Structured Semiconductors in Photocatalysis

Structured Semiconductors in Photocatalysis

Editors

Jorge Bedia
Carolina Belver

MDPI • Basel • Beijing • Wuhan • Barcelona • Belgrade • Manchester • Tokyo • Cluj • Tianjin

Editors

Jorge Bedia
Universidad Autonoma de Madrid
Madrid, Spain

Carolina Belver
Universidad Autónoma de Madrid
Madrid, Spain

Editorial Office
MDPI
St. Alban-Anlage 66
4052 Basel, Switzerland

This is a reprint of articles from the Special Issue published online in the open access journal *Catalysts* (ISSN 2073-4344) (available at: https://www.mdpi.com/journal/catalysts/special_issues/Photocatalysis_Semiconductors).

For citation purposes, cite each article independently as indicated on the article page online and as indicated below:

LastName, A.A.; LastName, B.B.; LastName, C.C. Article Title. *Journal Name* **Year**, *Volume Number*, Page Range.

ISBN 978-3-0365-8340-2 (Hbk)
ISBN 978-3-0365-8341-9 (PDF)

Contents

About the Editors

Jorge Bedia

Dr. Jorge Bedia is an Industrial Engineer and obtained his PhD in Chemical Engineering at the University of Malaga (Spain). He has been an Assistant Professor at the Chemical Engineering Department of the Autonomous University of Madrid (Spain) since 2013. His research interests include (i) the synthesis and characterization of carbon-based materials for adsorption and catalysis; (ii) synthesis and applications of MOFs; (iii) water purification by advanced oxidation processes, especially photocatalysis; (iv) gas-phase hydrodechlorination of chlorinated volatile organic compounds; and (v) separation and catalytic processes with ionic liquids and supported ionic liquids. Dr. Bedia is the co-author of around 110 journal publications (more than 4800 citations, H factor of 42), 3 book chapters, and 2 Spanish patents. He has presented over 150 works to national and international conferences with over 20 oral presentations. He has been involved in several research contracts with private companies and more than 20 research projects with different entities: European, national, integrated or cooperative with other countries (Russia, Mexico, Germany, Peru, USA). He is an Editor of the *Separation and Purification Technology* and *Catalysis Communications* journals. He is also a member of the Editorial Board of *Chemical Engineering* and Associate Editor of *Separation and Purification Technology*.

Carolina Belver

Dr. Carolina Belver is Associate Professor at the Department of Chemical Engineering at the Universidad Autónoma de Madrid (Spain). She received a B.Sc. in Chemistry from the Universidad Autónoma de Madrid in 1997 and a Ph.D. in 2004 from the University of Salamanca (Spain). She enjoyed competitive national postdoctoral contracts at the Catalysis and Petrochemistry Institute (Spain) (2005-2007) and the Material Science Institute of Madrid (2008-2011). Dr. Belver's specialization deals with the design, processing, and evaluation of novel heterostructures for applications in environmental remediation, mainly focused on heterogeneous catalysis, photocatalysis, and adsorption. She is the co-author of more than 100 referred publications and 13 book chapters (more than 5000 citations, H factor of 40), and in 2013 she received a Fulbright Fellowship recognizing her accomplishments as a researcher. Dr. Belver is the Editor-in-Chief of *Applied Catalysis B*, the Executive Editor of *Chemical Engineering*, and a member of the Editorial Advisory Boards of Materials Science for the journals *Energy* and *Catalysts*.

catalysts

MDPI

Editorial

Structured Semiconductors in Photocatalysis

Carolina Belver * and Jorge Bedia *

Departamento de Ingeniería Química, Universidad Autónoma de Madrid, Cantoblanco, 28049 Madrid, Spain
* Correspondence: carolina.belver@uam.es (C.B.); jorge.bedia@uam.es (J.B.)

1. Introduction

Photocatalysis appears as an interesting approach for different applications, with the possibility of using sunlight as a sustainable and renewable source of energy. This technology is based on the use of a semiconductor that can be excited by light with an energy higher than its band gap inducing the formation of energy-rich electron-hole pairs, which can be involved in redox reactions. Recent progress explored the chemical nature of structured semiconductors with the object to improve their electronic and optical properties, enhancing their photoresponse under different conditions. This Special Issue collects original research papers, reviews, and commentaries focused on the challenges for the design of structured semiconductors with photocatalytic applications, thus including synthesis, characterization of new photocatalysts, studies of activity and stability, and the mechanisms of photocatalytic reactions.

2. Structured Semiconductors in Photocatalysis

This Special Issue includes outstanding studies focused on structured semiconductors in photocatalysis for different applications. In this sense, Bobape et al. [1] synthesized CuO-TiO_2 nanocomposites using *C. Benghalensis* plant extracts. They analyzed the effect of the CuO to TiO_2 ratio on the morphological, optical, electrochemical, and photodegradation efficiency. The XRD data confirmed the tenorite structure of the CuO and the anatase phase of the TiO_2. The voltammogram of the CuO-TiO_2 30/70 electrode showed the highest response current density, suggesting a higher specific capacitance in this structured semiconductor. This sample also showed the highest photocatalytic degradation efficiencies against methylene blue, ciprofloxacin, and sulfisoxazole, with hydroxyl radicals being the primary species responsible for the photodegradation. In another study, Matussin et al. [2] synthesized CeO_2 and palladium–doped CeO_2 photocatalysts via the microwave-assisted synthesis method. The authors observed mixed phases of CeO_2/Ce_2O_3, although the Ce_2O_3 phase gradually disappeared upon doping with a higher percentage of Pd. The presence of Pd resulted in a decrease in the band gap energies, and the photoluminescence intensities were also quenched with Pd doping. Pd–CeO_2 NPs showed enhanced activities under visible light irradiation in the photodegradation of methylene blue and photoantibacterial activities. Chue et al. [3] prepared a hexagonal wurtzite ZnO photocatalyst via precipitation method. CuS nanoparticles and PbS quantum dots were loaded onto ZnO via a hydrothermal method to obtain a CuS/PbS/ZnO heterojunction photocatalyst. The CuS/PbS/ZnO photocatalyst showed significant absorption capabilities in the ultraviolet to near-infrared spectral regions, and effectively reduced the recombination of electron–hole pairs during a photocatalytic reaction. This catalyst demonstrated the best water splitting effect. Furthermore, after adding a 0.25 M mixed solution of Na_2S and Na_2SO_3 as the sacrificial reagent, the hydrogen production efficiency from water splitting reached 6654 μmol g^{-1} h^{-1} after 5 h. Tien et al. [4] synthesized a new binary MoS_2/Co_3O_4 nanohybrids. The heterojunction presented an S-scheme structure that acted as electron traps and promoted light absorption capacity for the degradation of methyl orange under visible light. The photocatalyst also showed excellent stability and recyclability over five

Citation: Belver, C.; Bedia, J. Structured Semiconductors in Photocatalysis. *Catalysts* **2023**, *13*, 1111. https://doi.org/10.3390/catal13071111

Received: 8 July 2023
Accepted: 12 July 2023
Published: 17 July 2023

consecutive cycles, without noticeable changes in the nanocomposite structure. The boosted photocatalytic degradation and redox activities of MoS_2/Co_3O_4 can be attributed to the created S-scheme heterostructure to facilitate the separation of and delay recombination of photoinduced charge carriers. In another study, Rezgui et al. [5] studied a heterogeneous photo-electro-Fenton process, which is an attractive technology for the removal of recalcitrant pollutants. To better exploit the presence of an irradiation source, a bifunctional catalyst with TiO_2 nanoparticles embedded into an iron–chitosan matrix was developed. The catalytic activity of the catalyst was improved by the optimization of the loaded TiO_2 content. The prepared composite catalysts based on TiO_2, Fe_3O_4, and chitosan were called TiO_2/Fe_3O_4-CS beads. The best catalyst with an optimal ratio $TiO_2/Fe = 2$ exhibited a high efficiency in the degradation and mineralization of chlordimeform insecticide. Under the optimum conditions, a real effluent loaded with 30 mg L^{-1} of the insecticide was efficiently treated, leading to $80.8 \pm 1.9\%$ TOC reduction after 6 h of treatment, with total removal of the pollutant after only 1 h. Hudandini et al. [6] loaded Ag onto ZnO by an ultrasonic spray pyrolysis system at different Ag contents (1, 5, and 10 wt%). An increase in the ZnO-Ag activity compared with pristine ZnO was observed at a carrier gas ratio of 0:1 with reaction rate constants of 0.0059 and 0.0025 min^{-1}, respectively, in the degradation of textile wastewater under UV light irradiation. Babyszko et al. [7] analyzed the modification of titanium dioxide with fumed silica. The SiO_2/TiO_2 photocatalysts were obtained by the sol-gel method and were then calcined under an argon atmosphere. Various SiO_2 weights (2–17.2 wt.%) were prepared. The modification of titanium dioxide with SiO_2 inhibited the increase in crystallite size of anatase and brookite during calcination and the decrease in specific surface area values. The photocatalytic activity was determined based on the decomposition of methylene blue under UV irradiation. All the obtained SiO_2/TiO_2 photocatalysts showed higher activity compared to the starting TiO_2. Dies et al. [8] prepared fluoride-doped TiO_2 (F-TiO_2) photocatalysts by an efficient and simple one-step synthesis and tested them in the UV-photo-degradation of methylene blue and bisphenol A. F-TiO_2 defeated commercial TiO_2, and almost complete pollutant removal was achieved within 30 min. The energy consumption was reduced as a result of the suitable reactor set-up, which reduced light scattering, and by the application of a long-pulse radiation procedure, where the lamp was switched off during periods of continued degradation. This enhanced the overall photocatalytic performance. Under these conditions, 80% of dye removal was attained within 15 min of radiation with an energy consumption of only 0.070 Wh min^{-1}, demonstrating a much better efficiency when compared to previously reported data. The catalyst was reusable, and its performance can be improved by the addition of H_2O_2. The results were validated by BPA degradation and the treatment of real wastewater with both pollutants. Harris et al. [9] investigated ZnO powders prepared by alkali precipitation using different $[Zn(acetate)_2(amine)_x]$ compounds to alter the particle size and aspect ratio. Slow precipitations from 95 °C solutions produced micron-scale particles with morphologies of hexagonal plates, rods, and needles, depending on the precursor used. Powders prepared at 65 °C with rapid precipitation yielded particles with minimal morphology differences, but the particle size was dependent on the precursor used. The smallest particles were produced using precursors that yielded crystals with low aspect ratios during high-temperature synthesis. Particles produced during rapid synthesis had sizes ranging from 21 to 45 nm. The materials prepared using precursors with less volatile amines were found to retain more organic material than ZnO produced using precursors with more volatile amines. The amount of organic material associated with the nanoparticles influenced the photocatalytic activity of the ZnO, with powders containing less organic material producing faster rate constants for the decolorizing of malachite green solutions under ultraviolet illumination, independent of particle size. $[Zn(acetate)_2(hydrazine)_2]$ produced ZnO with the fastest rate constant and was recycled five times for dye degradation studies that revealed minimal to no reduction in catalytic efficiency. Akitunde et al. [10] studied disinfection and photocatalytic degradation of organic contaminants using visible light-activated GCN/Ag_2CrO_4 nanocomposites. The organic pollutants studied were 2,4-

dichlorophenoxyacetic acid and methyl chlorophenoxy propionic acid present in Killex®, a commercially available herbicide, bovine serum albumin (BSA) protein, and SARS-CoV-2 spike protein. The disinfection experiments were conducted on wastewater secondary effluent. Overall, the results indicate that GCN/Ag_2CrO_4 nanocomposite is a promising photocatalyst in degrading organic pollutants and disinfecting microorganisms under visible light irradiation within a reasonable time. Lang et al. [11] evaluated a TiO_2–CNT–Ag ternary composite film synthesized using the plasma-enhanced chemical vapor deposition method by simultaneously feeding a carbon nanotube (CNT)/Ag suspension and titanium tetraisopropoxide gas. The performance of the TiO_2-composite film for the degradation of rhodamine 6G under simulated solar light irradiation was evaluated. The rate constant of the prepared TiO_2–CNT–Ag for rhodamine 6G degradation was approximately 1.8 times greater than that of the prepared TiO_2. This result indicates that the addition of CNT and Ag significantly improved the photocatalytic activity of the prepared films. This same research group prepared TiO_2–Ag nanoparticle composite films by plasma-enhanced chemical vapor deposition (PECVD) using a mixture of aerosolized $AgNO_3$, which was used as an Ag nanoparticle precursor, and titanium tetraisopropoxide, which acted as the TiO_2 precursor [12]. Notably, the use of PECVD enabled a low process temperature and eliminated the need for pre-preparing the Ag nanoparticles, thereby increasing the process efficiency. The photocatalytic activity of the deposited films was determined by assessing the degradation of methylene blue under UV light irradiation. The Ag ions were successfully reduced to metallic nanoparticles and were embedded in the TiO_2 film. The best photocatalytic activity was achieved for a 1 wt% Ag-loaded TiO_2 composite film, which was 1.75 times that of pristine TiO_2. Finally, Sanni et al. [13] studied the synthesis approach (thermal polyol and deposition–precipitation) regarding the dispersion of Ag/AgBr nanoparticles dispersed on activated carbon prepared from chemically impregnated pinecone to increase their photocatalytic efficiency on the degradation of tetracycline.

This Special Issue also includes two reviews. The first one is a work by Venezia et al. [14] that described the state of the art of the current design strategies and emerging trends to hierarchical catalysts. Nature mimicking implies the design of nanostructured materials, which can be assembled into a hierarchical structure, thus outperforming the features of the neat components because of their multiple length scale organization. This approach can be effectively exploited for the design of advanced photocatalysts with superior catalytic activity for energy and environment applications with considerable development in the recent six years. The review presented different synthesis strategies, including template-free structuring, and organic, inorganic, and hybrid templating. Furthermore, emerging approaches based on hybrid and bio-waste templating were also highlighted. Finally, a critical comparison among available methods was carried out based on the envisaged application. The second review by Sreedhar and Noh [15] described the advancements in solar desalination of seawater by various Ti_3C_2 MXene-based morphologies for freshwater generation. The study explained key features such as light absorption, reflection, multiple internal reflection, hydrophilicity, lower thermal conduction, light-to-heat generation, and salt rejection for achieving efficient desalination output throughout the visible and broadband region. Specifically, it explored the self-floating and salt rejection nature of various state-of-the-art 2D Ti_3C_2 MXene structures. Among the different morphologies, Ti_3C_2 MXene in the form of a membrane is believed to be a promising morphology that effectively desalinates seawater into freshwater.

It can be concluded that structured semiconductors have key characteristics for many kinds of applications and purposes [16]. Probably, due to their high tunability and versability, we will assist in the future with a fast development of these amazing materials that can be fundamental for many applications.

Author Contributions: C.B. and J.B. have equally contributed to the writing and revision of this editorial. All authors have read and agreed to the published version of the manuscript.

Acknowledgments: We would like to thank the authors, reviewers, and editorial team of *Catalysts* for the work devoted to this Special Issue.

Conflicts of Interest: The authors declare no conflict of interest.

References

1. Bopape, D.A.; Mathobela, S.; Matinise, N.; Motaung, D.E.; Hintsho-Mbita, N.C. Green Synthesis of CuO-TiO_2 Nanoparticles for the Degradation of Organic Pollutants: Physical, Optical and Electrochemical Properties. *Catalysts* **2023**, *13*, 163. [CrossRef]
2. Matussin, S.N.; Khan, F.; Harunsani, M.H.; Kim, Y.-M.; Khan, M.M. Effect of Pd-Doping Concentrations on the Photocatalytic, Photoelectrochemical, and Photoantibacterial Properties of CeO. *Catalysts* **2023**, *13*, 96. [CrossRef]
3. Chiu, M.-H.; Kuo, C.-C.; Huang, C.-W.; Yang, W.-D. Preparation of CuS/PbS/ZnO Heterojunction Photocatalyst for Application in Hydrogen Production. *Catalysts* **2022**, *12*, 1677. [CrossRef]
4. Tien, T.-M.; Chen, C.-H.; Huang, C.-T.; Chen, E.L. Photocatalytic Degradation of Methyl Orange Dyes Using Green Synthesized MoS_2/Co_3O_4 Nanohybrids. *Catalysts* **2022**, *12*, 1474. [CrossRef]
5. Rezgui, S.; Díez, A.M.; Monser, L.; Adhoum, N.; Pazos, M.; Sanromán, M.Á. Magnetic TiO_2/Fe_3O_4-Chitosan Beads: A Highly Efficient and Reusable Catalyst for Photo-Electro-Fenton Process. *Catalysts* **2022**, *12*, 1425. [CrossRef]
6. Hudandini, M.; Puri, N.R.; Winardi, S.; Widiyastuti, W.; Shimada, M.; Kusdianto, K. Photocatalytic Activity of ZnO/Ag Nanoparticles Fabricated by a Spray Pyrolysis Method with Different O_2:N_2 Carrier Gas Ratios and Ag Contents. *Catalysts* **2022**, *12*, 1374. [CrossRef]
7. Babyszko, A.; Wanag, A.; Sadłowski, M.; Kusiak-Nejman, E.; Morawski, A.W. Synthesis and Characterization of SiO_2/TiO_2 as Photocatalyst on Methylene Blue Degradation. *Catalysts* **2022**, *12*, 1372. [CrossRef]
8. Díez, A.M.; Núñez, I.; Pazos, M.; Sanromán, M.Á.; Kolen'ko, Y.V. Fluoride-Doped TiO_2 Photocatalyst with Enhanced Activity for Stable Pollutant Degradation. *Catalysts* **2022**, *12*, 1190. [CrossRef]
9. Harris, J.D.; Wade, E.A.; Ellison, E.G.; Pena, C.C.; Bryant, S.C.; McKibben, N.L.; Christy, A.J.; Laughlin, K.O.; Harris, A.E.; Goettsche, K.V.; et al. Zinc–Acetate–Amine Complexes as Precursors to ZnO and the Effect of the Amine on Nanoparticle Morphology, Size, and Photocatalytic Activity. *Catalysts* **2022**, *12*, 1099. [CrossRef] [PubMed]
10. Akintunde, O.O.; Yu, L.; Hu, J.; Kibria, M.G.; Hubert, C.R.J.; Pogosian, S.; Achari, G. Disinfection and Photocatalytic Degradation of Organic Contaminants Using Visible Light-Activated GCN/Ag_2CrO_4 Nanocomposites. *Catalysts* **2022**, *12*, 943. [CrossRef]
11. Lang, J.; Takahashi, K.; Kubo, M.; Shimada, M. Preparation of TiO_2-CNT-Ag Ternary Composite Film with Enhanced Photocatalytic Activity via Plasma-Enhanced Chemical Vapor Deposition. *Catalysts* **2022**, *12*, 508. [CrossRef]
12. Lang, J.; Takahashi, K.; Kubo, M.; Shimada, M. Ag-Doped TiO_2 Composite Films Prepared Using Aerosol-Assisted, Plasma-Enhanced Chemical Vapor Deposition. *Catalysts* **2022**, *12*, 365. [CrossRef]
13. Sanni, S.O.; Brink, H.G.; Viljoen, E.L. Influence of Synthesis Approach on Controlled Microstructures and Photocatalytic Properties of Ag/AgBr-Activated Carbon Composites on Visible Light Degradation of Tetracycline. *Catalysts* **2021**, *11*, 1396. [CrossRef]
14. Venezia, V.; Pota, G.; Silvestri, B.; Costantini, A.; Vitiello, G.; Luciani, G. Tailoring Structure: Current Design Strategies and Emerging Trends to Hierarchical Catalysts. *Catalysts* **2022**, *12*, 1152. [CrossRef]
15. Sreedhar, A.; Noh, J.-S. Advancements in Solar Desalination of Seawater by Various Ti_3C_2 MXene Based Morphologies for Freshwater Generation: A Review. *Catalysts* **2021**, *11*, 1435. [CrossRef]
16. Belver, C.; Bedia, J.; Gómez-Avilés, A.; Peñas-Garzón, M.; Rodriguez, J.J. Semiconductor Photocatalysis for Water Purification. In *Nanoscale Materials in Water Purification*; Elsevier: New York, NY, USA, 2018; p. 581. ISBN 978-012813927-1/978-012813926-4. [CrossRef]

catalysts

Article

Green Synthesis of CuO-TiO_2 Nanoparticles for the Degradation of Organic Pollutants: Physical, Optical and Electrochemical Properties

Dineo A. Bopape [1], Sarah Mathobela [1], Nolubabalo Matinise [2], David E. Motaung [3] and Nomso C. Hintsho-Mbita [1,*]

1 Department of Chemistry, University of Limpopo, Mankweng 0727, South Africa
2 Uneso-Unisa Africa Chair in Nanoscience and Nanotechnology, College of Graduate Studies Department, University of South Africa, Pretoria 0002, South Africa
3 Department of Physics, University of Limpopo, Mankweng 0727, South Africa
* Correspondence: nomso.hintsho-mbita@ul.ac.za

Abstract: CuO-TiO_2 nanocomposites were successfully synthesized using the *C. benghalensis* plant extracts. The effect of the composition of CuO to TiO_2 on the morphological, optical, electrochemical, and photodegradation efficiency in the composites was studied. SEM, XRD, UV-vis, FTIR, TGA, BET, and CV were used to characterize these materials. The XRD data reported the tenorite structure of the CuO and the anatase phase of the TiO_2. SEM showed the spherical morphologies for all the CuO-TiO_2 NPs, and these were also mesoporous in nature, as depicted by BET. The voltammogram of the CuO-TiO_2 30/70 electrode showed a higher response current density compared to the other two samples, suggesting a higher specific capacitance. Upon testing the photocatalytic efficiencies of the CuO-TiO_2 nanocomposites against methylene blue (MB), ciprofloxacin (CIP), and sulfisoxazole (SSX), the highest degradation of 94% was recorded for SSX using the CuO-TiO_2 30/70 nanocomposites. Hydroxyl radicals were the primary species responsible for the photodegradation of SSX, and the material could be reused once. The most active species in the photodegradation of SSX has been identified as OH•. From this study, it can be noted that the CuO-TiO_2 nanocomposites were more selective toward the degradation of antibiotics (sulfisoxazole and ciproflaxin) as compared to dyes (methylene blue).

Keywords: CuO-TiO_2 nanocomposites; green synthesis; photocatalysts; sulfisoxazole

Citation: Bopape, D.A.; Mathobela, S.; Matinise, N.; Motaung, D.E.; Hintsho-Mbita, N.C. Green Synthesis of CuO-TiO_2 Nanoparticles for the Degradation of Organic Pollutants: Physical, Optical and Electrochemical Properties. *Catalysts* **2023**, *13*, 163. https://doi.org/10.3390/catal13010163

Academic Editors: Jorge Bedia and Carolina Belver

Received: 3 November 2022
Revised: 12 December 2022
Accepted: 17 December 2022
Published: 10 January 2023

1. Introduction

Water pollution is one of South Africa's most serious environmental problems, adding to the water scarcity the country is facing [1–3]. Despite this, contaminants such as dyes and antibiotics are being detected in natural water sources, and the number of contaminants detected has increased in recent years as new emerging pollutants are reported each year [4,5]. The textile sector contributes significantly to water pollution in the environment by dumping dye-bearing waste effluents [6–9]. Even at low concentrations, the presence of these dyes (which can be anionic, cationic, or non-ionic) in water is highly apparent and undesired [10]. Humans can be exposed to these toxic dyes through direct or indirect consumption of polluted water via the food chain [11]. The after effect has been reported to be mutagenic to human health and damaging to the aquatic ecology [12].

Methylene blue (MB) is a cationic dye that has been classified as a toxic colorant [11]. This pollutant is non-biodegradable due to its complex aromatic composition, and it damages septic tanks during wastewater treatment [13].

Sulfisoxazole (SSX) is an endothelin receptor antagonist that protects retinal neurons against ischemia/reperfusion and lipopolysaccharide damage [14]. On the other hand, excessive usage of SSX may leave traces of the medicine in animal products [15]. However, wastewater treatment plants that treat pharmaceutical wastes transfer the SSX into freshwater environments [14]. SSX residues in the aquatic environment may cause allergic responses, drug resistance, and endocrine problems [16].

Ciprofloxacin (CIP) is the most commonly prescribed fluoroquinolone antibiotic [17,18]. CIP is effective against a wide range of Gram-negative and Gram-positive bacteria and is one of the emerging pharmaceutical contaminants found in water [17–19]. It is often found in the environment and has been demonstrated to be genotoxic and it is a primary metabolite of enrofloxacin, a popular fluoroquinolone used in veterinary medicine [18]. CIP is also difficult to completely remove using conventional treatment methods [18,19]. Thus, some of the CIP has remained in the water following treatment in concentrations measured in parts per million (g/mL) or even parts per billion (ng/mL), its unpredictable toxicological effects have sparked growing concern [20]. As a result, there is a strong need to develop materials that can remove various pollutants such as MB, SSX, and CIP from effluents before they are discharged into freshwater systems [13–19].

Several methods such as adsorption, electrocoagulation, and membrane filtration have been used for the elimination of these pollutants in wastewater streams [11,14,18]. Most of these approaches have drawbacks such as excessive sludge generation, formation of by-products, high operational costs, and poor selectivity, thus researchers are moving toward Advanced Oxidation Processes in particular the photocatalysis process as its been shown to be an efficient method for the degradation of multiple pollutants. However, in the treatment of organic MB, SSX, and CIP, these approaches fall short of the dye wastewater discharge criteria. Given the current environmental issues, finding a cost-effective and environmentally friendly technique to breaking down organic contaminants is critical [21–23].

Metal oxide semiconductor catalysts have received much press in recent years because of their capacity to convert solar energy and clean up pollution [24]. The process of photocatalytic degradation using metal oxides is frequently utilized in water treatment due to its efficacy in removing pollutants and has no secondary pollution [24,25]. Researchers are currently focusing on metal oxide nanostructured composites due to their unique features, such as a high surface to volume ratio and quantum confinement effects [22]. Various metal oxides such as ZnO, CuO, NiO, TiO_2, etc., are highly explored but nanosized transition metal oxides, such as copper oxide (CuO) and titanium oxide (TiO_2), have attracted scientific interest due to their potential applications in medicine, sensing, catalysis, etc. [22,24,26]. CuO has gained much attention among transition metal oxides because of its intriguing photochemical and photomagnetic capabilities [26]. It is a photoconductive p-type semiconductor with a narrow bandgap (1.2 eV) and has been used in a variety of applications in gas sensing, catalysis, super capacitors, and field emission, to name a few [24,27]. CuO has a monoclinic structure and is employed in a variety of device fabrication units depending on its properties [22]. It is one of the most important catalysts used to remove industrial effluents from the environment [21].

On the other hand, TiO_2 is an n-type semiconductor with a large band gap that ranges from 3.2 to 3.6 eV and it is widely used in cosmetics, coatings for paper and medical devices, as well as gas sensors [24]. Although TiO_2 is a promising material for photocatalytic applications, its wide bandgap energy of 3.2 eV limits its absorption to UV light at λmax = 365–387 nm [23,25,28]. It was recorded that TiO_2 can only utilize 5–8% of the UV light and for the vis light it was recorded to be relatively low [6,25,29]. One significant benefit is that when TiO_2 reacts with another metal oxide, it forms heterojunctions [22,24,29]. However, metal doping of TiO_2 has a low thermal stability and is prone to photo corrosion [25]. CuO has been frequently employed to enhance the photocatalytic properties of TiO_2 [25–27]. This is because it can reduce the photogenerated electrons and holes from recombining while simultaneously lowering the energy of the band gap [24–28]. The disadvantage with the synthesis of such CuO-TiO_2 nanocomposite, is the use of harmful chemicals. Thus, it is critical to synthesize CuO-TiO_2 nanocomposites using a simple, eco-friendly, and cost-effective method that does not require the use of hazardous reagents. In recent years, the synthesis of $CuO-TiO_2$ using biological methods such as plant extracts has sparked considerable interest in the field of nanotechnology [30]. Some of the required properties in these materials have been reported to have been added from bimetal oxides synthesized from plant extracts, microorganisms, and enzymes. Plant-mediated nanocomposites for example, can be used to degrade a wider range of pollutants, such as pharmaceutical and dyes strains, due to the phytochemicals they contain.

Various studies have been conducted on the photodegradation of organic pollutants using the green-derived CuO/TiO_2 nanocomposites. Lu et al. [29] prepared the CuO/TiO_2 nanocomposite. In their study, UV light was irradiated and 100% of phenol was degraded after 80 min. Approximately, 98 and 88% of MB and MO, respectively, was degraded by CuO/TiO_2 nanocomposite after 150 min irradiation by Khodadadi et al. [30]. Based on the previous studies discussed, the photocatalytic efficiency of TiO_2 were improved by coupling the material with CuO. Hence there is a need to prepare the $CuO-TiO_2$ nanocomposites as photocatalyst semiconductors.

In this study various ratios of $CuO-TiO_2$ nanocomposites were synthesized using the *Commelina benghalensis* (*C. benghalensis*) plant extracts for the first time. The plant is rich in phytochemicals that can act as trapping and capping agents during bimetallic synthesis [25]. The physical, optical, surface, and electrochemical properties were investigated. The effect of the different compositions of the photocatalysts were also explored for the degradation of various pollutants, methylene blue, ciprofloxacin, and sulfisoxazole.

2. Results and Discussions

2.1. FTIR Characterization

Figure 1a,b shows the IR spectrum of the *C. benghalensis* plant extract, CuO, TiO_2-NPs, and $CuO-TiO_2$ nanocomposites. An intense band at 1641 cm^{-1} for *C. benghalensis* extracts was noted. This corresponds to the amide band, which is caused by a carbonyl stretch and N-H deformation vibrations in the amide linkage of *C. benghalensis* proteins [31]. These could be from alkaloids and tannins, which were identified from this plant, both of which are known to be stabilizing and reducing agents [32,33]. For the CuO-NPs, a Cu-O bond vibration at around 516, 832, and 1052 cm^{-1}, which shows the formation of the CuO metal oxide, was noted [29,34]. The hydroxyl groups (–OH) from the *C. benghalensis* plant extract were responsible for the absorption bands at 1663 cm^{-1} (bending mode) and 3388 cm^{-1} (stretching mode), whereas the Cu-O stretching vibration mode was responsible for the formation of the absorption bands at 510, 832, and 1052 cm^{-1} [35,36]. Peaks associated with the bending and stretching vibrations of hydroxyl groups (OH) were seen in $CuO-TiO_2$ samples at ~1651 and 3356 cm^{-1}, respectively, and the Ti-O and Ti-O-Ti bands were also detected vibrating at 400–900 cm^{-1} [32]. In addition, when identifying the stretches of the composites (Figure 1b), a peak at 1376 cm^{-1} representing the stretching vibrations of Cu-O-Ti groups confirmed the presence of CuO structures on the surface of TiO_2 nanoparticles [37,38]. This verifies the development of a chemical link at the $CuO-TiO_2$ interface. The $CuO-TiO_2$ nanocomposite at the 50/50 composition had more intense peaks, which corresponded to the *C. benghalensis*, CuO, and TiO_2 peaks. The intensity of the peaks increased in the order $CuO-TiO_2$ 30–70 < $CuO-TiO_2$ 70–30 < $CuO-TiO_2$ 50–50. Table 1 shows the detailed peak formations of the compounds.

Table 1. The IR peaks depicted *for C. benghalensis* plant extract, *C. benghalensis*-mediated CuO, and TiO_2 nanoparticles.

Wavenumber	Compound	Correspondence	Functional Group	Refs.
3345 cm^{-1} 3356 cm^{-1} 3335 cm^{-1}	30/70 CuO-TiO_2 50/50 CuO-TiO_2 70/30 CuO-TiO_2	O-H (stretch) H- (bonded)	Phenols	[31–34]
3335 cm^{-1}	*C. benghalensis*	N-H (vibrations)	Amide proteins	[31,34]
2210 2335	TiO_2	C-H (stretch)	Alkanes	[35]
1925 2094	TiO_2 CuO	-C-C-	Stretch	[35]
1641 cm^{-1}	CuO TiO_2 *C. benghalensis*	(NH)C=O	II amines	[31–34]
1557 cm^{-1}	CuO	N-H (bending)	I amines	[35]
1361 cm^{-1}	CuO-TiO_2	Cu-O-Ti (stretch)	Metal oxides	[37,38]
1010 cm^{-1}	TiO_2	O-H (bending)	Alkyl halides	[35]
411 cm^{-1} 916 cm^{-1} 1010 cm^{-1}	TiO_2	Ti-O (bending) Ti-O-Ti (bending)	Alkyl halides	[36–38]
516 cm^{-1} 832 cm^{-1} 1052 cm^{-1}	CuO	Cu-O (stretch)	Metal oxide	[35]

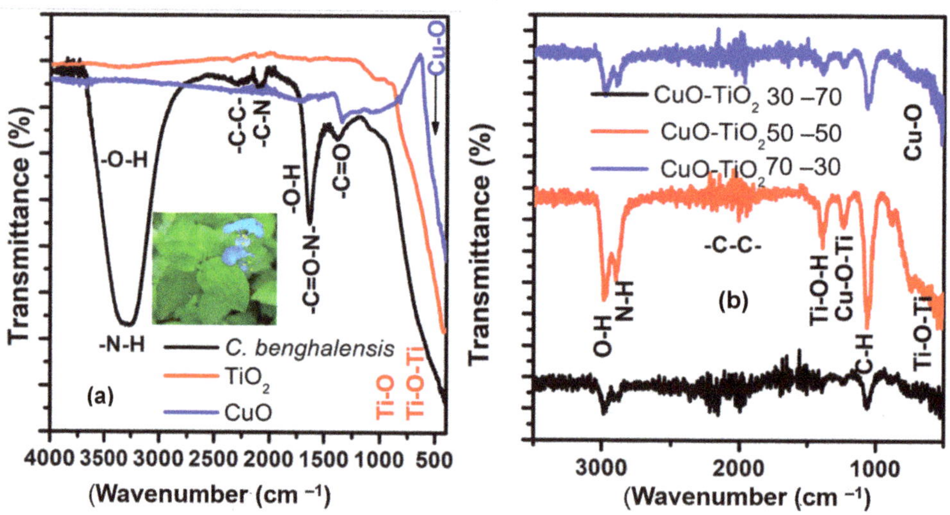

Figure 1. (**a**) FTIR diagram for the *C. benghalensis* extract, TiO_2, and CuO and the (**b**) CuO-TiO_2 nanocomposites with different compositions.

2.2. Optical Properties of CuO-TiO_2 Nanocomposites

The UV-vis absorption spectra of CuO-TiO_2 30/70, CuO-TiO_2 50/50, and CuO-TiO_2 70/30 were explored in Figure 2a to determine the properties of the material in the 200–800 nm spectrum. For all the CuO-TiO_2 nanocomposites, the absorption band was found to be 301 nm. The intensity of the absorption peak increased with the composition of CuO on the TiO_2 material. The adsorption edge of all the prepared samples was the same, indicating that CuO was deposited on the surface of TiO_2 rather than doped into the TiO_2 crystalline [39,40]. This absorbance band makes the CuO-TiO_2 materials better photocatalysts in the visible light region. Zhang et al. [39] obtained an absorption band at 380 nm, whereas Shi et al. [40] obtained it at 385 nm, and Yu et al. [41] at 388 nm. When compared to the literature, it shows that the absorbance peaks of all green-synthesized CuO-TiO_2 in this study were redshifted. This could be due to the difference in the synthetic process of the CuO-TiO_2 nanocomposites. In this study, a plant extract of *C. benghalensis* was utilized as a reducing agent, as opposed to the NaOH, $NaBH_4$, and $C_2H_6O_2$ that was used by Zhang et al. [39], Shi et al. [40], and Yu et al. [41], respectively. The band gap energies (E_g) of materials were determined by extrapolating and intersecting the linear portion of absorbance to the energy axis [28]. The E_g values for the CuO-TiO_2 30/70, CuO-TiO_2 50/50, and CuO-TiO_2 70–30 were found to be 3.76, 3.72, and 3.79 eV, respectively. The optical band gap energy (Eg) for all the CuO-TiO_2 nanocomposites was calculated based on the absorption spectrum of the samples according to the equation of E_g = 1240/wavelength [23]. From the literature, Gnanasekaran et al. [35] obtained a band gap of 2.98 eV, Muzakki et al. [36] recorded 3.01 eV, and Chowdhury et al. [27] obtained 3.17 eV. These studies used NaOH and urea as reducing agents. Therefore, in this study, maybe due to the phytochemicals present in the *C. benghalensis* plant extract, which was used as a reducing and capping agent, may have led to the higher band gap values of the CuO-TiO_2 nanocomposites.

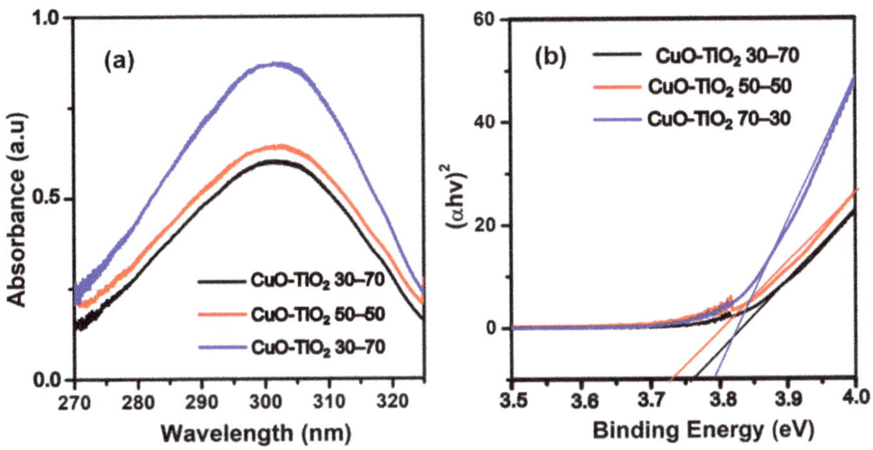

Figure 2. (**a**) UV-vis curves for CuO-TiO_2 (30/70, 50/50, and 70/30) and the corresponding (**b**) Tauc's plot.

2.3. Structural and Phase Composition Analysis

To obtain the structural properties, crystallite size, and phase composition of the nanocomposites, XRD analysis (Figure 3) was conducted. The XRD data of pure CuO were assigned with 2θ diffraction peaks at 35.6° (002), 38.9° (200), 48.65° (−202), 58.23° (202), 61.53° (−113), and 66.2° (−311) agreed to the JCPDS card no. 48-1548, which identified the tenorite structure of CuO [42–44]. The TiO_2 was identified by the characteristic diffraction peaks at 25.3° (101), 36.9° (103), 37.8° (004), 48.05° (200), 53.9° (105), 62.12° (213), 68.7° (116), 70.311° (220), 75.03° (215), and 76.02° (301) corresponding to the anatase phase of TiO_2 (according to the JCPD 21-1272) [44]. The relative broad peaks at 35.3°, 38.9°, and 48.37° were indexed to the diffraction of the TiO_2 planes (103), (004), and (200), and CuO planes (−103), (200) and (−202) demonstrating that TiO_2 and CuO coexist in the CuO-TiO_2 heterojunction [45]. This is reasonable given that the anatase TiO_2 lattice constants are identical to those of CuO. The intensity of the peaks of CuO and TiO_2 decreased in the CuO-TiO_2 bimetallic. It is interesting to note that the visibility of CuO peaks (plane −111 and 200) was more visible in the CuO-TiO_2 (70/30) than the (30/70) and the TiO_2 peaks (plane 101 and 200) were more intense in the CuO-TiO_2 (30/70), vice versa. This showed the difference in the composition of the two metal oxides. Then on the TiO_2-CuO (50/50), the peaks from both CuO and TiO_2 were more intense and visible as compared to those of CuO-TiO_2 (70/30) and CuO-TiO_2 (30/70). These data corroborate with the FTIR data because the presence of Ti-O, Cu-O, and C-H. N-H peaks were also intense on the FTIR data for the 50/50 composition. The Scherrer Equation (2) was used to calculate the crystalline sizes of materials, and the sizes were found to be 17.03, 17.13, 39.72, 47.2, and 39.14 nm for the TiO_2, CuO, and CuO-TiO_2 30/70, 50/50, and 70/30, respectively. This demonstrates that the formation of the heterojunctions led to an increased crystallite sizes [41].

$$D = K\lambda/\beta \cos\theta \quad (1)$$

where: K = Scherrer constant (i.e., 0.94),

λ = X-ray wavelength (0.154060 nm)
β = Full width at half-maximum of the (101) XRD peak
θ = The Bragg diffraction angle (degree).

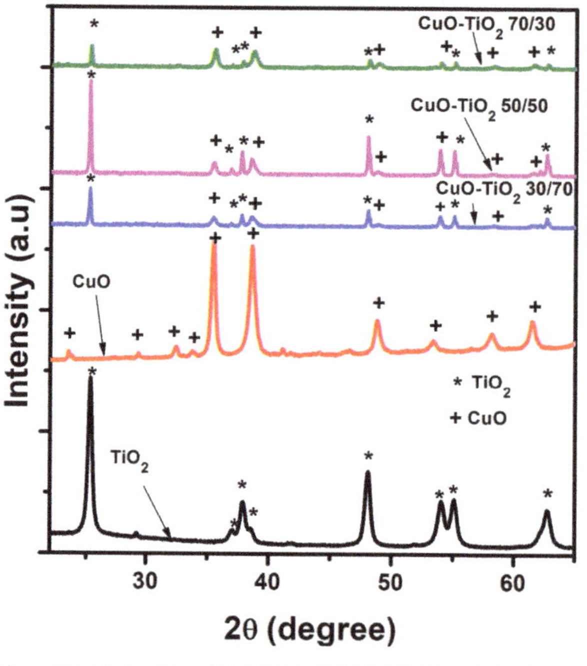

Figure 3. TiO_2, CuO, and CuO-TiO_2 (30/70, 50/50 and 70/30) XRD patterns.

2.4. Morphological Analysis of the Various Materials

Figure 4a,b shows the SEM images of the biosynthesized CuO nanoparticles. The materials are shown to be thick tubes in shape, as shown in the literature previously, with some aggregation, and have a particle size distribution of 50–250 nm, with dominant particle sizes being between 100 and 200 nm (Figure 4k) [24]. For the TiO_2, the SEM images in Figure 4c,d showed spherically shaped nanoparticles with a particle size distribution of (Figure 4l) 50–400 nm with dominating particles ranging in size from 100 to 300 nm. The lesser agglomeration that was observed on the TiO_2 may be due to the optimum concentration of the plant extract that was utilized [37]. Upon analyzing the three bimetallic materials, the more we increased the composition of CuO on TiO_2, the less spherical the composites became, which is evidence that Cu, Ti, and O species are strongly demonstrated to be homogeneously distributed throughout the entire selected area, and the formation of the p-n heterojunction between CuO and TiO_2 is revealed [40]. The particle sizes of the spherical nanocomposites increased as we increased the composition of CuO. The particle size distribution of CuO-TiO_2 30–70, 50–50, and 70–30 was 100–200 nm, 125–250, and 105–250, respectively. The larger particle sizes may be due to the agglomeration that is observed from CuO-TiO_2 30–70 to 70–30. To further confirm the formation of these materials, EDS analysis was conducted as reported in Figure S1a–e.

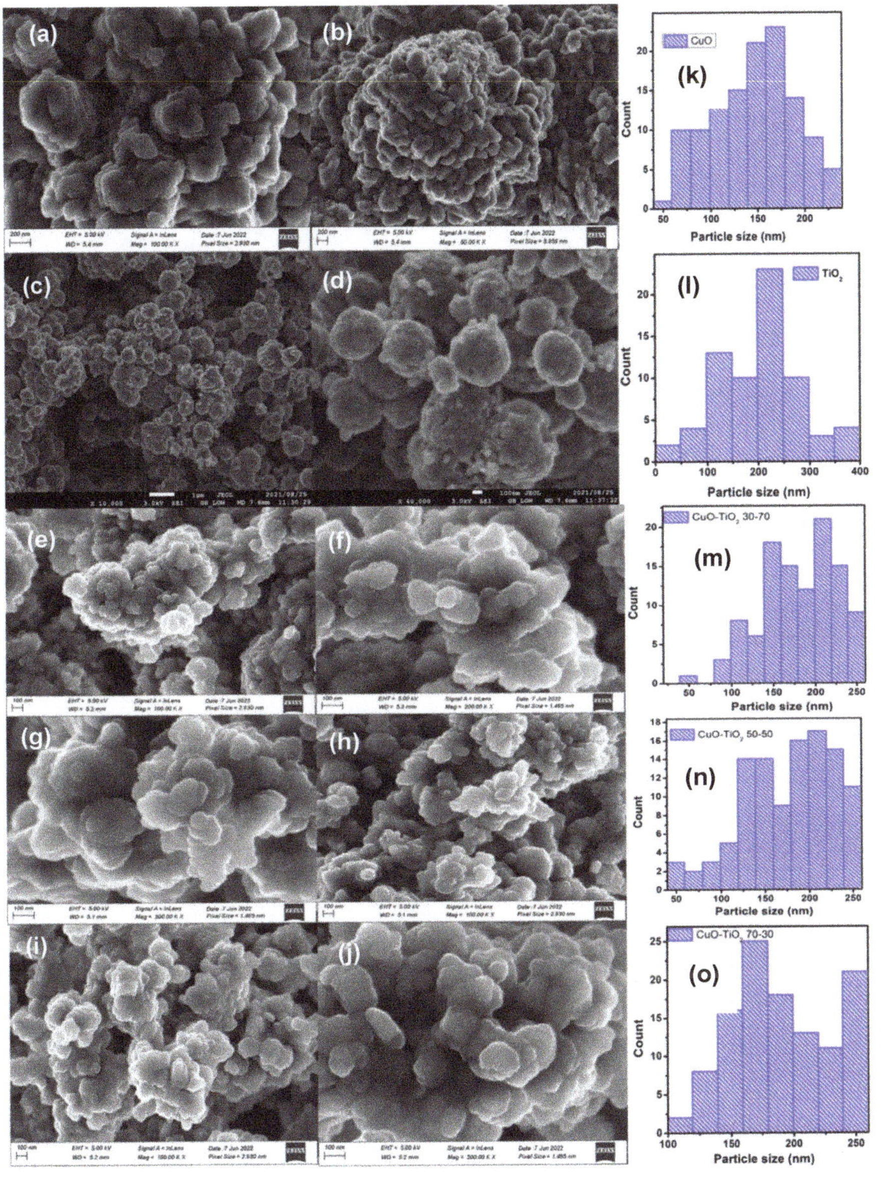

Figure 4. **(a–j)** SEM images for TiO_2, CuO, and CuO-TiO_2 (30/70, 50/50, and 70/30) and their **(k–o)** particle size distributions.

2.5. TGA and DTA Analysis

For thermal analysis studies on the CuO-TiO_2 nanocomposites (Figure 5a), two mass losses were observed arising from the elimination of water. All mass losses were linked to the endothermal transformation, according to the DTA curves (Figure 5b). At the molar ratios CuO-TiO_2 = 30:70, 50:50, and 70:30, total mass losses of the nanocomposites were 6.5%, 1.8%, and 8.3%, respectively. When compared to the TiO_2 reference sample, a 50% increase in copper (II) oxide content enhanced the thermal stability in the temperature range of 20–1000 °C. All of the materials studied had suitable thermal stability up to 1000 °C, which is in line with previous studies by Ashok et al. [42] where weight loss percentages were 9.0, 4.5, 3.0, and 2.5 for CT-5, CT-6, and CT-7, respectively. The 50/50 CuO-TiO_2 nanocomposite was the most thermally stable composite of all the materials

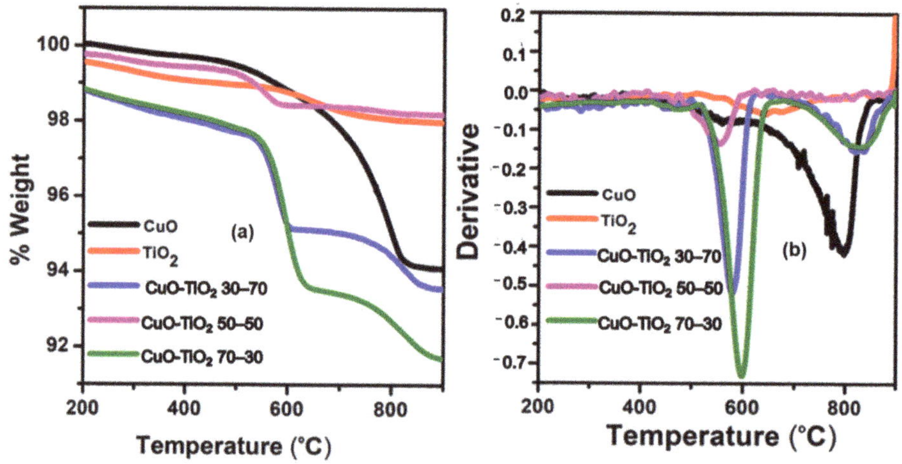

Figure 5. (**a**) TGA diagrams and (**b**) the derivatives diagrams of TiO_2, CuO, and CuO-TiO_2 (30/70, 50/50, and 70/30).

2.6. N_2 Adsorption-Desorption Studies, BJH Pore Volume, and BET Surface Area

The surface area, pore diameter, and pore volume analysis through N_2 adsorption/desorption isotherms for all the CuO-TiO_2 nanocomposites are demonstrated in Figure 6a–e. All the samples exhibited a type III hysteresis loop except for the TiO_2 material, which had an H1 hysteresis loop [43]. The BET surface areas of the photocatalysts generally decreased as the CuO loading increased. The BET surface area of the materials formed at the molar ratios CuO-TiO_2 30/70, 50/50, and 70/30 was 5.58, 5.74, and 3.22 m^2/g, respectively. On the other hand, the surface area of TiO_2 and CuO was 13.18 m^2/g and 1.613 m^2/g, respectively. All the materials, excluding TiO_2, were mesoporous in nature [44]. The mesoporous nature of these materials could have been influenced by the manner of the pore diameter distribution of copper (II) oxide [45].

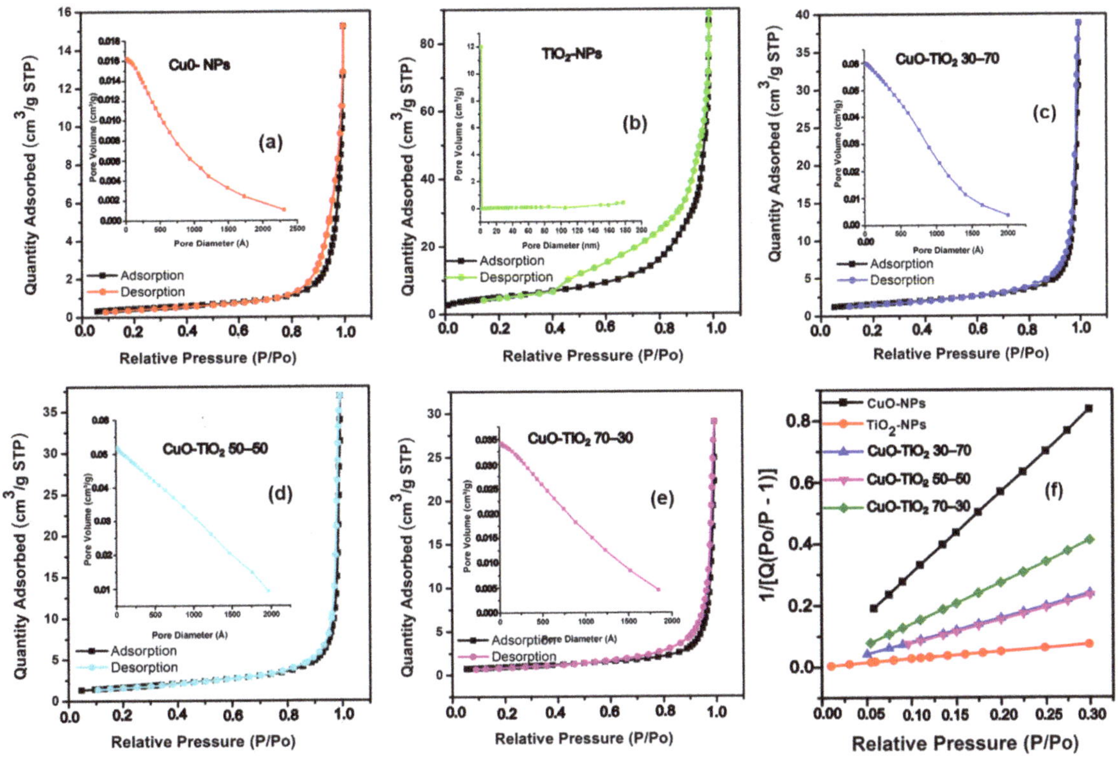

Figure 6. The adsorption-desorption isotherms (**a**–**e**) with the pore volume distribution curve (inset) and (**f**) the surface area diagrams for TiO_2, CuO, and CuO-TiO_2 (30/70, 50/50, and 70/30).

2.7. Electrochemical Characterization of the Composites

Cyclic and linear swept voltammetry were used to test the electrochemical properties to quickly determine information about the thermodynamics of redox processes and the electron transfer between an electrolyte and a modified working electrode. CV and linear curves of bare (unmodified) glassy carbon electrodes modified with different samples (S1–S3) at the scanning rate of 2 mV/s are displayed in Figure 7A,B. According to the graph observed, there was no electrochemical performance revealed by the bare GCE electrode. An electrochemical reaction was revealed after the modification of GCE with the three different samples noted due to the peak potential and high current density. All the CV curves for the modified electrode displayed a similar shape, a redox peak, but the different positions of the redox peaks were dependent on the type of sample. A couple of redox peaks and an increase in the current density can be seen clearly in the voltammogram of each sample. Voltammogram curves generally reveal redox peaks and high current density, indicating a pseudo-capacitance behavior [11,12]. An anodic peak appeared to the positive sweep around 0 to 0.2 V due to the oxidation of CuO, and cathodic peaks appeared to the negative ongoing from −0.2 to −0.3 due to the reduction of Cu^{2+}/Cu^{1+}. Noticeably, the voltammogram of the CuO-TiO_2 30/70 (S1) electrode showed a higher response current density compared to the other two samples, suggesting a higher specific capacitance, and it could be due to more titanium dioxide, as there is a synergistic effect that increases the electrochemical properties of copper oxide. In addition, this may be attributed to the high conductivity of the electrode due to the p-n junction between CuO and TiO_2 nanoparticles, which enhances the electrochemical activities. This implies that more TiO_2 nanoparticles added to the CuO, which is an n-type semiconductor, play an important role in increasing the electrochemical activity of the electrode to show the best response in terms of voltammetry. The electrochemical results can make the electrode to be the best candidate for electrochemical applications due to their best voltammetric response and possible photocatalytic applications. The specific capacitance was calculated from the area under the curve of cyclic voltammetry using Equation (3) and was found to be 164, 134, and 113 for samples 1, 2, and 3, respectively.

$$C_p = \frac{A}{2km\Delta V} \tag{2}$$

A—area under I-V curve, k—scan rate, m—active mass, ΔV—potential window.

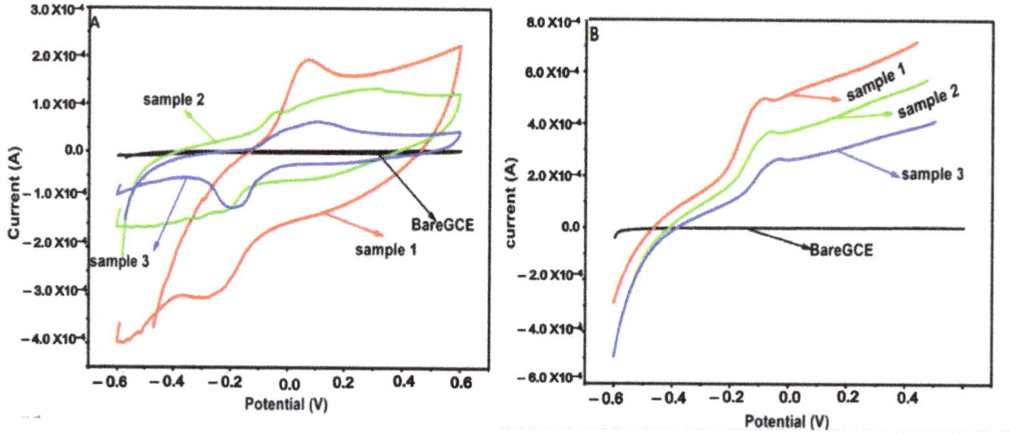

Figure 7. Cyclic (**A**) and linear (**B**) voltammetry curves of bare GCE and modified A: GCE/ CuO-TiO_2 30/70 (S1) B: GCE/ CuO-TiO_2 50/50 (S2): GCE/ CuO-TiO_2 70/30 (S3) in 0.5 M H_2SO_4 as an electrolyte.

Figure 8 reveals CV at different scan rates of 5–50 mV/s for three different samples. The cyclic voltammogram curves showed that the current density increased with increasing the scan rates; thus, the voltammetric current is directly proportional to the scan rate and gives an indication of the excellent rate capability of the modified electrode. There was a slight shift of redox peaks (oxidation and reduction) toward a more positive and negative potential, illustrating that the electrochemical kinetics are controlled by the diffusion process and displaying a rapid charge transfer [38].

The EIS was also used to study the electrochemical impedance properties of the samples (S1–S3). It is a prevailing tool to investigate the capacitive behavior of the samples. It is used to study the charge transfer resistance and diffusion process from the solution to the modified electrode surface and to the resistance of the electrolyte [14,45–47]. In Figure 9, the Nyquist plots of CuO-TiO_2 30/70 (S1), CuO-TiO_2 50/50 (S2), and CuO-TiO_2 70/30 (S3) that were modified in the glassy carbon electrode are shown. As seen from the graph, the results show that all samples revealed a semi-circle in the high-frequency range, which is related to the charge transfer resistance caused by faradic reactions. The intersection of the curve signifies solution resistance, which is an indication of the resistance of the electrolyte and modified electrode [48,49]. A straight line in the low-frequency resistance, which corresponds to Warburg impedance, indicates an ion diffusion process on the modified electrode and electrolyte [15,16]. The EIS data suggest that the CuO-TiO_2 electrode material showed faster charge transport ability, leading to suitable electrochemical properties. Therefore, the electrochemical results showed that the modified CuO-TiO_2 electrode (30:70) is an outstanding electrode material for electrochemical application due to its great voltammetric response, suitable electrochemical properties, and best charge transfer resistance.

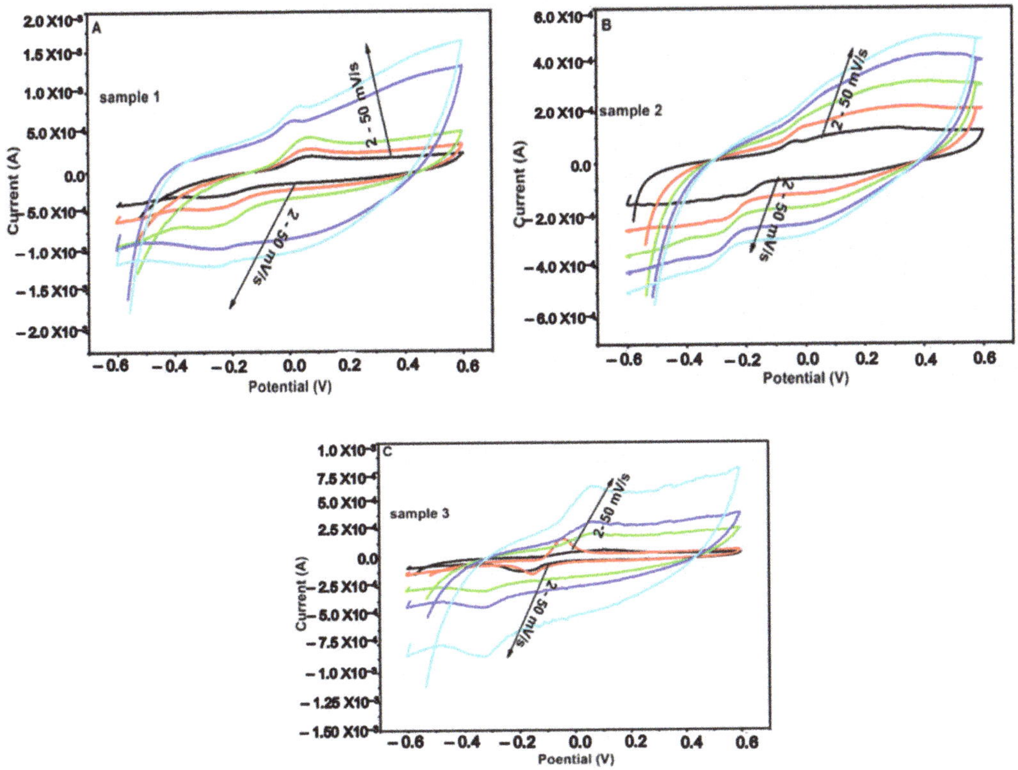

Figure 8. Cyclic voltammetric at different scan rates (2–50 mV/s). (**A**): GCE/ CuO-TiO_2 30/70 (S1) (**B**): GCE/ CuO-TiO_2 50/50 (S2) and (**C**) GCE/ CuO-TiO_2 70/30 (S3) in 0.5 M H_2SO_4 as an electrolyte.

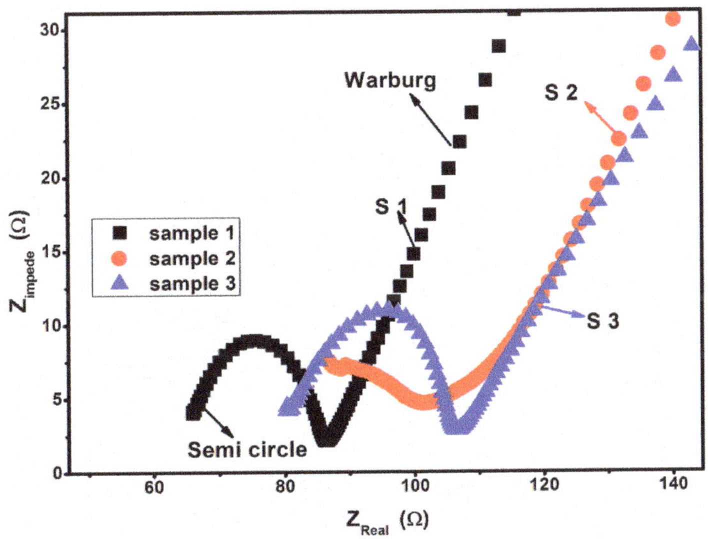

Figure 9. Nyquist plot of GCE/ CuO-TiO_2 30/70 (S1) GCE/ CuO-TiO_2 50/50 (S2): GCE/ CuO-TiO_2 70/30 (S3) in 0.5 M H_2SO_4 as an electrolyte.

2.8. Photodegradation of MB Dye Using the CuO-TiO_2 Nanocomposite

2.8.1. Photodegradation of MB UV Light

The green-synthesized CuO, TiO_2, CuO-TiO_2 30/70, 50/50, and 70/30 photocatalysts were used for the degradation of MB dye (Figure 10A). From the analysis, 18%, 65%, 17%, 33%, and 14% of the MB pollutant was degraded. It can be noted that there was a drastic drop in the degradation efficiencies of these materials, in particular with the 30/70 CuO-TiO_2 and the 70:30 CuO-TiO_2, which recorded 17% and 14% degradation. Sankar et al. [46], using the CuO photocatalyst, were able to degrade 50% of Coomassie brilliant blue R-250 after 240 min exposure to sunlight. Lingaraju et al. [50] utilized a green-synthesized CuO photocatalyst to degrade ~100% of trypan blue dye after 140 min irradiation under UV light. The green-synthesized CuO/TiO_2 prepared by Lu et al. [29] degraded 100% of phenol after 160 min exposure to UV light. Khodadadi et al. [30] synthesized CuO/TiO_2 nanocomposite, which was able to degrade 100% of MB and methyl orange (MO) after 10 min irradiation with UV light. It can be noted that for the green-derived CuO and TiO_2 NPs, for their improved degradation to occur, extended time periods (360 min) were noted even though in one study, the degradation was still only 50% using CuO. In addition, the choice of the pollutant could play another major role. Using CuO/TiO_2, for all the pollutants there was a complete degradation for all the mentioned pollutants, and thus it is the complete opposite of what has been obtained in our study, with a maximum degradation of 33% for MB, considering that optimum conditions from other similar studies were used. Thus, it is suspected that the phytochemicals from the various plant extracts used for the synthesis of these materials also play a major in their optical properties. In addition, their activity could be pollutant dependent; thus, it was important that we also investigate other pollutants, such as sulfisoxazole and ciprofloxacin, which are pharmaceutical pollutants. From the kinetics study (Figure 10B), it was noted that the CuO-TiO_2 30/70 nanocomposite had a better fitting than the first-order kinetics as the other two materials did not fit completely. The rates of reactions were 0.014, 0.0029, and 0.001 min^{-1} for the CuO-TiO_2 30/70, 50/50, and 70/30, respectively.

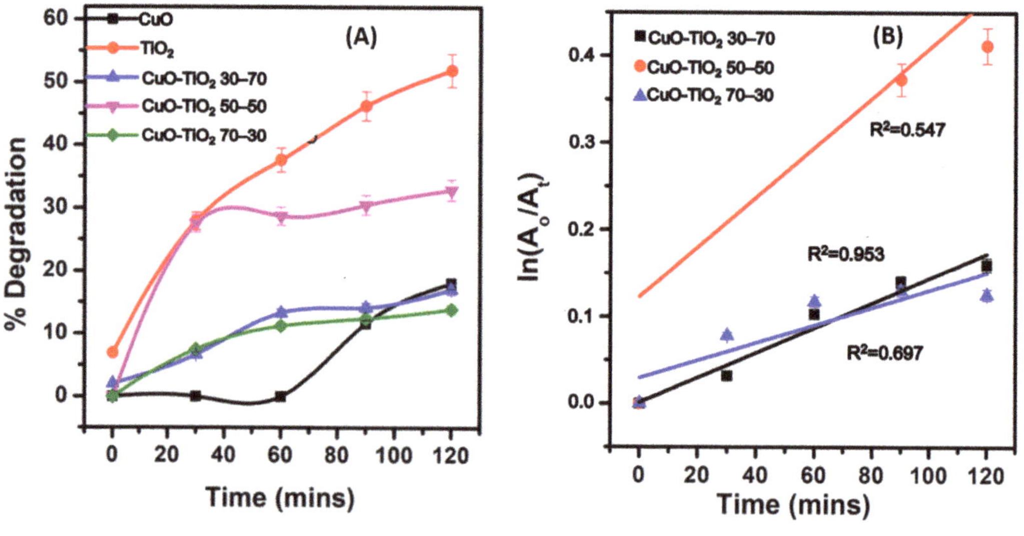

Figure 10. (**A**) Photodegradation of MB by CuO, TiO_2, and various ratios of CuO-TiO_2 and (**B**) kinetics study of the CuO-TiO_2.

2.8.2. Photodegradation of CIP and SSX Using UV light

For the degradation of antibiotics (CIP and SSX), similar conditions as with MB, but the concentration of SSX and CIP was 10 mg/L (Figures S2 and S3), and from Figure 11A, all the CuO-TiO_2 nanocomposite showed a high degradation efficiency against SSX than for CIP. The CuO-TiO_2 photocatalyst degraded 51% (30/70), 55% (50/50), and 67% (70/30) of CIP, while CuO and TiO_2 only degraded 58.2 and 62.7%, respectively. From the kinetics diagram (Figure 11B), the rates of reactions were 0.044, 0.0069, and 0.009 min^{-1} for the CuO-TiO_2 30/70, 50/50, and 70/30, respectively, meaning the rate of the reaction for CuO-TiO2 was faster compared to the other two materials. On the other hand, the percentage degradation of SSX by CuO-TiO_2 30/70, 50/50, and 70/30 was recorded to be 93.6%, 93.2%, and 91.2%, which is the highest for these composites compared to MB and CIP, respectively. While the degradation of CuO and TiO_2 was 44.8% and 50.2%, respectively. Comparing the results with previous studies, Mofokeng et al. [48] prepared a CuO/TiO_2@GCN, which was able to degrade 94% of 10 mg/mL ketoprofen (KP) after 30 min irradiation of visible light. Using UV light, Castaneda et al. [49] degraded 90% of caffeine within 3 h using 100 mg of CuO-TiO_2-F photocatalyst. Hajipour et al. [50] utilized 5 mg of TiO_2/CuO photocatalyst to degrade 80% of amoxicillin. The CuO-TiO_2-GO prepared by Cosma et al. [51] degraded 95% and 98% of amoxicillin and ciprofloxacin, respectively. The photodegradation process was conducted over 120 min. Li et al. [52] used green-synthesized TP-TiO_2 to degrade 90% of CIP in 60 min under the UV-vis light source. These nanocomposites are suitable photocatalysts for the degradation of pharmaceutical drugs.

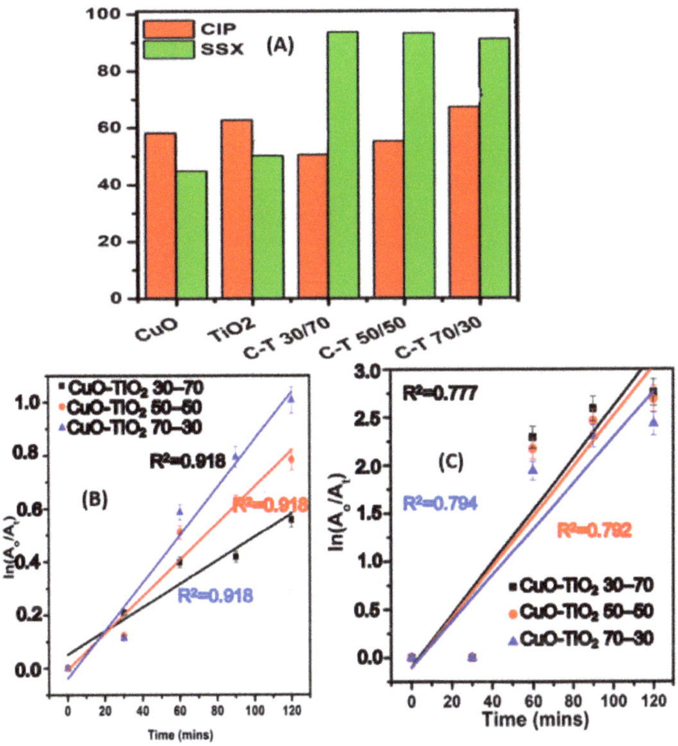

Figure 11. (**A**) Photocatalytic degradation of various materials against SSX and CIP, (**B**) kinetics of CIP, and (**C**) kinetics of SSX.

2.8.3. Effect of Trapping and Mechanism of Degradation

In an effort to test the cost-effectiveness of a material, reusability studies were conducted using the 30/70 CuO-TiO_2 NPs for the degradation of SSX. It is important for synthesized materials to not only have a high degradation efficiency but to ensure their material are cost-effective should they be upscaled. In this study, the 30/70 CuO-TiO_2

material was tested for its efficiency at least four times. As it can be noted (Figure 12B), the material was not reusable as the efficiency significantly dropped. Upon testing the species responsible for the degradation, it can be noted that the •OH species were the main active species for the degradation of SSX as the degradation dropped to 59% as these species were trapped during the photodegradation process. Other studies have also shown the •OH radicals to be the species responsible for their degradation.

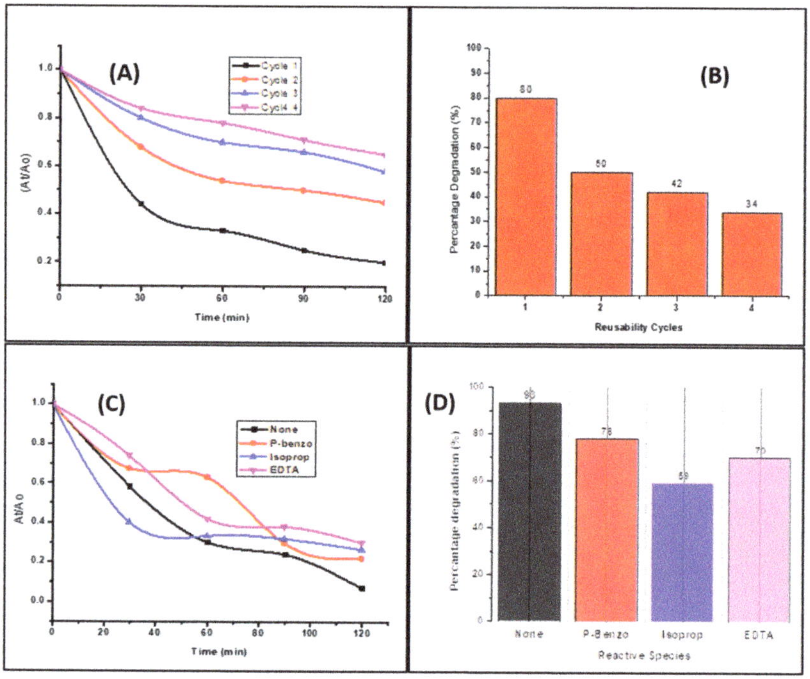

Figure 12. (**A**,**B**) Reusability studies, (**C**,**D**) effect of reactive.

On the basis of the observed photocatalytic activities and the characterizations shown above, a probable mechanism for the photooxidation of SSX is illustrated in Scheme 1. TiO_2 is a wide band gap semiconductor; thus, due to its limited access to the wider spectrum, low bandgap materials such as CuO are coupled with it to reduce the recombination rate and fast generation of radicals. Upon irradiating enough light on the photocatalysts, the electrons were excited and moved up to the conduction band of TiO_2. These photoexcited electrons reacted with the O_2 species that were adsorbed on the TiO_2 surface, thus forming the O_2 radical anion and the OH radical through the protonation process [43]. Therefore, the CuO, which is a low bandgap material (1.2 eV), assisted in the reduction of the holes and electron recombining. Furthermore, environmentally friendly materials are potentially formed.

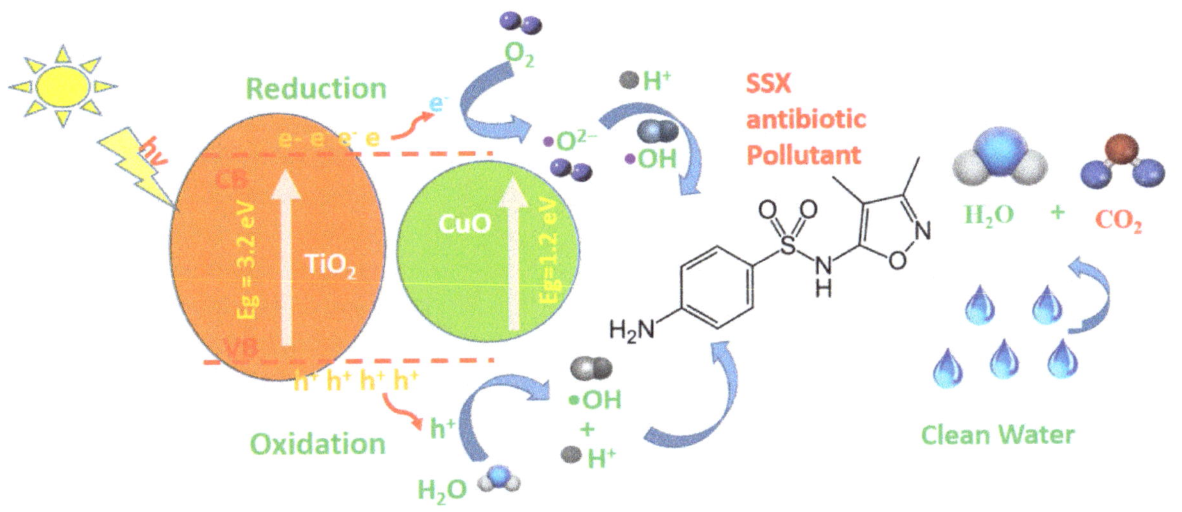

Scheme 1. For the mechanism of degradation.

3. Experimental Procedure

3.1. Chemicals and Materials

Analytical grade chemicals such as $Cu(NO_3)_2{\cdot}3H_2O$, propan-2-ol (C_3H_8O), ethylenediaminetetraacetic acid (EDTA), p-benzoquinone ($C_6H_4O_2$) and ciprofloxacin (CIP), methylene blue (MB), Titanium Fluoride (TiF_4) were purchased from Sigma Aldrich, Germany. The pharmaceutical antibiotic sulfisoxazole (SSX) was obtained from the organic group at the University of Limpopo's Department of Chemistry. Methylene blue dye was purchased from Merck chemicals, Germany. The *C. benghalensis* plant was harvested, dried, and ground at the University of Limpopo. No further purifications were performed in all the chemicals used.

3.2. Preparation of the Plant Extracts

The *C. benghalensis plants* were collected in the fields of University of Limpopo between September–November (yearly basis). The plant was washed with tap water, then the leaves and flowers were removed in order to work with only the roots and stem. The plant was dried (indoors) for 6–8 weeks. The dry plant was then crushed (using a blender) and stored in an airtight container. The plant extracts of *C. benghalensis* (See Scheme 2) were prepared in a 500 mL flask containing 250 mL of hot water. A total of 10 g *C. benghalensis* powder was added and further heated at 80 °C for 15 min. The resulting green-brownish extract was cooled (at room temperature) and then collected by vacuum filtration. The extracts were reserved in a refrigerator for further use.

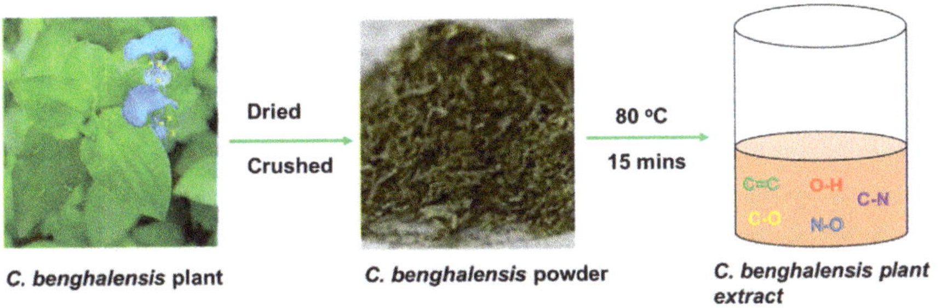

Scheme 2. *C. benghalensis* plant extract preparation.

3.3. Preparation of *C. benghalensis*-Mediated TiO_2 Nanoparticle Synthesis

Green-medisted TiO_2-NPs were prepared by mixing 50 mL of 0.5 M TiF_4 with an aliquot of 50 mL of 10 g *C. benghalensis* extract, respectively. The mixture was then heated for two hours at 90 °C until it turned brownish-yellowish color. The mixture was allowed to cool at room temperature before being centrifuged for 30 min at 4000 rpm. It was then washed and centrifuged in distilled water for another 15 min at 4000 rpm. After that, the solid precipitate was transferred to the crucible and calcined for 2 h in a 400 °C furnace before cooling. The product was gathered, crushed, and stored for later.

3.4. Preparation of *C. benghalensis*-Mediated CuO-NPs

An aliquot of 50 mL of *C. benghalensis* extract was added to 5 g of cupric nitrate ($Cu(NO_3)_2{\cdot}3H_2O$) and then boiled at 80 °C for 1 h until a brown-colored paste was observed. The paste was collected in a ceramic crucible and heated in an air furnace at 400 °C for 2 h. A black powder was obtained, and the product was carefully collected and packed for characterization.

3.5. Synthesis of CuO-TiO_2 Nanocomposite

The CuO-TiO_2 nanocomposites were synthesized (Scheme 3) by adding an aliquot of 50 mL of the *C. benghalensis* to the $Cu(NO_3)_2 3H_2O$ and TiF_4 at ratios of 70:30, 50:50, and 30:70, separately. The mixture was heated at 80 °C for an hour to produce a paste. The paste was then transferred into a crucible and calcined in a furnace for 2 h. The products were collected and crushed for further characterization and applications.

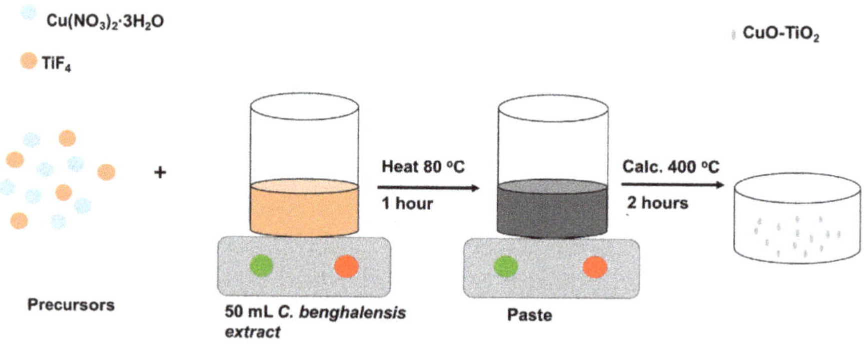

Scheme 3. Green synthesis of CuO-TiO_2 nanocomposite.

3.6. Characterization of CuO-TiO_2 Materials

FTIR, over a range of 4000–500 cm^{-1}, was used to study the stretching bond frequencies of the synthesized CuO-TiO_2 nanocomposites, phytochemical identification, and nanoparticle synthesis. The optical parameters of the generated materials, as well as the concentration of the dye and pharmaceutical component, were determined using a PerkinElmer Spectrum SP-UV 500VDB spectrophotometer. The bandgap energy (Eg) of CuO-TiO_2 nanocomposites was calculated using Tauc's plot. The crystallinity and phase composition of the materials were determined by scanning for 15 min in the range of 20–90 2θ degrees on a Bruker AXS D8 Advance X-ray Diffractometer (XRD) (Germany). Scanning electron microscopy (SEM), coupled with Image J, was used to determine the particle size distribution and morphology of the samples. The electron-dispersive X-ray (EDX) analysis was used for elemental identification and chemical composition of the nanocomposites. Pore sizes, specific surface area, and pore volume were measured at 196 °C using the Bruner–Emmet–Teller (BET) surface analyzer, with which the sample was heated to 150 °C for 2 h using N_2 gas flow. A Perkin Elmer Pyris Thermogravimetric Analyzer (TGA) was used to test the thermal stability, with a thermal analyzer, SDT Q600, heating 10 mg of the sample to 900 °C at a rate of 10 °C/min.

3.7. Electrochemical Experiment

Cyclic voltammetry (CV), linear swept voltammetry (LSV), and electrochemical impedance spectroscopy (EIS) techniques were evaluated to characterize the electrochemical properties of the samples, using an Autolab electrochemical workstation with three electrodes electrochemical cell, namely the modified glassy carbon electrode as a working electrode, Ag/AgCl as a reference and platinum wire as a counter electrode in 0.5 M H_2SO_4 as an electrolyte. Glassy carbon was modified with 3 different samples, namely CuO-TiO_2 (30/70), CuO-TiO_2 (50/50) CuO-TiO_2 (70/30). The samples were renamed as CuO-TiO_2 (30/70): sample 1 (S1), CuO-TiO_2 (50/50) sample 2, and (S2) CuO-TiO_2 (70/30): sample 3 (S3). The voltametric tests were recorded in the potential range of −0.6 V to 0.6 V at a different scan rate from 2 to 50 mV/s. Before the modification of the glassy carbon electrode, the electrode was polished with 1, 0.5, and 0.03 μm polishing powder and then rinsed with ethanol and distilled water after each polish.

3.8. Preparation of the Modified Electrode

An amount of 3 mg of different powders (S1–S3) was dissolved in 1000 μL of ethanol, and 3 μL of 5% Nafion solution was added; thereafter, the mixtures were ultra-sonicated to make a slurry. Then 3 μL of the solution was drop-coated on the surface area of the glassy carbon electrode and allowed to dry using an oven at 35 °C for 15 min. The electrochemical impedance spectroscopy was recorded over the frequency range of 0.1–100 kHz. Before each electrochemical measurement, the solution was purged using Ar gas for deoxygenation.

3.9. Photocatalytic Degradation of Dyes and Antibiotics

The 30 mg of CuO-TiO_2 powder was added in 300 mL of 20 ppm methylene blue (MB), then stirred in a photochemical reactor in the absence of light for 30 min to reach the adsorption-desorption equilibrium. Then, the mixture was exposed to UV light (450 W) for 2 h. Upon exposure to the light, at 30 min intervals, 3 mL of the samples were collected for analysis using the UV-vis spectrophotometer. The λ_{max} of 668 nm for MB was used to monitor the degradation over time. The efficiency of the degradation was calculated by the following equation:

$$\% \text{ Degradation } = \left(\frac{\text{Ao} - \text{Af}}{\text{Ao}}\right) \times 100\% \tag{3}$$

where:

A_o = is the initial absorbance at 0 min (adsorption-desorption).
A_f = is the final absorbance of 30, 60, 90, and 120 min of photodegradation

Similar steps were followed for the degradation of the antibiotic CIP and SSX, except upon analysis, λ_{max} values of 275 and 290 were used, and the concentration was 10 ppm.

3.10. Recyclability Studies of SSX

To determine both the stability and recyclability of the CuO-TiO_2 nanocomposites, then 4 cycle experiments were conducted following similar conditions for normal photocatalytic degradation experiments. After every cycle, the photocatalyst was thoroughly washed with distilled water and recovered by membrane filtration. Then, it was dried at 60 °C to obtain a powder-like material for the next cycle.

3.11. Effect of Scavengers on the Degradation of SSX

To determine the effect of scavengers on the degradation of the SSX, holes (h^+), hydroxyl radicals (OH·), and superoxide (O_2^-) were investigated during the photodegradation process. A measure of 2 mL (5 mmol) of EDTA, isopropyl alcohol, and p-benzoquinone was added, respectively, for each individual experiment. The photodegradation of similar conditions for normal photocatalytic degradation experiments was followed.

4. Conclusions

CuO-TiO_2 p-n heterostructured photocatalysts were prepared in this study and tested for the degradation of MB, CIP, and SSX pollutants under UV light irradiations. This research was motivated by the need for highly efficient photocatalysts for the degradation of pollutants in wastewater. The materials were found to be spherical in shape with varied particle sizes, thermally stable, and electrochemically active, which could potentially assist in the photodegradation of the organic pollutants. These materials also displayed high degradation potential against antibiotics CIP and SSX when exposed to UV light. The degradation efficiency of CIP and SSX were recorded to be 67.3% and 93.64%, respectively. Furthermore, it was shown that the heterostructured photocatalysts could not be reused. Additionally, the species that is most active in the photodegradation of SSX was identified as OH•. These heterojunctions were shown to be selective toward antibiotic degradation as compared to dyes.

Supplementary Materials: The following supporting information can be downloaded at: https://www.mdpi.com/article/10.3390/catal13010163/s1, Figure S1. (a) EDS of *C. Benhghanlensis* TiO_2; (b) EDS of *C. Benhghanlensis* CuO NPs; (c) EDS of *C. Benhghanlensis* $CuOTiO_2$(30/70); (d) EDS of *C. Benhghanlensis* $CuOTiO_2$(50/50); € EDS of *C. Benhghanlensis* $CuOTiO_2$(70/30); Figure S2: SSX photodegradation curves by (a) 30/70 CuO-TiO_2 (b) 50/50 CuO-TiO_2 (c) 70/30 CuO-TiO_2 (d) TiO_2 and (e) CuO nanomaterials; Figure S3: CIP photodegradation curves by (a) 30/70 CuO-TiO_2 (b) 50/50 CuO-TiO_2 (c) 70/30 CuO-TiO_2 (d) TiO_2 and (e) CuO nanomaterials.

Author Contributions: Conceptualization, N.H-M. and D.E.M.; methodology, D.A.B., S.M., N.M.; formal analysis, D.A.B., S.M., N.M.; investigation, D.A.B., N.M.; resources, N.C.H.-M. and D.E.M.; writing—original draft preparation, D.A.B., N.M.; writing—review and editing N.C.H.-M.; supervision, N.C.H.-M. and D.E.M.; funding acquisition, N.C.H.-M. and D.E.M. All authors have read and agreed to the published version of the manuscript.

Funding: The National Research Foundation and the DSI NRF Centre of Excellence in Strong Materials (grant no. R808-CoE Mbita) provided funding for this work and are greatly acknowledged.

Data Availability Statement: The data can be made available upon request from the authors.

Conflicts of Interest: The authors affirm that they have no known financial or interpersonal conflicts that would have appeared to have an impact on the research presented in this study.

References

1. Khoshbin, S.; Seyyedi, K. Removal of acid red 1 dye pollutant from contaminated waters by electrocoagulation method using a recirculating tubular reactor. *Lat. Am. Appl. Res.* **2017**, *47*, 101–105. [CrossRef]
2. Santhy, K.; Selvapathy, P. Removal of reactive dyes from wastewater by adsorption on coir pith activated carbon. *Bioresour. Technol.* **2006**, *97*, 1329–1336. [CrossRef] [PubMed]
3. Gómez, M.J.; Gómez-Ramos, M.M.; Agüera, A.; Mezcua, M.; Herrera, S.; Fernández-Alba, A.R. A new gas chromatography/mass spectrometry method for the simultaneous analysis of target and non-target organic contaminants in waters. *J. Chromatogr. A* **2009**, *1216*, 4071–4082. [CrossRef] [PubMed]
4. Aigbe, U.O.; Ukhurebor, K.E.; Onyancha, R.B.; Okundaye, B.; Pal, K.; Osibote, O.A.; Esiekpe, E.L.; Kusuma, H.S.; Darmokoesoemo, H. A facile review on the sorption of heavy metals and dyes using bionanocomposites. *Adsorpt. Sci. Technol.* **2022**, *2022*, 8030175. [CrossRef]
5. Arismendi, D.; Becerra-Herrera, M.; Cerrato, I.; Richter, P. Simultaneous determination of multiresidue and multiclass emerging contaminants in waters by rotating-disk sorptive extraction–derivatization-gas chromatography/mass spectrometry. *Talanta* **2019**, *201*, 480–489. [CrossRef] [PubMed]
6. Vishnu, D.; Dhandapani, B.; Authilingam, S.; Sivakumar, S.V. A comprehensive review of effective adsorbents used for the removal of dyes from wastewater. *Curr. Anal. Chem.* **2022**, *18*, 255–268. [CrossRef]
7. Sridhar, A.; Ponnuchamy, M.; Kapoor, A.; Prabhakar, S. Valorization of food waste as adsorbents for toxic dye removal from contaminated waters: A review. *J. Hazard. Mater.* **2022**, *424*, 127432. [CrossRef]

8. Aftab, S.; Shabir, T.; Shah, A.; Nisar, J.; Shah, I.; Muhammad, H.; Shah, N.S. Highly efficient visible light active doped ZnO photocatalysts for the treatment of wastewater contaminated with dyes and pathogens of emerging concern. *Nanomaterials* **2022**, *12*, 486. [CrossRef]
9. Zhao, X.; Baharinikoo, L.; Farahani, M.D.; Mahdizadeh, B.; Farizhandi, A.A.K. Experimental modelling studies on the removal of dyes and heavy metal ions using $ZnFe_2O4$ nanoparticles. *Sci. Rep.* **2022**, *12*, 5987. [CrossRef]
10. Ariaeenejad, S.; Motamedi, E.; Salekdeh, G.H. Highly efficient removal of dyes from wastewater using nanocellulose from quinoa husk as a carrier for immobilization of laccase. *Bioresour. Technol.* **2022**, *349*, 126833. [CrossRef]
11. Kang, S.; Zhao, Y.; Wang, W.; Zhang, T.; Chen, T.; Yi, H.; Rao, F.; Song, S. Removal of methylene blue from water with montmorillonite nanosheets/chitosan hydrogels as adsorbent. *Appl. Surf. Sci.* **2018**, *448*, 203–211. [CrossRef]
12. Siddiqui, S.I.; Manzoor, O.; Mohsin, M.; Chaudhry, S.A. Nigella sativa seed based nanocomposite-MnO_2/BC: An antibacterial material for photocatalytic degradation, and adsorptive removal of Methylene blue from water. *Environ. Res.* **2019**, *171*, 328–340. [CrossRef] [PubMed]
13. Duan, J.; Jian, H.; Dou, Q.; Shi, X.; Su, R. Indirect photodegradation of sulfisoxazole: Effects of environmental factors (CDOM, pH, salinity, HCO_3-, metal ions, halogen ions and NO_3-). *Mar. Pollut. Bull.* **2022**, *174*, 113320. [CrossRef] [PubMed]
14. Orachorn, N.; Bunkoed, O. Nanohybrid magnetic composite optosensing probes for the enrichment and ultra-trace detection of mafenide and sulfisoxazole. *Talanta* **2021**, *228*, 122237. [CrossRef] [PubMed]
15. Uhlemann, T.; Seidel, S.; Müller, C.W. Site-specific binding of a water molecule to the sulfa drugs sulfamethoxazole and sulfisoxazole: A laser-desorption isomer-specific UV and IR study. *Phys. Chem. Chem. Phys.* **2018**, *20*, 6891–6904. [CrossRef]
16. Girardi, C.; Greve, J.; Lamshöft, M.; Fetzer, I.; Miltner, A.; Schäffer, A.; Kästner, M. Biodegradation of ciprofloxacin in water and soil and its effects on the microbial communities. *J. Hazard. Mater.* **2011**, *198*, 22–30. [CrossRef]
17. Igwegbe, C.A.; Oba, S.N.; Aniagor, C.O.; Adeniyi, A.G.; Ighalo, J.O. Adsorption of ciprofloxacin from water: A comprehensive review. *J. Ind. Eng. Chem.* **2021**, *93*, 57–77. [CrossRef]
18. Ojo, B.O.; Arotiba, O.A.; Mabuba, N. Sonoelectrochemical degradation of ciprofloxacin in water on a Ti/$BaTiO_3$ electrode. *J. Environ. Chem. Eng.* **2022**, *10*, 107224. [CrossRef]
19. Xing, X.; Du, Z.; Zhuang, J.; Wang, D. Removal of ciprofloxacin from water by nitrogen doped TiO_2 immobilized on glass spheres: Rapid screening of degradation products. *J. Photochem. Photobiol. A Chem.* **2018**, *359*, 23–32. [CrossRef]
20. Soto-Robles, C.A.; Luque, P.A.; Gómez-Gutiérrez, C.M.; Nava, O.; Vilchis-Nestor, A.R.; Lugo-Medina, E.; Ranjithkumar, R.; Castro-Beltrán, A. Study on the effect of the concentration of Hibiscus sabdariffa extract on the green synthesis of ZnO nanoparticles. *Results Phys.* **2019**, *15*, 102807. [CrossRef]
21. Ngoepe, N.M.; Mathipa, M.M.; Hintsho-Mbita, N.C. Biosynthesis of titanium dioxide for the degradation of methylene blue and removal of bacteria. *Optik* **2020**, *224*, 165728. [CrossRef]
22. Jang, Y.J.; Simer, C.; Ohm, T. Comparison of zinc oxide nanoparticles and its nano-crystalline particles on the photocatalytic degradation of methylene blue. *Mater. Res. Bull.* **2006**, *41*, 67–77. [CrossRef]
23. Munyai, S.; Hintsho-Mbita, N.C. Green derived metal suphides as photocatalysts for wastewater treatment. A review. *Curr. Res. Green Sustain. Chem.* **2021**, *4*, 100163. [CrossRef]
24. Lv, Y.; Liu, J.; Zhang, Z.; Zhang, W.; Wang, A.; Tian, F.; Zhao, W.; Yan, J. Green synthesis of CuO nanoparticles-loaded ZnO nanowires arrays with enhanced photocatalytic activity. *Mater. Chem. Phys.* **2021**, *267*, 124703. [CrossRef]
25. Bopape, D.A.; Motaung, D.E.; Hintsho-Mbita, N.C. Green synthesis of ZnO: Effect of plant concentration on the morphology, optical properties and photodegradation of dyes and antibiotics in wastewater. *Optik* **2022**, *251*, 168459. [CrossRef]
26. Prabu, H.J.; Varghese, R.; Johnson, I.; Sundaram, S.J.; Raj, A.D.; Rajakrishnan, R.; Kuppusamy, P.; Sathya, R.; Kaviyarasu, K. Laser induced plant leaf extract mediated synthesis of CuO nanoparticles and its photocatalytic activity. *Environ. Res.* **2022**, *212*, 113295. [CrossRef]
27. Chowdhury, I.H.; Ghosh, S.; Basak, S.; Naskar, M.K. Mesoporous CuO-TiO_2 microspheres for efficient catalytic oxidation of CO and photodegradation of methylene blue. *J. Phys. Chem. Solids* **2017**, *104*, 103–110. [CrossRef]
28. Patil, B.N.; Taranath, T.C. *Limonia acidissima* L. leaf mediated synthesis of silver and zinc oxide nanoparticles and their antibacterial activities. *Microb. Pathog.* **2018**, *115*, 227–232. [CrossRef]
29. Lu, D.; Zelekew, O.A.; Abay, A.K.; Huang, Q.; Chen, X.; Zheng, Y. Synthesis and photocatalytic activities of a CuO/TiO_2 composite catalyst using aquatic plants with accumulated copper as a template. *RSC Adv.* **2019**, *9*, 2018–2025. [CrossRef]
30. Khodadadi, B.; Yeganeh Faal, A.; Shahvarughi, A. Tilia platyphyllos extract assisted green synthesis of CuO/TiO_2 nanocomposite: Application as a reusable catalyst for the reduction of organic dyes in water. *J. Appl. Chem. Res.* **2019**, *13*, 51–65.
31. Ibrahim, J.; Ajaegbu, V.C.; Egharevba, H.O. Pharmacognostic and phytochemical analysis of *Commelina benghalensis* L. *Ethnobot. Leafl.* **2010**, *2010*, 7.
32. Johnson, I.; Prabu, H.J. Green synthesis and characterization of silver nanoparticles by leaf extracts of *Cycas circinalis, Ficus amplissima, Commelina benghalensis* and *Lippia nodiflora*. *Int. Nano Lett.* **2015**, *5*, 43–51. [CrossRef]
33. Feng, Y.; Yan, X.; Liu, C.; Hong, Y.; Zhu, L.; Zhou, M.; Shi, W. Hydrothermal synthesis of CdS/Bi_2MoO_6 heterojunction photocatalysts with excellent visible-light-driven photocatalytic performance. *Appl. Surf. Sci.* **2015**, *353*, 87–94. [CrossRef]
34. Karim, K.M.R.; Jebi, M.A.P.; Ong, H.R.; Abdullah, H.; Tarek, M.; Cheng, C.K.; Khan, M.M.R. CuO-TiO_2 as a visible light responsive photocatalyst for the photoelectroreduction of CO_2 to methanol. *Natl. Conf. Postgrad. Res.* **2018**, 103–111.

35. Gnanasekaran, L.; Pachaiappan, R.; Kumar, P.S.; Hoang, T.K.; Rajendran, S.; Durgalakshmi, D.; Soto-Moscoso, M.; Cornejo-Ponce, L.; Gracia, F. Visible light driven exotic p (CuO)-n (TiO_2) heterojunction for the photodegradation of 4-chlorophenol and antibacterial activity. *Environ. Pollut.* **2021**, *287*, 117304. [CrossRef] [PubMed]
36. Muzakki, A.; Shabrany, H.; Saleh, R. Synthesis of ZnO/CuO and TiO_2/CuO nanocomposites for light and ultrasound assisted degradation of a textile dye in aqueous solution. *AIP Conf. Proc.* **2016**, *1725*, 020051.
37. Kubiak, A.; Bielan, Z.; Kubacka, M.; Gabała, E.; Zgoła-Grześkowiak, A.; Janczarek, M.; Zalas, M.; Zielińska-Jurek, A.; Siwińska-Ciesielczyk, K.; Jesionowski, T. Microwave-assisted synthesis of a TiO_2-CuO heterojunction with enhanced photocatalytic activity against tetracycline. *Appl. Surf. Sci.* **2020**, *520*, 146344. [CrossRef]
38. Zangeneh, H.; Farhadian, M.; Zinatizadeh, A.A. N (Urea) and CN (L-Asparagine) doped TiO_2-CuO nanocomposites: Fabrication, characterization and photodegradation of direct red 16. *J. Environ. Chem. Eng.* **2020**, *8*, 103639. [CrossRef]
39. Zhang, S.; Gong, X.; Shi, Q.; Ping, G.; Xu, H.; Waleed, A.; Li, G. CuO nanoparticle-decorated TiO_2-nanotube heterojunctions for direct synthesis of methyl formate via photo-oxidation of methanol. *ACS Omega* **2020**, *5*, 15942–15948. [CrossRef]
40. Shi, Q.; Qin, Z.; Yu, C.; Waheed, A.; Xu, H.; Gao, Y.; Abroshan, H.; Li, G. Experimental and mechanistic understanding of photo-oxidation of methanol catalyzed by CuO/TiO_2-spindle nanocomposite: Oxygen vacancy engineering. *Nano Res.* **2020**, *13*, 939–946. [CrossRef]
41. Yu, Y.H.; Chen, Y.P.; Cheng, Z. Microwave-assisted synthesis of rod-like CuO/TiO_2 for high-efficiency photocatalytic hydrogen evolution. *Int. J. Hydrogen Energy* **2015**, *40*, 15994–16000. [CrossRef]
42. Ashok, C.H.; Venkateswara Rao, K. Microwave-assisted synthesis of CuO/TiO_2 nanocomposite for humidity sensor application. *J. Mater. Sci. Mater. Electron.* **2016**, *27*, 8816–8825. [CrossRef]
43. Fouda, A.; Salem, S.S.; Wassel, A.R.; Hamza, M.F.; Shaheen, T.I. Optimization of green biosynthesized visible light active CuO/ZnO nano-photocatalysts for the degradation of organic methylene blue dye. *Heliyon* **2020**, *6*, e04896. [CrossRef] [PubMed]
44. Gou, X.; Guo, Z. Hybrid Hydrophilic–Hydrophobic CuO@ TiO2-Coated Copper Mesh for Efficient Water Harvesting. *Langmuir* **2019**, *36*, 64–73. [CrossRef]
45. Tayeb, A.M.; Hussein, D.S. Synthesis of TiO_2 nanoparticles and their photocatalytic activity for methylene blue. *Am. J. Nanomater.* **2015**, *3*, 57–63.
46. Sankar, R.; Manikandan, P.; Malarvizhi, V.; Fathima, T.; Shivashangari, K.S.; Ravikumar, V. Green synthesis of colloidal copper oxide nanoparticles using Carica papaya and its application in photocatalytic dye degradation. *Spectrochim. Acta Part A Mol. Biomol. Spectrosc.* **2014**, *121*, 746–750. [CrossRef]
47. Lingaraju, K.; Naika, H.R.; Manjunath, K.; Nagaraju, G.; Suresh, D.; Nagabhushana, H. Rauvolfia serpentina-mediated green synthesis of CuO nanoparticles and its multidisciplinary studies. *Acta Metall. Sin. (Engl. Lett.)* **2015**, *28*, 1134–1140. [CrossRef]
48. Mofokeng, L.E.; Hlekelele, L.; Moma, J.; Tetana, Z.N.; Chauke, V.P. Energy-efficient CuO/TiO_2@ GCN cellulose acetate-based membrane for concurrent filtration and photodegradation of ketoprofen in drinking and groundwater. *Appl. Sci.* **2022**, *12*, 1649. [CrossRef]
49. Castañeda, C.; Martínez, J.J.; Santos, L.; Rojas, H.; Osman, S.M.; Gómez, R.; Luque, R. Caffeine photocatalytic degradation using composites of NiO/TiO_2–F and CuO/TiO_2–F under UV irradiation. *Chemosphere* **2022**, *288*, 132506. [CrossRef]
50. Hajipour, P.; Eslami, A.; Bahrami, A.; Hosseini-Abari, A.; Saber, F.Y.; Mohammadi, R.; Mehr, M.Y. Surface modification of TiO_2 nanoparticles with CuO for visible-light antibacterial applications and photocatalytic degradation of antibiotics. *Ceram. Int.* **2021**, *47*, 33875–33885. [CrossRef]
51. Cosma, D.; Urda, A.; Radu, T.; Rosu, M.C.; Mihet, M.; Socaci, C. Evaluation of the Photocatalytic Properties of Copper Oxides/Graphene/TiO_2 Nanoparticles Composites. *Molecules* **2022**, *27*, 5803. [CrossRef]
52. Li, Y.; Fu, Y.; Zhu, M. Green synthesis of 3D tripyramid TiO_2 architectures with assistance of aloe extracts for highly efficient photocatalytic degradation of antibiotic ciprofloxacin. *Appl. Catal. B Environ.* **2020**, *260*, 118149. [CrossRef]

catalysts

MDPI

Article

Effect of Pd-Doping Concentrations on the Photocatalytic, Photoelectrochemical, and Photoantibacterial Properties of CeO_2

Shaidatul Najihah Matussin [1], Fazlurrahman Khan [2,3], Mohammad Hilni Harunsani [1], Young-Mog Kim [2,3,4] and Mohammad Mansoob Khan [1,*]

1 Chemical Sciences, Faculty of Science, Universiti Brunei Darussalam, Jalan Tungku Link, Gadong BE 1410, Brunei
2 Marine Integrated Biomedical Technology Center, The National Key Research Institutes in Universities, Pukyong National University, Busan 48513, Republic of Korea
3 Research Center for Marine Integrated Bionics Technology, Pukyong National University, Busan 48513, Republic of Korea
4 Department of Food Science and Technology, Pukyong National University, Busan 48513, Republic of Korea
* Correspondence: mmansoobkhan@yahoo.com or mansoob.khan@ubd.edu.bn

Abstract: Cerium oxide (CeO_2) can exhibit good photocatalytic and photoantibacterial activities. However, its light-harvesting property is rather limited due to its large band gap. In order to boost these properties, doping with metal ions can improve light absorption and charge mobility. In this report, CeO_2 and palladium−doped CeO_2 (Pd−CeO_2) NPs were synthesized via the microwave-assisted synthesis method. The structural, optical, and morphological studies of CeO_2 and Pd−CeO_2 NPs were carried out using various techniques. Mixed phases of CeO_2/Ce_2O_3 were observed in pure CeO_2 (S−CeO_2) and Pd−CeO_2. However, the Ce_2O_3 phase gradually disappeared upon doping with a higher percentage of Pd. Almost spherical particles were observed with average sizes between 6 and 13 nm. It was found that the incorporation of Pd reduced the particle size. Moreover, band gap energies of S−CeO_2 and Pd−CeO_2 NPs were reduced from 2.56 to 2.27 eV, and the PL intensities were also quenched with more Pd doping. The shifts in the conduction band and valence band were found to cause the reduction in the band gap energies of S−CeO_2 and Pd−CeO_2 NPs. In the case of photocatalytic degradation of methylene blue, photoelectrochemical, and photoantibacterial activities, Pd−CeO_2 NPs showed enhanced activities under visible light irradiation. Therefore, Pd−CeO_2 NPs have been shown to be a visible-light active material.

Keywords: cerium oxide; CeO_2; Pd−doped CeO_2; noble metal; oxygen vacancies; MB degradation; antibacterial studies

Citation: Matussin, S.N.; Khan, F.; Harunsani, M.H.; Kim, Y.-M.; Khan, M.M. Effect of Pd-Doping Concentrations on the Photocatalytic, Photoelectrochemical, and Photoantibacterial Properties of CeO_2. *Catalysts* **2023**, *13*, 96. https://doi.org/10.3390/catal13010096

Academic Editors: Jorge Bedia and Carolina Belver

Received: 1 December 2022
Revised: 22 December 2022
Accepted: 25 December 2022
Published: 3 January 2023

1. Introduction

Antibiotics have been used to treat bacterial infections that are major causes of chronic infections and mortality. However, studies found that the widespread use of antibiotics has led to the emergence of multidrug-resistant bacterial strains [1]. In general, the major groups of antibiotics that are currently in use have three bacterial targets, namely: cell wall synthesis, translational machinery, and DNA replication machinery [2]. However, bacterial resistance can develop at each mode of action. Nanoparticles (NPs) have demonstrated broad-spectrum antibacterial properties against Gram-positive and Gram-negative bacteria. Studies have revealed that NPs do not show antibiotic resistance mechanisms because NPs contact directly with the bacterial cell wall, which would be less prone to promoting resistance in bacteria. Therefore, NPs can be potentially used as an alternative antibiotic. According to research, the major processes underlying the antibacterial effects of NPs are disruption of the bacterial cell membrane, generation of ROS, penetration of bacterial cell membrane, and induction of intracellular antibacterial effects, including interactions with DNA and proteins [2,3].

Organic dyes, which generally consist of non-biodegradable, highly poisonous, and colored pigments, have been reported to be present and widely spread in industrial wastewater originating from the paper, textile, and apparel industries [4,5]. Even at low concentrations, dyes can be clearly seen and pollute various aquatic environments [6]. These dye-polluted effluents are harmful to living organisms [7]. Dyes such as methylene blue (MB) can also cause serious threats to humans, for example, abdominal disorders, respiratory problems, and digestive and mental disorders [8,9]. Hence, the removal of dyes from wastewater is important. Apart from that, redox reactions have shown promising water treatment activity as the reaction products can be conveniently separated and removed [10]. Owing to this, researchers have applied electrochemical systems to diverse functional applications, including photovoltaics [11], fuel cells [12], supercapacitors [13], as well as sensors [14]. It is reported that heterostructure exhibits interesting electrochemical properties owing to the possibility of a dual charge storage contribution from both materials [15]. Nevertheless, materials with high porosity and large surface areas are highly suitable for electrochemical applications [16].

Metal oxides such as TiO_2 [17], ZnO [18], Fe_2O_3 [19], CeO_2 [20], etc., have shown remarkable activities in biological and photocatalytic activities. Among these, CeO_2 has shown extensive industrial applications in medicine, catalysis, solid oxide fuel cells, luminescence, optical, and sensor technologies [21–25]. Despite being a highly abundant material, its redox ability to change between $Ce^{3+} \leftrightarrow Ce^{4+}$ has also attracted significant attention. CeO_2 has been synthesized using hydrothermal, sol-gel, precipitation, microwave, and green synthesis methods where multiple morphologies such as cube, rod, spherical, and flower-like particles have been produced [26–30]. In catalytic applications, the activity of CeO_2 primarily arises from the reduction of Ce^{4+} to Ce^{3+} and the formation of oxygen vacancies [31,32]. The modification in shape and size results in the formation of surface defects such as V_o, which endows it with the ability to exist in either the Ce^{3+} or Ce^{4+} state on the surface of the particle [33,34].

In addition, dopants with different ionic states incorporated into CeO_2 can generate more structural defects to gain charge neutrality, enhancing physical properties and biocompatibility. Many studies have emphasized a strong synergy between CeO_2 and noble metals, which can significantly enhance catalytic activities [35]. Among the noble metals, palladium is increasingly used in electrical equipment, dental materials, implanted medical devices and automobiles as catalysts [36]. The incorporation of Pd^{2+} into the CeO_2 crystal lattice has shown better catalytic activity than the palladium-supported CeO_2 catalyst [37]. It is believed that the unique properties of Pd-doped CeO_2 depend on the active components of palladium and the interaction between palladium and CeO_2 [38].

Therefore, in this study, CeO_2 (S−CeO_2) and Pd−doped CeO_2 (Pd−CeO_2) NPs were synthesized using a microwave-assisted synthesis method. To the best of the authors' knowledge, a study on Pd−CeO_2 with mixed phases of CeO_2 and Ce_2O_3 has not yet been explored. Moreover, studies on the photocatalytic degradation of methylene blue (MB) were conducted using synthesized-CeO_2 (S−CeO_2) and (0.5, 1, 3 and 5%) Pd−CeO_2 under visible light irradiation for 5 h. The photoelectrochemical studies, namely, linear sweep voltammetry (LSV) and electrochemical impedance spectroscopy (EIS), were performed under dark and visible light conditions using NaCl as the electrolyte. Moreover, the photoantibacterial/antibacterial properties of S−CeO_2 and (0.5, 1, 3 and 5%) Pd−CeO_2 NPs were also carried out for 24 h using bacterial strain *Staphylococcus aureus* (*S. aureus*) KCTC 1916 (Gram-positive) and *Pseudomonas aeruginosa* (*P. aeruginosa*) KCTC 1637 (Gram-negative).

2. Results and Discussion

2.1. Structural Analysis of CeO_2 and Pd−CeO_2 NPs

XRD provides a broad range of information related to the crystallographic nature and chemical structure of materials. The XRD patterns of the S−CeO_2 and Pd−CeO_2 NPs are shown in Figure 1a, with diffraction peaks in the range of 20° to 80°. The peak positions at about 2θ = 28.56°, 33.27°, 47.36°, 56.40°, and 59.10° of each sample correspond to the (111), (200), (220), (311), and (222) planes of CeO_2 cubic fluorite phase (JCPDS no. 00-004-0593) (Figure S1). In S−CeO_2, the diffraction peak of Ce_2O_3 was observed, suggesting that mixed-phase CeO_2/Ce_2O_3 (JCPDS no. 00-023-1048) was present. It was found that gradually, the peak corresponding to Ce_2O_3 decreased with higher Pd doping. Hence, the peak intensity at 28.56° was seen to increase. This suggests that Pd doping inhibited the formation of Ce_2O_3 and suggested that the incorporation of Pd^{2+} lowered the appearance of the Ce_2O_3 phase. The ionic radii of Pd^{2+} and Ce^{4+} are 86 and 101 pm, respectively. No other peaks were observed that contribute to the presence of PdO. Therefore, it can be concluded that Pd^{2+} was successfully incorporated into the CeO_2 lattice.

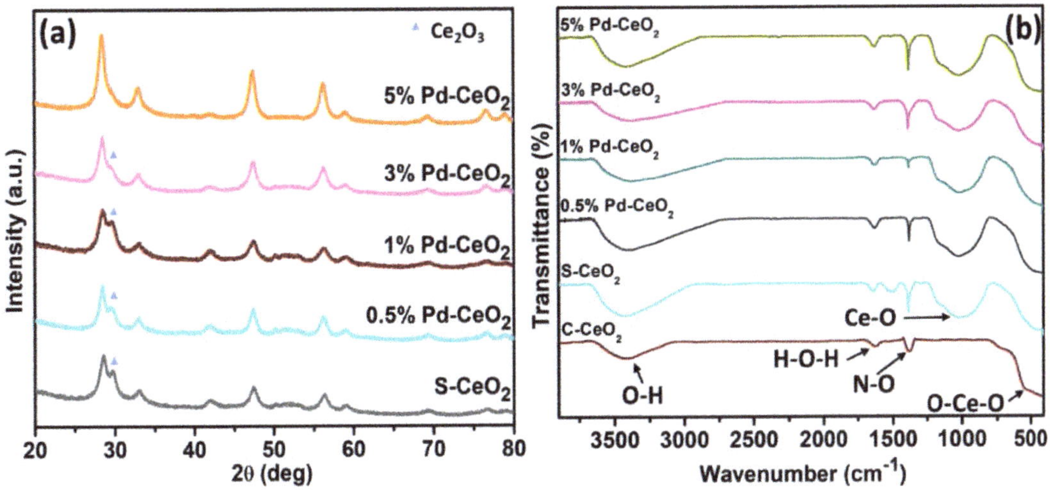

Figure 1. (**a**) XRD patterns and (**b**) FTIR-spectra of S−CeO_2 and (0.5, 1, 3, and 5%) Pd−CeO_2 NPs.

Furthermore, the average crystallite sizes were estimated using Debye-Scherrer's Formula (1):

$$D = \frac{0.9\lambda}{\beta cos\theta} \quad (1)$$

where β is the full width at half maximum (FWHM), in radians, of the peak for a given (*hkl*) value, λ = 1.5406 Å for the CuKα radiation used, and θ is the diffracting angle. The average crystallite sizes were found to be 34.67, 37.25, 16.25, 23.03, and 19.17 nm for S−CeO_2, 0.5%, 1%, 3%, and 5% Pd−CeO_2 NPs, respectively (Table 1).

Table 1. The average crystallite size (D), lattice parameter, cell volume, and average lattice strain of S−CeO_2 and Pd−CeO_2 NPs.

Sample	Average Crystallite Size (D, nm)	Lattice Parameter, *a* (Å)	Cell Volume (Å^3)	Average Lattice Strain
S−CeO_2	34.67	5.414	158.69	0.0022
0.5% Pd−CeO_2	37.25	5.408	158.16	0.0009
1% Pd−CeO_2	16.25	5.423	159.48	0.0004
3% Pd−CeO_2	23.03	5.409	158.25	0.0019
5% Pd−CeO_2	19.17	5.428	159.93	0.0022

The average lattice strain was also calculated using Equation (2):

$$\varepsilon = \frac{\beta_{hkl}}{4tan\theta} \quad (2)$$

The lattice parameters of $S{-}CeO_2$ and $Pd{-}CeO_2$ were observed to be comparable with the bulk CeO_2 (5.411 Å). The average lattice strain was decreased significantly when 0.5% and 1% Pd were incorporated. However, as more Pd^{2+} were incorporated into the lattice, it would cause higher lattice strain. This can also be seen that as more Pd was doped, the cell volume increased slightly.

2.2. Fourier Transform-Infrared Spectroscopy of CeO_2 and $Pd{-}CeO_2$ NPs

FT-IR spectroscopic studies of $S{-}CeO_2$ and $Pd{-}CeO_2$ NPs were carried out within the range of 450–4000 cm^{-1} at room temperature (Figure 1b). The broad band at approximately 3400 cm^{-1} for all samples is attributed to the stretching mode of absorbed O-H in the samples. A broad, intense peak at 450 cm^{-1} corresponds to the O-Ce-O stretching mode. At approximately 900 cm^{-1}, a broad peak was observed in $S{-}CeO_2$ and $Pd{-}CeO_2$ NPs, which corresponds to the bending of the intercalated Ce-O [39]. Moreover, a peak at about 1730 cm^{-1} was also observed in all samples, and it could be assigned to the H-O-H bending mode.

2.3. X-ray Photoelectron Spectroscopy

XPS was performed at room temperature to investigate the chemical state and the electronic structure of the elements in $S{-}CeO_2$ and 0.5 and 5% $Pd{-}CeO_2$ NPs (Figure 2). Figure 2a shows the complete survey scan spectra of the samples, which confirmed the presence of Pd 3*d*, O 1*s*, and Ce 3*d*. Figure 2b shows the six typical peaks for Ce 3*d*. The peaks positioned at 883.43, 886.80, 896.26, 898.89, 905.59, and 914.75 eV is characteristic of Ce^{4+} [40]. Peaks at 898.89, 905.59, and 914.75 eV correspond to $3d_{3/2}$, whereas peaks at 883.43, 886.80, and 896.26 eV correspond to $3d_{5/2}$ [41]. No obvious shift in the binding energy of Ce 3*d* spectra was observed.

Figure 2c shows the Pd 3*d* spectra of 0.5% $Pd{-}CeO_2$ and 5% $Pd{-}CeO_2$ NPs. As a result of the spin-orbit coupling, the Pd 3*d* spectra are split into two peaks of $3d_{5/2}$ and $3d_{3/2}$ at 335.03:335.03 eV and 340.40:340.31 eV, respectively. The XPS spectrum of O 1*s* can be seen in Figure 2d, in which all peaks exhibit two asymmetrical peaks, indicating the presence of O^{2-}, OH^-, and O^- at the surface of the nanostructures. The peak at higher binding energy, i.e., 531–532 eV, is attributed to O^{2-} vacancies and adsorbed -OH or H_2O, while the peak at lower binding energy, 529 eV, is attributed to the metal-oxygen binding [42]. $S{-}CeO_2$ showed a dominant peak at 531–532 eV, suggesting that O^{2-} vacancies or adsorbed -OH were mainly found in the sample. On the other hand, 0.5 and 5% $Pd{-}CeO_2$ showed a dominant peak at 529 eV [42]. The difference might be due to the formation of the Ce_2O_3 phase in $S{-}CeO_2$ and the gradual decrease of the Ce_2O_3 phase in both 0.5 and 5% $Pd{-}CeO_2$, as shown in XRD. The typical C 1*s* were observed in the spectra (Figure 2e), which were derived from the carbon coating used in the analysis. The atomic concentration of C 1*s*, O 1*s*, Ce 3*d*, and Pd 2*p* can be found in Table 2. The estimated Pd content in both 0.5% and 5% $Pd{-}CeO_2$ is lower than the expected value.

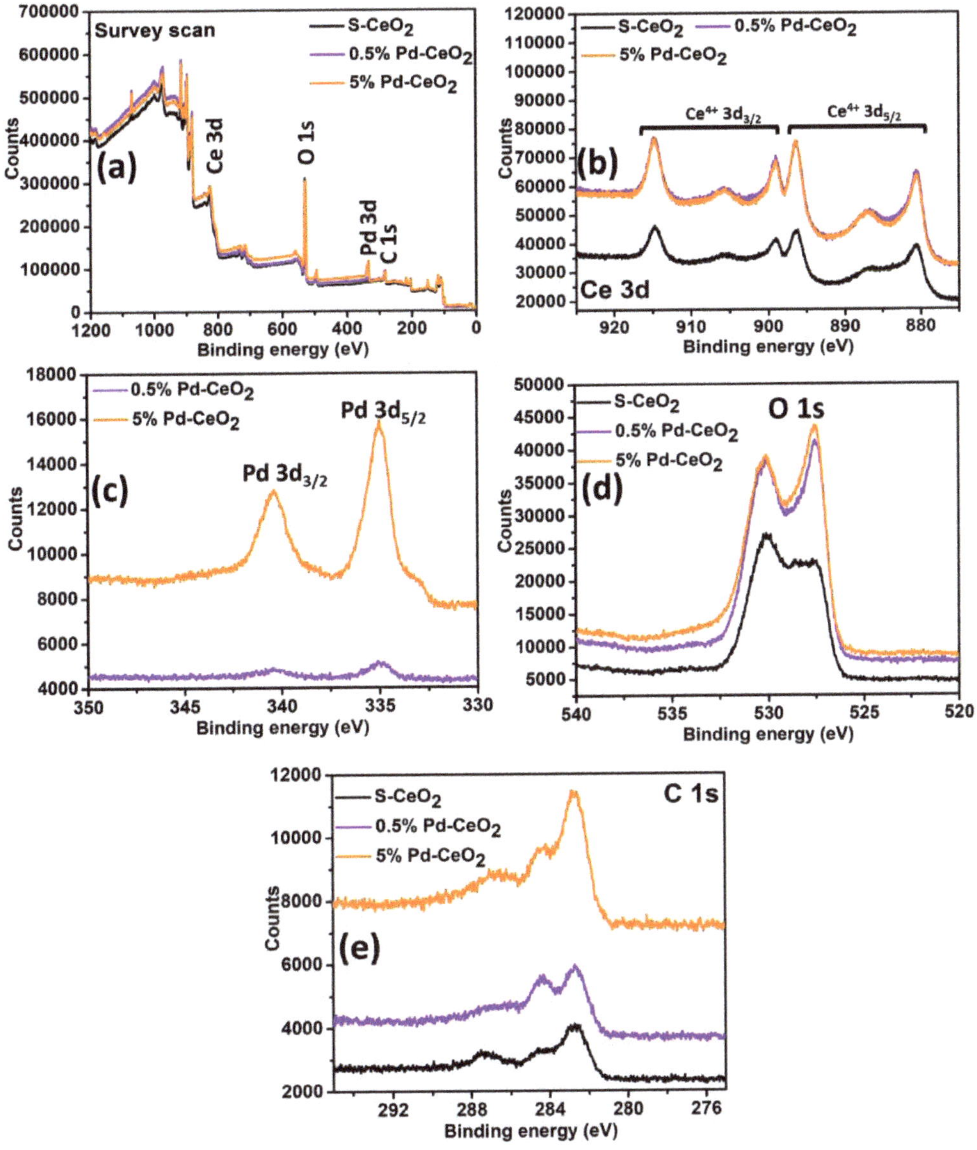

Figure 2. XPS spectra of S−CeO_2, 0.5% Pd−CeO_2, and 5% Pd−CeO_2: (**a**) Survey scan, (**b**) Ce 3*d*, (**c**) Pd 3*d*, (**d**) O 1s, and (**e**) C 1s.

Table 2. Atomic concentration of C 1s, O 1s, Ce 3*d*, and Pd 3*d* of S−CeO_2, 0.5% Pd−CeO_2, and 5% Pd−CeO_2.

Sample	Atomic Concentration (%)			
	C 1*s*	O 1*s*	Ce 3*d*	Pd 3*d*
S−CeO_2	12.4	74.1	13.5	-
0.5% Pd−CeO_2	12.6	72.1	15.1	0.2
5% Pd−CeO_2	19.0	66.3	12.8	1.9

The band gap reduction of S−CeO_2 and Pd−CeO_2 NPs might be due to the development of mid-gap states either above the valence band (VB) or below the conduction band (CB). Therefore, to study the cause of the reduction of band gap energy, VB-XPS was carried out for S−CeO_2 and Pd−CeO_2 NPs (Figure 3a).

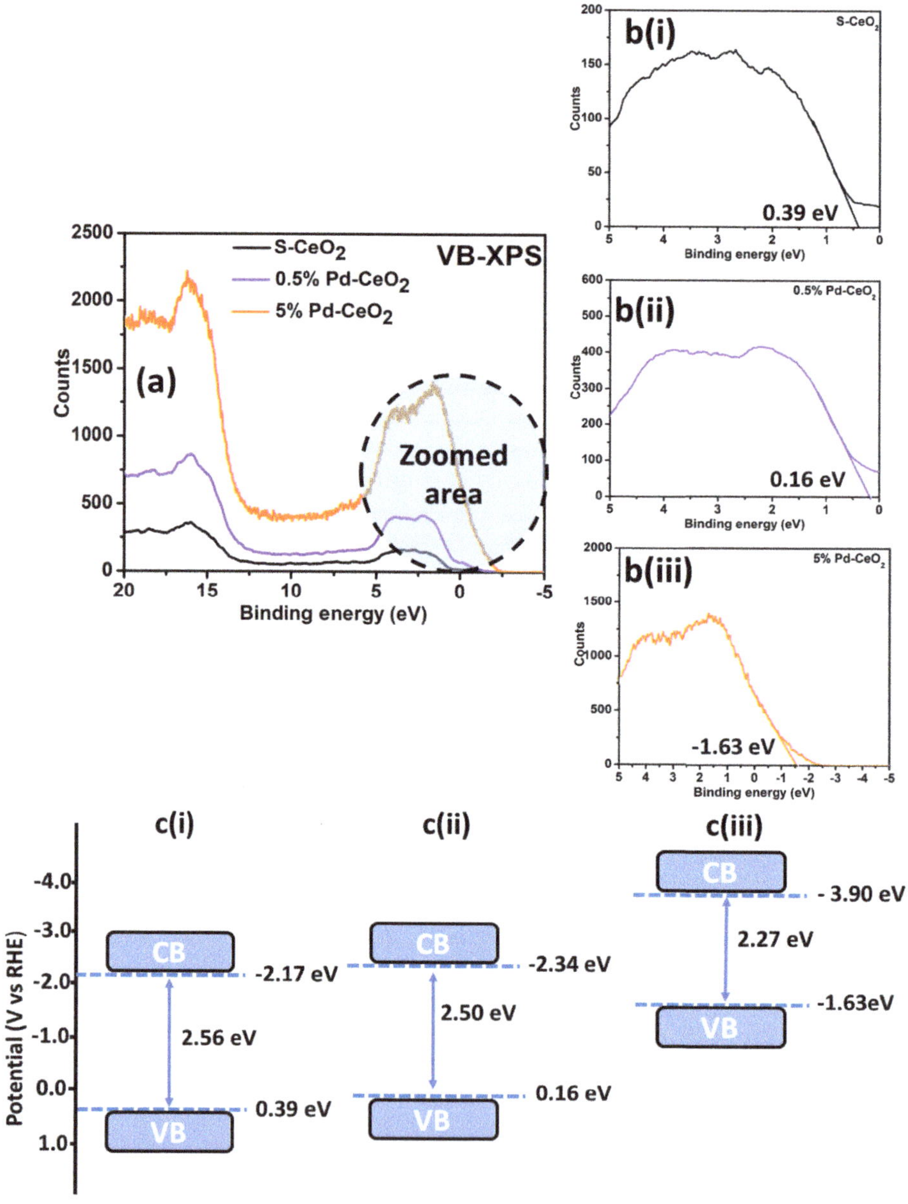

Figure 3. (**a**) Valence band XPS spectra of S−CeO_2, 0.5% Pd−CeO_2, and 5% Pd−CeO_2 NP; (**b**) zoomed valence band spectra of (**i**) S−CeO_2, (**ii**) 0.5% Pd−CeO_2, and (**iii**) 5% Pd−CeO_2 to estimate the band gap; and (**c**) proposed density of electronic states (DOS) for (**i**) S−CeO_2, (**ii**) 0.5% Pd−CeO_2, and (**iii**) 5% Pd−CeO_2 NPs.

The zoomed area of VB-XPS is shown in Figure 3b. The VB maximum of S−CeO_2, 0.5% Pd−CeO_2, and 5% Pd−CeO_2 NPs were found to be at 0.39, 0.16, and −1.63 eV, respectively (Figure 3b(i–iii)). The optical band gap energy obtained from the Tauc plot was 2.56, 2.50, and 2.27 eV for S−CeO_2, 0.5% Pd−CeO_2, and 5% Pd−CeO_2 NPs, respectively. Consequently, based on the materials' optical band gap and VB maxima, the CB minima would follow at −2.17, −2.34, and −3.90 eV for S−CeO_2, 0.5% Pd−CeO_2, and 5% Pd−CeO_2 NPs, respectively. The shifts of VB and CB result in the band gap energy reduction, as proposed in Figure 3c.

2.4. Optical Studies of CeO_2 and Pd−CeO_2 Using UV-vis-DRS and Photoluminescence Spectroscopy

UV-Vis DRS analysis is used to estimate the band gap energies of S−CeO_2 and Pd−CeO_2 NPs. The Tauc plot derived from the Kubelka–Munk function is shown in Figure 4a. The band gap energies of all samples were estimated using the Kubelka–Munk Equation (3):

$$F(R) = \left(\frac{(1-R)^2}{2R} \times hv\right)^{\frac{1}{2}} \tag{3}$$

where R is the measured absolute reflectance of the samples. The band gap was obtained from the plots of $[F(R)hv]^{1/2}$ versus hv, as the intercept of the extrapolated linear part of the plot at $[F(R)hv]^{1/2} = 0$, assuming that the absorption coefficient (α) is proportional to the Kubelka-Munk function F(R).

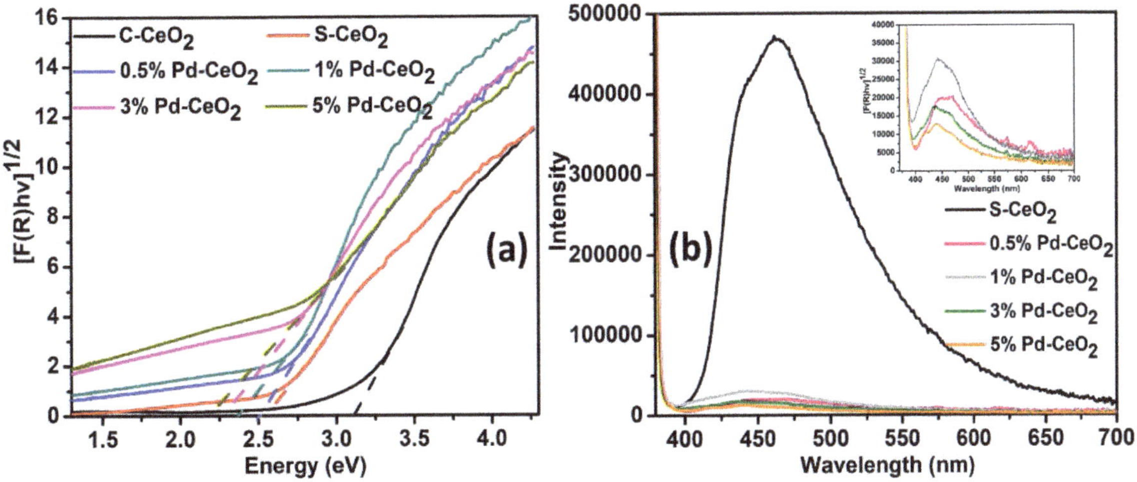

Figure 4. (**a**) Tauc plot derived from Kubelka-Munk function for band gap energy estimation, and (**b**) photoluminescence spectra of C−CeO_2, S−CeO_2, and Pd−CeO_2 NPs.

The band gap energy of C−CeO_2, S−CeO_2, 0.5% Pd−CeO_2, 1% Pd−CeO_2, 3% Pd−CeO_2, and 5% Pd−CeO_2 NPs were found to be decreasing (3.10 to 2.27 eV) as shown in Table 3. This shows that Pd doping is proven to decrease the band gap energy of the CeO_2. Pd doping is reported to help in reducing the energy difference between a CB and VB of CeO_2, which enhances the electronic conductivity of CeO_2. Moreover, oxygen vacancies (V_o) might be produced at the interface between the grains of Pd−CeO_2 NPs. Furthermore, the absorption edge of C−CeO_2, S−CeO_2, 0.5% Pd−CeO_2, 1% Pd−CeO_2, 3% Pd−CeO_2, and 5% Pd−CeO_2 NPs were greatly shifted into the visible region with the Pd doping as shown in Figure 4a.

Table 3. Band gap energy of C−CeO_2, S−CeO_2, and Pd−CeO_2 NPs.

Sample	Band Gap Energy (eV)
C−CeO_2	3.10
S−CeO_2	2.56
0.5% Pd−CeO_2	2.50
1% Pd−CeO_2	2.47
3% Pd−CeO_2	2.42
5% Pd−CeO_2	2.27

PL is used to evaluate the optical study of the crystals, defects on the surface and excitation fine structure of the semi-conducting materials. PL intensity describes the lifetime of electron relaxation from the VB to the CB, as well as the separation efficiency of photogenerated e^-/h^+ [43]. From Figure 4b, the PL intensity was decreased with more Pd doping concentration. This indicates that Pd doping can efficiently inhibit the e^-/h^+ recombination process due to the electron or hole trap levels (in this case, Pd^{2+}/Pd^{3+} or Pd^+/Pd^{2+}) in the band structure of CeO_2. Moreover, the reduction in peak intensity might be due to the formation of more oxygen defects, increased surface defects, as well as maximum separation of charge carriers [44].

2.5. Transmission Electron Microscopy Analysis of CeO_2 and Pd−CeO_2 NPs

Figure 5 exhibits the TEM, HR-TEM, and SAED images of S−CeO_2, 0.5%, and 5% Pd−CeO_2 NPs. Both S−CeO_2 and Pd−CeO_2 (Figure 5a,d and g) have almost spherical morphology. At the same time, a rod-like structure was observed for S−CeO_2 NPs only. The average particle sizes of S−CeO_2, 0.5% Pd−CeO_2, and 5% Pd−CeO_2 are about 13, 9, and 6 nm, respectively. This shows that the incorporation of Pd into the lattice of CeO_2 decreased the particle size. The *d*-spacing value of the lattice planes was also determined from the HR-TEM images (Figure 5b,e,h). The *d*-spacing values for S−CeO_2, 0.5% Pd−CeO_2, and 5% Pd−CeO_2 NPs are estimated to be around 0.3, 0.2 and 0.1 nm, which corresponds to the (111), (200), and (220) planes of a fluorite-structure of cubic CeO_2 [45]. This is in accordance with the XRD analysis (Section 2.1).

Moreover, the nano-crystallinity of S−CeO_2 and Pd−CeO_2 NPs were examined by selected area electron diffraction (SAED) analysis, as shown in Figure 5c,f and i. It was observed that S−CeO_2, 0.5%, and 5% Pd−CeO_2 exhibit four prominent broad rings, which can be attributed to the (111), (200), (220), and (311) reflections of the fluorite cubic CeO_2 structure. This observation is supported by the XRD patterns reported earlier (Section 2.1).

The EDX mapping images shown in Figure S2a–c confirmed the presence of Ce, O, and Pd in S−CeO_2, 0.5%, and 5% Pd−CeO_2 NPs, respectively. The percentage of Ce was seen to decrease with more Pd content. The percentage mass of Ce, O, and Pd can be found in Table S1. This shows that Pd has been successfully incorporated into the CeO_2 lattice.

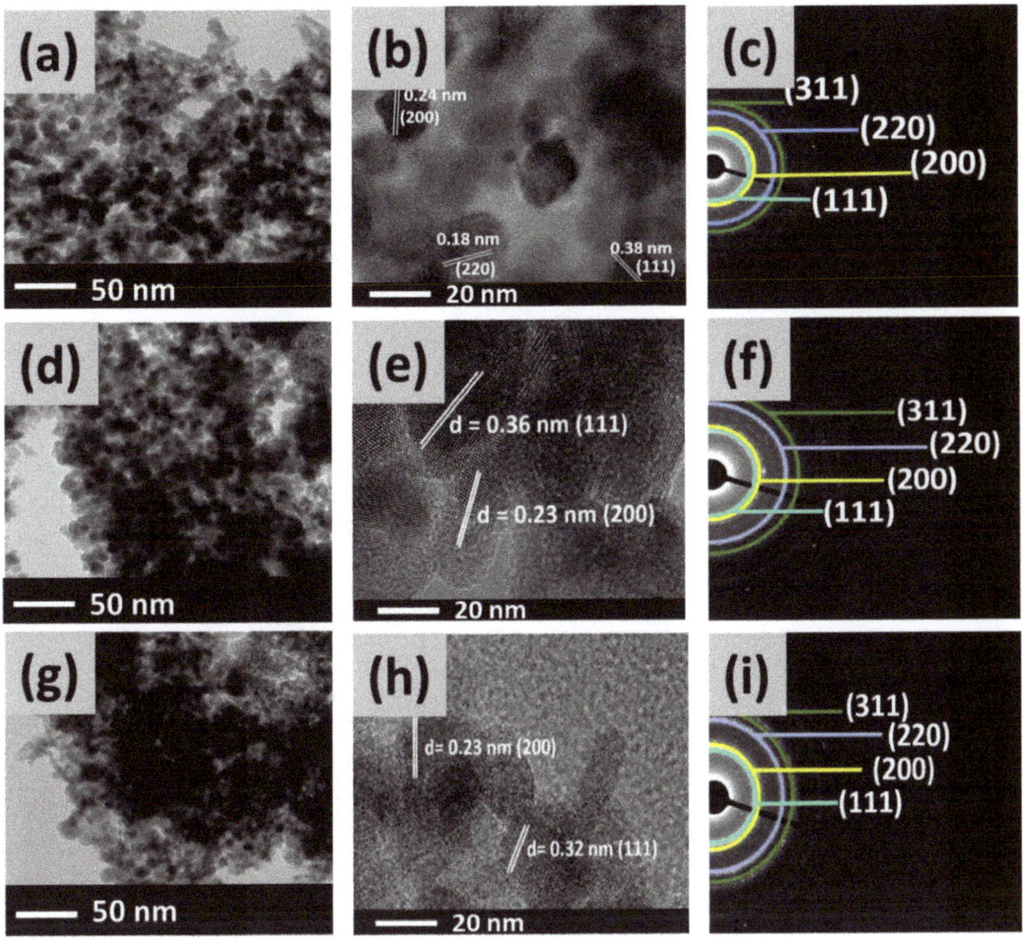

Figure 5. TEM, HR-TEM, and SAED patterns of (**a**–**c**) S−CeO_2, (**d**–**f**) 0.5% Pd−CeO_2, and (**g**–**i**) 5% Pd−CeO_2 NPs.

3. Applications

3.1. Photocatalytic Degradation of MB

The photocatalytic degradation of MB using S−CeO_2 and Pd−CeO_2 NPs was carried out under visible light irradiation in a total experiment duration of 5 h. The activity of C−CeO_2 against MB dye under visible light was also investigated to compare the photocatalytic efficiency with the synthesized materials. Figure 6a shows the average C/C_0 of the photocatalytic degradation of MB using C−CeO_2, S−CeO_2, and Pd−CeO_2 NPs, while the average percentage of the photocatalytic degradation of MB and the absorption spectra of MB within 5 h can be seen in Figures S3 and S4, respectively. Based on the study, adsorption-desorption equilibrium was reached in 30 min, in which C−CeO_2 showed a slight adsorption capacity. About 39–61% adsorption was reached for the synthesized materials suggesting that the addition of Pd effectively increases the affinity of MB molecules towards photocatalysts [46]. The enhanced adsorption is also driven by the increase in the surface area contributed by significantly reduced particle size [46].

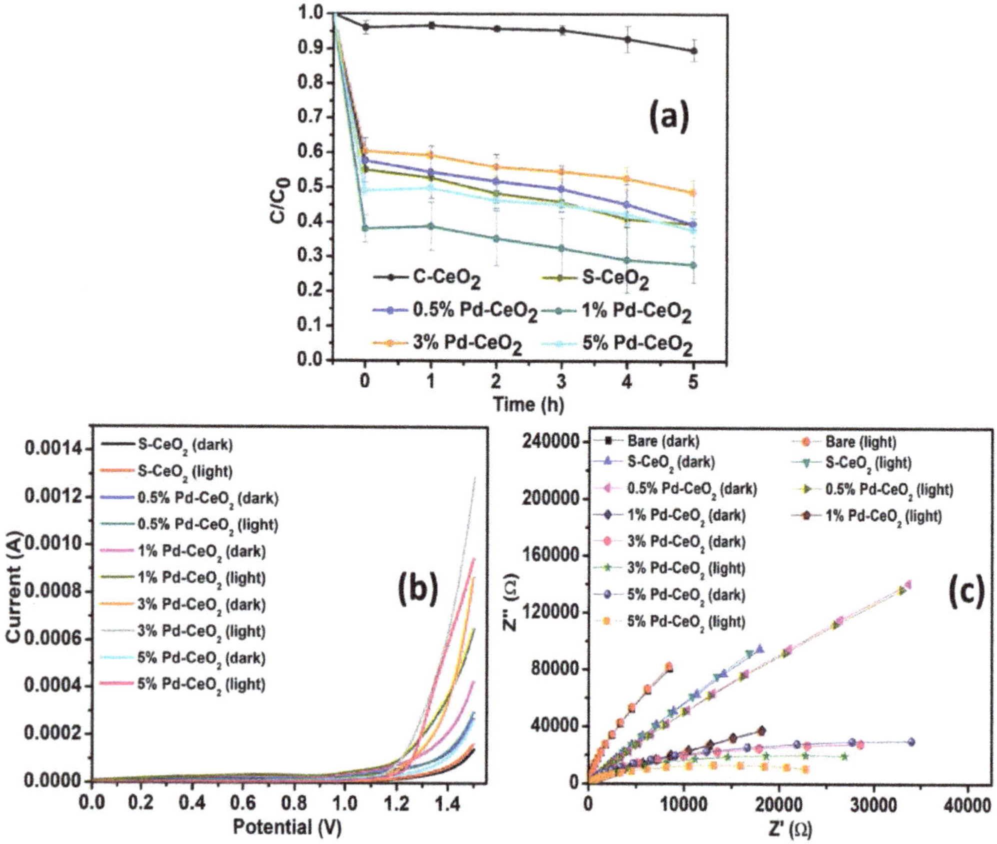

Figure 6. (**a**) The average C/C_0 of photocatalytic degradation of MB using C−CeO_2, S−CeO_2, and Pd−CeO_2 NPs under visible light irradiation, (**b**) LSV, and (**c**) EIS Nyquist plots of S−CeO_2, 0.5%, 1%, 3%, and 5% Pd−CeO_2 photoelectrode in the dark and under visible light irradiation.

The effectiveness of the photocatalysts in the photocatalytic degradation of MB was seen as follows: 1% Pd−CeO_2 > 5% Pd−CeO_2 > S−CeO_2 > 0.5% Pd−CeO_2 > 3% Pd−CeO_2 > C−CeO_2. A small response from C−CeO_2 might be due to the larger band gap energy, which was found to be 3.10 eV. Moreover, 3% Pd−CeO_2 and 0.5% Pd−CeO_2 also showed lower responses despite their lower band gap energies. This might be due to variations in particle size, which resulted in lower responses in the photocatalytic activity. In addition to that, a smaller surface area might also lead to less active sites present on the surface of the photocatalysts [47]. S−CeO_2 and 5% Pd−CeO_2 showed almost similar activities. The activity of S−CeO_2 was higher than some of the materials, which might be due to the mixed CeO_2/Ce_2O_3 phases, as shown in Figure 1a. The presence of both phases can ease the redox reaction between Ce^{3+} and Ce^{4+}, which causes the photocatalytic activities to increase [48–50]. Choudhury et al. also stated that an electron from Ce^{3+} can be transferred to adsorbed oxygen to form reactive oxygen species (ROS), i.e., superoxide radicals ($O_2^{\bullet-}$), whereas holes interact with water to form hydroxyl radicals ($^{\bullet}OH$) [51]. In addition, the e^- in the CB can be trapped by oxygen vacancies and then react with O_2 to form $O_2^{\bullet-}$. These radicals are involved in the degradation of MB into harmless products [52]. Recombination of photogenerated e^- and h^+ was restrained through the Ce^{3+}/Ce^{4+} redox and oxygen vacancies which finally improved the photocatalytic performance of the Pd−CeO_2 materials [53]. Moreover, smaller particle size as well as low band gap energy also effectively increased the photocatalytic activity [54]. Nevertheless, 5% Pd−CeO_2 showed a slightly better performance than S−CeO_2 due to its smaller particle size and band gap energy. Accordingly, 1% Pd−CeO_2 showed the highest response, which might be due to the optimum Pd loading. This shows that a further increase in doping does not enhance the photocatalytic degradation of MB [46]. Table 4 summarizes the average percentage of photocatalytic degradation of MB activities using S−CeO_2 and Pd−CeO_2 NPs under visible light irradiation.

Table 4. The average percentage MB dye degradation using C−CeO_2, S−CeO_2, and Pd−CeO_2 under visible light irradiation.

	C−CeO$_2$	S−CeO$_2$	0.5% Pd−CeO$_2$	1% Pd−CeO$_2$	3% Pd−CeO$_2$	5% Pd−CeO$_2$
0 h	3.87 ± 1.94	44.92 ± 3.60	42.21 ± 6.34	61.87 ± 3.97	39.57 ± 0.81	50.83 ± 1.52
1 h	3.33 ± 1.04	47.22 ± 2.06	45.58 ± 7.54	61.20 ± 6.92	40.67 ± 1.03	50.18 ± 0.80
2 h	4.19 ± 0.66	51.69 ± 2.77	48.19 ± 7.73	64.63 ± 7.92	44.07 ± 2.50	53.62 ± 2.72
3 h	4.57 ± 1.33	54.13 ± 1.52	50.35 ± 6.55	67.42 ± 8.64	45.36 ± 2.02	54.85 ± 2.15
4 h	7.12 ± 3.82	58.99 ± 2.29	54.72 ± 5.71	70.87 ± 9.48	47.31 ± 3.19	57.56 ± 3.31
5 h	10.28 ± 3.13	60.48 ± 3.76	60.33 ± 1.76	72.07 ± 5.34	51.20 ± 3.47	62.10 ± 4.27

3.2. Photoelectrochemical Studies of Pd−CeO_2 NPs

The electrochemical and photoelectrochemical studies of S−CeO_2 and Pd−CeO_2 NPs were analyzed using LSV and EIS. Figure 6b shows the LSV analysis carried out in the dark and under visible light at 100 mV/s in the range of −0.7–1.5 V. S−CeO_2 showed the lowest photocurrent, and 5% Pd−CeO_2 showed the highest photocurrent. As the doping percentage increases (Figure 6b), the photocurrent response also increases. Each material, respectively, showed a higher response under visible light irradiation as compared to their response in the dark, suggesting that the materials were light-responsive. The improvement in the photocurrent response in Pd−CeO_2 was attributed to the light absorption ability due to the creation of a mid-gap state which lowered the band gap energy [55]. The band gap energy of Pd−CeO_2 was narrowed with higher Pd doping, in which the valence electrons can be excited to the conduction band state by absorbing light. It is stated that the e^-/h^+ pairs will recombine unless they are separated quickly [56]. In general, a high photocurrent suggests that the material has a strong ability for the generation and transfer of the photoexcited charge carrier under light irradiation [57]. Therefore, in this case, 5% Pd−CeO_2 exhibits better electron-hole separation, and it might be excited easily by visible light.

EIS was performed in the dark and under visible light irradiation at 0.0 V with a frequency ranging from 1–10^6 Hz (Figure 6c). The EIS Nyquist plots show the charge transfer resistance and separation efficiency between the photogenerated electrons and holes [58]. In general, a small arc radius and low resistance indicate higher charge transfer efficiency [55]. S−CeO_2 showed the highest resistance, followed by 0.5% and 1% Pd−CeO_2 indicating the slow interfacial charge-transfer process. This response might be due to their high band gap energies. In each case, there was no significant decrease in the resistance under visible light, which suggests that the transfer efficiency of the photogenerated electrons and holes was not accelerated under light irradiation. On the other hand, 3% Pd−CeO_2 and 5% Pd−CeO_2 showed significant responses, especially under visible light irradiation. This suggests that lower resistances derived from both 3% and 5% Pd−CeO_2 show a faster interfacial charge-transfer process. Other factors can also facilitate charge separation and transfer efficiency, such as the presence of Ce^{3+} and oxygen vacancies [55].

3.3. Antibacterial Activities of Pd−CeO_2 NPs

The effect of bactericidal activities of S−CeO_2 NPs and Pd−CeO_2 NPs was examined against one Gram-positive bacteria (i.e., *S. aureus*) and another against Gram−negative pathogenic bacteria (i.e., *P. aeruginosa*). Since the synthesized NPs are light sensitive, the antibacterial activities were performed in the absence and presence of light. The antibacterial effect of Pd−CeO_2 and S−CeO_2 NPs under light and dark conditions is shown in Figure 7. The results showed that both S−CeO_2 NPs and Pd−CeO_2NPs exhibit concentration-dependent inhibitions of *S. aureus* growth under light or dark conditions (Figure 7a,b). At a high concentration of 2048 μg/mL in the dark, the growth inhibition of *S. aureus* cells was found to be higher (reduction of 1.3 log CFU) in the presence of 5% Pd−CeO_2 NPs as compared to the CeO_2 NPs (reduction of 1.2 log CFU) (Figure 7a). Similarly, in the presence of light at 2048 μg/mL, the antibacterial effect of 5% Pd−CeO_2 NPs was also found to be higher (reduction of 1.6 log CFU) as compared to the S−CeO_2 NPs (reductions 1.0 log CFU) (Figure 7b). The bactericidal effect of CeO_2 NPs was found to be higher in the presence of light as compared to dark conditions. This finding suggests that light enhances the bactericidal effect of CeO_2 NPs towards Gram-positive bacteria *S. aureus*.

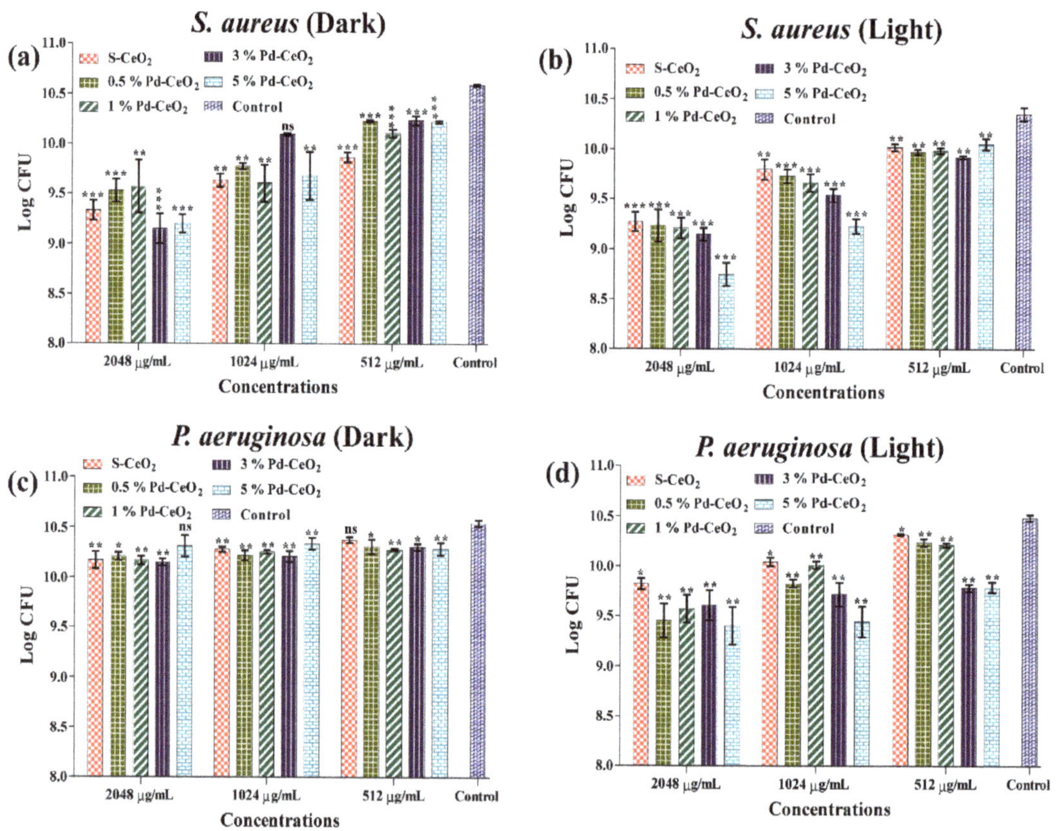

Figure 7. Log reduction of the *S. aureus* treated with different concentrations of S−CeO_2 NPs and Pd−CeO_2 NPs incubated under (**a**) dark and (**b**) light conditions. Log reduction of the *P. aeruginosa* treated with different concentrations of S−CeO_2 NPs and Pd−CeO_2 NPs incubated under (**c**) dark conditions and (**d**) light conditions, where ***, **, and * imply significance at $p < 0.0001$, $p < 0.01$, and $p < 0.05$, respectively, while ns means non-significant.

The antibacterial effect of S−CeO_2 NPs and Pd−CeO_2 NPs on *P. aeruginosa* was found to be different as compared to *S. aureus* (Figure 7c,d). There was no concentration-dependent inhibition of *P. aeruginosa* cell growth in the presence of light and dark conditions. The growth inhibition at the higher concentration (2048 μg/mL) and dark conditions was found to be non-significant in the presence of 5% Pd−CeO_2 NPs as compared to the S−CeO_2 NPs

(Figure 7c). The results in the presence of light were found to be significantly effective at all tested concentrations of S−CeO_2 NPs and Pd−CeO_2 NPs (Figure 7d).

In the presence of light and high concentration (2048 μg/mL), the 5% Pd−CeO_2 NPs showed a significant reduction of *P. aeruginosa* cells (1.0 log CFU) compared to the S−CeO_2 NPs (0.6 log CFU). Based on the findings of the present study, it is clear that light acts as a sensitizing agent, which results in the enhanced antibacterial activity of CeO_2 NPs as compared to dark conditions. Previous research has shown that UV light irradiation causes the formation of reactive oxygen species (ROS) such as $^{\bullet}OH$, $O_2^{\bullet -}$, and singlet oxygen (O^-) from metal-NPs [59–62]. This evidence can be explained in the current study by the possibility of ROS generation during the photocatalytic degradation of the MB experiment (Section 3.1). In addition, Pd-doping will also result in the enhanced antibacterial effect of CeO_2 NPs, and the effect was increased with the increasing concentration of Pd. Overall, these effects were found to be more effective towards Gram-positive bacteria *S. aureus* than the Gram-negative pathogen *P. aeruginosa*.

In general, the antibacterial activity of metal oxides is associated with the release of metal ions from metal oxides. It is reported that metal ions are involved in the destruction of the bacterial cell and membrane [63]. This could lead to the possibility of metal oxides penetrating the cell. In the case of CeO_2, CeO_2 dissociates into Ce^{4+} ions (most of the time under irradiation of visible light) and interacts with the bacterial cell and penetrates the cell [64]. This might lead to changes in cell integrity, lactate dehydrogenase leakage and cell death. Moreover, the generation of ROS is also responsible for antibacterial activity [65]. When CeO_2 is irradiated with light, e^- are promoted to CB and reacted with molecular oxygen to form $O_2^{\bullet -}$. The photogenerated h^+ reacts with H_2O molecules on the surface of CeO_2 to form $^{\bullet}OH$ radicals. These two radicals are believed to be contributing significantly to the destruction of the bacterial cell membrane [60].

4. Experimental

4.1. Chemicals, Bacterial Strain, Culture Media, and Growth Conditions

Cerium(III) nitrate hexahydrate 98% ($Ce(NO_3)_3{\cdot}6H_2O$) and commercial CeO_2 (C−CeO_2) were purchased from Sigma-Aldrich whereas, palladium(II) chloride anhydrous ($PdCl_2$) was obtained from Fluka. Water was purified using Aquatron (England) prior to use. For photocatalytic degradation of MB and photoelectrochemical studies, MB and NaCl were obtained from Merck. For electrode preparation, ethanol (95%), ethyl cellulose 48–49.5% (*w*/*w*) ethoxy basis, and alpha-terpineol were obtained from Daejung Chemicals and Metals Co., Ltd., Sigma-Aldrich and Merck, respectively. The bacterial strain *S. aureus* KCTC 1916 (Gram-positive) and *P. aeruginosa* KCTC 1637 (Gram-negative) were purchased from the Korean Collection for Type Cultures (KCTC, Daejeon, Korea). The bacteria were cultivated using Tryptic soy broth (TSB) and a TSB agar plate (Difco Laboratory Inc., Detroit, MI, USA). The growth temperature of the bacterial strain was 37 °C under aerobic conditions.

4.2. Instrumentations

CeO_2 and (0.5, 1, 3, and 5%) Pd−doped CeO_2 NPs were synthesized using a microwave-assisted method (Anton Paar Monowave 400, Graz, Austria). Fourier Transform-Infrared Spectroscopy (FT-IR) was used to identify the possible functional groups present in the synthesized CeO_2 and Pd−CeO_2 NPs using Shimadzu IRPrestige-21 FT-IR Spectrophotometer. The stretching frequencies obtained were plotted as %transmittance mode on the y-axis and wavenumber (cm^{-1}) on the x-axis from 450 to 4000 cm^{-1}. The morphology, structure, and elemental mapping were analyzed using field emission transmission electron microscopy (FE-TEM) and selected area electron diffraction (SAED) using JEM-F200 (JEOL Ltd., Tokyo, Japan). The determination of band gap energy of S−CeO_2 and Pd−CeO_2 NPs was investigated using UV-Vis diffuse reflectance spectroscopy (DRS) (Shimadzu, UV-2600, Tokyo, Japan). A photoluminescence (PL) study was carried out using F-7000 Fluorescence spectroscopy (Hitachi High Tech, Tokyo, Japan) with an excitation wavelength of 370 nm. X-ray photoelectron spectroscopy (XPS) and valence band (VB)-XPS were performed (Kratos Analytical, AXIS Nova, Manchester, UK). Photocatalytic activities of MB dye degradation were carried out using a photochemical reactor Toption (TOPT-V) having a 300 W Xenon lamp, and the photocatalytic degradation of MB was monitored using UV–visible spectrophotometer (Shimadzu UV-1900, Japan). Photoelectrochemical studies were performed using Autolab (MetroHm, Herisau, Switzerland) under dark and visible light irradiation (Simon FL30 LED Floodlight, 100 W, Jiangsu, China). The measurements were carried out and taken from NOVA software. Photoantibacterial studies of S$-$CeO$_2$ and Pd$-$CeO$_2$ NPs were investigated using a 96-well microplate in the presence of the LED light (One & One Plus OP-0303 Nape Slim LED Stand).

4.3. Microwave-Assisted Synthesis of Cerium Oxide Nanoparticles

CeO_2 NPs were synthesized using the microwave-assisted method. Briefly, 0.05 M of $Ce(NO_3)_3{\cdot}6H_2O$ solution (15 mL) was prepared in a microwave vessel. Exactly 2.4 mL of 1 M NaOH was added dropwise into the solution. Subsequently, the microwave reaction was carried out at 180 °C for 15 min at 850 W microwave power. The precipitate formed was centrifuged and washed three times with water before it was dried at 80 °C. The product was stored and labeled as S$-$CeO$_2$.

4.4. Microwave-Assisted Synthesis of Pd$-$CeO$_2$ Nanoparticles

Pd$-$CeO$_2$ NPs were synthesized using the same method as mentioned above. A 15 mL of 0.05 M $Ce(NO_3)_3$ solution was prepared. A specific amount of $PdCl_2$ was then added to prepare 0.5, 1, 3, and 5% Pd$-$CeO$_2$. Subsequently, 1 M NaOH was added dropwise into the solution. The microwave reaction was carried out at 180 °C for 15 min at 850 W microwave power. The precipitate was formed, centrifuged, and washed three times with water before it was dried at 80 °C. The products were stored and labeled as 0.5% Pd$-$CeO$_2$, 1% Pd$-$CeO$_2$, 3% Pd$-$CeO$_2$, and 5% Pd$-$CeO$_2$ for further use.

4.5. Electrode Preparation

S$-$CeO$_2$, 0.5%, 1%, 3%, and 5% Pd$-$CeO$_2$ electrodes were prepared using the doctor blade method. In brief, 25 mg of the respective sample was mixed with 0.5 mL ethanol and 0.5 mL α-terpineol. The mixture was sonicated for 10 min. Then, 25 mg of ethyl cellulose was added to the mixture and stirred at 80 °C for ~2 h to produce a thick paste. The paste was then spread on fluorine-doped tin oxide (FTO) glass electrode by 2 cm × 1 cm using a doctor blade. The glass electrode was dried at 80 °C for 24 h prior to use.

4.6. Photocatalytic Degradation of MB Dye

The photocatalytic degradation of MB dye using S$-$CeO$_2$ and Pd$-$CeO$_2$ NPs under visible light irradiation was monitored using UV-Vis spectroscopy. Exactly 10 mg of S$-$CeO$_2$ and 0.5, 1, 3, and 5% Pd$-$CeO$_2$ NPs were dispersed in 50 mL of 10 ppm MB dye solution. The sample mixture was stirred in the dark for 30 min to reach adsorption-desorption equilibrium. Then, the reaction tubes were irradiated with visible light (300 W) for 5 h in which the absorbance of the solution was measured every 1 h. The percentage of photocatalytic dye degradation of MB was obtained using Equation (4):

$$\%\text{ photocatalytic MB dye degradation} = \frac{\left(A_{blank} - A_{sample}\right)}{A_{blank}} \times 100 \quad (4)$$

where A_{blank} is the absorbance of MB only and A_{sample} is the absorbance of MB dye after photocatalytic degradation by the photocatalyst.

4.7. Photoelectrochemical Studies

The photoelectrochemical response of S$-$CeO$_2$ and Pd$-$CeO$_2$ NPs was examined through linear sweep voltammetry (LSV) and electrochemical impedance spectroscopy (EIS) experiments. The experiments were carried out under ambient conditions in the dark and under visible light irradiation (LED, 100 W) in 100 mL of a 1 M NaCl aqueous electrolyte solution. The distance between the light and reactor vial was 22 cm. The prepared glass electrode, Ag/AgCl electrode, and Pt electrode were used as the working electrode, reference electrode, and counter electrode, respectively. For each electrode, LSV and EIS were performed in the dark and later under visible light irradiation at 100 mV/s in the potential range of −0.7–1.5 V and at 0.0 V with a frequency ranging from 1–10^6 Hz, respectively.

4.8. Assays for Antibacterial Activities of Pd−CeO_2 Nanoparticles

A single colony from the agar plate of each bacterium was inoculated in the TSB and incubated for 12 h at 37 °C under shaking conditions (250 rpm). The seed culture was diluted into fresh TSB in order to achieve the final OD_{600} ~0.05. The diluted cell culture (300 µL) was placed in a 96-well microtiter plate and treated with different concentrations of S−CeO_2 and 0.5, 1, 3, and 5% Pd−CeO_2 NPs. The concentrations of these NPs ranged from 265 to 2048 µg/mL. Two sets of microtiter plates were prepared to contain different concentrations of NPs. One set of microplates was incubated in the dark for 24 h at 37 °C. The second set of the microplates was incubated in the presence of LED light that was set at the height of 27 cm for 24 h at 37 °C. After incubation, the cell culture (100 µL) was transferred into another 96-well microplate containing 200 µL sterile TSB. A two-fold serial dilution (up to 10^{-8} dilutions) of the cell culture was carried out. The serially diluted cell culture (100 µL) was spread-plated on the TSA plates. The TSA plates were incubated at 37 °C for 24 h, followed by the counting of the total colonies. The log colony forming unit (CFU) values of treated and untreated cells were calculated. All the experiments were carried out in triplicate.

5. Conclusions

S−CeO_2 and 0.5, 1, 3, and 5% Pd−CeO_2 NPs were successfully synthesized using the microwave-assisted method. Mixed phases of CeO_2/Ce_2O_3 were obtained in S−CeO_2 and 0.5, 1, and 3% Pd−CeO_2. However, there were no mixed phases observed in 5% Pd−CeO_2 NPs. Average crystallite sizes were found to be between 16.25–37.25 nm. Based on the TEM images, irregularly spherical shaped S−CeO_2 and Pd−CeO_2 NPs were observed with an average particle size between 6 and 13 nm. The band gap energy of S−CeO_2 decreased with the addition of higher Pd content, in which the band gap narrowing phenomena was illustrated through the DOS scheme. One percent of Pd−CeO_2 NPs showed enhanced responses under visible light irradiation in photocatalytic degradation of MB. Meanwhile, 5% Pd−CeO_2 showed enhanced responses in photoelectrochemical and photoantibacterial activities. Therefore, Pd−CeO_2 has shown visible light active properties and can be potentially used in photocatalysis and photoantibacterial applications.

Supplementary Materials: The following supporting information can be downloaded at: https://www.mdpi.com/article/10.3390/catal13010096/s1. Figure S1. Standards XRD patterns of CeO_2, Ce_2O_3, and PdO; Figure S2. EDX-mapping of Ce, O, and Pd elements in (a) S−CeO_2, (b) 0.5% Pd−CeO_2, and (c) 5% Pd−CeO_2 NPs; Figure S3. The average percentage of photocatalytic degradation of MB dye using C−CeO_2, S−CeO_2, and Pd−CeO_2 NPs under visible light irradiation; Figure S4. Absorbance spectra of photocatalytic degradation of MB using (a) C−CeO_2, (b) S−CeO_2 (c) 0.5%, (d) 1%, (e) 3%, and (f) 5% Pd−CeO_2 NPs; Table S1. Percentage mass of Ce, O, and Pd elements in S−CeO_2, 0.5% Pd−CeO_2, and 5% Pd−CeO_2 NPs using EDX-Mapping.

Author Contributions: S.N.M.: Methodology, Investigation, Data curation, and Writing—original draft. F.K.: Methodology, Investigation, Data curation, and Writing. M.H.H.: Supervision, Writing—review and editing. Y.-M.K.: Resources, Formal analysis. M.M.K.: Supervision, Conceptualization, Funding acquisition, Writing—review and editing. All authors have read and agreed to the published version of the manuscript.

Funding: The authors would like to acknowledge the FRC grant (UBD/RSCH/1.4/FICBF(b)/2022/046) received from Universiti Brunei Darussalam, Brunei Darussalam. This research was supported by

Basic Science Research Program through the National Research Foundation of Korea (NRF) grant funded by the Ministry of Education (2021R1A6A1A03039211 and 2022R1A2B5B01001998).

Conflicts of Interest: The authors declare no conflict of interest.

References

1. Sánchez-López, E.; Gomes, D.; Esteruelas, G.; Bonilla, L.; Lopez-Machado, A.L.; Galindo, R.; Cano, A.; Espina, M.; Ettcheto, M.; Camins, A.; et al. Metal-Based Nanoparticles as Antimicrobial Agents: An Overview. *Nanomaterials* **2020**, *10*, 292. [CrossRef] [PubMed]
2. Wang, L.; Hu, C.; Shao, L. The Antimicrobial Activity of Nanoparticles: Present Situation and Prospects for the Future. *Int. J. Nanomed.* **2017**, *2017*, 1227–1249. [CrossRef] [PubMed]
3. Kadiyala, U.; Kotov, N.A.; VanEpps, J.S. Antibacterial Metal Oxide Nanoparticles: Challenges in Interpreting the Literature. *Curr. Pharm. Des.* **2018**, *24*, 896–903. [CrossRef]
4. Chiu, Y.-H.; Chang, T.-F.; Chen, C.-Y.; Sone, M.; Hsu, Y.-J. Mechanistic Insights into Photodegradation of Organic Dyes Using Heterostructure Photocatalysts. *Catalysts* **2019**, *9*, 430. [CrossRef]
5. Thambiraj, S.; Sharmila, G.; Ravi Shankaran, D. Green Adsorbents from Solid Wastes for Water Purification Application. *Mater. Today Proc.* **2018**, *5*, 16675–16683. [CrossRef]
6. Daneshvar, N.; Salari, D.; Khataee, A.R. Photocatalytic Degradation of Azo Dye Acid Red 14 in Water: Investigation of the Effect of Operational Parameters. *J. Photochem. Photobiol. A Chem.* **2003**, *157*, 111–116. [CrossRef]
7. Rafiq, A.; Ikram, M.; Ali, S.; Niaz, F.; Khan, M.; Khan, Q.; Maqbool, M. Photocatalytic Degradation of Dyes Using Semiconductor Photocatalysts to Clean Industrial Water Pollution. *J. Ind. Eng. Chem.* **2021**, *97*, 111–128. [CrossRef]
8. Santoso, E.; Ediati, R.; Kusumawati, Y.; Bahruji, H.; Sulistiono, D.O.; Prasetyoko, D. Review on Recent Advances of Carbon Based Adsorbent for Methylene Blue Removal from Waste Water. *Mater. Today Chem.* **2020**, *16*, 100233. [CrossRef]
9. Khan, I.; Saeed, K.; Zekker, I.; Zhang, B.; Hendi, A.H.; Ahmad, A.; Ahmad, S.; Zada, N.; Ahmad, H.; Shah, L.A.; et al. Review on Methylene Blue: Its Properties, Uses, Toxicity and Photodegradation. *Water* **2022**, *14*, 242. [CrossRef]
10. Beitollahi, H.; van Le, Q.; Farha, O.K.; Shokouhimehr, M.; Tajik, S.; Nejad, F.G.; Kirlikovali, K.O.; Jang, H.W.; Varma, R.S. Recent Electrochemical Applications of Metal-Organic Framework-Based Materials. *Cryst. Growth Des.* **2020**, *20*, 7034–7064. [CrossRef]
11. Basu, K.; Benetti, D.; Zhao, H.; Jin, L.; Vetrone, F.; Vomiero, A.; Rosei, F. Enhanced Photovoltaic Properties in Dye Sensitized Solar Cells by Surface Treatment of SnO_2 Photoanodes. *Sci. Rep.* **2016**, *6*, 1–10. [CrossRef] [PubMed]
12. Li, L.; Zhu, B.; Zhang, J.; Yan, C.; Wu, Y. Electrical Properties of Nanocube CeO_2 in Advanced Solid Oxide Fuel Cells. *Int. J. Hydrogen Energy* **2018**, *43*, 12909–12916. [CrossRef]
13. Liang, S.; Wang, H.; Li, Y.; Qin, H.; Luo, Z.; Huang, B.; Zhao, X.; Zhao, C.; Chen, L. Rare-Earth Based Nanomaterials and Their Composites as Electrode Materials for High Performance Supercapacitors: A Review. *Sustain. Energy Fuels* **2020**, *4*, 3825–3847. [CrossRef]
14. Xie, W.; Liu, B.; Xiao, S.; Li, H.; Wang, Y.; Cai, D.; Wang, D.; Wang, L.; Liu, Y.; Li, Q.; et al. High Performance Humidity Sensors Based on CeO_2 Nanoparticles. *Sens. Actuators B Chem.* **2015**, *215*, 125–132. [CrossRef]
15. Manibalan, G.; Murugadoss, G.; Thangamuthu, R.; Ragupathy, P.; Kumar, M.R.; Mohan Kumar, R.; Jayavel, R. High Electrochemical Performance and Enhanced Electrocatalytic Behavior of a Hydrothermally Synthesized Highly Crystalline Heterostructure CeO_2@NiO Nanocomposite. *Inorg. Chem.* **2019**, *58*, 13843–13861. [CrossRef]
16. Nwachukwu, I.M.; Nwanya, A.C.; Osuji, R.; Ezema, F.I. Nanostructured Mn-Doped CeO_2 Thin Films with Enhanced Electrochemical Properties for Pseudocapacitive Applications. *J. Alloys Compd.* **2021**, *886*, 161206. [CrossRef]
17. Wang, H.; Liu, Y.; Li, M.; Huang, H.; Xu, H.M.; Hong, R.J.; Shen, H. Multifunctional TiO_2 nanowires-Modified Nanoparticles Bilayer Film for 3D Dye-Sensitized Solar Cells. *Optoelectron. Adv. Mater. Rapid Commun.* **2010**, *4*, 1166–1169. [CrossRef]
18. Rahman, A.; Harunsani, M.H.; Tan, A.L.; Ahmad, N.; Khan, M.M. Antioxidant and Antibacterial Studies of Phytogenic Fabricated ZnO Using Aqueous Leaf Extract of Ziziphus Mauritiana Lam. *Chem. Pap.* **2021**, *75*, 3295–3308. [CrossRef]
19. Aboutaleb, W.A.; El-Salamony, R.A. Effect of Fe_2O_3-CeO_2 Nanocomposite Synthesis Method on the Congo Red Dye Photodegradation under Visible Light Irradiation. *Mater. Chem. Phys.* **2019**, *236*, 121724. [CrossRef]
20. Naidi, S.N.; Khan, F.; Tan, A.L.; Harunsani, M.H.; Kim, Y.-M.; Khan, M.M. Photoantioxidant and Antibiofilm Studies of Green Synthesized Sn-Doped CeO_2 Nanoparticles Using Aqueous Leaf Extracts of Pometia Pinnata. *New J. Chem.* **2021**, *45*, 7816–7829. [CrossRef]
21. Su, Z.; Si, W.; Liu, H.; Xiong, S.; Chu, X.; Yang, W.; Peng, Y.; Chen, J.; Cao, X.; Li, J. Boosting the Catalytic Performance of CeO_2 in Toluene Combustion via the Ce-Ce Homogeneous Interface. *Environ. Sci. Technol.* **2021**, *55*, 12630–12639. [CrossRef] [PubMed]
22. Su, Z.; Yang, W.; Wang, C.; Xiong, S.; Cao, X.; Peng, Y.; Si, W.; Weng, Y.; Xue, M.; Li, J. Roles of Oxygen Vacancies in the Bulk and Surface of CeO_2 for Toluene Catalytic Combustion. *Environ. Sci. Technol.* **2020**, *54*, 12684–12692. [CrossRef] [PubMed]
23. Gnanam, S.; Gajendiran, J.; Ramana Ramya, J.; Ramachandran, K.; Gokul Raj, S. Glycine-Assisted Hydrothermal Synthesis of Pure and Europium Doped CeO_2 Nanoparticles and Their Structural, Optical, Photoluminescence, Photocatalytic and Antibacterial Properties. *Chem. Phys. Lett.* **2021**, *763*, 138217. [CrossRef]
24. Bača, L'.; Steiner, H.; Stelzer, N. Upconversion Luminescence and Optical Thermometry in Er^{3+}/Yb^{3+} Co-Doped CeO_2 for Space Application. *J. Alloys Compd.* **2019**, *774*, 418–424. [CrossRef]

25. Khan, M.M.; Rahman, A.; Matussin, S.N. Recent Progress of Metal-Organic Frameworks and Metal-Organic Frameworks-Based Heterostructures as Photocatalysts. *Nanomaterials* **2022**, *12*, 2820. [CrossRef]
26. dos Santos, A.P.B.; Dantas, T.C.M.; Costa, J.A.P.; Souza, L.D.; Soares, J.M.; Caldeira, V.P.S.; Araújo, A.S.; Santos, A.G.D. Formation of CeO_2 Nanotubes through Different Conditions of Hydrothermal Synthesis. *Surf. Interfaces* **2020**, *21*, 100746. [CrossRef]
27. Seong, G.; Dejhosseini, M.; Adschiri, T. A Kinetic Study of Catalytic Hydrothermal Reactions of Acetaldehyde with Cubic CeO_2 Nanoparticles. *Appl. Catal. A Gen.* **2018**, *550*, 284–296. [CrossRef]
28. Xu, B.; Zhang, Q.; Yuan, S.; Liu, S.; Zhang, M.; Ohno, T. Synthesis and Photocatalytic Performance of Yttrium-Doped CeO_2 with a Hollow Sphere Structure. *Catal. Today* **2017**, *281*, 135–143. [CrossRef]
29. Lian, J.; Liu, P.; Jin, C.; Liu, Q.Y.; Zhang, X.; Zhang, X. Flower-like CeO_2/CoO p-n Heterojuncted Nanocomposites with Enhanced Peroxidase-Mimicking Activity for l-Cysteine Sensing. *ACS Sustain. Chem. Eng.* **2020**, *8*, 17540–17550. [CrossRef]
30. Zheng, J.; Zhu, Z.; Gao, G.; Liu, Z.; Wang, Q.; Yan, Y. Construction of Spindle Structured CeO_2 Modified with Rod-like Attapulgite as a High-Performance Photocatalyst for CO_2 Reduction. *Catal. Sci. Technol.* **2019**, *9*, 3788–3799. [CrossRef]
31. Bezkrovnyi, O.S.; Lisiecki, R.; Kepinski, L. Relationship between Morphology and Structure of Shape-Controlled CeO_2 Nanocrystals Synthesized by Microwave-Assisted Hydrothermal Method. *Cryst. Res. Technol.* **2016**, *51*, 554–560. [CrossRef]
32. Khan, M.M.; Ansari, S.A.; Pradhan, D.; Han, D.H.; Lee, J.; Cho, M.H. Defect-Induced Band Gap Narrowed CeO_2 Nanostructures for Visible Light Activities. *Ind. Eng. Chem. Res.* **2014**, *53*, 9754–9763. [CrossRef]
33. Abbas, F.; Iqbal, J.; Jan, T.; Naqvi, M.S.H.; Gul, A.; Abbasi, R.; Mahmood, A.; Ahmad, I.; Ismail, M. Differential Cytotoxicity of Ferromagnetic Co Doped CeO_2 Nanoparticles against Human Neuroblastoma Cancer Cells. *J. Alloys Compd.* **2015**, *648*, 1060–1066. [CrossRef]
34. Matussin, S.N.; Rahman, A.; Khan, M.M. Role of Anions in the Synthesis and Crystal Growth of Selected Semiconductors. *Front. Chem.* **2022**, *10*, 881518. [CrossRef] [PubMed]
35. Liu, H.; Zhu, Y.; Ma, J.; Chen, C.; Cheng, P.; Zhang, S. Hydrothermal Synthesis of Pd-Doped CeO_2 Nanomaterials and Electrochemical Detection for Phenol. *J. Cryst. Growth* **2022**, *586*, 126626. [CrossRef]
36. Gad, S.C. Palladium. In *Encyclopedia of Toxicology*; Elsevier: Amsterdam, The Netherlands, 2014; pp. 751–753. [CrossRef]
37. Hegde, M.S.; Madras, G.; Patil, K.C. Noble Metal Ionic Catalysts. *Acc. Chem. Res.* **2009**, *42*, 704–712. [CrossRef]
38. Wang, B.; Weng, D.; Wu, X.; Ran, R. Modification of Pd–CeO_2 Catalyst by Different Treatments: Effect on the Structure and CO Oxidation Activity. *Appl. Surf. Sci.* **2011**, *257*, 3878–3883. [CrossRef]
39. George, S.E.; George, M.; Alex, J.; Joy, L.K.; Aravind, A.; Sajan, D.; Thakur, A.; Hussain, S.; Vinitha, G. Nonlinear Optical and Photocatalytic Dye Degradation of Co Doped CeO_2 Nanostructures Synthesized through a Modified Combustion Technique. *Ceram. Int.* **2020**, *46*, 13932–13940. [CrossRef]
40. Matussin, S.N.; Khan, M.M. Phytogenic Fabrication of CeO_2@SnO_2 Heterojunction Nanostructures for Antioxidant Studies. *Chem. Pap.* **2022**, *76*, 2071–2084. [CrossRef]
41. Naidi, S.N.; Harunsani, M.H.; Tan, A.L.; Khan, M.M. Green-Synthesized CeO_2 Nanoparticles for Photocatalytic, Antimicrobial, Antioxidant and Cytotoxicity Activities. *J. Mater. Chem. B* **2021**, *9*, 5599–5620. [CrossRef]
42. Tou, M.; Michalsky, R.; Steinfeld, A. Solar-Driven Thermochemical Splitting of CO_2 and In Situ Separation of CO and O_2 across a Ceria Redox Membrane Reactor. *Joule* **2017**, *1*, 146–154. [CrossRef] [PubMed]
43. Channei, D.; Nakaruk, A.; Jannoey, P.; Phanichphant, S. Preparation and Characterization of Pd Modified CeO_2 Nanoparticles for Photocatalytic Degradation of Dye. *Solid State Sci.* **2019**, *87*, 9–14. [CrossRef]
44. Syed Khadar, Y.A.; Balamurugan, A.; Devarajan, V.P.; Subramanian, R.; Dinesh Kumar, S. Synthesis, Characterization and Antibacterial Activity of Cobalt Doped Cerium Oxide (CeO_2:Co) Nanoparticles by Using Hydrothermal Method. *J. Mater. Res. Technol.* **2019**, *8*, 267–274. [CrossRef]
45. He, L.; Ren, Y.; Fu, Y.; Yue, B.; Edman Tsang, S.C.; He, H. Morphology-Dependent Catalytic Activity of Ru/CeO_2 in Dry Reforming of Methane. *Molecules* **2019**, *24*, 526. [CrossRef] [PubMed]
46. Cui, X.; Xu, W.; Xie, Z.; Dorman, J.A.; Gutierrez-Wing, M.T.; Wang, Y. Effect of Dopant Concentration on Visible Light Driven Photocatalytic Activity of Sn1-XAgxS2. *Dalton Trans.* **2016**, *45*, 16290–16297. [CrossRef]
47. Xing, J.; Jiang, H.B.; Chen, J.F.; Li, Y.H.; Wu, L.; Yang, S.; Zheng, L.R.; Wang, H.F.; Hu, P.; Zhao, H.J.; et al. Active Sites on Hydrogen Evolution Photocatalyst. *J. Mater. Chem. A Mater.* **2013**, *1*, 15258. [CrossRef]
48. Khan, M.E.; Khan, M.M.; Cho, M.H. Ce^{3+}-Ion, Surface Oxygen Vacancy, and Visible Light-Induced Photocatalytic Dye Degradation and Photocapacitive Performance of CeO_2-Graphene Nanostructures. *Sci. Rep.* **2017**, *7*, 5928. [CrossRef]
49. Kumar, K.M.; Mahendhiran, M.; Diaz, M.C.; Hernandez-Como, N.; Hernandez-Eligio, A.; Torres-Torres, G.; Godavarthi, S.; Gomez, L.M. Green Synthesis of Ce^{3+} Rich CeO_2 Nanoparticles and Its Antimicrobial Studies. *Mater. Lett.* **2018**, *214*, 15–19. [CrossRef]
50. Hu, Q.; Huang, B.; Li, Y.; Zhang, S.; Zhang, Y.; Hua, X.; Liu, G.; Li, B.; Zhou, J.; Xie, E.; et al. Methanol Gas Detection of Electrospun CeO_2 Nanofibers by Regulating Ce^{3+}/Ce^{4+} Mole Ratio via Pd Doping. *Sens. Actuators B Chem.* **2020**, *307*, 127638. [CrossRef]
51. Choudhury, B.; Chetri, P.; Choudhury, A. Oxygen Defects and Formation of Ce^{3+} Affecting the Photocatalytic Performance of CeO_2 Nanoparticles. *RSC Adv.* **2014**, *4*, 4663–4671. [CrossRef]
52. Khan, I.; Saeed, K.; Ali, N.; Khan, I.; Zhang, B.; Sadiq, M. Heterogeneous Photodegradation of Industrial Dyes: An Insight to Different Mechanisms and Rate Affecting Parameters. *J. Environ. Chem. Eng.* **2020**, *8*, 104364. [CrossRef]

53. Yang, C.; Yang, J.; Duan, X.; Hu, G.; Liu, Q.; Ren, S.; Li, J.; Kong, M. Roles of Photo-Generated Holes and Oxygen Vacancies in Enhancing Photocatalytic Performance over CeO_2 Prepared by Molten Salt Method. *Adv. Powder Technol.* **2020**, *31*, 4072–4081. [CrossRef]
54. Li, Y.F.; Liu, Z.P. Particle Size, Shape and Activity for Photocatalysis on Titania Anatase Nanoparticles in Aqueous Surroundings. *J. Am. Chem. Soc.* **2011**, *133*, 15743–15752. [CrossRef] [PubMed]
55. Khan, M.M.; Ansari, S.A.; Pradhan, D.; Ansari, M.O.; Han, D.H.; Lee, J.; Cho, M.H. Band Gap Engineered TiO_2 Nanoparticles for Visible Light Induced Photoelectrochemical and Photocatalytic Studies. *J. Mater. Chem. A Mater* **2014**, *2*, 637–644. [CrossRef]
56. Li, W.; Li, M.; Xie, S.; Zhai, T.; Yu, M.; Liang, C.; Ouyang, X.; Lu, X.; Li, H.; Tong, Y. Improving the Photoelectrochemical and Photocatalytic Performance of CdO Nanorods with CdS Decoration. *Cryst. Eng. Comm.* **2013**, *15*, 4212. [CrossRef]
57. Lu, X.; Zheng, D.; Zhang, P.; Liang, C.; Liu, P.; Tong, Y. Facile Synthesis of Free-Standing CeO_2 Nanorods for Photoelectrochemical Applications. *Chem. Commun.* **2010**, *46*, 7721. [CrossRef]
58. Ansari, S.A.; Khan, M.M.; Ansari, M.O.; Kalathil, S.; Lee, J.; Cho, M.H. Band Gap Engineering of CeO_2 Nanostructure Using an Electrochemically Active Biofilm for Visible Light Applications. *RSC Adv.* **2014**, *4*, 16782–16791. [CrossRef]
59. Zhang, W.; Li, Y.; Niu, J.; Chen, Y. Photogeneration of Reactive Oxygen Species on Uncoated Silver, Gold, Nickel, and Silicon Nanoparticles and Their Antibacterial Effects. *Langmuir* **2013**, *29*, 4647–4651. [CrossRef]
60. Li, Y.; Zhang, W.; Niu, J.; Chen, Y. Mechanism of Photogenerated Reactive Oxygen Species and Correlation with the Antibacterial Properties of Engineered Metal-Oxide Nanoparticles. *ACS Nano* **2012**, *6*, 5164–5173. [CrossRef]
61. Li, Y.; Zhang, W.; Niu, J.; Chen, Y. Surface-Coating-Dependent Dissolution, Aggregation, and Reactive Oxygen Species (ROS) Generation of Silver Nanoparticles under Different Irradiation Conditions. *Environ. Sci. Technol.* **2013**, *47*, 10293–10301. [CrossRef]
62. Brunet, L.; Lyon, D.Y.; Hotze, E.M.; Alvarez, P.J.J.; Wiesner, M.R. Comparative Photoactivity and Antibacterial Properties of C_{60} Fullerenes and Titanium Dioxide Nanoparticles. *Environ. Sci. Technol.* **2009**, *43*, 4355–4360. [CrossRef] [PubMed]
63. Azizi-Lalabadi, M.; Ehsani, A.; Divband, B.; Alizadeh-Sani, M. Antimicrobial Activity of Titanium Dioxide and Zinc Oxide Nanoparticles Supported in 4A Zeolite and Evaluation the Morphological Characteristic. *Sci. Rep.* **2019**, *9*, 65. [CrossRef] [PubMed]
64. Sharma, G.; Prema, D.; Venkataprasanna, K.S.; Prakash, J.; Sahabuddin, S.; Devanand Venkatasubbu, G. Photo Induced Antibacterial Activity of CeO_2/GO against Wound Pathogens. *Arab. J. Chem.* **2020**, *13*, 7680–7694. [CrossRef]
65. Shams, S.; Khan, A.U.; Yuan, Q.; Ahmad, W.; Wei, Y.; Khan, Z.U.H.; Shams, S.; Ahmad, A.; Rahman, A.U.; Ullah, S. Facile and Eco-Benign Synthesis of Au@Fe_2O_3 Nanocomposite: Efficient Photocatalytic, Antibacterial and Antioxidant Agent. *J. Photochem. Photobiol. B* **2019**, *199*, 111632. [CrossRef]

catalysts

MDPI

Article

Preparation of CuS/PbS/ZnO Heterojunction Photocatalyst for Application in Hydrogen Production

Ming-Huan Chiu [1], Cheng-Ching Kuo [1], Chao-Wei Huang [2] and Wein-Duo Yang [1,*]

[1] Department of Chemical and Materials Engineering, National Kaohsiung University of Science and Technology, Kaohsiung 807, Taiwan
[2] Department of Engineering Science, National Cheng Kung University, Tainan 701, Taiwan
* Correspondence: ywd@nkust.edu.tw; Tel.: +886-7-381-4526 (ext. 15116)

Abstract: A hexagonal wurtzite ZnO photocatalyst was prepared via a precipitation method. CuS nanoparticles (NPs) and PbS quantum dots (QDs) were loaded onto ZnO via a hydrothermal method to obtain a CuS/PbS/ZnO heterojunction photocatalyst. The CuS/PbS/ZnO photocatalyst obtained via the abovementioned method has significant absorption capabilities in the ultraviolet to near-infrared spectral regions, and effectively reduced the recombination of electron–hole pairs during a photocatalytic reaction. Electron microscope images showed that in the CuS/PbS/ZnO photocatalyst prepared at 130 °C, the particle size of the PbS QDs was approximately 5.5–5.7 nm, and the bandgap determined from the Tauc plot was 0.84 eV; this catalyst demonstrated the best water splitting effect. Furthermore, after adding a 0.25 M mixed solution of Na_2S and Na_2SO_3 as the sacrificial reagent in photocatalysis for 5 h, the hydrogen production efficiency from water splitting reached 6654 μmol g^{-1} h^{-1}.

Keywords: heterojunction; CuS/PbS/ZnO; photocatalyst; hydrogen production; sacrificial reagent

Citation: Chiu, M.-H.; Kuo, C.-C.; Huang, C.-W.; Yang, W.-D. Preparation of CuS/PbS/ZnO Heterojunction Photocatalyst for Application in Hydrogen Production. *Catalysts* **2022**, *12*, 1677. https://doi.org/10.3390/catal12121677

Academic Editors: Jorge Bedia and Carolina Belver

Received: 29 October 2022
Accepted: 12 December 2022
Published: 19 December 2022

1. Introduction

With industrial development and rapid population growth, humanity has accelerated the consumption of the Earth's resources, especially fossil fuels and forest resources. Some scholars have pointed out that, among the energy sources on Earth, petroleum can be used for about 40 years, natural gas for 60 years, and coal for 200 years. The concentration of carbon dioxide in the atmosphere due to the heavy usage of fossil fuels has reached a record high, exacerbating the greenhouse effect and causing rapid melting of the Arctic ice cap and global climate change. These phenomena are extremely detrimental to the sustainable development of human civilization. Therefore, all countries worldwide are actively undertaking research and development to secure new energy resources, with the aim of discovering an energy system to replace petroleum.

The calorific value of hydrogen is very high. Hydrogen can generate 142 million joules of heat energy per kilogram, which is three times that of gasoline and three and a half times that of natural gas. Hydrogen combustion can generate higher energy density, along with water, and will not produce greenhouse gases such as carbon dioxide, rendering it a form of clean energy. If hydrogen can be produced sustainably, combined with safe and efficient storage, transportation, and utilization, a platform for the sustainable use of hydrogen energy can be established to transform the current "petroleum economy" society into a clean and zero-carbon "hydrogen economy" society.

Solar energy can be regarded as an inexhaustible resource. The average light energy reaching the Earth is approximately 1.7×10^{17} W, and 1% of this energy is equivalent to providing the world's 7.4 billion people with 200 days of energy consumption. If this solar energy can be effectively harnessed, the efficiency of energy usage can be improved, and huge progress can be made towards realizing the vision of establishing low-carbon nations and sustainable development. In recent years, the utilization of semiconductor photocatalysts for converting solar energy has become an important research direction. Semiconductor photocatalysts are used for water splitting to produce hydrogen [1,2], as well as in photocatalysis [3], the degradation of organic pollutants [4], and biomass energy conversion [5], all of which can be used to find reliable and stable solutions for environmental issues.

The current shortage of traditional energy requires the emergence of alternative energy sources. Among them, hydrogen energy is one of the high-profile energy sources, and there are more and more reports on the production of H_2. The advantage of this work is that the materials used are not expensive, which reduces the cost of synthesis. The photocatalyst synthesis method is a hydrothermal method, and the process is simple.

Photocatalytic water splitting using solar energy to produce hydrogen was first proposed by Japanese scholars Fujishima and Honda in 1972. The titanium dioxide electrode of the anode absorbs light, and its valence band electrons excite the conduction band. Electrons come to the platinum electrode along the circuit to generate hydrogen. The holes on the titanium dioxide of the anode react with water to generate oxygen. Concurrently, hydrogen gas is generated on the platinum cathode; this phenomenon is the well-known Honda–Fujishima effect [6].

The method of electrochemically producing H_2 from H_2O is to put the cathode, anode and membrane in water and apply voltage to anode to produce H^+, which then reacts with the electrons on the cathode to produce H_2. Moreover, the strategy of photochemical production of H_2 is to place a material that can absorb light energy in water, and through the irradiation of sunlight, the electrons of the material jump from the valence band to the conduction band, generating electron–hole pairs, and H^+ reacts to produce H_2. These two methods of producing H_2 are the methods that are currently being discussed and researched, and it is expected that they can replace the traditional methods of producing H_2 in the future. These two methods can be linked together. In fact, photocatalyst degradation, photocatalyst hydrogen production, solar cells, etc., are all combined and used together [7].

In photocatalytic systems, the main function of the photocatalyst is to absorb the solar energy of different wavelengths to conduct reactions. Photocatalytic materials, such as TiO_2 [8], ZnO [9], CdS [10], and $BiVO_4$ [11], have been developed for hydrogen production via water splitting. As ultraviolet radiation only accounts for 4% of sunlight, the wavelength of visible light accounts for a much larger percentage. In order to further extend the absorption range of the photocatalyst to the visible region of sunlight and adjust the redox potential to increase the efficacy of hydrogen production, different metal ions such as Pt, Au, Cu, and other transition metals can be added to [12–14] or combined with other semiconductor molecular structures to form heterojunction composite photocatalytic compounds [15–17], such as CdS@ZnO, CuS/ZnO, ZnO/ZnS, etc., which have heterojunction photocatalytic potential. In this manner, the original bandgap of the semiconductor can be altered to achieve effective electron–hole separation and enhance the efficiency of hydrogen production. For the CuS material, the reduction potential is −0.48 eV and the oxidation potential is +1.18 eV, while the reduction potential of water is 0 eV and the oxidation potential is +1.23 eV. It is revealed that the reduction potential of CuS is lower than that of water. When CuS receives energy, it is suitable for the reduction reaction of water to generate hydrogen; the energy gap of CuS is 1.66 eV, which belongs to the visible light absorption range. However, due to the poor optical stability of semiconductors with narrow bandgaps, the high rate of electron–hole recombination in the photocatalytic hydrogen production process leads to poor hydrogen production efficiency. In order to improve the hydrogen production efficiency, in their research, Wang et al. added Na_2S and Na_2SO_3 as sacrificial reagents to effectively reduce electron–hole recombination [18]. In 1996, Alivisatos reported on the use of quantum dots (QDs) to control and adjust the size of semiconductor materials [19] in order to improve the solar absorption and bonding problems encountered with narrow bandgap semiconductors, and increase the photocatalytic hydrogen production efficiency.

The PbS QDs and CuS loaded onto ZnO in this study can absorb light in the near-infrared range during the photocatalytic reaction, and thus the photocatalyst is not limited by the range of UV and visible wavelength. The size of the PbS QDs can also be adjusted by changing the light wave absorption position of the photocatalyst.

When the conduction band potential of the photocatalyst is lower than the reduction potential of water (0 V vs. NHE, pH = 0), and the valence potential is higher than the oxidation potential of water (1.23 V vs. NHE, pH = 0), the photocatalyst absorbs light energy greater than the bandgap value, resulting in electron–hole pair separation. The electron transitions to the conduction band to reduce the hydrogen ions in the water to hydrogen, and the holes left in the valence band oxidize the oxygen ions in the water to generate oxygen; this process is termed photolysis by the semiconductor photocatalyst. Due to its bandgap of 3.2 eV, TiO_2 can only absorb in the UV region of the solar spectrum: its photoelectric conversion efficiency is limited, and it is subject to electron–hole recombination, which affects the overall photocatalytic performance.

Nandi et al. loaded ZnO NPs and CdS NPs onto a CuS photocatalyst and adjusted the energy band relationship of the photocatalyst through the heterostructure, which effectively improved the efficiency of photocatalytic degradation [20]. Chang et al. prepared dendritic PbS@CuS photocatalysts by a hydrothermal method and used NaCl, Na_2S, and Na_2SO_3 as sacrificial reagents for photocatalytic hydrogen production, achieving a hydrogen production efficiency of 1736 μmol h^{-1} g^{-1} [3]. Shi et al. first synthesized PbS with QDs and then prepared a PbS@ZnO/graphene oxide photocatalyst. Through the multiple exciton generation of PbS QDs, a good cooperative effect with graphene oxide was produced, which effectively facilitated electron generation and electron–hole separation and improved the efficiency of photocatalytic hydrogen production [21]. Liu et al. prepared a (ZnS-PbS)/Au/TiO_2 photocatalyst via a hydrothermal method, where the loaded PbS QDs enhanced the absorption range of the photocatalyst from the near-infrared zone to the visible light region. Moreover, 20 wt% methanol was used as a sacrificial reagent for hydrogen production, which afforded an efficiency of 5011 μmol h^{-1} g^{-1} [22]. Wang et al. electrochemically modified a ZnO rod/reduced graphene (rGO)/CdS composite with CuS at room temperature to obtain a good heterostructure material. These studies show that CuS nanoparticles play a key role in enhancing the visible light response, and exhibit excellent catalytic performance. A visible-light photocatalytic H_2 production rate of 1073 μmol h^{-1} g^{-1} was obtained with a CuS-ZnO/rGO/CdS heterostructure containing 0.23% CuS and 1.62% CdS [23].

Thus far, many ZnO-based substrates have been prepared by doping with transition metals or rare earth metal nanoparticles to reduce the electron–hole recombination rate and extend the light response into the visible region [24,25]. The generation of composites with other semiconductors has also been used to form heterostructures that combine bandgaps and different properties to enhance photocatalytic activity. However, if energy can be absorbed in the near-infrared region, the efficiency of photocatalytic hydrogen production could be further improved [26].

The aforementioned research results demonstrate that the photocatalytic efficiency of photocatalysts can be improved by loading a narrow bandgap semiconductor on the photocatalyst substrate to form a heterojunction. Therefore, in this study, CuS NPs and PbS QDs were loaded onto the surface of a ZnO photocatalyst to improve the bandgap relationship of the photocatalyst by utilizing sulfide with a narrow bandgap and increasing the absorption range of the photocatalyst in the visible light region and the near-infrared region, so as to improve the photocatalytic hydrogen production efficiency.

2. Results and Discussion

2.1. Analysis of Material Properties

The prepared heterojunction photocatalyst was analyzed using transmission electron microscopy (TEM), as is shown in Figure 1. Figure 1a shows that ZnO prepared via the precipitation method has a rectangular shape and a particle size of approximately 323.8 nm. The sulfide CuS NPs prepared via the hydrothermal method at 130 °C had a size of approximately 35.7 nm, and presented a sheet-like morphology similar to the shape reported by Saranya et al. [27]. The particle size of the PbS QDs was approximately 5.5–5.7 nm, and the particle shape was approximately spherical, which is similar to that of the PbS QDs prepared by Liu and colleagues [22].

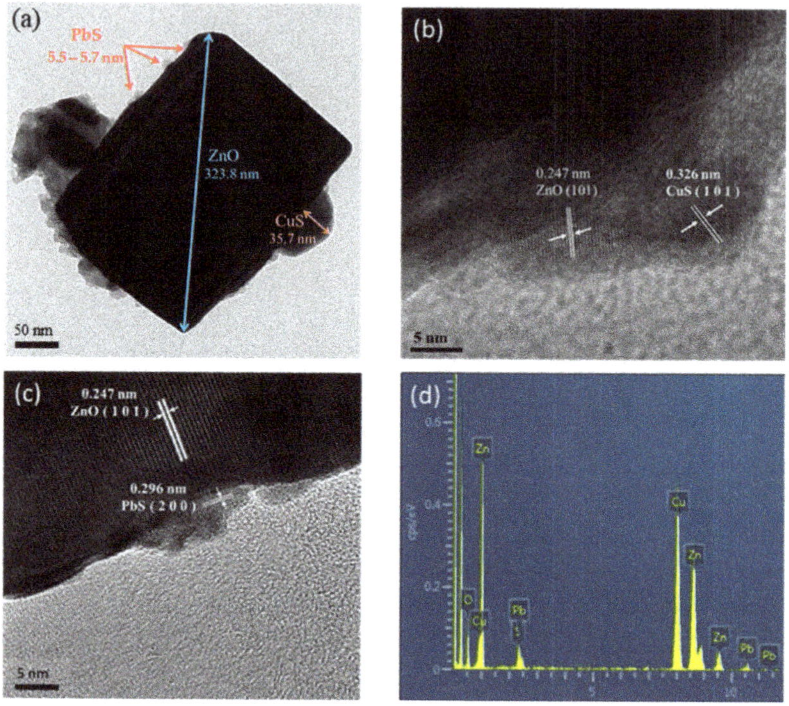

Figure 1. TEM analysis of CuS/PbS/ZnO heterojunction photocatalyst. (**a**) Analytical images of CuS/PbS/ZnO photocatalyst, (**b**) image analysis of the lattice spacing for CuS NPs loaded onto ZnO, (**c**) image analysis of PbS QDs and ZnO lattice spacing, and (**d**) EDX analysis of CuS/PbS/ZnO photocatalyst.

The high magnification TEM image of the photocatalyst (Figure 1b) shows a lattice spacing of 0.247 nm for ZnO, corresponding to the (101) crystal face of hexagonal wurtzite; the lattice spacing of the CuS (101) crystal plane was 0.326 nm. From Figure 1c, the lattice spacing of the (200) plane of PbS QDs was 0.296 nm, which is consistent with the results of Sun and Chang et al. [3,7] and is consistent with the XRD results presented hereinafter. The TEM image of the prepared CuS/PbS/ZnO photocatalyst was analyzed by energy-dispersive X-ray spectroscopy (EDX) (as shown in Figure 1d), whereby Cu, Pb, S, Zn, and O were clearly observed, in addition to C originating from the carbon-coated copper mesh itself. Thus, it was confirmed that the CuS NPs and PbS QDs were successfully loaded onto the ZnO photocatalyst.

The XRD data for the prepared CuS/PbS/ZnO photocatalyst are shown in Figure 2. The characteristic peaks of hexagonal wurtzite ZnO were observed at 2 θ values of 31.8°, 34.5°, 36.3°, 47.6°, 56.6°, 62.9°, 66.4°, 68.0°, and 69.1°, corresponding to the (100), (002), (101), (102), and (110) crystal faces; the data are consistent with JCPDS card no. 36-1451 [28]. The characteristic peaks of the loaded CuS NPs were observed at 27.3°, 27.8°, 29.4°, 31.9°, 48.1°, 52.7°, and 59.1°, corresponding to the (100), (102), (103), (110), (108), and (116) crystal faces, consistent with JCPDS card no. 06-0464 [29]. The characteristic peaks of PbS were apparent at 26.0°, 30.1°, 43.1°, 51.0°, 53.5°, 62.6°, 68.9°, 70.9°, and 78.9°, corresponding to the (111), (200), (220), (311), (222), (400), (331), (420), and (422) crystal faces, consistent with

JCPDS card no. 05-0592 [3]. Thus, CuS NPs and PbS QDs were successfully loaded onto the ZnO photocatalysts at 100 °C, 130 °C, 150 °C, and 170 °C by the hydrothermal method.

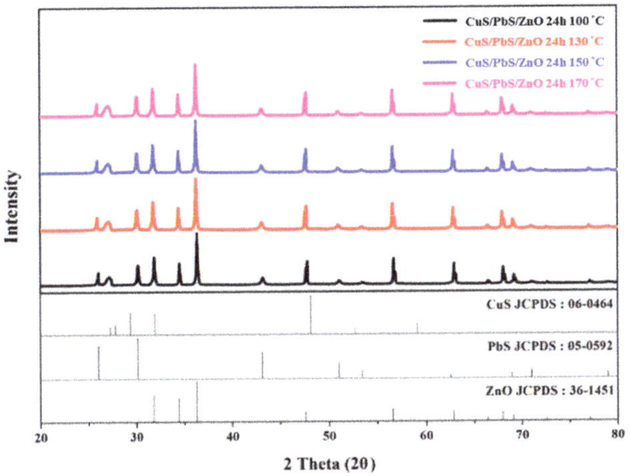

Figure 2. XRD analysis of CuS/PbS/ZnO heterojunction photocatalysts prepared under different hydrothermal temperatures.

As is shown in Figure 2, the characteristic peaks of the loaded CuS at 31.9° and 48.0° overlapped with the characteristic peaks of ZnO at 31.8° and 47.5°, which intensified the peaks. The characteristic peaks of the loaded PbS at 62.6°and 68.9° overlapped with those of ZnO at 62.9° and 69.1°, leading to higher intensities. This indicates the presence of CuS and PbS in the photocatalyst. Moreover, these peaks were more intense for the CuS/PbS/ZnO photocatalyst prepared at higher hydrothermal temperature because the formation of sulfides was more complete under the hydrothermal environment [30].

2.2. Spectral Properties of Prepared Heterostructure Photocatalysts

Table 1 shows ZnO, PbS/ZnO, CuS/ZnO, and CuS/PbS/ZnO photocatalysts incident at a single wavelength light to detect the absorbance of the photocatalysts. It can be found that single composition of ZnO photocatalyst absorbs UV (300–350 nm), but ZnO does not absorb from VIS. to NIR (400–1800 nm). Comparing PbS/ZnO and CuS/ZnO, it can be seen that ZnO-loaded PbS QDs can enhance the absorption of 400–1800 nm; while the ZnO-loaded CuS can enhance the absorption of 350–1800 nm, the above two carriers can improve the absorption of ZnO. Further comparison can be found that for the wavelength of 350–600 nm, CuS is more improved; as for PbS QDs, it is more improved at 800–1800 nm.

Table 1. Monochromatic light source to the contribution of each component in the photocatalyst.

Single Wave Length Light (nm) / Photocatalyst	300	350	400	600	800	1000	1400	1800
	Absorbance of the Photocatalysts (a.u.)							
ZnO	0.408	0.615	0.000	0.000	0.000	0.000	0.000	0.000
PbS/ZnO	0.408	0.616	0.345	0.746	0.861	0.838	0.913	0.860
CuS/ZnO	0.408	0.625	0.665	0.754	0.627	0.627	0.587	0.546
CuS/PbS/ZnO	0.408	0.626	0.445	0.787	0.870	0.885	0.904	0.858

The UV-VIS-NIR spectra of the CuS/PbS/ZnO photocatalysts prepared at various hydrothermal temperatures are shown in Figure 3, and the dotted line in Figure 3 represents the position of the maximum absorption wavelength of sulfide quantum dots synthesized by different hydrothermal temperatures, which can confirm the quantum confinement effect of quantum dots. When the hydrothermal temperature is lower, the smaller size of synthesized quantum dots is produced and spectrum present the blue-shift of the absorption wavelength is occurred. The CuS/PbS/ZnO photocatalysts prepared at all the hydrothermal temperatures had significant absorption bands from the near-infrared to the visible light region. The photocatalysts prepared at 100 °C and 130 °C absorbed more strongly at wavelengths of 380–760 nm. In comparison, the photocatalysts prepared at 150 °C and 170 °C had weaker absorption in the visible spectral region. The maximum absorption wavelength was shifted from 1435 nm for the sample prepared at 170 °C to 817 nm for the congener prepared at 100 °C, which suggests that the absorption of the photocatalyst prepared using the lower-temperature hydrothermal conditions shifted towards the shorter wavelength. Hence, it was concluded that the sulfide QDs prepared at lower temperatures had a smaller particle size, and the bandgap was widened due to the quantum confinement effect, which resulted in a blue-shift of the absorption [31].

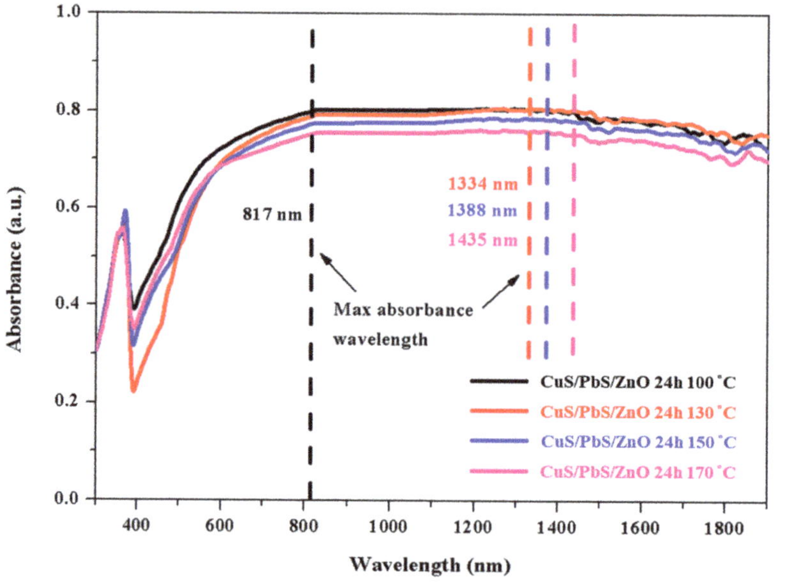

Figure 3. UV-VIS-NIR spectra of CuS/PbS/ZnO photocatalysts prepared at various hydrothermal temperatures. Dotted lines indicate the maximum absorption wavelength of photocatalysts.

The bandgaps of the sulfide QDs and ZnO photocatalysts were obtained via an analysis of the Tauc plots. Figure 4 shows the Tauc plot for the CuS/PbS/ZnO photocatalysts prepared at different hydrothermal temperatures. The CuS/PbS/ZnO photocatalyst had two bandgaps: a lower bandgap of PbS, and a higher bandgap of ZnO. As can be deduced from Figure 4, the bandgap of ZnO increased from 2.95 eV to 3.10 eV and the bandgap of PbS QDs decreased from 0.71 eV to 0.55 eV for the CuS/PbS/ZnO photocatalysts prepared in the hydrothermal temperature range of 100–170 °C. The bandgap of the PbS QDs prepared at a hydrothermal temperature of 130 °C was approximately 0.84 eV, which corresponds to PbS QDs of approximately 5.5 nm in the TEM image in Figure 1a. The relationship between the particle size and bandgap of PbS QDs is consistent with the study of Liang and co-workers [32]. Guchhait et al. also indicated that smaller PbS QDs particles should have a larger bandgap, which could shift the absorption from the near-infrared region to the visible light region, resulting in a blue-shift phenomenon [33].

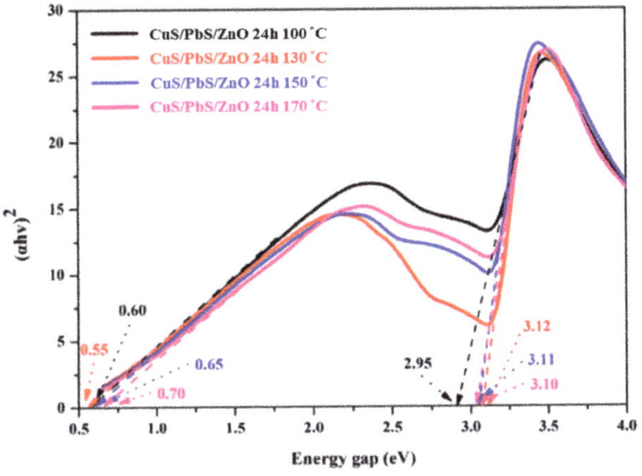

Figure 4. Tauc plot of CuS/PbS/ZnO photocatalysts prepared at various hydrothermal temperatures.

Electron–hole pair recombination in the photocatalyst was investigated by PL spectral analysis. When the photocatalyst absorbs light energy, the electrons will transition from the ground state to the excited state, and the electrons will fall back to the ground state and generate fluorescence. This process is termed electron–hole pair recombination. The degree of electron–hole recombination in the photocatalyst determines the fluorescence intensity [29].

Figure 5 shows the PL spectrum of the prepared photocatalyst. Figure 5a shows the PL spectrum of the ZnO photocatalysts loaded with different sulfides. Fluorescence was generated in the wavelength range of approximately 400 nm to 800 nm, and the strongest fluorescence intensity was at a wavelength of 600 nm. Non-loaded ZnO produced the strongest fluorescence intensity, at a wavelength of 600 nm, whereas the fluorescence intensity decreased significantly for the ZnO photocatalyst loaded with the CuS NPs and PbS QDs, indicating that the loaded sulfide QDs effectively reduced the electron–hole recombination rate of the photocatalyst [34]. In comparison with the PbS/ZnO and CuS/ZnO photocatalysts, the fluorescence intensity generated by the CuS/PbS/ZnO photocatalyst was the lowest, which indicates that when CuS NPs and PbS QDs were simultaneously loaded onto the ZnO photocatalyst, the sulfide QDs and ZnO photocatalysts approached each other more closely, which effectively reduced the recombination of electron–hole pairs [35]. Yendrapati et al. [29] pointed out that the higher the peak value of the PL, the stronger the fluorescence intensity generated by the photocatalyst, corresponding to a faster electron–hole recombination rate. In contrast, the lower the peak value of the PL, the weaker the fluorescence intensity generated by the photocatalyst, indicating a slower electron–hole recombination rate.

Figure 5b shows the PL spectrum of the CuS/PbS/ZnO photocatalysts prepared at different hydrothermal temperatures. The CuS/PbS/ZnO photocatalysts prepared at different hydrothermal temperatures also emitted the maximum fluorescence intensity at 600 nm. The fluorescence intensity of the photocatalysts prepared at temperatures from 170 to 130 °C decreased gradually, where the photocatalyst prepared at 130 °C had the lowest fluorescence intensity. This is because the particle size of the sulfide QDs prepared at a lower hydrothermal temperature was smaller, and the bandgap of the sulfide QDs increased due to the quantum confinement effect, which effectively facilitated the separation of the electron–hole pairs and reduced the incidence of recombination. Notably, the highest PL peak intensity was observed for the photocatalyst prepared at 100 °C. This might be due to the hydrothermal temperature being too low, the crystallinity of the sulfide QDs being low, or crystal defects, which may have led to a high electron–hole pair recombination rate [36].

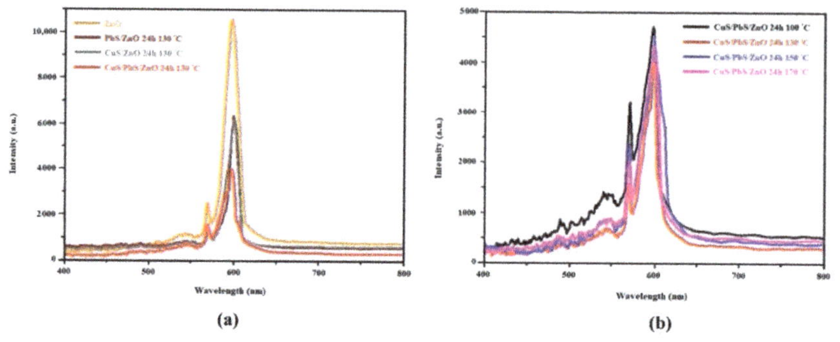

Figure 5. PL spectra of the prepared photocatalysts at 579 nm as the excitation wavelength (**a**) Various sulfides supported on ZnO photocatalyst, and (**b**) CuS/PbS/ZnO photocatalysts prepared at different hydrothermal temperatures.

2.3. Hydrogen Production by Water Splitting with Photocatalysts

Figure 6 shows the hydrogen production efficiency of the respective photocatalysts in a 0.25 M aqueous mixed solution of Na_2S and Na_2SO_3 as sacrificial reagents. The photocatalytic efficiency of the photocatalyst was studied by irradiating the aqueous solution containing the photocatalyst and sacrificial reagents with light generated by a solar simulator for 5 h. Figure 6a shows that the hydrogen production of the ZnO photocatalyst loaded with sulfide QDs was better than that of the non-loaded ZnO photocatalyst, and the hydrogen production achieved with the respective photocatalysts increased with time. CuS and PbS of the CuS/ZnO and PbS/ZnO photocatalysts facilitated energy absorption from the near-infrared to visible light region. PbS and CuS are prone to photocorrosion; thus, the hydrogen production rates of the CuS/ZnO and PbS/ZnO photocatalysts were lower. As is shown in Figure 6b, the hydrogen production achieved with the ZnO photocatalyst without any loading was 203 μmol g^{-1} h^{-1}, whereas the hydrogen production of the CuS/PbS/ZnO photocatalyst was 6654 μmol g^{-1} h^{-1}, which is 32 times that of the non-loaded ZnO photocatalyst. The CuS/PbS/ZnO photocatalyst had a higher hydrogen production rate, because the simultaneous loading of PbS QDs and CuS NPs not only facilitated absorption in the near-infrared and visible light regions, but the prepared ternary CuS/PbS/ZnO heterojunction photocatalyst was also more efficient. The effective electron–hole separation reduced the incidence of recombination [35].

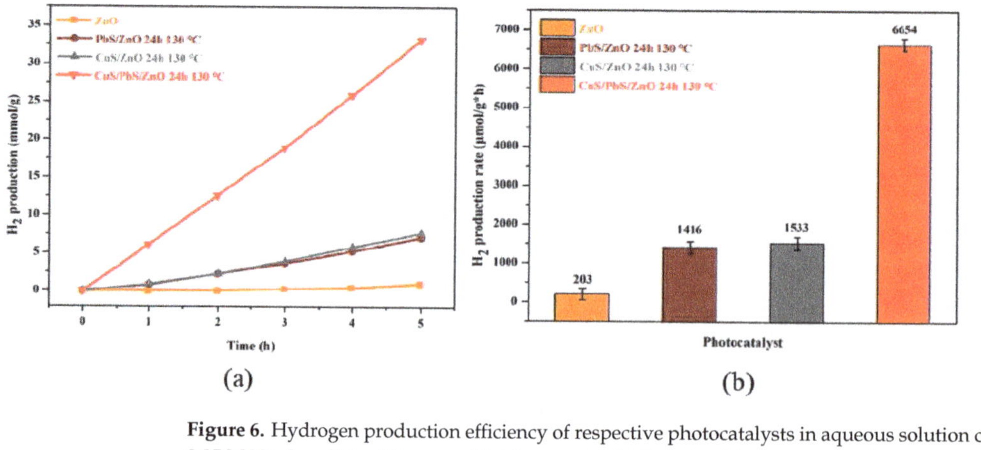

Figure 6. Hydrogen production efficiency of respective photocatalysts in aqueous solution containing 0.25 M Na_2S and Na_2SO_3 as sacrificial reagents. (**a**) Relationship between hydrogen production and time. (**b**) Hydrogen production rate in μmol g^{-1} h^{-1}.

Wang et al. reported that CuS/ZnS nanomaterials exhibit high visible light-induced H_2 generation activity. The H_2 generation rate increases with increasing Cu^{2+} ions. However, as with other cocatalysts, when the maximum amount of Cu^{2+} is reached (above 7 mol %), the hydrogen evolution rate decreases significantly. This is due to light shielding by excess CuS, which reduces the number of active sites on the surface [37].

Figure 7 shows the hydrogen production experiment of loading different contents of CuS NPs on PbS/ZnO and adding Na_2S and Na_2SO_3 mixture (0.25 M) as a sacrificial reagent. It exhibits that as the amount of CuS NPs loaded onto PbS/ZnO increases, the hydrogen production will increase. When the CuS NPs loading is 2 wt.%, the hydrogen production is the highest, and when the loading reaches 3 wt.% and 4 wt.% the hydrogen production of the photocatalyst decreased. The reason is that the over-loaded CuS NPs shield the specific surface area of PbS/ZnO, greatly reducing the active sites and light absorption of PbS/ZnO.

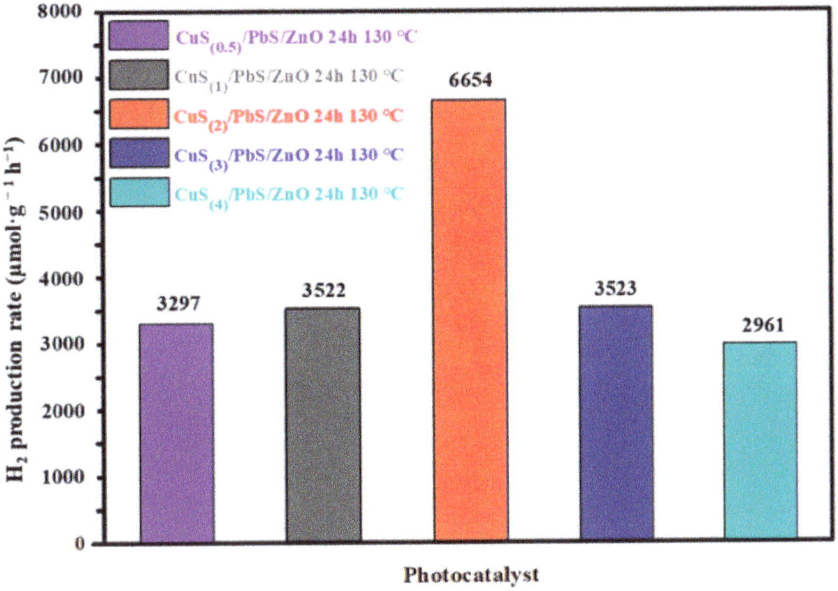

Figure 7. Rate diagram of hydrogen production on PbS/ZnO loaded with different contents of CuS NPs.

TON specifies the maximum use that can be made of a catalyst for a special reaction under defined reaction conditions by the number of molecular reactions or reaction cycles occurring at the reactive center up to the decay of activity.

The calculation formula of TON:

$$\text{Hydrogen production rate } (\mu\text{mole} \times \text{min}^{-1})/\text{amounts of photocatalyst } (\mu\text{mole})$$

In order to obtain the hydrogen production effect of a single-component photocatalyst and heterogeneous photocatalyst (CuS/PbS/ZnO) in a monochromatic light environment, the authors designed the following experiments:

50 μmole of CuS, PbS, ZnO and CuS/PbS/ZnO were used for the hydrogen production experiment under the monochromatic light source environment between 300 nm and 1200 nm (0.25 M mixed solution of Na_2S and Na_2SO_3 as the sacrificial reagent). The relevant data collection Tables 2 and 3 are as follows.

Table 2. Evaluation of hydrogen production rate by single and heterojunction photocatalysts at monochromatic light.

Single Wave Length Light (nm) / Photocatalyst	350	600	800	1000	1200
	Hydrogen Production Rate (μmol min^{-1})				
ZnO	224	0	0	0	0
PbS	0	212	234	188	201
CuS	0	197	220	207	194
CuS/PbS/ZnO 24 h 130 °C	271	346	375	394	446

Table 3. Evaluation of single and heterojunction photocatalyst activities at monochromatic light.

Single Wave Length Light (nm) / Photocatalyst	350	600	800	1000	1200
	Turnover Number, TON (min^{-1})				
ZnO	4.48	0	0	0	0
PbS	0	4.24	4.68	3.76	4.02
CuS	0	3.94	4.40	4.14	3.88
CuS/PbS/ZnO 24 h 130 °C	5.42	6.92	7.50	7.88	8.92

Note: The photocatalyst amounts of CuS, PbS, ZnO and CuS/PbS/ZnO are all 50 μmole.

From the table 'Evaluation of single and heterojunction photocatalyst activities at monochromatic light', it can be seen that the TON of each component of a single-component photocatalyst is smaller than that of heterojunction photocatalyst under the environment of monochromatic light, which can also explain that the heterojunction synthesized by hydrothermal method photocatalyst has better hydrogen production rate.

Figure 8 shows hydrogen production in the presence of the CuS/PbS/ZnO photocatalysts prepared at different hydrothermal temperatures. The photocatalytic reaction proceeded in the presence of the photocatalyst without the addition of a sacrificial reagent (as is shown in Figure 8a), where the photocatalyst prepared at 130 °C afforded the best hydrogen production. The hydrogen production curves of the photocatalysts prepared at hydrothermal temperatures in the range of 150–170 °C were similar. This is due to the quantum confinement effect, because the particles of sulfide QDs prepared at lower hydrothermal temperature (130 °C) were smaller, which increased the bandgap of the sulfide QDs and effectively separated the electron–hole pairs. However, the particle size of the sulfide QDs prepared at higher hydrothermal temperatures (150 and 170 °C) was relatively larger; thus, the hydrogen production was lower than that achieved with the photocatalysts prepared at 130 °C. Moreover, after five hours of the photocatalytic process, the hydrogen production was lower for the photocatalyst prepared at a hydrothermal temperature of 100 °C. This might be because the hydrothermal temperature was low, and the crystallinity of the prepared sulfide QDs was low; thus, it was easy to generate crystal defects, leading to the recombination of the electron–hole pairs. This result is consistent with the previous PL analysis.

Wang et al. [38] pointed out that adding a mixed solution of Na_2S and Na_2SO_3 to the aqueous solution of the photocatalytic system as a sacrificial reagent effectively reduced electron–hole recombination in the photocatalyst. Na_2S generates S^{2-}, which is more unstable than the sulfide photocatalyst, causing the electron–hole pairs to oxidize S^{2-} to a greater extent whilst protecting the sulfide photocatalyst from photocorrosion. However, the oxidation of S^{2-} produces polysulfide Sn^{2-} (Equation (1)), which affects the efficiency of hydrogen production. The added Na_2SO_3 will dissociate to give $SO_3{}^{2-}$ and convert Sn^{2-} back to S^{2-} (Equation (2)); thus, the photocatalyst is able to produce hydrogen. The efficiency is unaffected and $SO_3{}^{2-}$ also reacts with the electron–hole pairs to sweep them away and prevent recombination (Equation (3)). The overall reaction formula with the mixed sacrificial reagents is represented by Equation (4). The reaction mechanism in the presence of Na_2S mixed with Na_2SO_3 as sacrificial reagents is as follows:

$$2S^{2-} + 2h^+ \rightarrow S_2{}^{2-} \tag{1}$$

$$S_2{}^{2-} + SO_3{}^{2-} \rightarrow S_2O_3{}^{2-} + S^{2-} \tag{2}$$

$$SO_3{}^{2-} + h^+ + 2OH^- \rightarrow SO_4{}^{2-} + 2H^+ \tag{3}$$

$$SO_3{}^{2-} + S^{2-} + 2h^+ \rightarrow S_2O_3{}^{2-} \tag{4}$$

In this study, 0.25 M of Na_2S and Na_2SO_3 mixed solution was added as a sacrificial reagent to aid the photocatalyst in removing electron–holes during the photocatalytic reaction, so as to reduce the electron–hole recombination rate (as is shown in Figure 8b).

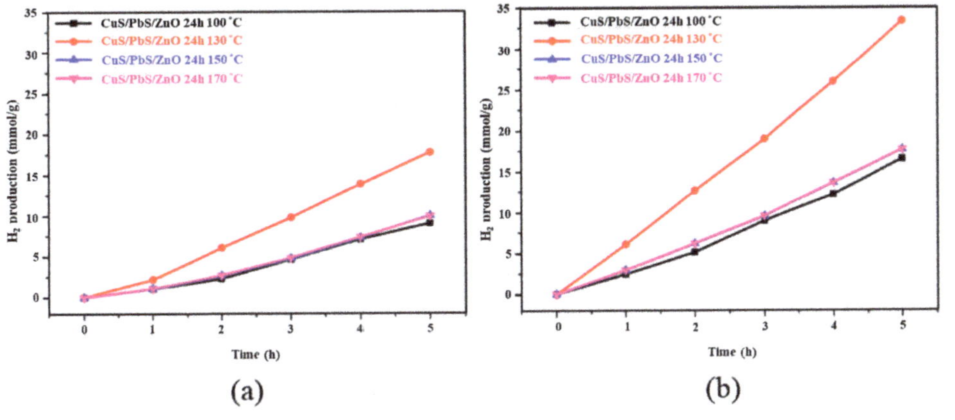

Figure 8. Analysis of hydrogen production in the presence of CuS/PbS/ZnO photocatalysts prepared at different hydrothermal temperatures. (**a**) No addition of Na_2S and Na_2SO_3 mixed sacrificial reagent. (**b**) Addition of 0.25 M Na_2S and Na_2SO_3 mixed sacrificial reagent.

After adding the sacrificial reagent (Na_2S mixed with Na_2SO_3), the hydrogen production of the photocatalyst was significantly improved. In comparison with the photocatalyst without the added sacrificial reagents, the hydrogen production increased greatly (by 77–88%), which proves that adding Na_2S and Na_2SO_3 as sacrificial reagents effectively reduced electron–hole recombination in the photocatalyst and improved the hydrogen production efficiency of the photocatalyst.

An SEM analysis of the photocatalyst (Figure 9) showed that the hexagonal wurtzite ZnO (in the shape of a rectangle) was still visible in the samples treated at hydrothermal temperatures of 100–150 °C. As the hydrothermal temperature increased, agglomeration gradually occurred, and was most prominent in the photocatalyst prepared at the hydrothermal temperature of 170 °C. The specific surface area/pore volume/pore size of the CuS/PbS/ZnO photocatalysts prepared at different hydrothermal temperatures was determined through Brunauer–Emmett–Teller (BET) analysis (see Table 4). The surface area decreased from 33.2 $m^2\ g^{-1}$ for the photocatalyst prepared at 100 °C to 23.0 $m^2\ g^{-1}$ for the congener prepared at 170 °C, indicating that a higher hydrothermal temperature led to a lower specific surface area of the photocatalyst. This is because the photocatalyst powders prepared at higher hydrothermal temperatures agglomerated easily [39], which is consistent with the SEM analysis. The larger the surface area of the photocatalyst, the more reaction sites there are, and the better the photocatalytic hydrogen production efficiency of the photocatalyst [40]. Through the abovementioned analysis of the photocatalytic hydrogen production, it was discovered that the crystallinity of the loaded sulfide QDs also affected the hydrogen production efficiency of the photocatalyst.

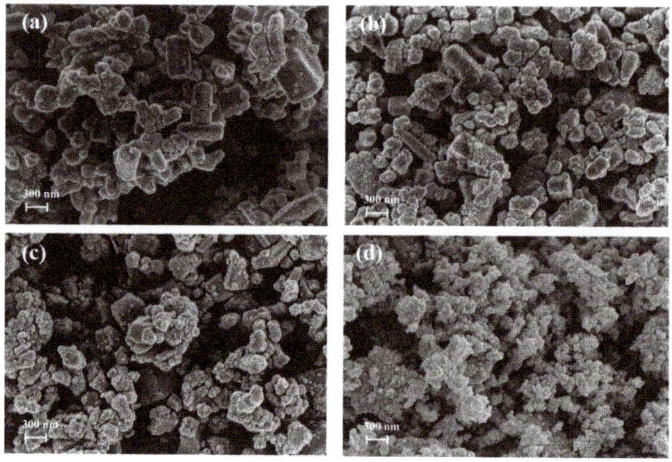

Figure 9. SEM images of CuS/PbS/ZnO photocatalysts prepared at different hydrothermal temperatures. (**a**) 100 °C, (**b**) 130 °C, (**c**) 150 °C, and (**d**) 170 °C.

Table 4. BET surface area, pore volume, and pore size of CuS/PbS/ZnO.

Photocatalyst	BET Surface Area ($m^2\ g^{-1}$)	Pore Volume ($cm^3\ g^{-1}$)	Pore Size (nm)
CuS/PbS/ZnO 100 °C	28.98	0.21	29.05
CuS/PbS/ZnO 130 °C	33.20	0.26	30.79
CuS/PbS/ZnO 150 °C	27.57	0.23	33.13
CuS/PbS/ZnO 170 °C	23.09	0.23	39.37

The CuS/PbS/ZnO photocatalyst, prepared at a hydrothermal temperature of 130 °C in an autoclave, was irradiated with a solar simulator for photocatalytic reaction for 5 h, the solution was centrifugated to collect the photocatalyst, and was reused 10 times for hydrogen production analysis (as shown in Figure 10), during which the photocatalyst still maintained a hydrogen production rate of 78%. Thus, the prepared CuS/PbS/ZnO photocatalyst exhibits good stability and reusability, and it is speculated that when the mixture of water and sacrificial reagent is replaced, it is not performed in a dark environment, so the photo-corrosion of sulfides is caused, and when the photocatalyst is recycled and the photocatalytic reaction is performed again, the amount of hydrogen produced is dropped.

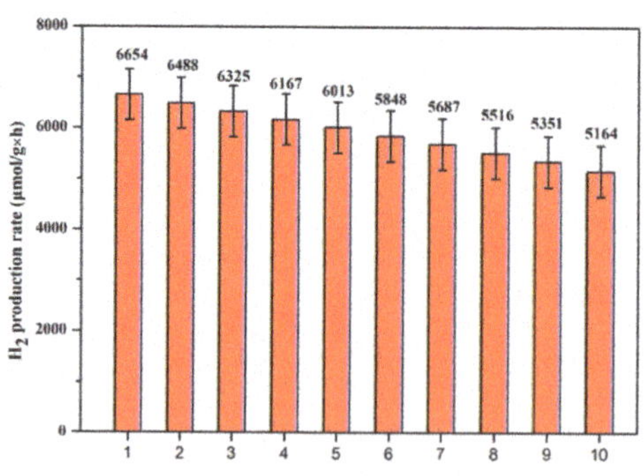

Figure 10. Hydrogen production efficiency of CuS/PbS/ZnO photocatalyst prepared at a hydrothermal temperature of 130 °C, under photocatalytic reaction with solar simulator irradiation for 5 h.

3. Materials and Methods

3.1. Experimental Reagents

The laboratory-grade chemicals used in this study and their purity were as follows: zinc acetate (Showa, 99%, Tokyo, Japan), copper acetate (Alfa Aesar, 99%, Heysham, UK), lead acetate (Alfa Aesar, 99%, Heysham, UK), thioacetamide (Alfa Aesar, 99%, Essex, MA, USA), sodium sulfide (Acros, 98+%, Hel, Belgium), sodium sulfite (Showa, 97%, Tokyo, Japan), sodium hydroxide (Showa, 97%, Tokyo, Japan), and ammonia (Showa, 28%, Tokyo, Japan).

3.2. Synthesis of ZnO Nanomaterials

The ZnO used in this study was prepared by the precipitation method. Firstly, 3 g of $Zn(CH_3COO)_2{\cdot}2H_2O$ was weighed and dissolved in 100 mL of deionized (DI) water. The solution was then heated to 80 °C and stirred at 300 rpm with a magnetic stirrer for 30 min. At the same time, 20 mL of dilute NH_3 solution was added dropwise at a rate of 1–2 mL/min, and stirring was continued for 2 h. After stirring, the solution was allowed to stand at room temperature, after which the precipitate was filtered by ultrasonic vibration for 30 min. The precipitate was then washed with DI water and C_2H_5OH several times and freeze-dried for 8 h. After drying, the ZnO photocatalyst powder was collected in a sample bottle and covered with aluminum foil to avoid exposure to external light sources.

3.3. Synthesis of CuS/PbS/ZnO Photocatalyst by Hydrothermal Method

An appropriate amount of $Pb(CH_3COO)_2{\cdot}2H_2O$ and $Cu(CH_3COO)_2{\cdot}H_2O$ was weighed and dissolved in DI water to form an acetate solution, which was then mixed with the ZnO photocatalyst powder synthesized above and stirred uniformly. Excess CH_3CSNH_2 (2.5 times the mixture of lead acetate and copper acetate) was added at a rate of 1–2 mL/min, stirred for 2 h, and then ultrasonically shaken for 30 min. After completion of the reaction, 1 M NaOH was added to adjust the pH of the solution to 11, and the solution was poured into a Teflon-lined container, placed in an autoclave, and heated to 100, 130, 150, or 170 °C for 24 h via the hydrothermal method. After heating, the mixture was left to cool to room temperature, and the precipitate was washed several times with DI water and C_2H_5OH and freeze-dried for 8 h. After drying, the CuS/PbS/ZnO photocatalyst powder was obtained and stored in a sample bottle wrapped with aluminum foil to avoid exposure to external light. Figure 11 shows the bandgap correlation diagram of the prepared CuS/PbS/ZnO heterojunction photocatalyst.

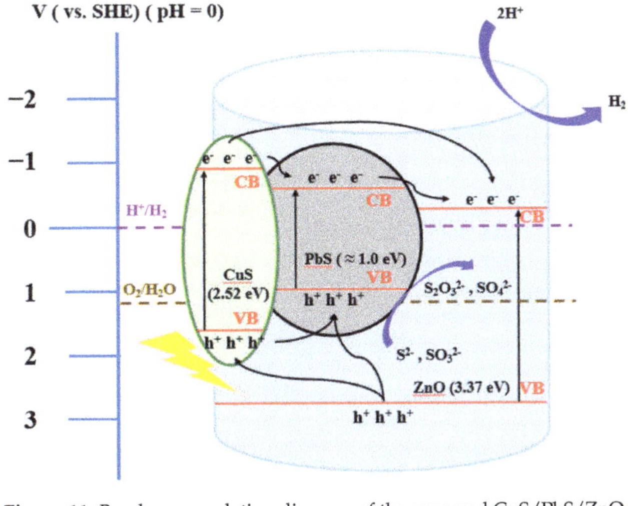

Figure 11. Bandgap correlation diagram of the prepared CuS/PbS/ZnO heterojunction photocatalysts.

3.4. The Degree of Recombination of Photocatalyst

The degree of recombination of the photocatalyst is the electron–hole recombination rate of the prepared photocatalyst, which is measured via fluorescence spectrometer photoluminescence (PL). When the photocatalyst is irradiated with a light source with a wavelength of 579 nm, the excited photocatalyst generates electrons in the conduction band, and then falls back to the conduction band to generate fluorescence. In this way, the recombination rate of the electron–hole pairs of the photocatalyst was judged by the fluorescence intensity of each sample.

3.5. Photocatalytic Hydrogen Production

The photocatalytic water splitting hydrogen production reactor system used in this study was a semi-open type. The simulated light source was a 300 W solar simulator, and an AM 1.5 G filter was installed as the illumination light source. The distance between the light source and the photocatalytic reactor was 20 cm. The main body of the reactor was made of light-transmitting quartz glass. There were four holes on the reactor, two of which were the inlet and outlet holes for helium gas, and two of which were the pressure detection hole and the sampling port. CuS/PbS/ZnO photocatalyst powder (50 mg) was placed in the reactor, and the photocatalytic reaction was carried out in pure water to study the effect of hydrogen production via water splitting. The results were compared those obtained by adding 0.25 M sacrificial reagent (mixed solution of Na_2S and Na_2SO_3). After adding the 0.25 M sacrificial reagent (mixed solution of Na_2S and Na_2SO_3) and allowing the photocatalytic reaction to proceed for 5 h, the solution was centrifuged at 6000 rpm to filter out the photocatalyst. Subsequently, under similar conditions, the water splitting experiment to produce hydrogen was repeated 10 times.

In the photocatalytic decomposition of water for hydrogen production, the gas in the reactor was extracted by a gas needle every hour and passed through a gas chromatograph (GC, YL Instrument 6500GC system) connected to a thermal conductivity detector (TCD) to analyze the hydrogen content. During the process, helium was used as the carrier gas, where the gas flow rate was 30 mL/min. The sample was injected into the injection port at a temperature of 180 °C; the column type was molecular-sieves, and the temperature was set at 60 °C. CuS/PbS/ZnO photocatalyst is used on organic compounds (such as a five-membered ring system), but it may be cracked due to high temperature in the process of illumination. However, it is believed that photocatalyst research can be carried out in the ring catalyst system in the future. [41].

4. Conclusions

CuS NPs and PbS QDs were successfully loaded onto ZnO photocatalysts by a hydrothermal method, and the resulting CuS/PbS/ZnO heterojunction photocatalyst was applied to photocatalytic hydrogen production under sunlight irradiation. The sulfide-loaded PbS QDs and CuS NPs on the ZnO photocatalyst enhanced the absorption of the photocatalyst in the near-infrared region to the visible light region. The quantum confinement effect of the sulfide QDs widened the bandgap of the photocatalyst, and indeed reduced electron–hole recombination, thus effectively improving the photocatalytic hydrogen production efficiency of the photocatalyst. The CuS/PbS/ZnO photocatalyst prepared at a hydrothermal temperature of 130 °C had better efficiency, where the hydrogen production amounted to 6654 μmol g^{-1} h^{-1} when a 0.25 M aqueous mixture of Na_2S and Na_2SO_3 was added as a sacrificial reagent.

Author Contributions: W.-D.Y. conceived and designed the experiments; C.-C.K. performed the experiments; M.-H.C. and C.-W.H. analyzed the data; M.-H.C. wrote the paper. All authors have read and agreed to the published version of the manuscript.

Funding: The authors thank the Ministry of Science and Technology of Taiwan for financial support under grant no. MOST 110-2221-E-151-040.

Conflicts of Interest: The authors declare no conflict of interest.

References

1. Hermesmann, M.; Müller, T.E. Green, Turquoise, Blue, or Grey? Environmentally Friendly Hydrogen Production in Transforming Energy Systems. *Prog. Energy Combust. Sci.* **2022**, *90*, 100996. [CrossRef]
2. Okonkwo, P.C.; Barhoumi, E.M.; Mansir, I.B.; Emori, W.; Bhowmik, H. Effect of electrode material on the hydrogen production using a low-cost home-made alkaline electrolyzer. *Vacuum* **2022**, *198*, 110878. [CrossRef]
3. Chang, C.J.; Lin, Y.G.; Chen, J.K.; Huang, C.Y.; Hsieh, S.C.; Wu, S.Y. Ionic liquid/surfactant-hydrothermal synthesis of dendritic PbS@CuS core-shell photocatalysts with improved photocatalytic performance. *Appl. Surf. Sci.* **2021**, *546*, 149106. [CrossRef]
4. Frank, S.N.; Allen, J. Heterogeneous photocatalytic oxidation of cyanide and sulfite in aqueous solutions at semiconductor powders. *J. Phys. Chem.* **1977**, *15*, 1484–1488. [CrossRef]
5. Chiarello, G.L.; Aguirre, M.H.; Selli, E. Hydrogen production by photocatalytic steam reforming of methanol on noble metal-modified TiO_2. *J. Catal.* **2010**, *20*, 182–190. [CrossRef]
6. Fujishima, A.; Honda, K. Electrochemical photolysis of water splitting at a semiconductor electrode. *Nature* **1972**, *5358*, 37–38. [CrossRef]
7. Sun, L.; Koh, Z.Y. PbS quantum dots embedded in a ZnS dielectric matrix for bulk heterojunction solar cell applications. *Adv. Mater.* **2013**, *33*, 4598–4604. [CrossRef]
8. Li, X.; Sun, Z.; Bao, Y.; Xia, X.; Tao, T.; Homewood, K.P.; Li, R.; Gao, Y. Comprehensively improved hydrogen sensing performance via constructing the facets homojunction in rutile TiO2 hierarchical structure. *Sens. Actuators B* **2022**, *350*, 130869. [CrossRef]
9. Xu, Y.; Li, H.; Sun, B.; Qiao, P.; Ren, L.; Guohui, T.; Jiang, B.J.; Pan, K.; Zhou, W. Surface oxygen vacancy defect-promoted electron-hole separation for porous defective ZnO hexagonal plates and enhanced solar-driven photocatalytic performance. *Chem. Eng. J.* **2020**, *379*, 122295. [CrossRef]
10. You, J.; Wang, L.; Bao, W.; Yan, A.; Guo, R. Synthesis and visible-light photocatalytic properties of BiOBr/CdS nanomaterials. *J. Mater. Sci.* **2021**, *56*, 6732–6744. [CrossRef]
11. Abd-Rabboh, H.S.M.; Benaissa, M.; Hamdy, M.S.; Ahmed, M.A.; Glal, M. Synthesis of an efficient, and recyclable mesoporous $BiVO_4/TiO_2$ direct Z-scheme heterojunction by sonochemical route for photocatalytic hydrogen production and photodegradation of rhodamine B dye in the visible region. *Opt. Mater.* **2021**, *114*, 110761. [CrossRef]
12. Tentu, R.D.; Basu, S. Photocatalytic water splitting for hydrogen production. *Curr. Opin. Electrochem.* **2017**, *1*, 56–62. [CrossRef]
13. Sclafani, A.; Herrmann, J.M. Influence of metallic silver and of platinum-silver bimetallic deposits on the photocatalytic activity of titania (anatase and rutile) in organic and aqueous media. *J. Photochem. Photobiol. A* **1998**, *113*, 181–188. [CrossRef]
14. Brezova, V.; Blazkova, A.; Karpinsky, L.; Ceppan, M. Phenol decomposition using Mn^+/TiO_2 photocatalysts supported by the sol-gel technique on glass fibres. *J. Photochem. Photobiol. A* **1997**, *109*, 177–183. [CrossRef]
15. Yang, G.; Yan, W.; Zhang, Q.; Shen, S.; Dingd, S.J. One-dimensional CdS/ZnO core/shell nanofibers via single-spinneret electrospinning: Tunable morphology and efficient photocatalytic hydrogen production. *Nanoscale* **2013**, *24*, 12432–12439. [CrossRef]
16. Goma, P.; Hachisuka, K.; Katsumata, H.; Suzuki, T.; Funasaka, K.; Kaneco, S. Photocatalytic hydrogen production with CuS/ZnO from aqueous $Na_2S + Na_2SO_3$ solution. *Int. J. Hydrogen Energy* **2013**, *38*, 8625–8630.
17. Zhao, X.; Feng, J.; Liu, J.; Lu, J.; Shi, W.; Yang, G.; Wang, G.; Feng, P.; Cheng, P. Metal-Organic Framework-Derived ZnO/ZnS Heteronanostructures for Efficient Visible-Light-Driven Photocatalytic Hydrogen Production. *Adv.Sci.* **2018**, *4*, 1700590. [CrossRef]
18. Inoue, T.; Watanabe, T.; Fujishma, A.; Honda, K.I. Suppression of Surface Dissolution of CdS Photoanode by Reducing Agents. *J. Electrochem. Soc.* **1977**, *124*, 719. [CrossRef]
19. Alivisatos, A.P. Semiconductor clusters, nanocrystals, and quantum dots. *Science* **1996**, *271*, 933–937. [CrossRef]
20. Nandi, P.; Das, D. ZnO/CdS/CuS heterostructure: A suitable candidate for applications in visible-light photocatalysis. *J. Phys. Chem. Solids* **2022**, *160*, 110344. [CrossRef]
21. Shi, X.F.; Xia, X.Y.; Cui, G.W.; Deng, N.; Zhao, Y.Q.; Zhuo, L.H.; Tang, B. Multiple exciton generation application of PbS quantum dots in ZnO@PbS/graphene oxide for enhanced photocatalytic activity. *Appl. Catal. B* **2015**, *163*, 123–128. [CrossRef]
22. Liu, R.; Feng, S.Y.; Dong, C.D.; Tsai, K.C.; Yang, W.D. Enhancing hydrogen evolution of water splitting under solar spectra using Au/TiO_2 heterojunction photocatalysts. *Int. J. Hydrogen Energy* **2021**, *46*, 28462–28473. [CrossRef]
23. Wang, X.W.; Li, Q.C.; Xu, H.P.; Gan, L.; Ji, X.F.; Lui, H.; Zhang, R.G. CuS-modified ZnO rod/reduced graphene oxide/CdS heterostructure for efficient visible-light photocatalytic hydrogen generation. *Int. J. Hydrogen Energy* **2020**, *45*, 28394–28403. [CrossRef]
24. Pahlevanpour, G.; Bashiri, H. Kinetic Monte Carlo simulation of hydrogen production from photocatalytic water splitting in the presence of methanol by 1 wt% Au/TiO_2. *Int. J. Hydrogen Energy* **2022**, *26*, 12975–12987. [CrossRef]
25. Patil, A.B.; Jadhav, B.D.; Bhoir, P. Optical band gap modification of Ce/ZnO for visible light photocatalytic H_2 production from aqueous methanol solution. *Opt. Mater.* **2021**, *121*, 111503. [CrossRef]
26. Conejeros, S.; Moreira Ide, P.; Alemany, P.; Canadell, E. Nature of Holes, Oxidation States, and Hypervalency in Covellite (CuS). *Inorg. Chem.* **2014**, *53*, 12402–12406. [CrossRef]
27. Saranya, M.; Nirmala Grace, A. Hydrothermal synthesis of CuS nanostructures with different morphology. *J. Nano Res.* **2012**, *18*, 43–51. [CrossRef]
28. Wei, J.; Wei, S.; Chang, N.; Wang, H.; Zhang, J.; Shao, W. Synthesis of PbS quantum dots/ZnO nanorods/Ag nanowires composite photocatalysts with enhanced photocatalytic activity. *Mater. Lett.* **2021**, *285*, 129130. [CrossRef]
29. Yendrapati, T.P.; Gautam, A.; Bojja, S.; Pal, U. Formation of ZnO@CuS nanorods for efficient photocatalytic hydrogen generation. *Sol. Energy* **2020**, *196*, 540–548. [CrossRef]

30. Guo, S.; Chen, W.; Li, M.; Wang, J.; Liu, F.; Cheng, J.P. Effect of reaction temperature on the amorphous-crystalline transition of copper cobalt sulfide for supercapacitors. *Electrochim. Acta* **2018**, *271*, 498–506. [CrossRef]
31. Shalahuddin Al Ja'farawy, M.; Kusumandari; Purwanto, A.; Widiyandari, H. Carbon quantum dots supported zinc oxide (ZnO/CQDs) efficient photocatalyst for organic pollutant degradation-A systematic review. *Environ. Nanotechnol. Monit. Manag.* **2022**, *18*, 100681. [CrossRef]
32. Liang, Y.; Novet, T.; Thorne, J.E.; Parkinson, B.A. Photosensitization of ZnO single crystal electrodes with PbS quantum dots. *Phys. Status Solidi A* **2014**, *9*, 1954–1959. [CrossRef]
33. Guchhait, A.A.; Rath, K.; Pal, A.J. To make polymer: Quantum dot hybrid solar cells NIR-active by increasing diameter of PbS nanoparticles. *Sol. Energy Mater. Sol. Cells* **2011**, *2*, 651–656. [CrossRef]
34. Hong, M.; Zhang, L.; Fang, H.; Feng, X.; Li, Z. Surface engineering of CdS quantum dots modified $SiO_2@C_3N_4$ nanospheres for effective photocatalytic hydrogen evolution. *Mater. Sci. Semicond. Process.* **2021**, *136*, 106134. [CrossRef]
35. He, B.; Bie, C.; Fei, X.; Cheng, B.; Yu, J.; Ho, W.; Al-Ghamdi, A.A.; Wageh, S. Enhancement in the photocatalytic H_2 production activity of CdS NRs by Ag_2S and NiS dual cocatalysts. *Appl. Catal. B* **2021**, *288*, 119994. [CrossRef]
36. Chen, W.; Wang, Y.; Liu, S.; Gao, L.; Mao, L.; Fan, Z.; Shangguan, W.; Jiang, Z. Non-noble metal Cu as a cocatalyst on TiO_2 nanorod for highly efficient photocatalytic hydrogen production. *Appl. Surf. Sci.* **2018**, *445*, 527–534. [CrossRef]
37. Lee, S.; Chang, C.J. Recent Progress on Metal Sulfide Composite Nanomaterials for Photocatalytic Hydrogen Production. *Catalysts* **2019**, *9*, 457. [CrossRef]
38. Wang, M.; Shen, S.; Li, L.; Tang, Z.; Yang, J. Effects of sacrificial reagents on photocatalytic hydrogen evolution over different photocatalysts. *J. Mater. Sci.* **2017**, *9*, 5155–5164. [CrossRef]
39. Anuchai, S.; Phanichphant, S.; Tantraviwat, D.; Pluengphon, P.; Bovornratanaraks, T.; Inceesungvorn, B. Low temperature preparation of oxygen-deficient tin dioxide nanocrystals and a role of oxygen vacancy in photocatalytic activity improvement. *J. Colloid Interface Sci.* **2018**, *512*, 105–114. [CrossRef]
40. Yavuz, C.; Erten-Ela, S. Solar light-responsive α-Fe_2O_3/CdS/g-C_3N_4 ternary photocatalyst for photocatalytic hydrogen production and photodegradation of methylene blue. *J. Alloys Compd.* **2022**, *908*, 164584. [CrossRef]
41. Shiri, P. An overview on the copper-promoted synthesis of five-membered heterocyclic systems. *Appl. Organomet. Chem.* **2020**, *34*, 5. [CrossRef]

catalysts

MDPI

Article

Photocatalytic Degradation of Methyl Orange Dyes Using Green Synthesized MoS_2/Co_3O_4 Nanohybrids

Tsung-Mo Tien *, Chao-Hsiang Chen, Chen-Tang Huang and Edward L. Chen *

Coastal Water and Environment Center, College of Hydrosphere Science, National Kaohsiung University of Science and Technology, Kaohsiung City 81157, Taiwan
* Correspondence: tmtien@nkust.edu.tw (T.-M.T.); edwardljchen@gmail.com (E.L.C.)

Abstract: In this work, a new binary MoS_2/Co_3O_4 nanohybrids was successfully fabricated and the chemical structures, morphologies, electrochemical and optical characterizations were carried out. In addition, heterojunction nanoparticles present in S-scheme structures act as electron traps and promote light absorption capacity for the degradation of Methyl orange (MO) with visible-light activity. MoS_2/Co_3O_4 nanohybrids suggested excellent photocatalytic performance compared to bare MoS_2 and Co_3O_4, where 95.6% of MO was degraded within 170 min, respectively. The results also showed excellent stability and recyclability over five consecutive cycles, without noticeable changes in the nanocomposite structure. The boosted photocatalytic degradation and redox activities of MoS_2/Co_3O_4 can be attributed to the created S-scheme heterostructure to facilitate the separation of and to delay recombination of photoinduced charge carriers. We believe that this strategy of exploiting nanohybrid photocatalysts has great potential in the field of environmental catalysis and diverse applications.

Keywords: photodegradation; nanohybrids; MoS_2/Co_3O_4; methyl orange (MO); environment remediation

Citation: Tien, T.-M.; Chen, C.-H.; Huang, C.-T.; Chen, E.L. Photocatalytic Degradation of Methyl Orange Dyes Using Green Synthesized MoS_2/Co_3O_4 Nanohybrids. *Catalysts* **2022**, *12*, 1474. https://doi.org/10.3390/catal12111474

Academic Editors: Jorge Bedia and Carolina Belver

Received: 25 October 2022
Accepted: 15 November 2022
Published: 18 November 2022

1. Introduction

Environmental contamination and energy demand represent a worldwide concern. To overcome these problems, nanomaterials are an alternative to address environmental aspects of wastewater treatment contamination and energy generation by organic compounds [1–3]. The immoderate emptying of waste toxic organic dyes, such as methyl orange (MO), into soils and water pools counteracts environmental protection and safety. Consequently, it is necessary to treat the dyeing wastewater with practical and harmless methods [4,5]. It is well-known that solar light is a favorite choice of clean energy for decomposing pollutants due to its great redox activity and energy density only makes H_2O [6]. The utilization of visible light is a rather effective process as it adopts abundantly available solar light. To date, the photocatalytic system, as a green and direct reach, has been found to be effective in environmental treatment, especially in the removal of organic pollutants. Photocatalytic degradation with visible light activity is employed to excite the semiconductor material creating electron (e^-) and hole (h^+) pairs for the degradation of organic compounds. To date, different semiconductors have been adopted as photocatalysts in environmental applications [7]. Photocatalytic removal pollutants have attracted sufficient exploration attention and are considered a hopeful system to solve the energy crisis and environmental issues. However, the low absorption of sunlight, high electron–hole recombination, and photo-decay effects are still limiting to the acquisition of an efficient photocatalyst with visible light activity. [8]. Accordingly, it is necessary to fabricate and acquire nano-heterojunction photocatalysts with great solar-to-redox reaction conversion performance.

Recently, heterostructured photocatalysts have attracted researchers' attention due to superior features such as improved electron–hole pair separation performance [9]. The transition-metal dichalcogenides (TMDs) have obtained large attention owing to a series of interesting excellent properties [10]. Between these materials, molybdenum disulfide (MoS_2) has great charge carrier density, high conductivity, and a controllable energy band gap, which is valuable and effective for optimizing environmental remediation and improving photocatalytic water splitting capability with visible light irradiation [11]. As a kind of normal TMDs, MoS_2 has a bandgap energy of about 1.7~2.7 eV and proper valence conduction band potentials that own great photocatalytic efficiency and can employ an immense section of solar light [12]. Unfortunately, the poor light harvesting ability, inactive removal pollutants performance, and quick recombination rate of photo-excited charge carriers are the natural disadvantages that inhibit the useful utilization of MoS_2 [13]. In addition, Co_3O_4 also exhibits obvious stability under the photocatalytic system. In addition, Co_3O_4 possesses the matched band gap potential with MoS_2 to develop an efficient step-scheme (S-scheme) heterostructure because the valence band (VB) of Co_3O_4 is close to the conduction band (CB) of MoS_2 based on previous works [12]. However, the Co_3O_4 catalyst absorbs primarily visible light and possesses a wide band gap energy, which prohibits their photocatalytic efficiency. Among these heterostructures, MoS_2 and Cobalt oxide (Co_3O_4) photocatalysts exhibit various superior properties for instance remarkable direct bandgaps, excellent mechanical properties, outstanding optical property as well as distinguished photocatalytic efficiency [14–16]. On the other hand, bare MoS_2 and Co_3O_4 for direct photocatalytic water splitting or removal of contaminants are quite challenging because of their poor operation and high electron and hole recombination rate [17,18]. In addition, this heterojunction not only improved H^+ adsorption and promoted migration and transport of H^+ to the surface reactive sites, but further suppress the stacking or aggregation of materials [19]. To solve this issue, a common method is to develop nano-heterostructure between Co_3O_4 and MoS_2 under visible light irradiation to improve their visible light activity. To overcome these barriers, substantial promotions have been made in the reasonable formation and design of MoS_2/Co_3O_4-based nanohybrids with well-established interfaces and nanostructures [20]. Constructing heterojunction photocatalysts can assist the charge transfer between the diverse photocatalysts and enhance the separation of charge carriers. Therefore, the introduction of new-form nano-heterojunction photocatalysts with structural variety, band gap energy, and better visible-light utilization is superior in this investigate area [21,22].

In the present study, a new MoS_2/Co_3O_4 nanohybrid photocatalyst was fabricated by a facile hydrothermal route for powerful photocatalytic activity. The green–hydrothermal technique is energy-saving, cost-effective, environmentally friendly, and a green process compared with other synthesis methods [23,24]. To our knowledge, there is no report on the photocatalytic degradation of MoS_2/Co_3O_4 nanohybrids with visible light irradiation. A potential S-scheme heterojunction can be constructed between MoS_2 and Co_3O_4 while Schottky junction developed at the MoS_2/Co_3O_4 interface, respectively, which connected to enhance the charge carries separation ability. With the optimized amount of Co_3O_4 loading in the composite, the MoS_2/Co_3O_4 nano-heterojunction displayed a great removal MO rate of 95.6%, which was approximately 2.1 and 2.3 folds higher than that of bulk MoS_2 and Co_3O_4 samples, respectively. MoS_2/Co_3O_4 nanohybrids demonstrated higher photocatalytic efficiency than MoS_2 and Co_3O_4 on the removal of MO dye with visible light irradiation. This system opened up a new route for the controlled fabrication of MoS_2-based heterojunction and also improved their organic compound photodegradation activity. The further the surface reactive positions created by the huger specific surface area, the higher the redox efficiency due to the farther negative conduction capacity, and decreased recombination degree of photogenerated electrons and holes eventually resulted in improved photocatalytic performance.

2. Results and Discussion

XRD analysis was further collected to check the crystal phase form and structural composition of the as-fabricated photocatalysts, as exhibited in Figure 1. During the hydrothermal process, the apparent feature peaks of MoS_2 were weaker while several new feature peaks appeared at 31.2°, 36.8°, 44.9° and 59.4° of the MoS_2/Co_3O_4 photocatalysts, which are related with the (220), (311), (400) and (511) diffraction face of Co_3O_4 (JCPDS No. 42–1467) [15], respectively. The MoS_2 materials exhibits a series of primary feature peaks at 14.3°, 32.9°, 39.8°, 44.5°, 50.2°, 50.7°, and 60.7°, which assigned to the (002), (100), (103), (006), (105), and (110) crystal planes (JCPDS: 37-1492), respectively, and no other impurity peaks were perceived [11]. Its value indicating that there is no altered signal in the XRD analysis of the MoS_2/Co_3O_4 nanocomposites compared to the MoS_2 and Co_3O_4 samples, suggesting that the presented Co_3O_4 content is fewer, and the dispersion is greater, and there are no other impurity peaks or feature peak shift. In addition, when the reaction solutions concentration of Co_3O_4 was increased from 5 wt% to 20 wt%, the (311) feature peak intensity of Co_3O_4 became higher, which indicated an enlarge in the dosage of Co_3O_4 in the heterojunction. In particular, contrast with the XRD analysis of Co_3O_4, the overall diffraction intensities of MoS_2/Co_3O_4 heterojunctions were altered at 36.8° with (311) diffraction face (see Figure 1 inset). In addition, the feature peaks of MC-X (X = 5 wt%, 10 wt%, 15 wt%, and 20 wt%) are comparable to the feature XRD analysis of MoS_2 without additional impurity peaks, which suggests that the nanohybrids photocatalyst further possess coherent crystal structure [25,26].

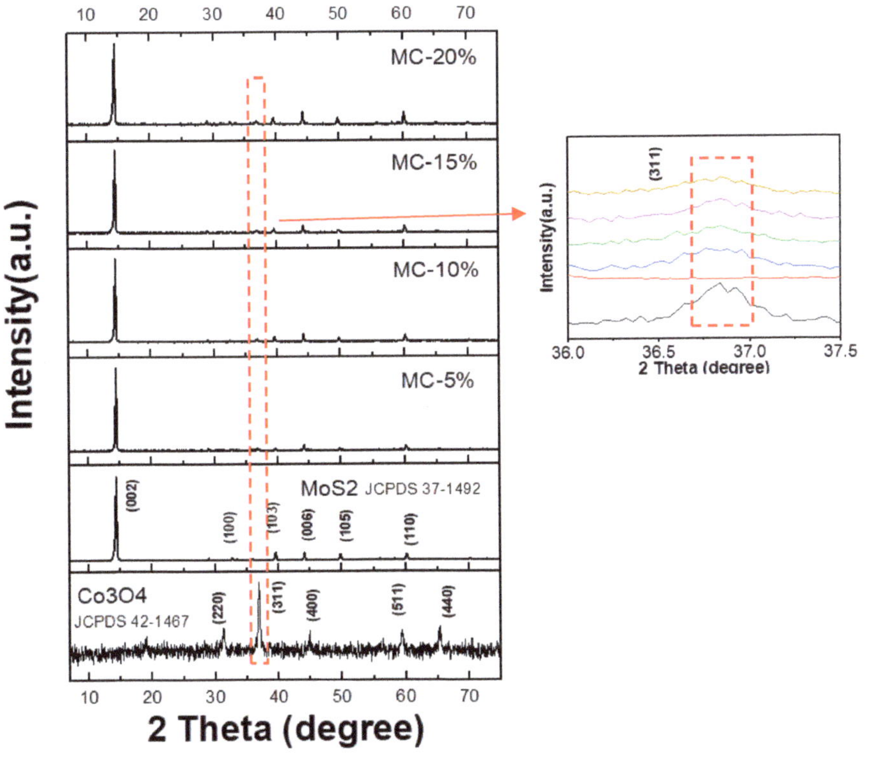

Figure 1. XRD patterns of MoS_2, Co_3O_4, MC-5%, MC-10%, MC-15%, and MC-20% samples, inset shows the intensity of (311) XRD peak.

In order to explore the development step of morphologies after hydrothermal process, SEM and TEM were characterized. As exhibited in Figure 2a,b, apparently large-sized bulk MoS_2 and Co_3O_4 can be found, respectively. It is obtained that the microstructure of the photocatalyst is sphere-like and plate-like with approximately 5 μm equivalent size.

However, the MoS_2 and Co_3O_4 possess a high grade of collection, which intensely suppress the surface of the active location of the photocatalysts. As displayed in Figure 2c, the MoS_2/Co_3O_4 heterojunction presents a nanomaterial with an irregular surface, which gives it a greater specific surface area. Therefore, this reflection was verified in the specific surface area examination test. It is pronounced that despite the size of MC-X being immensely decreased, the crystals of MoS_2 still retain great crystallinity, as presented by the (002) major peak observed in the XRD patterns (Figure 1). This nanohybrid of MoS_2/Co_3O_4 serves a sufficient area for the efficient load of Co_3O_4, which can mainly disperse MoS_2/Co_3O_4 nanohybrids and decrease aggregation. As displayed in Figure 2d, the MoS_2/Co_3O_4-20% heterojunction displays an aggregated form with a dimension of approximately 4–10 nm. From Figure 2e, the lattice fringe of 0.26 nm could be apparently noticed, which corresponds to the (002) plane of hexagonal MoS_2. The fringe spacing of 0.32 nm is assigned to the (311) plane of Co_3O_4. During further testing, it could be found that the MoS_2/Co_3O_4-20% nanohybrids were eventually constructed. The element mappings exhibit that the Mo, S, and Co elements are consistently distributed in the MoS_2/Co_3O_4-20% nanohybrids (Figure 2f–i). The construction of the heterostructure of MoS_2/Co_3O_4 is broadly supposed to possess a definite effect on promoting the charge carrier's separation ability of photocatalytic. The consistent distribution of Mo, S, and Co elements can confirm the effective formation of MoS_2/Co_3O_4 nanohybrids.

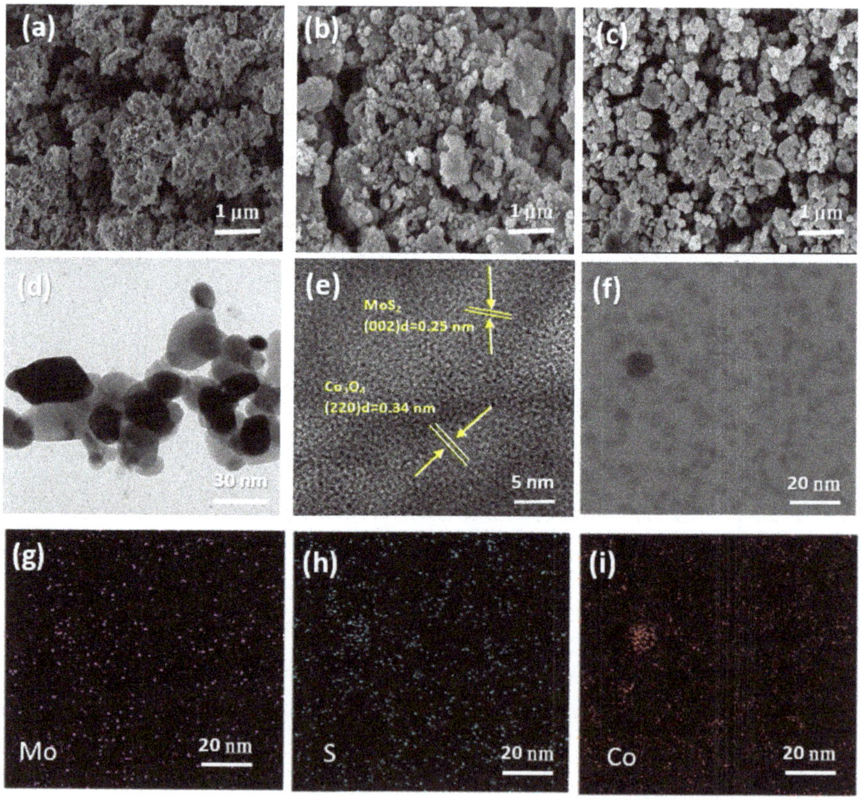

Figure 2. SEM images (**a**) MoS_2 (**b**) Co_3O_4 (**c**) MoS_2/Co_3O_4, (**d**) TEM and (**e**) HRTEM images of MoS_2/Co_3O_4, (**f–i**) Elemental mapping images of MoS_2/Co_3O_4 nanohybrids with elemental distribution of Mo, S, and Co.

The UV-vis absorption spectra are generally adopted to check the visible light absorption capability of the photocatalyst, and the analysis of the as-prepared MoS_2/Co_3O_4 heterojunction is exhibited in Figure 3. As can be observed, the two optical band gaps of MoS_2/Co_3O_4-X nanocomposites are 1.56~1.69 eV (*Eg*1) and 2.31~2.61 eV (*Eg*2), which are in good agreement with the MoS_2 and Co_3O_4 energy band gap. As MoS_2/Co_3O_4 photocatalyst is a photo-response sample, it possessed a wide wavelength absorption ability. *Eg*1 is joined to the onset of O(II)-Co(III) excitations [13–15]. The maximum response wavelength of MoS_2/Co_3O_4 is extended to 550 nm. Then, it presents a narrower band gap (*Eg*2) of 2.31~2.61 eV after adding Co_3O_4, which is favorable to improve visible light application and photoexcited charge carriers. In addition, the UV-vis spectra of the MoS_2/Co_3O_4 nanohybrids are relativity close. The construction for this case is that the minor added dosage of Co_3O_4 does not generate an apparent alteration to the light absorption efficiency of the MoS_2/Co_3O_4 nanohybrids. The MC-20% nanohybrids display better light absorption and the absorption range has a minor red shift in the range of 550–680 nm. It could be assigned to the cooperation between MoS_2 and Co_3O_4. Hence, the Co_3O_4 materials may be a significant reason to enlarge the light absorption of the MoS_2/Co_3O_4 nanohybrids. The band gap energy (*Eg*) of the corresponding photocatalyst was received via the transformed Kubelka–Munk function [27,28]. As exhibited in Figure 3b, the *Eg*2 (*Eg*1) values of MC-5%, MC-10%, MC-15%, and MC-20% are about 2.61 (1.56) eV, 2.51 (1.61) eV, 2.45 (1.64) eV, and 2.31 (1.69) eV.

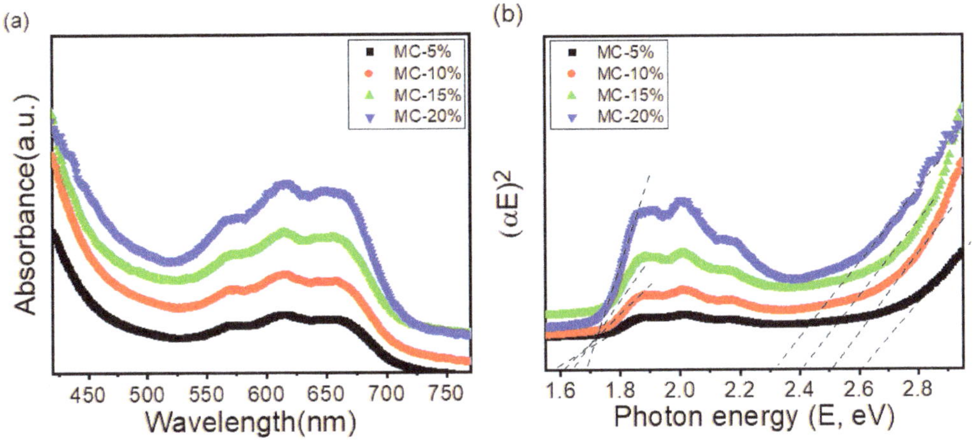

Figure 3. (**a**) UV–vis DRS absorption spectrum and (**b**) Tauc plots of as-prepared nanocomposites.

Then, X-ray photoelectron spectroscopy (XPS) is validated to recognize surface information of the fabricated samples. Figure 4a displays the survey XPS spectra. As further observed, MC-20% is constructed of Mo, S, Co and O elements, which involve the integral elements of MoS_2/Co_3O_4 nanohybrids. There are no other impurity peaks emerging in survey spectra, suggesting that there are no other contaminate photocatalysts during hydrothermal route. The high-resolution spectrum of Mo 3d (Figure 4b) could be divided into two peaks, which centered are at ~229.6 eV and ~232.8 eV, referring to Mo 3d 5/2 and Mo 3d 3/2 [10]. Correspondingly, two feature peaks in the high-resolution S 2p XPS spectra of MC-20% centered at nearly 162.6 eV and nearly 163.6 eV (Figure 4c) are referring to S 2p3/2 and S 2p1/2, respectively [11]. As shown in Figure 4d, binding energies of ~780.1 eV and ~795.5 eV referring to Co 2p 3/2 and Co 2p 1/2 for MC-20% nanohybrids, respectively [14]. The O 1s spectra of MC-20% (Figure 4e) binding energies at 529.8 eV and 531.8 eV, referring to the O^{2-} in the Co_3O_4 and the adsorbed oxygen [15]. These results suggested that the MoS_2/Co_3O_4 nanohybrids were well fabricated with a strong interaction between MoS_2 and Co_3O_4. The elements concerned with the formation of

MoS_2/Co_3O_4 nanohybrids were confirmed by the EDS spectra. Four noticeable peaks were obtained at nearly 0.8 keV, which was owing to the presence of oxygen atom, and other three peaks arising at 0.8 keV and 6.9 keV related to Co, and 2.3 keV, are related to Mo and S as exhibited in Figure 4f. The results confirmed the successful fabrication of MoS_2/Co_3O_4 nanohybrids. The Brunauer-Emmett-Teller (BET) specific surface areas of MoS_2, Co_3O_4, and MC-20% are evaluated to be ~35.2, 24.6, and 68.7 $m^2 g^{-1}$, respectively (Table 1). The main aperture size of MoS_2, Co_3O_4, and MC-20% are ~38.4, 48.5 and 8.9 nm, respectively. The pore volumes of MoS_2, Co_3O_4, and MC-20% are ~0.193 $cm^3 g^{-1}$, 0.124 $cm^3 g^{-1}$and 0.372 $cm^3 g^{-1}$, respectively. BET of MC-20% was more than 2 times than that of MoS_2 and Co_3O_4, demonstrating the introduced Co_3O_4 had a critical influence on raising the specific surface area of MoS_2.

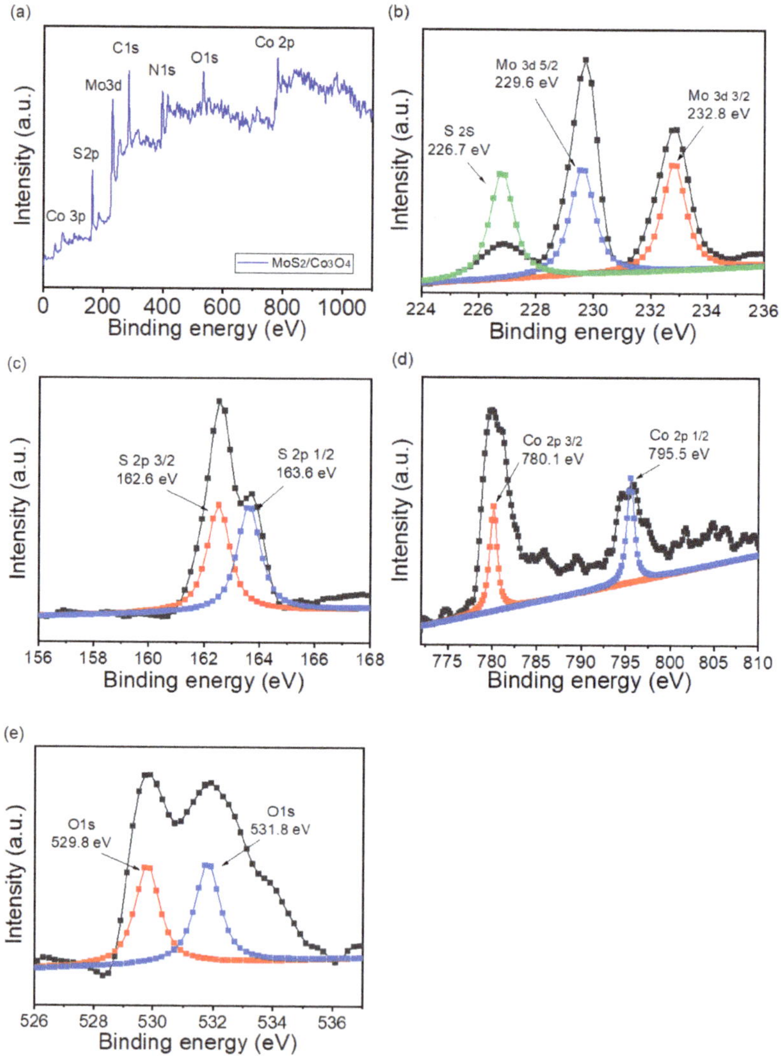

Figure 4. *Cont.*

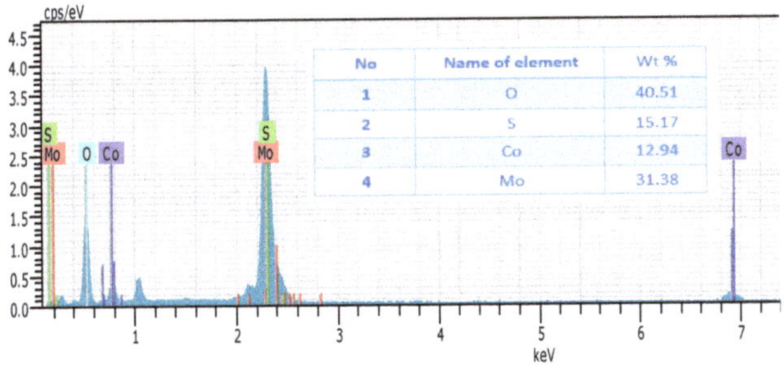

(f)

Figure 4. The surface chemical station analysis of MC-20%, (**a**) Survey spectra, High resolution XPS spectra of (**b**) Mo 3d spectra, (**c**) S 2p spectra, (**d**) Co 2p spectra, and (**e**) O 1s spectra, (**f**) EDS analysis of MC-20% nanohybrids.

Table 1. The physical adsorption parameters of the photocatalysts of MoS_2, Co_3O_4, and MC-20% were measured three times.

Samples	Specific Surface Area (m^2/g, BET) [a]	Total Pore Volume (cm^3/g, BET) [b]	Average Pore Diameter (nm, BJH) [b]
MoS_2	35.2 ± 5	0.193 ± 0.02	38.4 ± 5
Co_3O_4	24.6 ± 5	0.124 ± 0.02	48.5 ± 5
MC-20%	68.7 ± 5	0.372 ± 0.02	8.9 ± 3

[a] Received from BET analysis. [b] Relative pressure (P/P_0) was 0.99.

Figure 5a exhibits the decay in absorbance peak intensity of 447 nm with extending irradiation time. It displays that the absorbance ability of MO was continuously removed with visible light activity in the presence of the as-fabricated photocatalysts. In addition, the absorbance of MO decreased significantly with the increase of irradiation time by using MC-20% nanohybrids as primarily photocatalyst under solar irradiation was recorded by UV–vis absorption spectroscopy, which proved the destruction of MO molecules. The photocatalytic activity of MO by as-fabricated MoS_2, Co_3O_4 and MC-X materials was evaluated with and without visible light activities (Figure 5b). The control experiments displayed the capability values for the self-photocatalytic of MO pollutants under same light condition. This result suggests that the degradation of MO pollutants is owing to the photodegradation activity in the as-fabricated samples. On the other hand, the individual MoS_2 and Co_3O_4 samples could degrade about 45.8 and 41.4% of MO under 140 min of visible light radiation. The photodegradation capability of the photocatalysts followed the sequence: MC-20% nanohybrids > MC-15% nanohybrids > MC-10% nanohybrids > MC-5% nanohybrids > MoS_2 > Co_3O_4. This poor photodegradation performance could be assigned to the high recombination of electrons and holes in the bulk component [29]. Noticeably, MC-20% nanohybrids described excellent photodegradation performance of MO pollutants (~95.6%) than bare MoS_2 or Co_3O_4, which was about 2.3 folds than bare Co_3O_4. The major process of photocatalytic reaction for the composite magnetic photocatalyst MoS_2/Co_3O_4 based on the basic principles and studies reported in the literature can be summarized as follows:

$$MoS_2 + h\nu \rightarrow MoS_2(e^-_{CB} + h^+_{VB}) \quad (1)$$

$$Co_3O_4 + h\nu \rightarrow Co_3O_4(e^-_{CB} + h^+_{VB}) \quad (2)$$

$$O_2 + e^-_{CB} \rightarrow \bullet O_2^- \quad (3)$$

$$MoS_2(e^-_{CB}) \rightarrow Co_3O_4(e^-_{CB}) \quad (4)$$

$$2H^+ + \bullet O_2^- + 2e^-_{CB} \rightarrow H_2O_2 \quad (5)$$

$$Co_3O_4(h^+_{VB}) \rightarrow MoS_2(h^+_{VB}) \quad (6)$$

$$2H_2O + h^+_{VB} \rightarrow O_2 + 4H^+ \tag{7}$$

$$\text{Composite } h^+ + MO \rightarrow MO^* \text{ (Oxidation)} \tag{8}$$

$$MO + \text{Composite } h^+ \text{or } \bullet O_2^- \rightarrow \text{Degradation products} \tag{9}$$

$$MO + HO^\bullet \rightarrow MO(HO^\bullet) \rightarrow \text{Intermediates} + HO^\bullet \rightarrow CO_2 + H_2O \tag{10}$$

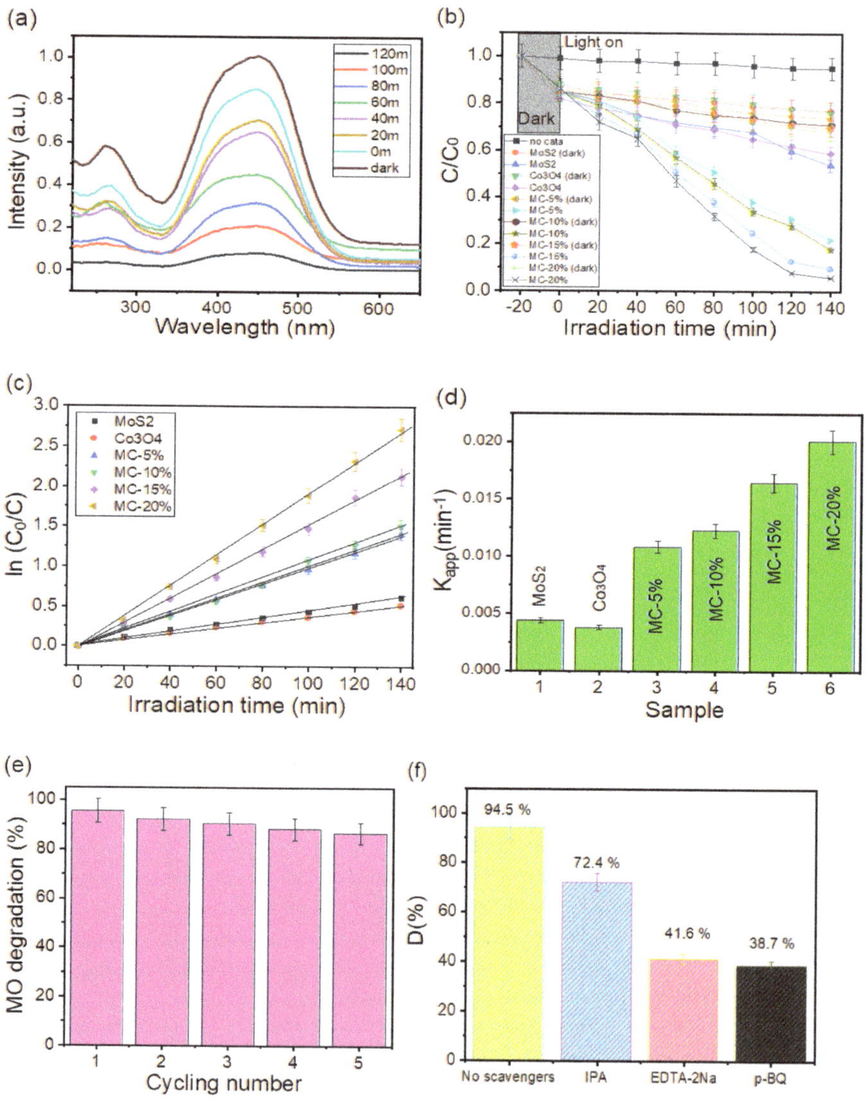

Figure 5. *Cont.*

(g)

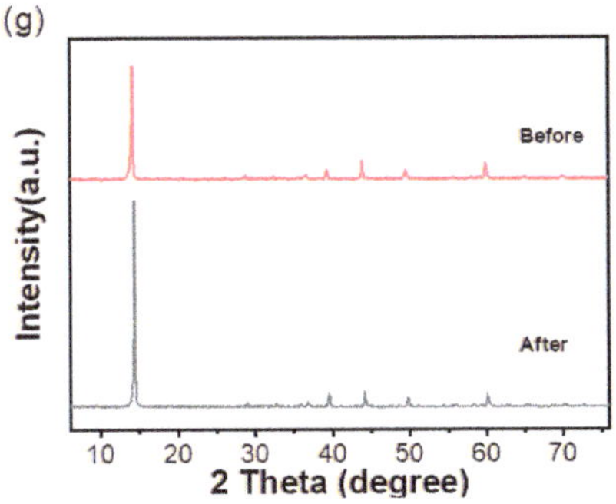

Figure 5. Photocatalytic study of MO degradation, (**a**) time-dependent UV−Vis absorption spectra, (**b**) conversion of MO over various photocatalysts, (**c**) pseudo−first-order kinetics study of the degradation, (**d**) degradation rate constant of as−fabricated samples, (**e**) recycling and (**f**) radical scavenging activity during the decomposition of MO over MC−20% under visible light, (**g**) Comparison of XRD analysis of MC−20% samples before and after photocatalytic.

In addition, the photocatalytic reaction kinetics of an MO removal test was calculated to a pseudo first order equation model in order to check the photocatalytic activity. The linear relationship between ln (Co/C) vs. irradiation time (t) is exhibited in Figure 5c and the removal rate constant (K_{app}) is displayed in Figure 5d. The MC-20% nanohybrids were measured with a higher K_{app} value compared to the other synthesized samples. For effective utility purposes, the stability and reusability of MC nanohybrids were estimated by cycling runs. After every test, the sample was centrifuged, washed and dried before operating the next run. As shown in Figure 5e, the photodegradation capability of the MC-20% sample was 95.6%, 92.3%, 90.5%, 88.3%, and 86.8% for first, second, third, fourth, and fifth test, respectively. The noticed reduction of the photocatalytic activity of MC-20% can be referred to an adsorbed MO and its evolution products on the surface of MC-20% nanohybrids, thus prohibiting the active sites and visible light from approaching the photocatalyst [30]. The exploration of the photodegradation mechanism of the nanohybrids acted as a critical factor in the effective utilization of the heterojunction photocatalyst. Accordingly, the free radical trapping tests were performed, in which p-BQ (1,4-benzoquinone), IPA (2-propanol) and EDTA-2Na were adopted as scavenging free radical ($^{\bullet}O_2^-$, $^{\bullet}OH$ and h^+) agents [31]. It could be observed from the analysis results in Figure 5f that when various scavenging free radical agents were further joined, the photocatalytic efficiency of the MC-20% sample was influenced to altering degrees. The MC-20% sample was highly responsive to p-BQ and EDTA-2Na, implying that $^{\bullet}O_2^-$ and h^+ were the major active species during the photodegradation activity, and $^{\bullet}OH$ only acted as an assisting active species. Consequently, these results demonstrate that the formation of MoS_2/Co_3O_4 with a proper ratio (20 wt%) is able to receive the better photocatalytic efficiency of MoS_2/Co_3O_4 in the nano-heterostructure. The stability and repeatability of the sample are important indexes to evaluate the performance of photocatalysts. XRD analysis was performed after the fourth recycling experiment. Taking MC-20% photocatalyst as an example, its activity did not change significantly after five cycles, which can still degrade 86.8% of MO (Figure 5g). This result clearly suggests the stability of the fabricated MoS_2/Co_3O_4 NC photocatalyst. In addition, Table 2 displays the photocatalytic performance of relevant heterojunction in previous reports [31–36]. By contrast, it is observed that the MoS_2/Co_3O_4 nanomaterial we prepared has remarkable benefits for organic compounds of high concentration in the photocatalytic process.

Table 2. Comparison of photocatalytic activity of some photocatalysts for removal MO dye under visible light irradiation.

Catalysts	Pollutants Volume	Concentration	Lamp	Mass of Catalysts	Time (min)	Degradation Rate %	Reference
ZnO-TiO_2/SO_4^{2-}	25 mL	20 mg L^{-1}	halide	20 mg	72	90	[31]
CoO_x/g-C_3N_4	35 mL	10 mg L^{-1}	Xenon	35 mg	180	92	[32]
ZnPcs	3 mL	3.3×10^{-5} M	White LED	10 mg	720	93	[33]
SbSI MRs	50 mL	30 mg L^{-1}	halide	25 mg	10	78	[34]
Sn-ZnO/GO	1000 mL	50 mg L^{-1}	Xenon	100 mg	120	87	[35]
CdS	150 mL	10 mg L^{-1}	Xenon	500 mg	180	94	[36]
MoS_2/Co_3O_4	50 mL	20 mg L^{-1}	halide	10 mg	170	96	Present work

For confirm the presented S-scheme heterojunction, electron spin resonance (ESR) is adopted to recognize the free radicals produced in the system of MoS_2, Co_3O_4 and MC-20% nanohybrids. Applying DMPO as a spin capture agent, DMPO-$^{\bullet}O_2^-$ and DMPO-$^{\bullet}$OH are observed with light conditions. It could be found in Figure 6a that the MC-20% photocatalyst displays the higher feature signal of DMPO-$^{\bullet}O_2^-$, while the feature signal of DMPO-$^{\bullet}O_2^-$of MoS_2 and Co_3O_4 is nearly undetectable. It exhibits that the construction of heterojunction is more favorable to the production of $^{\bullet}O_2^-$, which could be referred to the rapider migration and separation of photoexcited charge carriers in heterojunction catalysts [36]. Furthermore, the DMPO-$^{\bullet}$OH feature signal peaks of MoS_2, Co_3O_4 and MC-20% could be found in Figure 6b, of which MC-20% possess the highest signal peak. Hence, it could be resulted that MC-20% nanohybrids can generate more $^{\bullet}O_2^-$ and $^{\bullet}$OH. The coincidence of $^{\bullet}O_2^-$ and $^{\bullet}$OH radicals further verify that the charge carrier separation mechanism is S-scheme. The charge transfer routes are analyzed by photoelectrochemical measurements. Electrochemical impedance spectroscopy (EIS) with visible light activity is exhibited in Figure 6c, and the smaller semi-circular curve in Nyquist plots expresses the lesser charge carrier transport resistance. Apparently, the lowest charge carrier transport resistance in MC-20% imply the greatest efficient charge carrier transport routes [37], which corresponds with the outstanding PHE capability of MC-20% nanohybrids. As exhibited in Figure 6d, all the MC-20% nanohybrids electrodes exhibit a great transient photocurrent response than that bare MoS_2 and Co_3O_4 behaves at the highest intensity, implying the great efficient charge separation. Room-temperate PL spectra are carried out to check the charge carrier dynamics behavior. As displayed in Figure 6**e**, the apparently reduced PL intensity of MC-20% also proves the prohibited electrons and holes recombination, which is in good approval with the above results. Furthermore, the performance of the charge carrier is further explored under photocatalytic redox reactions. The efficient separation of photo-induced charge carriers is a required situation for enhancing the capability of photodegradation [38]. It is also found that the poorer the peak PL intensity, the lower the possibility of electron and hole pairs recombination rate, which is more favorable to the development of the pollutant removal reaction. This is owing to the evidence that the formation of nano-heterojunction improves the quick separation and transport of photo-excited charge carriers, hence, greatly suppressing the recombination rate of photo-induced charge carriers and enhancing the performance of charge carrier separation.

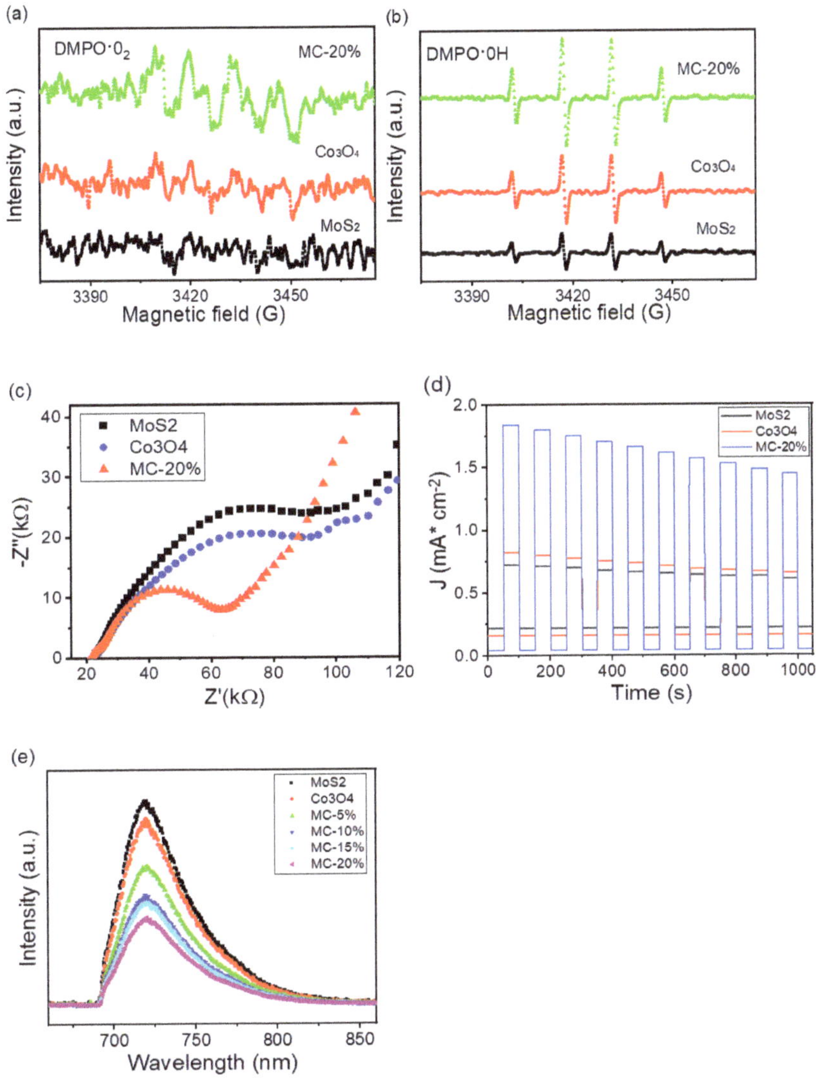

Figure 6. DMPO spin−trapping ESR spectra for (**a**) $^{\bullet}O_2-$ and (**b**) $^{\bullet}OH$ with visible light for MoS_2, Co_3O_4 and MC−20%, (**c**) Electrochemical impedance spectra (EIS) of MoS_2, Co_3O_4, and MoS_2/Co_3O_4 in 0.5 M Na_2SO_4 solution with visible light activity, and (**d**) Transient photocurrent densities of MoS_2, Co_3O_4, and MoS_2/Co_3O_4 electrodes in 0.5 M Na_2SO_4 solution with visible light, (**e**) PL spectra under the excitation wavelength of 580 nm of MoS_2, Co_3O_4, MC−5%, MC−10%, MC−15% and MC−20%.

Based on the above data analysis and results, a photocatalytic activity mechanism of MO removal on MC nanohybrids is recommended, as exhibited in Scheme 1. Following the introduction of Co_3O_4 materials on the surface of MoS_2, a firm heterostructure is organized. During exposure to visible light, photocatalysis activity has various efficiencies to the incident visible light in varied wavelength ranges. In this work, MC nanohybrids can absorb a broad range of incident visible light with the same conditions, thus generating more photoexcited charge carriers. More significantly, the system of photocatalytic reaction of nanohybrids design is higher than that of single-component design, implying in an improved interaction among reactant and photocatalyst. Ultimately, the excellent hydrogen evolution and pollutants removal promote the nanohybrids to have a higher charge carrier separation behavior [39,40]. In the meantime, a built electric field from Co_3O_4 to MoS_2 induces the downward band curving of MoS_2 and upward band curving of Co_3O_4 at the heterojunction interface, respectively. During visible light irradiation, both MoS_2 and Co_3O_4 could be photo-induced to produce electrons and holes. The electric field, curved band edges and the coulomb interaction notably drive the CB electrons of MoS_2 to rebuild with the VB holes of Co_3O_4, leading to the practical conservation of the high-oxidative holes in the VB of MoS_2 and high-reductive electrons in the CB of Co_3O_4, which significantly promotes the photo-response efficiency of MoS_2/Co_3O_4. This facile S-scheme system can efficiently weaken the recombination rate of photoinduced charge carriers in the heterojunction and improve the recombination of limited photo-generated electron and hole pairs, hence increasing the excellent redox activity of the photocatalyst. Finally, the enhanced photodegradation efficiency could be attributed to the increased visible light absorption ability, more reactive sites and powerful transfer of photo response charge carriers. In addition, the interface influence is larger than the efficiency of the photogenerated hole and electron to recombine. The electrons of the conduction band of MoS_2 contribute to the pollutant removal reduction reaction. Then, the holes of Co_3O_4 are oxidized in the valence band by the photocatalytic process. This novel S-scheme nanocomposite possesses better redox ability [19,26,27], which apparently enhances the capability of photocatalytic pollutants removal. The CB and VB of MC with visible light activity, are induced to create photoexcited electrons and holes, respectively. On the other hand, the photo-response charge carriers in MoS_2 and Co_3O_4 will recombine rapidly, implying a weakening in photocatalytic capability. Although, owing to the presence of Co_3O_4 cocatalyst, photoexcited electrons will rapidly migrate to MoS_2, and also further water reduction so hydrogen evolution will appear. Furthermore, the presence of MoS_2 can weaken over-potential and serve more active sites for photocatalytic and can further enlarge the visible light absorption of Co_3O_4.

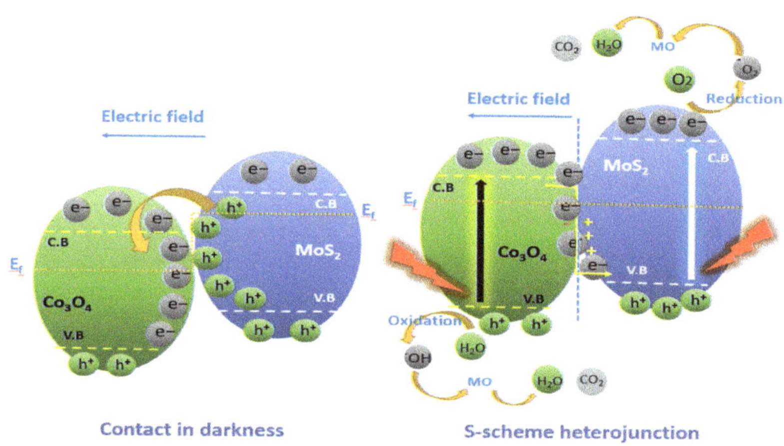

Scheme 1. The possible photocatalytic mechanism of nanohybrids.

3. Materials and Methods

Sodium molybdenum oxide dehydrates ($Na_2MoO_4{\cdot}2H_2O$) (≥99.5%), L-cysteine ($C_3H_7NO_2S$), Cobalt chloride hexahydrate ($CoCl_2{\cdot}6H_2O$, ≥98%), Methyl orange (MO, ≥85%), Triton X-100, Potassium hydroxide (KOH, ≥85%), 1-Methyl-2-pyrrolidone (NMP, ≥99%), absolute ethanol (CH_3CH_2OH, ≥95%), and acetone (CH_3COCH_3, ≥99.5%) were used from Sigma-Aldrich (St. Louis, MO, USA). Deionized water was utilized to produce solutions to be examined. All the chemicals were adopted directly without purification.

Typically, 1.2 mmol of $Na_2MoO_4 \cdot 2H_2O$ and 2.5 mmol the L-cysteine was used to 40 mL of deionized (DI) water to develop a homogeneous solution via stirring and ultrasonic treatment for 30 min. $CoCl_2 \cdot 6H_2O$ (2.5 mmol) was added in 40 mL distilled water and a certain amount of surfactant (1%, *w*/*w*, Triton X-100) was added subsequently. Subsequently, the homogeneous solution was transformed into a Teflon-lined stainless-steel autoclave (100 mL) and reacted at 180 °C for 20 h. After cooling down to room temperature naturally, the pristine dark precipitate was obtained by centrifugation, washed three times with deionized water, NMP, absolute ethanol and acetone, and dried at 70 °C in a vacuum oven overnight. A series of MoS_2/Co_3O_4 photocatalysts was provided by increasing various doses of Co_3O_4 (5 wt%~20 wt%); the individual samples are denoted as MC-X (X = 5 wt%, 10 wt%, 15 wt%, and 20 wt%). Finally, the nanocomposite is prepared via the hydrothermal method, as shown in Scheme 2.

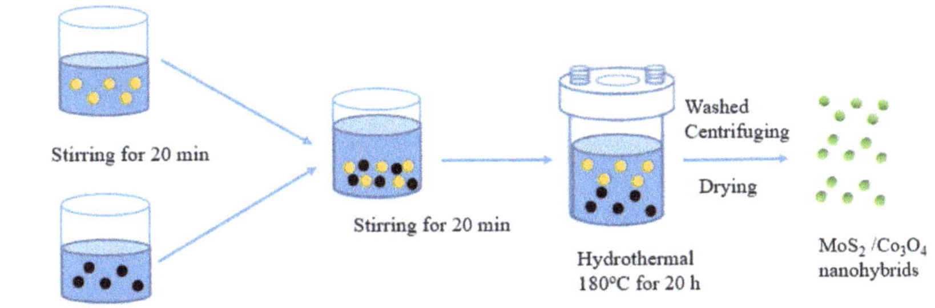

Scheme 2. Schematic illustration of the formation route of the MoS_2/Co_3O_4 nanohybrids.

X-ray diffraction patterns (XRD) were determined by a Bruker D8 Advanced diffractometer (Billerica, MA, USA) at 5–75° with a scanning step of 20°/min at room temperature. Hitachi UV-4100 UV–visible spectrophotometer and Edinburgh Analytical Instruments F-7000 (Tokyo, Japan) were adopted to check the UV–vis diffusive reflectance spectra and photoluminescence (PL) spectra with an excitation wavelength of 620 nm. The microstructure and morphology of different samples also were tested by employing scanning electron microscopy (SEM) micrographs (ZEISS AURIGA, Oberkochen, Germany) and transmission electron microscopy (TEM) experiment (JEOL JEM-2100F, Tokyo, Japan). Electron paramagnetic resonance (EPR) spectra were observed on a Bruker A300 (Munich, Germany) at room temperature under atmospheric conditions. A CH Instruments electrochemical workstation with a standard three-electrode system was recorded to carry out electrochemical measurements in 0.2 M Na_2SO_4 solution. Furthermore, the saturated Pt electrode and Hg/Hg_2Cl_2 electrode were used as the counter electrode and reference electrode, respectively. The working electrode was used according to the next procedure: the as-prepared photocatalysts (20 mg) were dispersed in 10 mL of DI water to obtain a homogeneous suspension with a consecutive ultrasonic process. The chemical states and surface species were tested by X-ray photoelectron spectroscopy (XPS, Thermo Fisher Scientific, Waltham, MA, USA).

Photodegradation of MO pollutants over various MC-X photocatalysts was performed at room temperature with visible light activity. A 350 W Xe lamp was employed as the visible-light source. A total of 20 mg of samples were added in 100 mL MO aqueous solution (20 mg L^{-1}). The above reaction solution was further stirred under the dark for 20 min to organize absorption-desorption equilibrium between photocatalysts and MO compounds. Under light activity, a definite quantity of reaction solution was carried out from the combination solution at contributed intervals, followed by centrifugation to eliminate all the materials. The concentration alters of MO pollutants were obtained via UV-vis spectroscopy under photocatalytic activity.

4. Conclusions

In this study, MoS_2/Co_3O_4 nanohybrids were fabricated adopting a facile hydrothermal process. The prepared nanohybrids exhibited better capability for the degradation of MO dye from aqueous solution. The maximum MO photodegradation performance approached by the nanohybrids was 95.6%, and the reaction rate constants was 0.021. The MoS_2/Co_3O_4 nanohybrids with 20 wt% loading of Co_3O_4 on MoS_2 nanoparticles, displays excellent photocatalytic response for MO dye removal as compared to the bare photocatalysts and other nano-heterojunctions. The 20 wt% MoS_2/Co_3O_4 nanohybrids were the optimal photocatalysts through photoreduction and photodegradation activity greater than that of bare MoS_2 and Co_3O_4. The result demonstrates that the formation of MoS_2/Co_3O_4 nanohybrids is conducive to the production of excellent photo response carriers and the enhancement of visible light absorption. In addition, the fabrication of nanohybrids effectively inhibits the recombination rate of photo response electrons and holes in the photocatalytic activity. Thereby, this research paves way for the facile formation of nanohybrids system for diverse environmental fields.

Author Contributions: Writing—original draft, Methodology, T.-M.T.; Writing—review and editing, E.L.C.; Data curation, Formal analysis, C.-H.C.; Funding acquisition, T.-M.T.; Validation, C.-T.H.; Resources, T.-M.T.; Formal analysis, data collection, E.L.C.; Supervision, Investigation E.L.C.; Supervision, Project administration. T.-M.T. All authors have read and agreed to the published version of the manuscript.

Funding: This research received no external funding.

Data Availability Statement: Not available.

Acknowledgments: The MOST and the NKUST are gratefully acknowledged for their general support. The authors gratefully acknowledge the use of HRTEM equipment belonging to the Instrument Center of National Cheng Kung University.

Conflicts of Interest: The authors declare no conflict of interest.

References

1. Gaim, Y.T.; Yimanuh, S.M.; Kidanu, Z.G.; Ashebir, M.E. Enhanced Photocatalytic Degradation of Amoxicillin with Mn-Doped Cu_2O under Sunlight Irradiation. *J. Compos. Sci.* **2022**, *6*, 317. [CrossRef]
2. Alam, M.W.; Azam, H.; Khalid, N.R.; Naeem, S.; Hussain, M.K.; BaQais, A.; Farhan, M.; Souayeh, B.; Zaidi, N.; Khan, K. Enhanced Photocatalytic Performance of Ag_3PO_4/Mn-ZnO Nanocomposite for the Degradation of Tetracycline Hydrochloride. *Crystals* **2022**, *12*, 1156. [CrossRef]
3. Wang, X.; Zhou, J.; Zhao, S.; Chen, X.; Yu, Y. Synergistic effect of adsorption and visible-light photocatalysis for organic pollutant removal over $BiVO_4$/carbon sphere nanocomposites. *Appl. Surf. Sci.* **2018**, *453*, 394. [CrossRef]
4. Das, T.K.; Remanan, S.; Ghosh, S.; Ghosh, S.K.; Das, N.C. Efficient synthesis of catalytic active silver nanoparticles illuminated cerium oxide nanotube: A mussel inspired approach. *Environ. Nanotechnol. Monit. Manag.* **2021**, *15*, 100411. [CrossRef]
5. Remanan, S.; Padmavathy, N.; Rabiya, R.; Ghosh, S.; Das, T.K.; Bose, S.; Sen, R.; Das, N.C. Converting Polymer Trash into Treasure: An Approach to Prepare MoS_2 Nanosheets Decorated PVDF Sponge for Oil/Water Separation and Antibacterial Applications. *Ind. Eng. Chem. Res.* **2020**, *59*, 20141. [CrossRef]
6. Hojamberdiev, M.; Kadirova, Z.C.; Gonçalves, R.V.; Yubuta, K.; Matsushita, N.; Teshima, K.; Hasegawa, M.; Okada, K. Reduced graphene oxide-modified Bi_2WO_6/BiOI composite for the effective photocatalytic removal of organic pollutants and molecular modeling of adsorption. *J. Mol. Liq.* **2018**, *268*, 715. [CrossRef]
7. Chen, L.J.; Chuang, Y.J.; Chen, C. Surface passivation assisted lasing emission in the quantum dots doped cholesteric liquid crystal resonating cavity with polymer template. *RSC Adv.* **2014**, *4*, 18600. [CrossRef]
8. Alnassar, M.A.; Alshehri, A.; Narasimharao, K. Visible Light Active Magnesium Silicate–Graphitic Carbon Nitride Nanocomposites for Methylene Blue Degradation and Pb^{2+} Adsorption. *Catalysts* **2022**, *12*, 1256. [CrossRef]
9. Yadav, A.A.; Hunge, Y.M.; Kang, S.W. Visible Light-Responsive CeO_2/MoS_2 Composite for Photocatalytic Hydrogen Production. *Catalysts* **2022**, *12*, 1185. [CrossRef]
10. Chen, L.J.; Tsai, M.L.; Chuang, Y.; Chen, C.W.; Dong, C.D. Construction of carbon nanotubes bridged MoS_2/ZnO Z-scheme nanohybrid towards enhanced visible light driven photocatalytic water disinfection and antibacterial activity. *Carbon* **2022**, *196*, 877. [CrossRef]
11. Pant, B.; Park, M.; Park, S.J. $MoS_2/CdS/TiO_2$ ternary composite incorporated into carbon nanofibers for the removal of organic pollutants from water. *Inorg. Chem. Commun.* **2019**, *102*, 113–119. [CrossRef]

12. Chen, Z.; Li, X.; Wu, Y.; Zheng, J.; Peng, P.; Zhang, X.; Duan, A.; Wang, D.; Yang, Q. S-scheme $Cs_2AgBiBr_6/Ag_3PO_4$ heterojunction with efficient photocatalysis performance for H_2 production and organic pollutant degradation under visible light. *Sep. Purif. Technol.* **2022**, *295*, 121250. [CrossRef]
13. Zia, A.; Naveed, A.B.; Javaid, A.; Ehsan, M.F.; Mahmood, A. Facile Synthesis of $ZnSe/Co_3O_4$ Heterostructure Nanocomposites for the Photocatalytic Degradation of Congo Red Dye. *Catalysts* **2022**, *12*, 1184. [CrossRef]
14. Wang, L.; Wang, L.; Shi, Y.; Zhu, J.; Zhao, B.; Zhang, Z.; Ding, G.; Zhang, H. Fabrication of Co_3O_4-Bi_2O_3-Ti catalytic membrane for efficient degradation of organic pollutants in water by peroxymonosulfate activation. *J. Colloid Interface Sci.* **2022**, *607*, 451. [CrossRef] [PubMed]
15. Chen, L.J.; Chen, C.W.; Huang, C.P.; Chuang, Y.; Nguyen, T.B.; Dong, C.D. A visible-light sensitive MoSSe nanohybrid for the photocatalytic degradation of tetracycline, oxytetracycline, and chlortetracycline. *J. Colloid Interface Sci.* **2022**, *616*, 67. [CrossRef]
16. Zhu, C.; Zhang, L.; Jiang, B.; Zheng, J.; Hu, P.; Li, S.; Wu, M.; Wu, W. Fabrication of Z-scheme Ag_3PO_4/MoS_2 composites with enhanced photocatalytic activity and stability for organic pollutant degradation. *Appl. Surf. Sci.* **2016**, *377*, 99–108. [CrossRef]
17. Kumar, S.; Maivizhikannan, V.; Drews, J.; Krishnan, V. Fabrication of nanoheterostructures of boron doped ZnO-MoS_2 with enhanced photostability and photocatalytic activity for environmental remediation applications. *Vacuum* **2019**, *163*, 88–98. [CrossRef]
18. Zhong, Y.; Wang, F.; Liang, C.; Guan, Z.; Lu, B.; He, X.; Yang, W. ZnO@MoS_2 Core–Shell Heterostructures Enabling Improved Photocatalytic Performance. *Appl. Sci.* **2022**, *12*, 4996. [CrossRef]
19. Dharman, R.K.; Shejale, K.P.; Kim, S.Y. Efficient sonocatalytic degradation of heavy metal and organic pollutants using CuS/MoS_2 nanocomposites. *Chemosphere* **2022**, *305*, 135415. [CrossRef]
20. Chen, L.J.; Chuang, Y.J. Directly electrospinning growth of single crystal Cu_2ZnSnS_4 nanowires film for high performance thin film solar cell". *J. Power Sources* **2013**, *241*, 259. [CrossRef]
21. Chen, L.J.; Chuang, Y.J. Diethylenetriamine assisted synthesis and characterization of stannite quaternary semiconductor $Cu_2ZnSnSe_4$ nanorods by self-assembly. *J. Cryst. Growth* **2013**, *376*, 11. [CrossRef]
22. Zhao, L.; Yu, T.; Yang, B.; Guo, H.; Liu, L.; Zhang, J.; Gao, C.; Yang, T.; Wang, M.; Zhang, Y. Wastewater Purification and All-Solid Z-Scheme Heterojunction ZnO-C/MnO_2 Preparation: Properties and Mechanism. *Catalysts* **2022**, *12*, 1250. [CrossRef]
23. Chen, L.J.; Lee, C.R.; Chuang, Y.J.; Wu, Z.H.; Chen, C. Synthesis and Optical Properties of Lead-Free Cesium Tin Halide Perovskite Quantum Rods with High-Performance Solar Cell Application. *J. Phys. Chem. Lett.* **2016**, *7*, 5028. [CrossRef] [PubMed]
24. Chen, L.J. Synthesis and optical properties of lead-free cesium germanium halide perovskite quantum rods. *RSC Adv.* **2018**, *18*, 18396–18399. [CrossRef] [PubMed]
25. Zhao, H.; Zhang, J.; Zheng, H. Facile preparation of dual Z-scheme Bi_2O_3/g-C_3N_4/$Ag_6Si_2O_7$ photocatalyst for highly efficient visible-light photocatalytic degradation of organic pollutants. *Appl. Surf. Sci.* **2020**, *527*, 146901. [CrossRef]
26. Wang, J.; Zhang, Q.; Deng, F.; Luo, X.; Dionysiou, D.D. Rapid toxicity elimination of organic pollutants by the photocatalysis of environment-friendly and magnetically recoverable step-scheme $SnFe_2O_4/ZnFe_2O_4$ nano-heterojunctions. *Chem. Eng. J.* **2020**, *379*, 122264. [CrossRef]
27. Lu, M.; Javed, M.; Javed, K.; Tan, S.; Iqbal, S.; Liu, G.; Khalid, W.B.; Qamar, M.A.; Alrbyawi, H.; Pashameah, R.A.; et al. Construction of a Well-Defined S-Scheme Heterojunction Based on Bi-$ZnFe_2O_4$/S-g-C_3N_4 Nanocomposite Photocatalyst to Support Photocatalytic Pollutant Degradation Driven by Sunlight. *Catalysts* **2022**, *12*, 1175. [CrossRef]
28. Shehzad, N.; Zafar, M.; Ashfaq, M.; Razzaq, A.; Akhter, P.; Ahmad, N.; Hafeez, A.; Azam, K.; Hussain, M.; Kim, W.Y. Development of $AgFeO_2$/rGO/TiO_2 Ternary Composite Photocatalysts for Enhanced Photocatalytic Dye Decolorization. *Crystals* **2020**, *10*, 923. [CrossRef]
29. Tien, T.M.; Chung, Y.J.; Huang, C.T.; Chen, E.L. WSSe Nanocomposites for Enhanced Photocatalytic Hydrogen Evolution and Methylene Blue Removal under Visible-Light Irradiation. *Materals* **2022**, *15*, 5616. [CrossRef]
30. Tien, T.M.; Chung, Y.J.; Huang, C.T.; Chen, E.L. Fabrication of WS_2/WSe_2 Z-Scheme Nano-Heterostructure for Efficient Photocatalytic Hydrogen Production and Removal of Congo Red under Visible Light. *Catalysts* **2022**, *12*, 852. [CrossRef]
31. You, J.; Zhang, L.; He, L.; Zhang, B. Photocatalytic degradation of methyl orange on ZnO–TiO_2/SO_4^{2-} heterojunction composites. *Opt. Mater.* **2022**, *131*, 112737. [CrossRef]
32. Chen, L.; Zhang, M.; Wu, J.; Zheng, X.; Liao, S.; Ou, B.; Tian, L. In situ embedment of CoO_x on g-C_3N_4 as Z scheme heterojunction for efficient photocatalytic degradation of methyl orange and phenol under visible light. *J. Alloys Compd.* **2022**, *927*, 167047. [CrossRef]
33. Sindelo, A.; Britton, J.; Lanterna, A.E.; Scaiano, J.C.; Nyokong, T. Decoration of glass wool with zinc (II) phthalocyanine for the photocatalytic transformation of methyl orange. *J. Photochem. Photobiol. A Chem.* **2022**, *432*, 114127. [CrossRef]
34. Wang, R.; Wang, Y.; Zhang, N.; Lin, S.; He, Y.; Yan, Y.; Hu, K.; Sun, H.; Liu, X. Synergetic piezo-photocatalytic effect in SbSI for highly efficient degradation of methyl orange. *Ceram. Int.* **2022**, *48*, 31818–31826. [CrossRef]
35. Oyewo, O.A.; Ramaila, S.; Mavuru, L.; Onwudiwe, D.C. Enhanced photocatalytic degradation of methyl orange using Sn-ZnO/GO nanocomposite. *J. Photochem. Photobiol.* **2022**, *11*, 100131. [CrossRef]
36. Guan, J.; Long, Z.; Li, Q.; Han, J.; Du, H.; Wang, P.; Zhang, G. Citric acid modulated preparation of CdS photocatalyst for efficient removal of Cr (VI) and methyl orange. *Opt. Mater.* **2021**, *121*, 111604. [CrossRef]
37. Uma, K.; Muniranthinam, E.; Chong, S.; Yang, T.C.K.; Lin, J.H. Fabrication of Hybrid Catalyst ZnO Nanorod/α-Fe_2O_3 Composites for Hydrogen Evolution Reaction. *Crystals* **2020**, *10*, 356. [CrossRef]

38. Zhang, N.; Peng, S.; Liu, Z.; Li, Y.; Huang, J.; Li, J.; Wan, H.; Zhou, S.; Gao, Z.; Chen, T. Ag NPs decorated on the magnetic Fe_3O_4@PDA as efficient catalyst for organic pollutants removal and as effective antimicrobial agent for microbial inhibition. *J. Alloys Compd.* **2022**, *928*, 167257. [CrossRef]
39. Vadivel, S.; Fujii, M.; Rajendran, S. Novel S-scheme 2D/2D $Bi_4O_5Br_2$ nanoplatelets/g-C_3N_5 heterojunctions with enhanced photocatalytic activity towards organic pollutants removal. *Environ. Res.* **2022**, *213*, 113736. [CrossRef]
40. Chen, L.J.; Chuang, Y.J. Quaternary Semiconductor Derived and formation mechanism by non-vacuum Route from Solvothermal Nanostructures for High-Performance application. *Mater. Lett.* **2013**, *91*, 372. [CrossRef]

catalysts

Article

Magnetic TiO_2/Fe_3O_4-Chitosan Beads: A Highly Efficient and Reusable Catalyst for Photo-Electro-Fenton Process

Soumaya Rezgui [1,2,*], Aida M. Díez [1], Lotfi Monser [2], Nafaa Adhoum [2], Marta Pazos [1] and M. Ángeles Sanromán [1]

1 BIOSUV Research Group, INTECX Building, University of Vigo, Lagoas-Marcosende Campus, 36310 Vigo, Spain
2 Research Unit in Electrochemistry, Materials and Environment (UR16ES02), IPEIK, University of Kairouan, Kairouan 3100, Tunisia
* Correspondence: soumaya.rezgui@insat.rnu.tn

Abstract: Heterogeneous photo-electro-Fenton process is an attractive technology for the removal of recalcitrant pollutants. To better exploit the presence of an irradiation source, a bifunctional catalyst with TiO_2 nanoparticles embedded into an iron–chitosan matrix was developed. The catalytic activity of the catalyst was improved by the optimization of the loaded TiO_2 content. The prepared composite catalysts based on TiO_2, Fe_3O_4 and chitosan were called TiO_2/Fe_3O_4-CS beads. The best catalyst with an optimal ratio TiO_2/Fe = 2 exhibited a high efficiency in the degradation and mineralization of chlordimeform (CDM) insecticide. Under the optimum conditions (concentration of catalyst equal to 1 g L^{-1} and applied current intensity equal to 70 mA), a real effluent doped with 30 mg L^{-1} of CDM was efficiently treated, leading to 80.8 ± 1.9% TOC reduction after 6 h of treatment, with total removal of CDM after only 1 h.The generated carboxylic acids and minerals were identified and quantified. Furthermore, the stability and reusability of the developed catalyst was examined, and an insignificant reduction in catalytic activity was noticed for four consecutive cycles of the photo-electro-Fenton process. Analyses using SEM, XRD and VSM showed a good stability of the physicochemical properties of the catalyst after use.

Keywords: photo-electro-Fenton process; chlordimeform; wastewater treatment; bifunctional catalyst

Citation: Rezgui, S.; Díez, A.M.; Monser, L.; Adhoum, N.; Pazos, M.; Sanromán, M.Á. Magnetic TiO_2/Fe_3O_4-Chitosan Beads: A Highly Efficient and Reusable Catalyst for Photo-Electro-Fenton Process. *Catalysts* **2022**, *12*, 1425. https://doi.org/10.3390/catal12111425

Academic Editors: Jorge Bedia and Carolina Belver

Received: 16 October 2022
Accepted: 9 November 2022
Published: 13 November 2022

1. Introduction

In recent decades, the extensive use of pesticides to improve agricultural production has led to an increased risk of water pollution. In fact, these pollutants can be classified as persistent and extremely toxic organic substances. The ability of these harmful pollutants to easily bioaccumulate even at very low concentrations represents a serious issue that can cause problems for the environment, human health and living organisms [1]. Consequently, the removal of pesticides from water and wastewater has become a great global challenge.

In recent years, electrochemical advanced oxidation processes (EAOPs) have been considered an option for the removal of toxic and persistent organic micropollutants from wastewater [2–4]. One of the most powerful and attractive techniques among EAOPs is the electro-Fenton (EF) process, whereby hydrogen peroxide (H_2O_2) is generated at a cathode by O_2 reduction (Equation (1)), and a ferrous ion or iron oxide catalyst is added to the effluent [5]. The EF process is very effective in the mineralization of organic pollutants due to the production of strongly oxidizing hydroxyl radicals through the Fenton reaction (Equation (2)) [6]. However, this technology is limited by a few factors, such as the rapid accumulation of Fe(III) ions and the possibility of their complexation with hydroxyl ions and oxidation products during treatment [7].

$$O_2 + 2\,H^+ + 2\,é \rightarrow H_2O_2 \quad (1)$$

$$H_2O_2 + Fe^{2+} \rightarrow Fe^{3+} + OH^- + \bullet OH \quad (2)$$

The presence of UV irradiation has two specific aims: the regeneration of Fe(II) ions and the degradation of the formed complexes until their mineralization due to an additional production of •OH [8]. This process is called the photo-electro-Fenton (PEF) process, and it leads to great performance in the mineralization of organic pollutants [9,10].

However, the application of UV irradiation could increase the operational cost, but it can be justified by the use of a photocatalyst, such as TiO_2, to increase its economic cost effectiveness. The application of a ferromagnetic-TiO_2 composite material is widely used for photocatalysis due to its efficiency and magnetic properties [11,12].

In fact, in a heterogeneous system, many solid catalysts containing iron, such as Fe, Fe_2O_3, Fe_3O_4 and FeOOH, were successfully used for the treatment of recalcitrant organic pollutants in water [13–17]. Among iron oxides, the immense popularity of Fe_3O_4 as a catalyst originates from its broad application potential due to high saturation magnetization, easy handling, relatively low cost, non-toxicity and environmentally friendly character [18,19]. Titanium oxide (TiO_2) has also shown widespread photocatalytic application in the field of wastewater treatment due to its unrivalled properties of non-toxicity, easy UV activation, chemical stability and availability [20]. Despite all these outstanding properties, TiO_2 deployment for photocatalytic application has witnessed drawbacks due to its large energy band gap of about 3.2 eV [20,21]. In addition, one of the problems hindering TiO_2 use is its agglomeration in aqueous solution, which requires a post-filtration treatment [21]. In order to overcome the problem of the band gap of TiO_2, several studies revealed that the presence of Fe_3O_4 enhances the photocatalytic activity of TiO_2 by decreasing the charge carrier recombination, with a band gap of the composite material around 2.5 eV. This is caused by the generated Z-scheme on the composite TiO_2/Fe_3O_4 where the recombination between holes and electrons on TiO_2 is restrained by the addition of another valence band (Fe_3O_4). Indeed, Fe-species have been reported as an outstanding alternative for creating Z-scheme photocatalytic heterojunction systems [22]. Moreover, this modified TiO_2 nanoparticle can be separated from water by means of an external magnetic field [11,23]. However, the use of magnetic separation is complicated in real water applications. Therefore, the use of a catalytic support, such as chitosan, for both metal oxides (Fe_3O_4 and TiO_2) not only avoids the agglomeration issue, but it also facilitates catalyst recovery, protects ferroparticles from oxidation and extends their storage life [24].

Therefore, in the present study, we innovatively synthesized the TiO_2/Fe_3O_4-CS catalytic beads through a green co-precipitation method in only one step. The co-precipitation method involves the dispersal of a mixture of chitosan and metals into an alkaline solution to form nanoparticles strongly bound to chitosan with attractive catalytic and magnetic properties for easy catalyst recovery and reuse [24–28]. TiO_2/Fe_3O_4-CS was used as a novel nano-structured heterogeneous catalyst for the oxidation of CDM insecticide by an efficient cyclic PEF process; therefore, Fe(II), within the structure of Fe_3O_4, can activate the Fenton system [29], and TiO_2 acts as a photocatalyst to better exploit the use of UV irradiation. In the PEF system, an electrochemical reactor was used for the electrogeneration of H_2O_2, and the catalyst was added to a second photocatalytic reactor. The effects of operational parameters were studied, such as the TiO_2 loading in the beads, catalyst dosage and the current intensity applied. Under optimal conditions, aqueous wastewater doped with 30 mg L^{-1} of CDM was treated. The carboxylic acids and inorganic ions generated during the treatment process were identified and quantified. Finally, the electric energy consumed was estimated, and the stability of the best catalyst in terms of CDM and TOC removal efficiencies was studied.

2. Results

2.1. Optimization of Operational Parameters

2.1.1. Performance Comparison of Several Processes for the Removal of CDM

Initially, some control experiments were performed. To begin with, the adsorption assays of the developed catalytic beads showed no CDM adsorption after 4 h of treatment on the synthesized catalytic beads (data not shown). Then, the low efficiencies of photoelectrolysis–H_2O_2 (electrogenerated H_2O_2 + UV-LED) and photocatalysis (1 g of TiO_2/Fe_3O_4-CS catalyst + UV-LED) systems revealed, respectively, 70.6 ± 5.3% and 13.2 ± 4.4% of CDM removal yield after 1 h of electrolysis (Figure 1) along with only 18.2 ± 3.3% and 4.4 ± 1.6% of TOC abatement after 4 h of treatment (Table 1). However, the PEF process running at I = 50 mA and pH_i = 3, in the presence of 1 g L^{-1} TiO_2/Fe_3O_4-CS, exhibited higher efficiency, with total degradation of CDM after only 1 h and TOC abatement equal to 69.6 ± 2.7% after 4 h of treatment (Table 1).

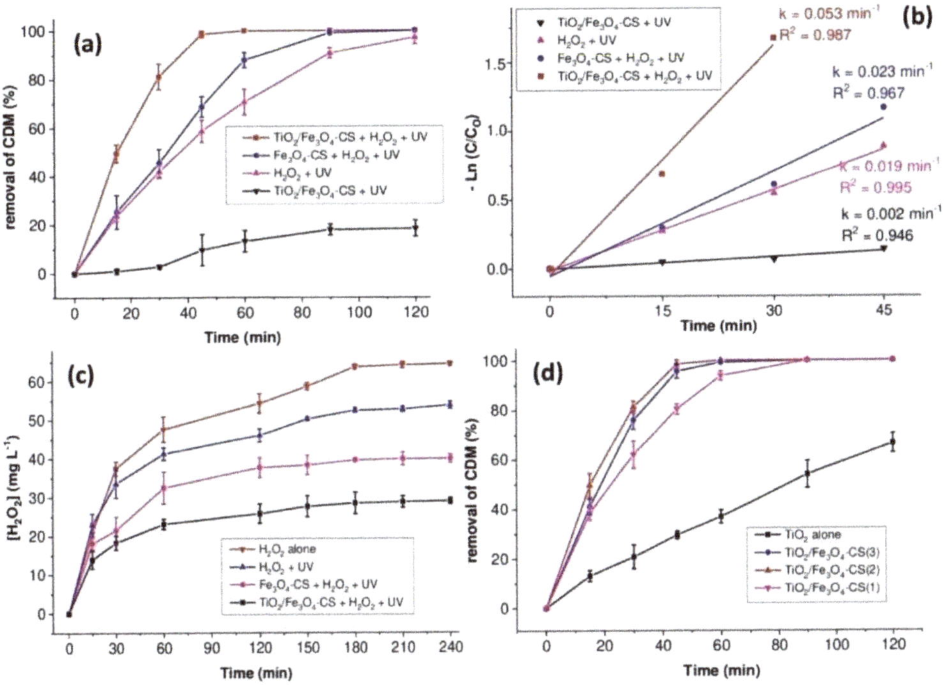

Figure 1. Performance comparison of several processes and effect of the addition of TiO_2 to magnetic chitosan beads: (**a**,**d**) % removal of CDM; (**b**) −Ln (C/C0) vs. time; (**c**) [H_2O_2] vs. time: I = 50 mA in the presence of 1 g L^{-1} of catalyst and at pH_i = 3.

2.1.2. Influence of TiO_2 in Magnetic Chitosan Beads

As found previously in the PEF system, the composite catalyst TiO_2/Fe_3O_4-CS showed higher catalytic activity, with total degradation of CDM after only 1 h and TOC abatement of around 70% after 4 h of treatment. However, in the presence of Fe_3O_4-CS catalytic beads, only 87.9 ± 3.1% CDM removal yield and TOC abatement was achieved, not exceeding 33.9 ± 4.1%. The higher photocatalytic activity of TiO_2/Fe_3O_4-CS beads could be attributed to the acceleration of electron mobility in the $Fe^{III/II}/TiO_2$ system [12,30] and the synergistic combination of PEF with photocatalysis. It is widely reported that the combination of TiO_2 and Fe_3O_4, with their different band gaps, can suppress the electron–hole recombination, enhancing the photocatalytic activity, and thus, the PEF process efficiency, thanks to the generation of the Z-scheme [11,22,30–32]. Moreover, the degradation of CDM by the PEF process was found to follow a pseudo-first-order kinetic (Figure 1b), which is probably related to the steady •OH concentration throughout the treatment process [33,34]. To evaluate the catalytic decomposition of the oxidant, the concentration of H_2O_2 was monitored during treatment with and without the developed catalysts, and the results are presented in Figure 1c. In the absence of the catalyst and UV-LED lamp, the accumulated amount of H_2O_2 after 4 h of electrolysis was estimated at 64.4 ± 0.5 mg L^{-1}. In the presence of UV-LED irradiation, the concentration of H_2O_2 decreased to 53.8 ± 0.9 mg L^{-1}, which means that the amount of H_2O_2 decomposed by photolysis after 4 h of electrolysis was around 16.6%. As expected, by adding the Fe_3O_4-CS and TiO_2/Fe_3O_4-CS catalytic beads to the (H_2O_2 + UV-LED) system, the concentration of the oxidant decreased significantly to 39.9 ± 1.1 mg L^{-1} and 29 ± 0.8 mg L^{-1}, respectively, indicating that approximately 38% and 55% of the H_2O_2 amount was decomposed. These findings were in agreement with our previous results, and they confirmed the best catalytic activity of TiO_2/Fe_3O_4-CS beads because an increase in the decomposition rate of H_2O_2 is related to an increase in the production yield of •OH radicals [35]. Therefore, it is clear that the incorporation of TiO_2 into magnetic CS beads favors an effective usage of H_2O_2, allowing great improvement to the PEF process efficiency.

Table 1. Summary table of mineralization rates.

Parameter	Assay	Rate of TOC Removal after 4 h of Treatment
Performance comparison of several processes	TiO_2/Fe_3O_4-CS(2) + UV-LED	4.4 ± 1.6
	H_2O_2 + UV-LED (Without catalyst)	18.2 ± 3.3
	TiO_2/Fe_3O_4-CS(2) + H_2O_2 + UV-LED	69.6 ± 2.7
Influence of TiO_2 in magnetic chitosan beads	Fe_3O_4-CS + H_2O_2 + UV-LED	33.9 ± 4.1
	TiO_2/Fe_3O_4-CS(2) + H_2O_2 + UV-LED	69.6 ± 2.7
Effect of TiO_2 loading content into magnetic chitosan beads	Molar ratio TiO_2/Fe = 1	38.2 ± 3.5
	Molar ratio TiO_2/Fe = 2	69.6 ± 2.7
	Molar ratio TiO_2/Fe = 3	54.2 ± 5.6
	Catalytic beads of TiO_2	24.5 ± 1.9
Effect of catalyst dosage	0.25 g L^{-1}	40.9 ± 4.1
	0. 5 g L^{-1}	60.8 ± 3.2
	1 gL^{-1}	69.6 ± 2.7
	1.5 g L^{-1}	70.2 ± 1.7
	2 g L^{-1}	58.2 ± 6.2
Effect of current intensity	50 mA	68.6 ± 4.9
	70 mA	76.9 ± 3.7
	100 mA	77.7 ± 1.2

2.1.3. Effect of TiO_2 Loading Content into Chitosan Beads

In order to increase the catalytic activity of TiO_2/Fe_3O_4-CS beads, the effect of TiO_2 content in the catalyst was studied. For this purpose, the molar ratio TiO_2/Fe was varied during the preparation of the catalyst from 1 to 3 (to facilitate a designation of the developed catalysts, the molar ratio values between the brackets were added to TiO_2/Fe_3O_4-CS). The results showed that a molar ratio equal to 2 (TiO_2/Fe_3O_4-CS(2)) was optimal, leading to total degradation of CDM after 1 h (Figure 1d) and TOC removal yield of around 70% after 4 h of treatment (Table 1), against 38.2 ± 3.5% and 54.2 ± 5.6%, respectively, for the molar ratios TiO_2/Fe = 1 and 3. On the other hand, the comparison of the photocatalytic activities of TiO_2/Fe_3O_4-CS(2) and TiO_2-CS (Figure 1d) showed a remarkable difference in their catalytic performance, which was greater for the composite catalyst. In fact, the CDM degradation rate did not exceed 35% after 1 h of treatment using TiO_2-CS beads, with low TOC abatement estimated at 24.5 ± 1.9% after 4 h of treatment. Therefore, it is obvious that the Fe(II) ions in TiO_2/Fe_3O_4-CS catalytic beads notably improve their catalytic performance by the Fenton reaction with H_2O_2, so that •OH radicals are produced. Numerous studies have reported the high photocatalytic activity of the mixture of TiO_2/Fe_3O_4 compared to pure TiO_2, and they explained this improvement by the fast photogenerated electron transfer between Fe_3O_4 and TiO_2, which can effectively reduce electron/hole recombination [12,33,36,37].

To sum up, a molar ratio TiO_2/Fe = 2 is optimal both for the removal of the pollutant and its mineralization. In fact, varying the molar ratio from 1 to 2, the photocatalytic activity of the catalytic beads increased through the increase in TiO_2 active sites. The improvement in pollutant degradation and mineralization by the increase in TiO_2 load in the iron catalysts has already been reported by many research studies, and it is related to the enhancement of the electron transfer reducing electron–hole recombination [32,38]. For a molar ratio $TiO_2/Fe = 3$, the catalytic performance of the system decreased due to the decrease in active iron sites in favor of the photocatalytic sites of TiO_2, which has a wide band gap energy, causing a rapid recombination of the electron–hole pairs [39]. It is well known that the fast recombination of charge carriers significantly lowers the photocatalytic performance [24,40]. Thus, a TiO_2/Fe molar ratio equal to 2 was optimal for the preparation of TiO_2/Fe_3O_4-CS beads, and the catalyst TiO_2/Fe_3O_4-CS(2) was selected thereafter.

The photodegradation of pesticides using TiO_2/Fe_3O_4-based materials as catalysts has been poorly reported. Compared to the efficiency of TiO_2/Fe_3O_4 composite catalysts listed in Table 2, TiO_2/Fe_3O_4-CS showed good catalytic activity. Nevertheless, the reactor set-up, such as the irradiation source, can greatly affect the overall efficiency of the process [41].

Table 2. The comparison of photodegradation efficiency of TiO_2/Fe_3O_4-CS catalytic beads with other TiO_2/Fe_3O_4-based materials against different pesticides.

Catalysts	Pesticides	Degradation Efficiency (%)	Refs.
Fe_3O_4-TiO_2/reduced graphene oxide	Atrazine	99% within 40 min	[42]
N-doped TiO_2@SiO_2@Fe_3O_4 nanocomposite	Paraquat	98.7% within 180 min	[43]
Bare 3D-TiO_2/magnetic biochar dots	Diazinon	98.5% within 30 min	[44]
TiO_2/Fe_3O_4-CS	Chlordimeform	100% removal within 60 min	Present work

2.1.4. Effect of Catalyst Dosage

Catalyst dosage was another parameter affecting the heterogeneous Fenton reaction [45]. The effect of catalyst dosage on the degradation of CDM by the PEF process was studied by applying a current intensity of 50 mA at $pH_{initial} = 3$. It is clear from Figure 2a that the increase in TiO_2/Fe_3O_4-CS(2) catalytic beads from 0.25 g L^{-1} to 1 g L^{-1} increased the rate of CDM removal, and an almost total degradation was obtained for all assays after 1 h of treatment, with an improvement in TOC abatement, respectively, from $40.9 \pm 4.1\%$ to $69.6 \pm 2.7\%$ (Table 1). For a concentration equal to 1.5 g L^{-1}, there was a trivial enhancement in CDM and TOC removal efficiencies. However, the removal yields of CDM and TOC were decreased for a high concentration equal to 2 g L^{-1}. Indeed, an increase in the concentration of catalytic beads in solution can inhibit the penetration of light into the photocatalytic reactor [46,47]. Additionally, an increase in the active catalytic sites can cause secondary reactions with •OH radicals, thus producing less reactive oxidizing species (Equation (3)), which can reduce the efficiency of the process for the mineralization of organic pollutants [25].

$$Fe^{2+} + \bullet OH \rightarrow Fe^{3+} + OH^- \quad (3)$$

2.1.5. Effect of the Current Intensity

The current intensity applied is a key parameter in the electrochemical Fenton technologies because it is the driving force for the reduction in oxygen, leading to the generation of H_2O_2 at the cathode, and it affects the regeneration of Fe(II) [48]. In order to study the effect of the current intensity applied on CDM degradation and its mineralization by the PEF process, several experiments were carried out using different current intensities in the range from 50 mA to 100 mA in the presence of 1 g of TiO_2/Fe_3O_4-CS(2) and at initial acid pH (Figure 2b). The results showed that an increase in the current intensity from 50 mA to 70 mA leads to total degradation of the pollutant after only 1 h, with almost 10% improvement in TOC abatement (Table 1). This finding can be mainly attributed to the fast production of H_2O_2 at a higher current and the fast regeneration of Fe(II) enhancing the production of •OH radicals [49].

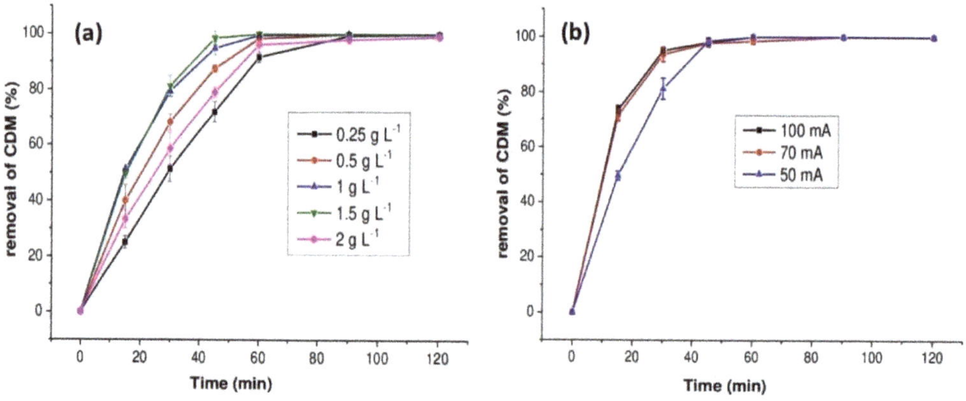

Figure 2. (**a**) Effect of catalyst dosage, (**b**) Effect of current intensity.

As seen in Figure 2b and Table 1, no further increase in the removal efficiencies was observed for the current intensity applied beyond 70 mA. This behavior can indicate that parasitic reactions, such as the four-electron reduction in O_2 with H_2O formation, as well as the decomposition and hydrogenation of H_2O_2, would take place when the current increased beyond a certain value [49–52]. On the other hand, working at higher current intensities can generate a degradation of our organic catalytic support "CS". In fact, several studies have shown the degradation of CS in the presence of an irradiation source and high concentrations of H_2O_2 [53,54]. An optimal current intensity equal to 70 mA, which ensured high treatment efficiency and good catalyst stability, was therefore established for the upcoming experiments.

To confirm the role of •OH radicals in the mineralization of the organic pollutant, isopropanol was used as a •OH scavenger. The degradation of CDM was totally inhibited in the presence of isopropanol. Thus, •OH should be the main active species for the degradation of CDM by the PEF process.

2.2. Treatment of Wastewater Doped with Chlordimeform by Photo-Electro-Fenton Process

To evaluate the applicability of the PEF process using TiO_2/Fe_3O_4-CS(2) as a catalyst for the removal of CDM in a more complex matrix than ultrapure water, an experiment was carried out on secondary treated wastewater kindly given by the municipal wastewater treatment plant in the northwest of Spain, whose physiochemical characteristics are summarized in Table 3.

Table 3. Physicochemical properties of the collected wastewater.

Parameter	Value
Total organic carbon TOC (mg L^{-1})	52.7
Chemical oxygen demand DCO (mg L^{-1})	35
Biological oxygen demand BOD_5 (mg L^{-1})	2
Conductivity (mS cm^{-1})	3.33
pH	7.31
Turbidity (mg L^{-1})	12

Under optimal experimental conditions ($[CDM]_i$ = 30 mg L^{-1}, $[TiO_2/Fe_3O_4$-CS(2)] = 1 g L^{-1}, and I = 70 mA), total degradation of CDM was achieved after 1 h, and TOC abatement reached 50.6 ± 5.1% after 1 h of electrolysis; then, it increased slowly to attain 80.8 ± 1.9% after 6 h of treatment (Figure 3b).

Furthermore, a slight decrease in the efficiency of the system was noticed compared to previous results when ultrapure water was used as the matrix (Figure 2b). This finding can be explained, according to zazouli et al. [55], by the presence of organic matter and inorganic anions in wastewater. In fact, the anions (X^-) can react with the •OH to produce $X^{-\bullet}$ radicals, which are less reactive.

In addition to the aqueous matrix influence, it should be noted that EEC is another big concern, which should be considered for sustainable development [56]. The results showed that global EEC was low compared to some literature values on the treatment of real wastewater by electrochemical processes [57–59] around 1.7 kWh m^{-3} and 1.4 kWh m^{-3}, respectively, for real wastewater and ultrapure water matrices. However, it should be noted that the difference ($\simeq$0.3 kWh m^{-3}) is probably due to the presence of degradable compounds in the treated wastewater matrix, whose oxidation requires a little more energy.

2.3. Evaluation of Organic Acids and Minerals Produced during Treatment

Studies of organic pollutants' mineralization by advanced oxidation processes show that oxidation by •OH radicals leads to the formation of carboxylic acids [60,61]. To better assess the oxidative capacity of our PEF process, an analysis of carboxylic acids was performed in order to identify and quantify them during treatment. The evolution of the identified acids during the mineralization of CDM is presented in Figure 3c. All the acids produced reached their maximum concentrations after around 2 h. Then, they decreased to a zero value after 6 h of treatment, indicating the deep mineralization of all detected acids. The concentration of oxalic acid during the treatment process was low (almost zero), which can be explained by the high reactivity of oxalic acid with Fe(III) and the formation of Fe(III)–oxalate complexes that are easily photolysed in the presence of UV light and H_2O_2 oxidant [62]. However, the other acids detected were malonic, succinic, formic and glycolic acids, appearing at the beginning of treatment. These results showed the effectiveness of our PEF process in removing the aliphatic compounds known for their resistance to oxidation.

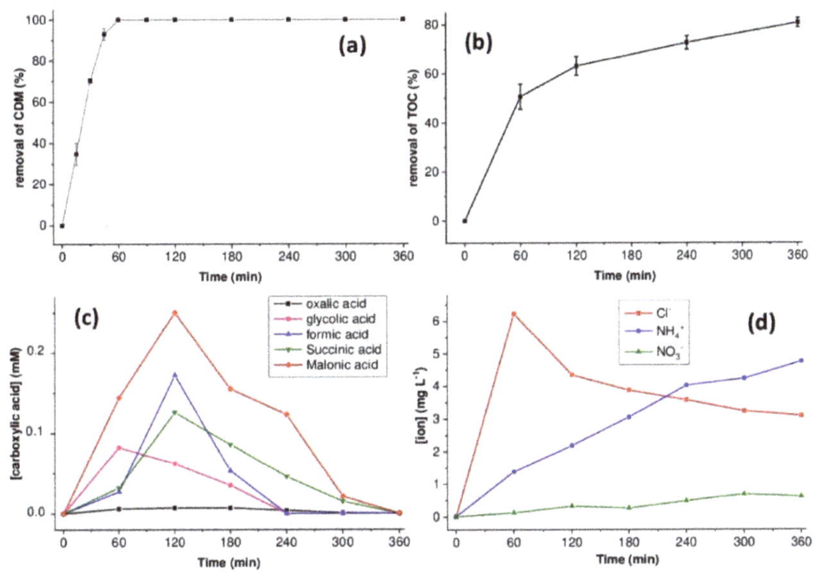

Figure 3. Treatment of a real effluent: (**a**) % removal of CDM; (**b**) % removal of TOC; (**c**) [carboxylic acids] vs. time (**d**) [ions] vs. time: I = 70 mA in the presence of 1 g L^{-1} of TiO_2/Fe_3O_4-CS(2) and at $pH_{initial}$ = 3.

On the other hand, considering the CDM molecule is composed of a chlorine atom and two nitrogen atoms, its treatment by an advanced oxidation process leads to the production of minerals. The evolution of these ions was monitored by ion chromatography. The obtained results (Figure 3d) showed that the chloride ions reached a maximum concentration of 6.22 mg L^{-1} after 1 h; this concentration presented the total amount initially present in the parent molecule. This finding was in agreement with several studies, which have shown that the release of chloride ions is rapid using a BDD anode and that the dechlorination of aromatic compounds occurs before the opening reaction of the aromatic cycle [63–65]. Then, the concentration of chloride ions decreased to approximately half (3.05 mg L^{-1}) after 6 h of treatment, which was in agreement with the results found by Mhemdi et al. [64] who studied the effect of the anode material (platinum and BDD) on the evolution of chlorides during the mineralization of 2-chlorobenzoic acid by an electrochemical oxidation process. The results showed that, for platinum, the concentration of chloride ions reached a maximum after 2 h and then remained constant. Using the BDD anode, a different behavior of chloride ions was observed. They accumulated rapidly after 1 h; then, their concentration decreased due to their oxidation to chlorine (Cl_2), which transformed into hypochlorite ions $HClO^-$ by hydrolysis. Consequently, the decrease in the accumulated chloride concentration during treatment could be explained by the oxidizing power of the BDD anode and the synthesized catalysts, causing the transformation of Cl^- ions into Cl_2.

Furthermore, the concentration of nitrate ions (NO_3^-) after 6 h was estimated at 0.68 mg L^{-1}. However, the accumulated concentration of ammonium ions (NH_4^+) was high, and it was around 4.75 mg L^{-1}. The accumulated concentrations of nitrate and ammonium ions after 6 h of treatment represent approximately 90% of the amount of nitrogen initially contained in the molecule of CDM. Consequently, one can confirm that CDM is mineralized during PEF application.

2.4. Stability of the Catalyst

2.4.1. Catalytic Stability

The recycling and reusability of a heterogeneous catalyst are important parameters for economic and environmental considerations [66]. The TOC abatement obtained using the same TiO_2/Fe_3O_4-CS(2) beads for four consecutive runs is depicted in Figure 4. After each run, the catalyst was washed with distilled water and dried at ambient temperature. The results showed that TOC removal remained almost constant for four cycles of reuse. The slight decrease ($\simeq$6.7%) observed in the fourth cycle could be attributed to the modification of the physicochemical properties of the catalyst's surface by the effect of UV-LED irradiation and the presence of H_2O_2 as an oxidant. In addition, possible leaching of Fe ions in the reaction medium was checked. The leached amount of Fe after each run did not exceed 1.7 mg L^{-1}, corresponding to 0.5% of the initial metal content in catalytic beads. These results revealed an excellent structural stability of the catalytic beads and minimal leaching of iron according to the environmental standards demanded by the European Union (<2 mg L^{-1}) [67].

2.4.2. Characterization of Catalyst before and after Use

SEM Analysis

The SEM images of TiO_2/Fe_3O_4-CS(2) beads before and after use are illustrated in Figure 5. As shown, the surface of the beads is smooth, indicating that iron and TiO_2 are attached to CS. The use of high magnification highlighted the observation of TiO_2 nanoparticles on the surface of the beads. The comparison of images in Figure 5a,b showed that, after four consecutive cycles, the surface of the beads was slightly damaged by oxidizing conditions and UV-LED irradiation, which explained the leaching of the metal.

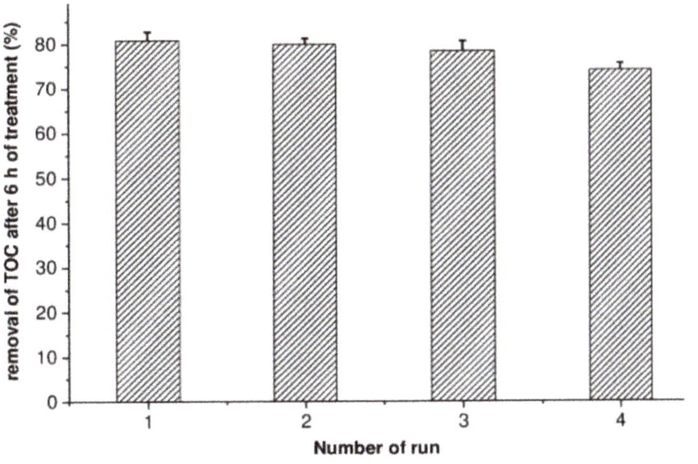

Figure 4. Stability of the catalytic activity of TiO_2/Fe_3O_4-CS(2) beads.

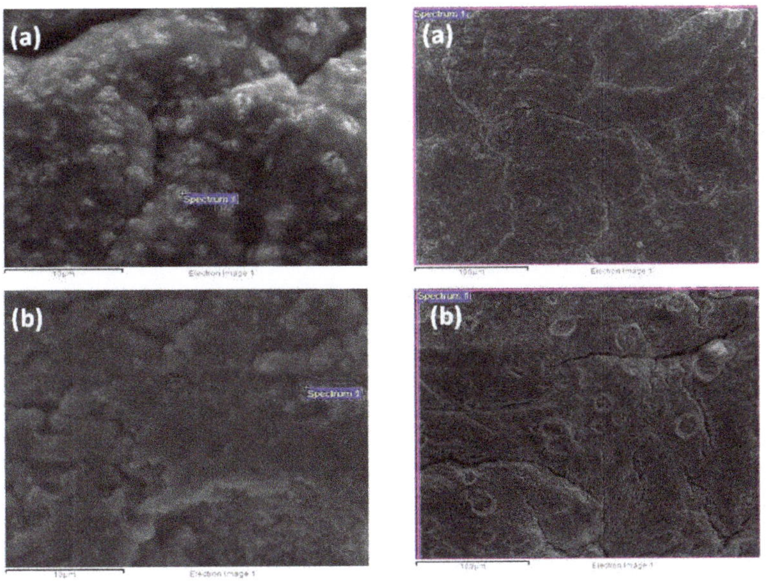

Figure 5. Morphology of catalyst: (**a**) TiO_2/Fe_3O_4-CS(2) before use, (**b**) TiO_2/Fe_3O_4-CS(2) after use.

XRD Analysis

In order to determine the crystalline structure of the TiO_2/Fe_3O_4-CS(2) beads, an XRD analysis was performed. In the XRD patterns (Figure 6a), the peak positions at 2Theta = 30.14°, 35.59°, 43.47°, 53.87°, 57.41° and 62.65° corresponded, respectively, to the Miller indices of (220), (311), (400), (422), (511) and (440) of the crystalline structure of Fe_3O_4, and the peak positions at 2Theta = 25.29°, 37.78°, 38.59°, 48.01°, 55.03°, 68.74°, 70,25°, 75.01°, 82.62° could be attributed to the diffraction planes of (110), (101), (004), (220), (105), (400), (220), (215) and (312) of the crystalline structure of TiO_2. The obtained peaks are well matched with standard data JCPDS-ICDD card N°:00-019-0629 for the Fe_3O_4 magnetite and 00-021-1272 for the TiO_2 anatase.

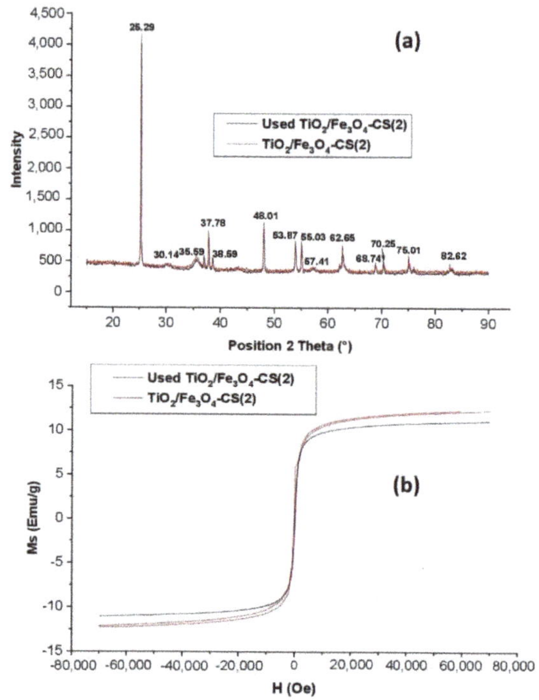

Figure 6. (**a**) X-ray diffraction patterns of TiO_2/Fe_3O_4-CS(2) beads, (**b**) Magnetization curves of the fresh and used TiO_2/Fe_3O_4-CS(2) beads.

The analysis of the catalytic beads used allowed us to obtain a diffractogram similar to that of fresh beads. In fact, indexing the pattern exhibited the crystal structure of the TiO_2 anatase (ICDD 00-021-1272). However, the peak positions at 2Theta = 30.14°, 35.59°, 43.47°, 48.01°, 55.03°, 57.41° and 62.65° corresponded, respectively, to the Miller indices of (220), (311), (400), (421), (422) (511) and (440) of the crystalline structure of maghemite Fe_2O_3 (ICDD 00-039-1346) instead of magnetite Fe_3O_4. Fe_2O_3 had the same spinel structure as magnetite, but it only contained iron in the trivalent Fe(III) state [68], which indicated the oxidation of Fe_3O_4. Furthermore, using the Debye–Scherrer equation, the grain size of TiO_2 was estimated at 64.5 nm, while the size of Fe_3O_4 was around 11.6 nm, indicating that the beads exhibit superparamagnetic behavior [69].

Magnetic Properties Analysis

The magnetization curve of the TiO_2/Fe_3O_4-CS(2) beads revealed a saturation intensity equal to 11.40 emu/g (Figure 6b). This low value is due to the presence of non-magnetic materials (CS and TiO_2) [70]. The magnetization intensity of the beads used was almost the same, and it was estimated at 10.48 emu g^{-1}, which means that the beads have a good stability of the magnetic properties after four consecutive cycles.

3. Materials and Methods

3.1. Chemical Products

All chemicals were of analytical–laboratory grade and applied without further purification. Chlordimeform, chitosan, sodium hydroxide, ferrous sulfate iron (II), iron (III) chloride, titanium oxide and potassium titanium oxide oxalate dehydrate were purchased from Sigma-Aldrich (Madrid, Spain). Acetic acid, sulfuric acid and nitric acid were supplied by Analar Normapur (Radnor, PA, USA). Acetonitrile (HPLC-grade) was purchased from Fisher Scientific (Loughborough, UK). Ultrapure water obtained through reverse osmosis technology (Basic 360) was utilized throughout all the experiments.

3.2. Preparation of TiO_2/Fe_3O_4-Chitosan Magnetic Beads

An eco-friendly, low-cost and simple approach was used to synthesize the composite catalysts. Magnetic TiO_2/Fe_3O_4-CS beads were prepared via a precipitation method using sodium hydroxide (1 M) as a precipitating agent. First, 2% CS gel solution was prepared by dissolving 1 g of CS in acetic acid (1%) under stirring at room temperature. After the total dissolution of CS flakes, 5 mmol of iron salts at a molar ratio Fe^{3+}:Fe^{2+} = 2:1 was added. Then, the TiO_2 nanoparticles were blended in the iron salts–CS gel solution. The amount of TiO_2 varied from 5 mmol, 10 mmol to 15 mmol, which corresponds, respectively, to the molar ratios TiO_2/Fe equal to 1, 2 and 3. By adding TiO_2, the color of the solution changed from orange to white, as shown in Figure 7A. Then, the mixture was dropped through a syringe into the hardening sodium hydroxide solution to create spherical CS gel beads. The beads were washed several times with deionizer water to remove any residual alkali and dried in an oven at 50 °C.

Figure 7. (**A**) Chitosan gel/Fe_3O solutions. (**B**) Wet chitosan beads at a molar ratio TiO_2/Fe: (**1**) 0; (**2**) 1; (**3**) 2 and (**4**) 3. (**C**) Superparamagnetic behavior of the TiO_{24}-CS beads at a molar ratio TiO_2/Fe = 2.

As seen in Figure 7B, the obtained beads are black for those containing iron only, and they take on a greenish coloration in the presence of TiO_2. All the prepared beads exhibited magnetic behavior in the presence of an external magnetic field (Figure 7C). The TiO_2-CS beads were prepared following the same alkaline co-precipitation method without addition of iron salts.

3.3. CDM Removal Assays

The PEF process of CDM degradation was performed in a cyclic mode, as shown in Figure 8. The process was composed of two reactors connected in series, allowing the treatment of (400 mL, 30 mgL^{-1}) CDM solution. A recirculation flow (200 mL min^{-1}) was set using a pump to connect the two reactors.

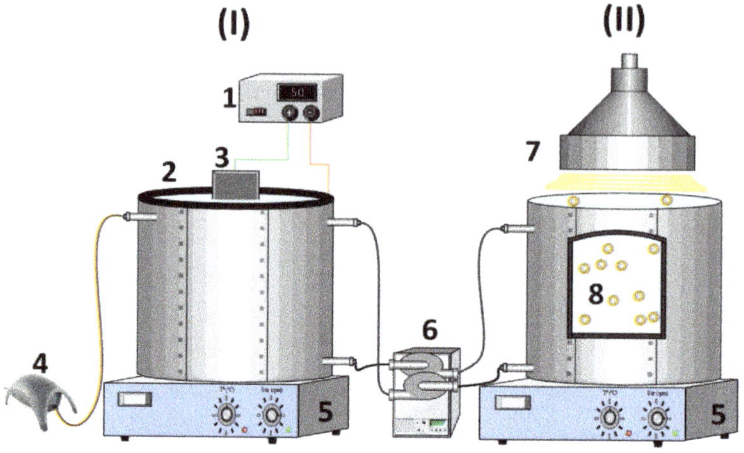

Figure 8. Cyclic electro-photocatalytic reactors: (**I**) electrochemical reactor: (**1**) power supply, (**2**) carbon felt electrode, (**3**) BDD electrode, (**4**) air supply, (**5**) magnetic stirrer, (**6**) peristaltic pump; (**II**) photocatalytic reactor: (**5**) magnetic stirrer, (**7**) UV-LED lamp, (**8**) catalytic beads.

The first reactor is electrochemical, composed of a glass beaker with a capacity of 250 mL, which permits the production of H_2O_2 by the reduction of dissolved oxygen on the cathode surface, which is a carbon felt (19.5 × 6 × 0.3 cm, Mersen, Barcelona, Spain) placed around the inner wall of the cylindrical cell, while the anode is a boron-doped diamond (BDD: 5 × 2.5 × 0.2 cm, Neocoat S.A.) placed in the middle of the cell. The system was run under a galvanostatic mode using an E 3512 A generator (Agilent, Santa Clara, CA, USA).

Na_2SO_4 was added previously as a support electrolyte at a concentration of 0.01 M, and the pH was adjusted to 3 using sulfuric acid solution. Air bubbling was maintained 15 min before the start of the reaction and through the treatment in order to saturate the medium with oxygen.

The second reactor is a cylindrical glass cell. A low-consumption UV-LED lamp (365 nm, 40 W, 550 lumens) from Luckylight electronics was placed above it, emitting at a wavelength equal to 365 nm. The catalyst was added into the photocatalytic reactor to promote its activation and avoid its electrochemical degradation, especially in the presence of a powerful oxidizing BDD anode in the first reactor.

To highlight the contribution of (photoelectrolysis + H_2O_2) and photocatalysis processes to the degradation of CDM, assays were conducted, respectively, without catalytic beads and in the absence of a current. Likewise, to evaluate the adsorption process contribution, another assay was conducted in the absence of UV irradiation and a current.

3.4. Analytical Methods

3.4.1. Determination of CDM Concentration

During experiments, the samples were collected and filtered prior to analysis through a 0.45 μm pore-size cellulose acetate membrane. CDM was quantified by HPLC (Agilent 1260 equipped with UV detector) with a C18 reverse-phase (4.6 mm × 250 mm, 5 μm; Agilent) column. The diode array detector was set at a fixed wavelength equal to 240 nm. The eluent was water/acetonitrile (60/40), with a flow rate of 1 mL min^{-1}.

3.4.2. Determination of Carboxylic Acids Concentrations

To identify and quantify the carboxylic acids generated during electrolysis, an HPLC was used with a diode array detector fixed at 206 nm. A Rezex™ ROA-Organic Acid H^+ (8%) (300 × 7.8 mm, i.e., 8 μm) column was used and placed in an oven at 60 °C. The eluent was 0.005 N H_2SO_4 solution pumped at a flow rate of 0.5 mL min^{-1}. The identification of carboxylic acids was based on a comparison of the retention times with those of pure standards.

3.4.3. Determination of Ions Concentrations

The ions generated were quantitatively followed by an ion chromatography system DIONEX ICS-3000. The separation of ions was performed by a Metrosep A Supp 5 column (4.0 × 250 mm). The eluent was 3.2 mmol L^{-1} Na_2CO_3 and 1 mmol L^{-1} $NaHCO_3$ at a flow rate of 0.7 mL min^{-1}. The limit of quantification (LOQ) of all chromatographic methods was 0.1 mg L^{-1}.

3.4.4. Total Organic Carbon Measurements

The total organic carbon (TOC) was measured via catalytic high-temperature combustion by multi N/C 3100 equipment (Analytic Jena, Germany) coupled with a non-dispersive infrared detector. The percentage removal of TOC was calculated using the following equation:

$$\% \text{ removal of TOC} = \frac{TOC_0 - TOC_t}{TOC_0} \times 100 \quad (4)$$

with TOC_0 and TOC_t representing the initial TOC and that at instant t.

3.4.5. Determination of Fe Concentration

The Fe concentration was determined by Inductively Coupled Plasma ICP (model: Optima 4300 DV Perkin Elmer Instruments). To obtain a metal solution from the heterogeneous catalyst, an acid digestion method was carried out using a concentrated nitric acid solution with the beads and placed on an autoclave at 121°C (Danish standard DS250).

3.4.6. Determination of H_2O_2 Concentration

The concentration of H_2O_2 was determined by a colorimetric method using a (Thermo Electron Corporation Helios) spectrophotometer because a titanium oxalate complexing agent can react with H_2O_2, producing a yellow peroxo–titanium complex, which absorbs at λ_{max} = 400 nm [71].

3.4.7. Characterization of the Synthesized Catalysts

The surface morphology of the catalytic beads was observed using a scanning electron microscope (JEOL JSM-6700F). The crystalline structure of the obtained catalysts was determined by X-Ray Diffraction (XRD, X'Pert PRO MPD). Finally, the Physical Properties Measurement System equipment (PPMS Ever Cool-II 9T) with a Vibrating Sample Magnetometer (VSM) at 298 K was used to explore the magnetic properties of the catalysts.

3.5. Specific Energy Consumption

The electric energy consumption (EEC) per unit volume of treated solution (kWh m^{-3}) was calculated according to the following equation [72]:

$$EEC\left(kWh\ m^{-3}\right) = \frac{I \times E \times t}{V} \quad (5)$$

where I is the current applied (A), E is the average cell voltage (V), t is the electrolysis time (h), and V is the solution volume (L or m^3).

4. Conclusions

In summary, magnetic TiO_2/Fe_3O_4-CS beads were synthesized following an easy in situ co-precipitation approach. The molar ratio TiO_2/Fe was studied, and the enhanced catalytic activity of TiO_2/Fe_3O_4-CS, with a molar ratio equal to 2, was probably due to the reduced recombination of charge carriers on the surface of the catalyst.

The performance of the PEF process using TiO_2/Fe_3O_4-CS(2) as a photocatalyst for the treatment of real wastewater doped with CDM insecticide was evaluated under optimal experimental conditions. Complete CDM removal was attained in 1 h, and more than 80% TOC abatement was achieved after 6 h of treatment, with simple carboxylic acids as the main by-product.

The catalytic activity of TiO_2/Fe_3O_4-CS(2) was satisfactorily validated in four consecutive cycles, and a slight decrease was obtained between the first and the fourth runs. Post-reaction catalyst characterization showed a high stability of magnetic properties despite the oxidation of Fe_3O_4 to Fe_2O_3, which is known for its good catalytic activity, sometimes similar to Fe_3O_4. Thus, the slight decrease in the catalytic activity could be attributed to the leaching of metals caused by the effect of the oxidizing conditions and UV-LED irradiation on catalytic beads.

This study offered a simple approach for constructing an eco-friendly, simple recovery and efficient bifunctional catalyst for advanced oxidation processes for treating recalcitrant organic pollutants in wastewater.

Author Contributions: Conceptualization, S.R.; Methodology, S.R. and A.M.D.; Validation, M.P. and M.Á.S.; Resources, M.P. and M.Á.S.; Writing—Original Draft Preparation, S.R.; Writing—Review and Editing, S.R., A.M.D., L.M., N.A., M.P. and M.Á.S.; Visualization, A.M.D., M.P. and M.Á.S.; Supervision, L.M., N.A., M.P. and M.Á.S.; Funding Acquisition, A.M.D., N.A., M.P. and M.Á.S. All authors have read and agreed to the published version of the manuscript.

Funding: This research was funded by Xunta de Galicia, grant number (ED481B 2019/091). This research was also funded by Project PID2020-113667GB-I00 464 funded by MCIN/AEI /10.13039/501100 011033. The authors are grateful to Xunta de Galicia for funding the researcher Aida Maria Díez Sarabia (ED481B 2019/091). This work was also financially supported by the Research Unit on Electrochemistry, Materials and Environment (UR16ES02) at the University of Kairouan in Tunisia.

Data Availability Statement: Data is contained within the article.

Conflicts of Interest: The authors declare no conflict of interest.

References

1. Malakootian, M.; Shahesmaeili, M.A.; Fraji, M.; Amiri, H.; Silva Martinez, S. Advanced oxidation processes for the removal of organophosphorus pesticides in aqueous matrices: A systematic review and meta-analysis. *Process. Saf. Environ. Prot.* **2020**, *134*, 292–307. [CrossRef]
2. Torres, N.H.; de Oliveira Santiago Santos, G.; Ferreira, L.F.R.; Américo-Pinheiro, J.H.P.; Eguiluz, K.I.B.; Salazar-Banda, G.R. Environmental aspects of hormones estriol, 17β-estradiol and 17α-ethinylestradiol: Electrochemical processes as next-generation technologies for their removal in water matrices. *Chemosphere* **2021**, *267*, 128888. [CrossRef] [PubMed]
3. Trellu, C.; Olvera Vargas, H.; Mousset, E.; Oturan, N.; Oturan, M.A. Electrochemical technologies for the treatment of pesticides. *Curr. Opin. Electrochem.* **2021**, *26*, 100677. [CrossRef]
4. Rajasekhar, B.; Venkateshwaran, U.; Durairaj, N.; Divyapriya, G.; Nambi, I.M.; Joseph, A. Comprehensive treatment of urban wastewaters using electrochemical advanced oxidation process. *J. Environ. Manag.* **2022**, *266*, 110469. [CrossRef] [PubMed]
5. Heidari, Z.; Pelalak, R.; Alizadeh, R.; Oturan, N.; Shirazian, S.; et Oturan, M.A. Application of Mineral Iron-Based Natural Catalysts in Electro-Fenton Process: A Comparative Study. *Catalysts* **2021**, *11*, 1–18. [CrossRef]
6. Ganzenko, O.; Trrellu, C.; Oturan, N.; Huguenot, D.; Péchaud, Y.; Van Hullebusch, E.D.; Oturan, M.A. Electro-Fenton treatment of a complex pharmaceutical mixture: Mineralization efficiency and biodegradability enhancement. *Chemosphere* **2020**, *253*, 126659. [CrossRef] [PubMed]
7. Hussain, S.; Aneggi, E.; et Goi, D. Catalytic activity of metals in heterogeneous Fenton-like oxidation of wastewater contaminants: A review. *Environ. Chem. Lett.* **2021**, *19*, 2405–2424. [CrossRef]
8. Guo, W.; Li, T.; Chen, Q.; Wan, J.; Zhang, J.; Wu, B.; Wang, Y. The roles of wavelength in the gaseous toluene removal with OH from UV activated Fenton reagent. *Chemosphere* **2021**, *275*, 129998. [CrossRef]
9. Babuponnusami, A.; Muthukumar, K. Advanced oxidation of phenol: A comparison between Fenton, electro-Fenton, sono-electro-Fenton and photo-electro-Fenton processes. *J. Chem. Eng.* **2012**, *183*, 1–9. [CrossRef]
10. Babuponnusami, A.; Muthukumar, K. Removal of phenol by heterogenous photo electro Fenton-like process using nano-zero valent iron. *Sep. Purif. Technol.* **2012**, *98*, 130–135. [CrossRef]
11. Chu, A.C.; Sahu, R.S.; Chou, T.H.; Shih, Y. Magnetic Fe_3O_4@TiO_2 nanocomposites to degrade bisphenol A, one emerging contaminant, under visible and long wavelength UV light irradiation. *J. Environ. Chem. Eng.* **2021**, *9*, 105539. [CrossRef]
12. Sun, Q.; Hong, Y.; Liu, Q.; Dong, L. Synergistic operation of photocatalytic degradation and Fenton process by magnetic Fe_3O_4 loaded TiO_2. *Appl. Surf. Sci.* **2018**, *430*, 399–406. [CrossRef]

13. Mukhtar, F.; Munawar, T.; Nadeem, M.S.; Rehman, M.N.; Khan, S.A.; Koc, M.; Batool, S.; Hasan, M.; Iqbal, F. Dual Z-scheme core-shell PANI-CeO_2-Fe_2O_3-NiO heterostructured nanocomposite for dyes remediation under sunlight and bacterial disinfection. *Environ. Res.* **2022**, *215*, 114140. [CrossRef] [PubMed]
14. Mukhtar, F.; Munawar, T.; Nadeem, M.S.; Khan, S.A.; Koc, M.; Batool, S.; Hasan, M.; Iqbal, F. Enhanced sunlight-absorption of Fe_2O_3 covered by PANI for the photodegradation of organic pollutants and antimicrobial inactivation. *Adv. Powder Technol.* **2022**, *33*, 103708. [CrossRef]
15. Goswami, A.; Jiang, J.-Q.; Petri, M. Treatability of five micro-pollutants using modified Fenton reaction catalyzed by zero-valent iron powder (Fe(0)). *J. Environ. Chem. Eng.* **2021**, *9*, 105393. [CrossRef]
16. Wang, Y.; Fang, J.; Crittenden, J.C.; Shen, C. Novel RGO/α-FeOOH supported catalyst for Fenton oxidation of phenol at a wide pH range using solar-light-driven irradiation. *J. Hazard. Mater.* **2017**, *239*, 321–329. [CrossRef]
17. Luo, H.; Zeng, Y.; He, D.; Pan, X. Application of iron-based materials in heterogeneous advanced oxidation processes for wastewater treatment: A review. *J. Chem. Eng.* **2021**, *407*, 127191. [CrossRef]
18. Dudchenko, N.; Pawar, S.; Perelshtein, I.; Fixler, D. Magnetite Nanoparticles: Synthesis and Applications in Optics and Nanophotonics. *Materials* **2022**, *15*, 2601. [CrossRef]
19. Nguyen, M.D.; Tran, H.-V.; Xu, S.; Lee, T.R. Fe_3O_4 Nanoparticles: Structures, Synthesis, Magnetic Properties, Surface Functionalization, and Emerging Applications. *Appl. Sci.* **2021**, *11*, 11301. [CrossRef]
20. Gopinath, K.P.; Madhav, V.M.; Krishnan, A.; Malolan, R.; Rangarajan, G. Present applications of titanium dioxide for the photocatalytic removal of pollutants from water: A review. *J. Environ. Manag.* **2020**, *270*, 110906. [CrossRef]
21. Mukhtar, F.; Munawar, T.; Nadeem, M.S.; Rehman, M.N.; Batool, S.; Hasan, M.; Riaz, M.; Rehman, K.; Iqbal, F. Highly efficient tri-phase TiO_2-Y_2O_3-V_2O_5 nanaocomposite: Structural, optical, photocatalyst and antibacterial studies. *J. Nanostruct. Chem.* **2022**, *12*, 547–564. [CrossRef]
22. Liu, C.; Dai, H.; Tan, C.; Pan, Q.; Hu, F.; Peng, X. Photo-Fenton degradation of tetracycline over Z-sheme Fe-g-C_3N_4/Bi_2WO_6 heterojunctions: Mechanism insight, degradation pathways and DFT calculation. *Appl. Catal. B* **2022**, *310*, 121326. [CrossRef]
23. Sun, L.; Zhou, Q.; Mao, J.; Ouyang, X.; Yuan, Z.; Song, X.; Gong, W.; Mei, S.; Xu, W. Study on Photocatalytic Degradation of Acid Red 73 by Fe_3O_4@TiO_2 Exposed (001) Facets. *Appl. Sci.* **2022**, *12*, 3574. [CrossRef]
24. Rezgui, S.; Díez, A.M.; Monser, L.; Adhoum, N.; Pazos, M.; et Sanromán, M.A. $ZnFe_2O_4$-chitosan magnetic beads for the removal of chlordimeform by photo-Fenton process under UVC irradiation. *J. Environ. Manag.* **2021**, *283*, 111987. [CrossRef] [PubMed]
25. Rezgui, S.; Amrane, A.; Fourcade, F.; Assadi, A.; Monser, L.; Adhoum, N. Electro-Fenton catalyzed with magnetic chitosan beads for the removal of Chlordimeform insecticide. *Appl. Catal. B* **2018**, *226*, 346–359. [CrossRef]
26. Lee, M.; Chen, B.-Y.; Den., W. Chitosan as a Natural Polymer for Heterogeneous Catalysts Support: A. Short Review on Its Applications. *Appl. Sci.* **2015**, *5*, 1272–1283. [CrossRef]
27. Zhang, W.; Jia, S.; Wu, Q.; Wu, S.; Ran, J.; Lui, Y.; Hou, J. Studies of the magnetic field intensity on the synthesis of chitosan-coated magnetite nanocomposites by co-precipitation method. *Mater. Sci. Eng. C* **2012**, *32*, 381–384. [CrossRef]
28. Donadel, K.; Felisberto, M.D.V.; Fávere, V.T.; Rigoni, M.; Batistela, N.J.; Laranjeira, C.M.C. Synthesis and characterization of the iron oxide magnetic particles coated with chitosan biopolymer. *Mater. Sci. Eng. C* **2008**, *28*, 509–514. [CrossRef]
29. Moura, F.C.C.; Araujo, M.H.; Costa, R.C.C.; Ardisson, J.D.; Macedo, W.A.A.; Lago, R.M. Efficient use of Fe metal as an electron transfer agent in a heterogeneous Fenton system based on Fe^0/Fe_3O_4 composites. *Chemosphere* **2005**, *60*, 1118–1123. [CrossRef]
30. Sun, Q.; Leng, W.; Li, Z.; Xu, Y. Effect of surface Fe_2O_3 clusters on the photocatalytic activity of TiO_2 for phenol degradation in water. *J. Hazard. Mater.* **2012**, *229–230*, 224–232. [CrossRef]
31. Afzal, S.; Julkapli, N.M.; Mun, L.K. Visible light active TiO_2/CS/Fe_3O_4 for nitrophenol degradation: Studying impact of TiO_2, CS and Fe_3O_4 loading on the optical and photocatalytic performance of nanocomposite. *Mater. Sci. Semicond. Process.* **2021**, *131*, 105891. [CrossRef]
32. Díez, A.M.; Pazos, M.; Sanromán, M.A. Synthesis of magnetic-photo-Fenton catalyst for degradation of emerging pollutant. *Catal. Today* **2019**, *328*, 267–273. [CrossRef]
33. Li, Y.; Cheng, H. Chemical kinetic modeling of organic pollutant degradation in Fenton and solar photo-Fenton processes. *J. Taiwan Inst. Chem. Eng.* **2021**, *123*, 175–184. [CrossRef]
34. Burbano, A.A.; Dionysiou, D.D.; Suidan, M.T.; Richardson, T.L. Oxidation kinetics and effect of pH on the degradation of MTBE with Fenton reagent. *Water Res.* **2016**, *39*, 107–118. [CrossRef] [PubMed]
35. Molamahmood, H.V.; Geng, W.; Wei, Y.; Miao, J.; Yu, S.; Shahi, A.; Chen, C.; Long, M. Catalyzed H_2O_2 decomposition over iron oxides and oxyhydroxides: Insights from oxygen production and organic degradation. *Chemosphere* **2022**, *291*, 133037. [CrossRef]
36. Li, Q.; Kong, H.; Jia, R.; Shao, J.; He, Y. Enhanced catalytic degradation of amoxicillin with TiO_2–Fe_3O_4 composites via a submerged magnetic separation membrane photocatalytic reactor (SMSMPR). *RSC Adv.* **2019**, *9*, 12538–12546. [CrossRef]
37. Li, Q.; Kong, H.; Li, P.; Shao, J.; He, Y. Photo-Fenton degradation of amoxicillin via magnetic TiO_2-graphene oxide-Fe_3O_4 composite with a submerged magnetic separation membrane photocatalytic reactor (SMSMPR). *J. Hazard. Mater.* **2019**, *373*, 437–446. [CrossRef]
38. Bai, X.; Lyu, L.; Ma, W.; Ye, Z. Heterogeneous UV/Fenton degradation of bisphenol A catalyzed by synergistic effects of $FeCo_2O_4$/TiO_2/GO. *Environ. Sci. Pollut. Res.* **2016**, *23*, 22734–22743. [CrossRef]
39. Nwe, T.S.; Sikong, L.; Kokoo, R.; Khangkhamano, M. Photocatalytic activity enhancement of Dy-doped TiO_2 nanoparticles hybrid with TiO_2 (B) nanobelts under UV and fluorescence irradiation. *Curr. Appl. Phys.* **2016**, *20*, 249–254. [CrossRef]

40. Nam, Y.; Lim, J.H.; Ko, K.C.; Lee, J.K. Photocatalytic activity of TiO_2 nanoparticles: A theoretical aspect. *J. Mater. Chem. A* **2019**, *7*, 13833–13859. [CrossRef]
41. Enesca, A.; Isac, L. The Influence of Light Irradiation on the Photocatalytic Degradation of Organic Pollutants. *Materials* **2020**, *11*, 2494. [CrossRef] [PubMed]
42. Boruah, P.K.; Das, M.R. Dual responsive magnetic Fe_3O_4-TiO_2/graphene nanocomposite as an artificial nanozymes for the colorimetric detection and photodegradation of pesticide in an aqueous medium. *J. Hazard. Mater.* **2020**, *385*, 121516. [CrossRef] [PubMed]
43. Pourzad, A.; Sobhi, H.R.; Behbahani, M.; Esrafili, A.; Kalantary, R.R.; Kermani, M. Efficient visible light-induced photocatalytic removal of paraquat using N-doped TiO_2@SiO_2@Fe_3O_4 nanocomposite. *J. Mol. Liq.* **2020**, *299*, 112467. [CrossRef]
44. Zahedifar, M.; Seyedi, N. Bare 3D-TiO_2/magnetic biochar dots (3D-TiO_2/BCDs MNPs): Highly efficient recyclable photocatalyst for diazinon degradation under sunlight irradiation. *Phys. E Low-Dimens. Syst. Nanostruct.* **2022**, *139*, 115151. [CrossRef]
45. Hua, Y.; Wang, S.; Xiao, J.; Cui, C.; Wang, C. Preparation and characterization of Fe_3O_4/gallic acid/graphene oxide magnetic nanocomposites as highly efficient Fenton catalysts. *RSC Adv.* **2017**, *7*, 28979–28986. [CrossRef]
46. Kasiri, M.B.; Aleboyeh, H.; Aleboyeh, A. Degradation of Acid Blue 74 using Fe-ZSM5 zeolite as a heterogeneous photo-Fenton catalyst. *Appl. Catal. B* **2008**, *84*, 9–15. [CrossRef]
47. Tekbaş, M.; Yatmaz, H.C.; Bektaş, N. Heterogeneous photo-Fenton oxidation of reactive azo dye solutions using iron exchanged zeolite as a catalyst. *Microporous Mesoporous Mater.* **2008**, *115*, 594–602. [CrossRef]
48. Jiang, B.; Niu, Q.; Li, C.; Oturan, N.; Oturan, M.A. Outstanding performance of electro-Fenton process for efficient decontamination of Cr(III) complexes via alkaline precipitation with no accumulation of Cr(VI): Important roles of iron species. *Appl. Catal. B* **2020**, *272*, 119002. [CrossRef]
49. Hammouda, S.B.; Amrane, A.; Fourcade, F.; Assadi, A.; Adhoum, N.; Monser, L. Effective heterogeneous electro-Fenton process for the degradation of a malodorous compound indole using iron loaded alginate beads as a reusable catalyst. *Appl. Catal. B* **2016**, *182*, 47–58. [CrossRef]
50. Liu, D.; Zhang, H.; Wei, Y.; Liu, B.; Lin, Y.; Li, G.; Zhang, F. Enhanced degradation of ibuprofen by heterogeneous electro-Fenton at circumneutral pH. *Chemosphere* **2018**, *209*, 998–1006. [CrossRef]
51. Chmayssem, A.; Taha, S.; Hauchard, D. Scaled-up electrochemical reactor with a fixed bed three-dimensional cathode for electro-Fenton process: Application to the treatment of bisphenol A. *Electrochim. Acta* **2017**, *225*, 435–442. [CrossRef]
52. Nidheesh, P.V.; Gandhimathi, R. Trends in electro-Fenton process for water and wastewater treatment: An overview. *Desalination* **2012**, *299*, 1–15. [CrossRef]
53. Kang, B.; Dai, Y.; Zhang, H.; Chen, D. Synergetic degradation of chitosan with gamma radiation and hydrogen peroxide. *Polym. Degrad. Stab.* **2007**, *92*, 359–362. [CrossRef]
54. Wang, S.-M.; Huang, Q.-Z.; Wang, Q.-S. Study on the synergetic degradation of chitosan with ultraviolet light and hydrogen peroxide. *Carbohydr. Res.* **2005**, *340*, 1143–1147. [CrossRef]
55. Zazouli, M.A.; Ghanbari, F.; Yousefi, M.; Madihi-Bidgoli, S. Photocatalytic degradation of food dye by Fe_3O_4–TiO_2 nanoparticles in presence of peroxymonosulfate: The effect of UV sources. *J. Environ. Chem. Eng.* **2017**, *5*, 2459–2468. [CrossRef]
56. Maktabifard, M.; Zaborowska, E.; Makinia, J. Achieving energy neutrality in wastewater treatment plants through energy savings and enhancing renewable energy production. *Rev. Environ. Sci. Biotechnol.* **2018**, *17*, 655–689. [CrossRef]
57. Ghazouani, M.; Bousselmi, L.; Akrout, H. Combined electrocoagulation and electrochemical treatment on BDD electrodes for simultaneous removal of nitrates and phosphates. *J. Environ. Chem. Eng.* **2020**, *8*, 104509. [CrossRef]
58. Ghanbari, F.; Moradi, M.A. comparative study of electrocoagulation, electrochemical Fenton, electro-Fenton and peroxi-coagulation for decolorization of real textile wastewater: Electrical energy consumption and biodegradability improvement. *J. Environ. Chem. Eng.* **2015**, *3*, 499–506. [CrossRef]
59. Gerek, E.E.; Yılmaz, S.; Koparal, A.S.; Gerek, Ö.N. Combined energy and removal efficiency of electrochemical wastewater treatment for leather industry. *J. Water Process. Eng.* **2019**, *30*, 100382. [CrossRef]
60. Flores, N.; Sirés, I.; Garrido, J.A.; Centellas, F.; Rodriguez, R.M.; Cabot, P.L.; Brillas., E. Degradation of trans-ferulic acid in acidic aqueous medium by anodic oxidation, electro-Fenton and photoelectro-Fenton. *J. Hazard. Mater.* **2016**, *319*, 3–12. [CrossRef]
61. Hammami, S.; Oturan, N.; Bellakhal, N.; Dachraoui, M.; Oturan, M.A. Oxidative degradation of direct orange 61 by electro-Fenton process using a carbon felt electrode: Application of the experimental design methodology. *J. Electroanal. Chem.* **2007**, *610*, 75–84. [CrossRef]
62. Aplin, R.; Feitz, A.J.; Waite, T.D. Effect of Fe(III)-ligand properties on effectiveness of modified photo-Fenton processes. *Water Sci. Technol.* **2001**, *44*, 23–30. [CrossRef] [PubMed]
63. Zazou, H.; Oturan, N.; Zhang, H.; Hamdani, M.; Oturan, M.A. Comparative study of electrochemical oxidation of herbicide 2,4,5-T: Kinetics, parametric optimization and mineralization pathway. *Sustain. Environ. Res.* **2017**, *27*, 15–23. [CrossRef]
64. Mhemdi, A.; Oturan, M.A.; Oturan, N.; Abdelhédi, R.; Ammar, S. Electrochemical advanced oxidation of 2-chlorobenzoic acid using BDD or Pt anode and carbon felt cathode. *J. Electroanal. Chem.* **2013**, *709*, 111–117. [CrossRef]
65. Skoumal, M.; Arias, C.; Cabot, P.L.; Centellas, F.; Garrido, J.A.; Rodriguez, R.M.; Brillas, E. Mineralization of the biocide chloroxylenol by electrochemical advanced oxidation processes. *Chemosphere* **2008**, *71*, 1718–1729. [CrossRef]
66. Arai, M.; Zhao, F. Metal Catalysts Recycling and Heterogeneous/Homogeneous Catalysis. *Catalysts* **2015**, *5*, 868–870. [CrossRef]

67. Thomas, N.; Dionysiou, D.D.; Pillai, S.C. Heterogeneous Fenton catalysts: A review of recent advances. *J. Hazard. Mater.* **2021**, *404*, 124082. [CrossRef]
68. Rusevova, K.; Kopinke, F.-D.; Georgi, A. Nano-sized magnetic iron oxides as catalysts for heterogeneous Fenton-like reactions—Influence of Fe(II)/Fe(III) ratio on catalytic performance. *J. Hazard. Mater.* **2012**, *241–242*, 433–440. [CrossRef]
69. Gijs, M.A.M. Magnetic bead handling on-chip: New opportunities for analytical applications. *Microfluid. Nanofluid.* **2004**, *1*, 22–40. [CrossRef]
70. Li, Y.; Zhang, M.; Guo, M.; Wang, X. Preparation and properties of a nano TiO_2/Fe_3O_4 composite superparamagnetic photocatalyst. *Rare Metals* **2009**, *28*, 423–427. [CrossRef]
71. Almuaibed, A.M.; Townshend, A. Flow spectrophotometric method for determination of hydrogen peroxide using a cation exchanger for preconcentration. *Anal. Chim. Acta* **1994**, *295*, 159–163. [CrossRef]
72. Martínez-Huitle, C.A.; Brillas, E. Decontamination of wastewaters containing synthetic organic dyes by electrochemical methods: A general review. *Appl. Catal. B* **2009**, *87*, 105–145. [CrossRef]

catalysts

MDPI

Article

Photocatalytic Activity of ZnO/Ag Nanoparticles Fabricated by a Spray Pyrolysis Method with Different O_2:N_2 Carrier Gas Ratios and Ag Contents

Meditha Hudandini [1], Nurdiana Ratna Puri [2], Sugeng Winardi [2], Widiyastuti Widiyastuti [2], Manabu Shimada [1] and Kusdianto Kusdianto [2,*]

1 Chemical Engineering Program, Graduate School of Advanced Science and Engineering, Hiroshima University, 4-1, Kagamiyama 1-Chome, Higashi-Hiroshima 739-8527, Hiroshima, Japan

2 Chemical Engineering Department, Institut Teknologi Sepuluh Nopember, Kampus ITS Sukolilo, Surabaya 60111, Indonesia

* Correspondence: kusdianto@chem-eng.its.ac.id

Abstract: Wastewaters of the textile industry, e.g., those generated in Gresik, Indonesia, are a possible threat to the environment and should be treated before disposal. Photodegradation is a more promising method to overcome this problem than conventional methods such as biodegradation. ZnO is widely used for photodegradation due to its unique physical and chemical properties and stability. In this study, Ag was loaded onto ZnO, which is non-toxic and inexpensive, can improve the electron–hole separation, and has a significant catalytic potential. Pristine ZnO and ZnO-Ag nanoparticles were fabricated by an ultrasonic spray pyrolysis system at different Ag contents (1, 5, and 10 wt%). The carrier gas ratio (O_2:N_2) was also changed (1:0, 1:2, 1:1, 2:1, and 0:1) to examine its effects on the nanoparticle characteristics. The nanoparticle characteristics were examined using X-ray diffraction (XRD), scanning electron microscopy (SEM), transmission electron microscopy (TEM), and Brunauer, Emmett, and Teller (BET) specific surface area. The results were interpreted in relation to photodegradation under UV light irradiation. An increase in the ZnO-Ag activity compared with pristine ZnO was observed at a carrier gas ratio of 0:1 with reaction rate constants of 0.0059 and 0.0025 min^{-1}, respectively.

Keywords: chemical vapor deposition (CVD); gas-phase; photodecomposition; specific surface area; nanocomposite

Citation: Hudandini, M.; Puri, N.R.; Winardi, S.; Widiyastuti, W.; Shimada, M.; Kusdianto, K. Photocatalytic Activity of ZnO/Ag Nanoparticles Fabricated by a Spray Pyrolysis Method with Different O_2:N_2 Carrier Gas Ratios and Ag Contents. *Catalysts* **2022**, *12*, 1374. https://doi.org/10.3390/catal12111374

Academic Editors: Jorge Bedia and Carolina Belver

Received: 3 October 2022
Accepted: 3 November 2022
Published: 5 November 2022

1. Introduction

The growth in the textile industry with its large water consumption impacts the environment because wastewater effluents contain large quantities of organic pollutions [1]. These compounds have highly toxic effects on living organisms [2]. Advanced oxidation processes (AOP) by photocatalysts have been considered for the treatment of dye-polluted wastewaters [1,3,4]. The advantages of AOP include (i) fast degradation of pollutants, (ii) complete mineralization of harmful organic wastes, (iii) operation at ambient temperature and pressure, and (iv) reduction of the toxicity of organic compounds [5]. The process of AOP mainly involves the generation of reactive hydroxyl radicals and its further functionalization to degrade organic compounds. Currently, most research focuses on photocatalytic oxidation using materials such as semiconductors [1,6].

ZnO (band gap energy = 3.37 eV) can degrade harmful organic compounds [3]. It is a semiconductor that is low-cost, is environmentally safe, and has good physical and chemical stability [2,7]. In addition, it has a high surface area and high photosensitivity [8]. However, the fast recombination of the excited electrons (e−) and holes (h+) due to its large band gap energy significantly limits the utilization of ZnO [7,8]. Nevertheless, several studies reported several methods, such as doping with metals, e.g., Au [8,9], Ag [8,10,11], and Pt [10], to increase the functionality of ZnO. These metals can suppress the recombination of electrons and holes by acting as an electron sink for the excited electrons from the valence band of ZnO [8]. In this study, ZnO was loaded with Ag because of its photocatalytic potential, low-cost, and non-toxicity [2]. ZnO-Ag for photodegradation of organic pollutants has also been extensively studied because of its good performance [11].

Nanocomposite fabrication can be obtained by several methods in the liquid or gas phase. However, the liquid-phase method involves more steps than the gas-phase method. Gas-phase methods include spray pyrolysis, sputtering, and electron-beam evaporation [7]. In this study, ultrasonic spray pyrolysis is used for nanocomposite synthesis. This is a single-step fabrication process that results in a better dispersion of the produced nanomaterials, which further enhances its activity [12]. This method consists of generation of the droplets by a nebulizer followed by solvent evaporation, precursor chemical reaction, and densification of the formed particles [9,13].

Here, ultrasonic spray pyrolysis is used to fabricate pristine ZnO and ZnO-Ag nanoparticles. The process was optimized to produce desirable nanoparticles for the degradation of organic compounds. The nanoparticle formation or characteristics of the fabricated ZnO nanoparticles is affected by several parameters such as the precursor concentration [14], carrier gas flow rate [14,15], type of carrier gas used [16], and furnace temperature [14,17]. Several studies have used spray pyrolysis with different carrier gases. Different carrier gases generate different structures and properties of the nanocomposites. Air and nitrogen as carrier gases generated ZnO nanoparticles with different morphologies and resistivities [16]. Other studies that used materials such as ZnO/Mn, fabricated using oxygen, nitrogen, or air as the carrier gas, showed that the carrier gas type affected the magnetic characteristics of the nanocomposites, which is significantly affected by lower system temperatures [18]. Furthermore, other semiconductors, such as α-Ga_2O_3, was fabricated using different carrier gases. A carrier gas containing a high oxygen concentration generated higher crystallinity and better film quality [19]. Previous studies have shown the effect of different carrier gases used toward the characteristics of the produced nanoparticles. These effects may be further attributed to the reaction conditions based on the type of gas used. Air and oxygen create an oxidizing atmosphere for the reactions, whereas gases such as Ar or N_2 provide an inert reaction atmosphere. Thus, these two types of carrier gases produce different characteristics of the nanoparticles [13].

To the best of our knowledge, the effect of the carrier gas on the characteristics and photocatalytic activity of ZnO-Ag nanocomposites has not been explored in detail. This work provides a close examination of the characteristics of the produced nanoparticles fabricated using an ultrasonic spray pyrolysis method at different carrier gas ratios (oxygen to nitrogen). The morphology was observed using scanning electron microscopy (SEM) and transmission electron microscopy (TEM), and the elemental analysis was observed using energy dispersive spectroscopy (EDS). Furthermore, the crystallinity and crystallite size were investigated by X-ray diffraction (XRD), and the specific surface area (SBET) was studied by the nitrogen adsorption isotherm using the Brunauer Emmett and Teller (BET) method. The correlation of these characteristics with the carrier gas ratio (O_2:N_2) and the Ag contents (0, 1, 5, and 10 wt%) was investigated. Furthermore, the photocatalytic degradation of organic compounds in actual textile wastewaters using the produced nanoparticles was examined under UV light irradiation.

2. Results and Discussion

2.1. Morphological Analysis of the Produced Nanoparticles

Figure 1a–h shows the morphology of ZnO and ZnO-Ag nanoparticles synthesized by ultrasonic spray pyrolysis. The figure shows crumpled spherical secondary submicron particles. No significant morphological change is observed at different carrier gas ratios and Ag contents. The cumulative size distribution (Figure 1i,j) of the secondary submicron nanoparticles was also measured. The increase in the Ag content does not affect the size distribution; however, the diameters of the secondary particles slightly increased with the increase in the nitrogen carrier gas ratio. The formation of these nanoparticles occurred during the evaporation of the solvent droplets in the aerosolized precursor followed by compression and decomposition of the precursor in the tubular furnace [20]. These morphological results agree well with previous studies. The formation of the nanoparticles and the obtained morphology are affected by several parameters such as the evaporation rate of the solvent and concentration of the precursor [12,21]. Upon evaporation, the confinement force, which compresses the droplets within the pathway, affects the morphology and size of the nanoparticles. The confinement force is proportional to the evaporation rate, which is significantly affected by the furnace temperature. A furnace temperature of 400 °C is sufficiently high to cause a significant water (solvent)-evaporation rate and the crumpled spherical structure of the nanoparticles. Furthermore, the precursor concentration is proportional to the confinement force and particle size. However, the confinement force is inversely proportional to the particle size. Thus, appropriate concentrations should be used to produce the desired morphology and size of nanoparticles.

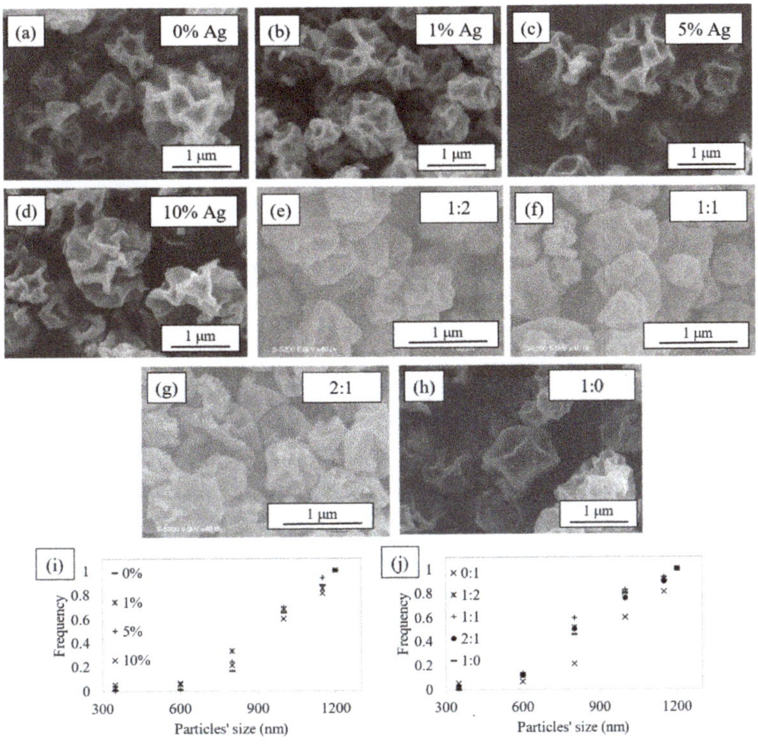

Figure 1. Morphology of (**a**) pristine ZnO, (**b–d**) ZnO-Ag with Ag content of (**b**) 1 wt%, (**c**) 5 wt%, (**d**) 10 wt% fabricated at an O_2:N_2 ratio of 0:1, (**e–h**) ZnO-Ag with Ag content of 10 wt% fabricated at different O_2:N_2 ratios, and (**i**,**j**) a cumulative size distribution of the nanoparticles.

TEM images (Figure 2a,b) show that the crumple-shaped spheres consisted of smaller primary nanoparticles. The EDS results (Figure 2c,d) indicated the presence of Zn, O, and Ag. In pristine ZnO, the ZnLα, ZnKα, ZnKβ, and OKα peaks confirm the presence of ZnO. The AgLα peak confirms the presence of Ag nanoparticles. The CuKα and CuKβ peaks correspond to the TEM grid that was used for observations in TEM measurements, and no impurities are observed in the samples.

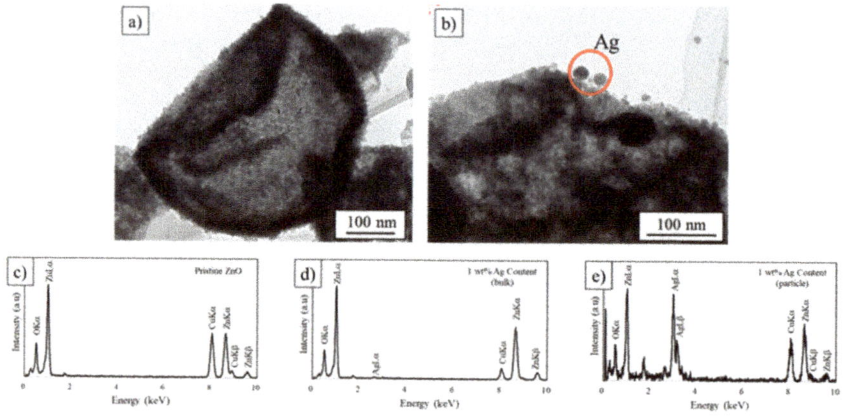

Figure 2. TEM-EDS of (**a**,**c**) Pristine ZnO and (**b**,**d**,**e**) ZnO-Ag nanoparticles at Ag content of 1 wt%.

2.2. Analysis of the Crystal Structure of the Produced Nanoparticles

The crystallization of the nanoparticles was observed using selected area electron diffraction (SAED) of ZnO-Ag (Figure 3a). The figure shows rings corresponding to the (100), (101), (102), and (110) lattice planes of wurtzite ZnO. The high-resolution TEM (HRTEM) image (Figure 3b) confirms the presence of the Ag nanoparticles and with a (111) crystal plane.

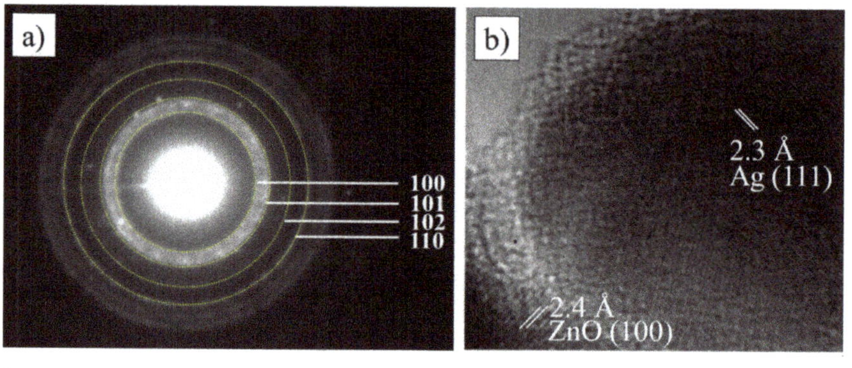

Figure 3. (**a**) SAED and (**b**) HR-TEM of ZnO-Ag nanoparticles at a Ag content of 1 wt%.

Hexagonal ZnO was identified by the XRD peaks (Figure 4) at $2\theta = 31.5°$, 33.9°, 35.8°, and 56.5° that correspond to the (100), (002), (101), and (110) crystal planes, respectively [6]. Zn peaks are observed with high nitrogen ratios (Figure 4a). During fabrication, zinc acetate dihydrate was used as the precursor, which decomposes to form Zn vapor at 400 °C, and when enough oxygen is present, ZnO is formed [22]. The Ag peak could only be observed at high Ag contents, either for oxygen or nitrogen-rich ratios, with the FCC peak of Ag observed at $2\theta = 37.82°$ for the (111) crystal plane [6,23]. These XRD peaks prove the existence and successful formation of crystallized ZnO and Ag nanoparticles from the decomposition of zinc acetate dihydrate and silver nitrate by spray pyrolysis for both nitrogen- and oxygen-rich ratios.

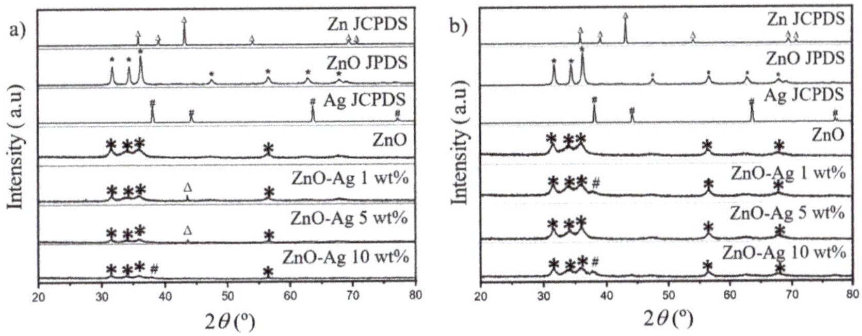

Figure 4. Crystalline phase of ZnO and ZnO-Ag fabricated at a carrier gas (O_2:N_2) ratio of (**a**) 0:1 and (**b**) 1:0. (symbols correspond to Δ Zn, * ZnO, and # Ag).

The crystallite size of ZnO was calculated using Scherrer's Equation (1). Figure 5 indicates that the crystallite size in a nitrogen-rich system that has not significantly changed with the addition of Ag to the ZnO nanoparticles. The crystallite size continues to increase with the increase in the oxygen ratio of the carrier gas. However, at a higher oxygen ratio, the crystallite size of pristine ZnO and the nanocomposite decreases. This is in accordance with the report by Li, et al. [24], such that for annealed ZnO under air, N_2, O_2 and vacuum following the grain size for air > N_2 > O_2 > vacuum, the existence of oxygen in the system was reported to suppress the grain growth by pinning the grain boundaries. At higher oxygen ratios, however, pristine ZnO nanoparticles have a slightly lower crystallite size than the ZnO-Ag nanocomposites. In this study, the presence of Ag nanoparticles affects the increase in the crystallite size of ZnO nanoparticles, especially at high oxygen ratios in the carrier gas. The increase in ZnO size with the addition of Ag in the oxygen-rich system can be explained by the more prominent effects of Ag doping toward ZnO in the oxygen, which may influence the crystal of ZnO nanoparticles [25]. Instead of following the significant decrease in size of pristine ZnO in higher oxygen ratios, the addition of Ag gave a higher crystallite size compared to pristine ZnO.

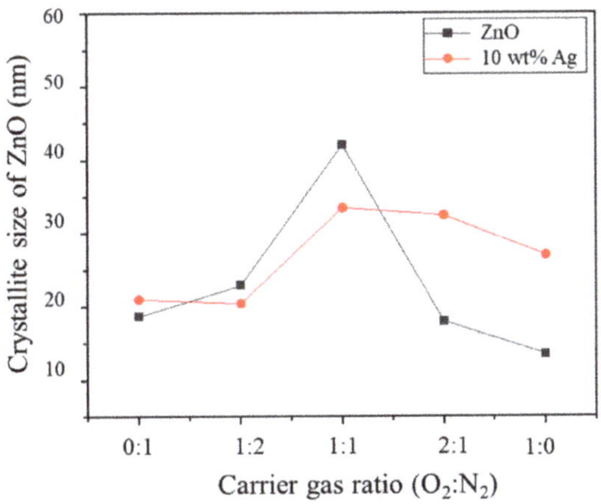

Figure 5. Crystallite size of ZnO at different carrier gas ratios and Ag contents.

2.3. *Surface Area of the Produced Nanoparticles*

The SBET of the nanoparticles was obtained using a nitrogen adsorption BET method. The surface area of the nanoparticles can significantly affect their photocatalytic activity. Tightly packed structure of the nanoparticles has a high surface area and free volume, which significantly increase their photocatalytic activity [21]. High surface area increases the contact between the active sites of the nanoparticles with pollutants, which can optimize the degradation of the organic compounds [12]. Figure 6 indicates that, except for samples with 5 wt% Ag content, SBET decreases with the increase in the oxygen ratio. The largest SBET is measured at a Ag content of 10 wt% using only nitrogen as the carrier gas. In general, a pure nitrogen carrier gas exhibits a larger SBET than that of samples produced with the mixture or pure oxygen carrier gas. However, this is not consistent with the size change of the secondary nanoparticles, of which the size increase should have decreased SBET. This difference in the results can be attributed to the crumpled sphere shape of the nanoparticles developed in this study.

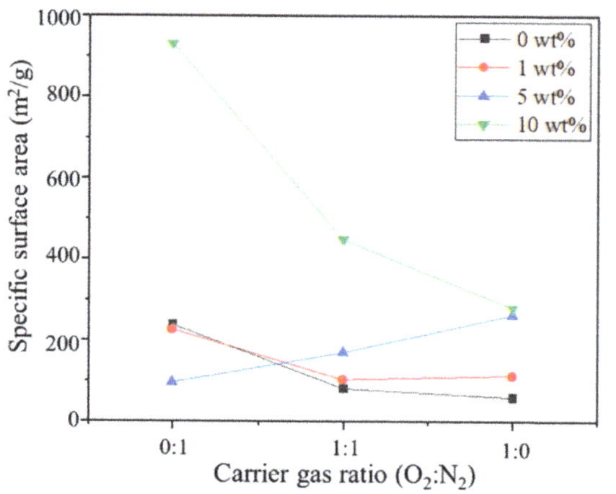

Figure 6. Specific surface area of ZnO and ZnO-Ag at different Ag contents and carrier gas (O_2:N_2) ratios.

Furthermore, the difference in SBET can be explained by the mechanism of the nanoparticle growth in the tubular furnace, in which the carrier gas atmosphere is expected to change the average temperature in the chamber and the evaporation rate of the solution [16]. Nitrogen has lower thermal conductivity than oxygen, which results in a lower evaporation rate, leading to the formation of non-aggregated nanoparticles. Aggregated nanoparticles tend to have lower surface area than non-aggregated nanoparticles.

2.4. Photocatalytic Activity of Pristine ZnO and ZnO-Ag at Different Carrier Gas Ratios

The photocatalytic activity of pristine ZnO and ZnO-Ag nanoparticles was studied for the degradation of textile wastewater under UV light irradiation. The process involves the decomposition of organic pollutants in the wastewater to less harmful compounds or minerals [26]. Figure 7 illustrates the effect of Ag content and type of carrier gas on the degradation reaction rate constant. Different Ag contents (0–10 wt%) and carrier gas ratios exhibited similar tendencies of rate constants, where the pure nitrogen carrier gas has the highest photocatalytic activity in all samples. As previously discussed, the carrier gas used in the system may affect several characteristics of the produced nanoparticles. The increase in the nitrogen ratio in the system increases the nanoparticle SBET. High SBET values can enhance the photocatalytic activity. Zn metal is also observed in the nanoparticles of the systems with higher nitrogen ratios. The synergetic effect of Zn can suppress the electron–hole recombination and further enhance the photocatalytic activity compared to pristine ZnO nanoparticles [27]. At 10 wt%, the effect of the carrier gases on the photocatalytic activity is more pronounced than at lower Ag contents. However, the reason for this phenomenon is still not clear. Thus, the Ag content and carrier gas type have a significant effect on the photocatalytic activity of the developed nanoparticles.

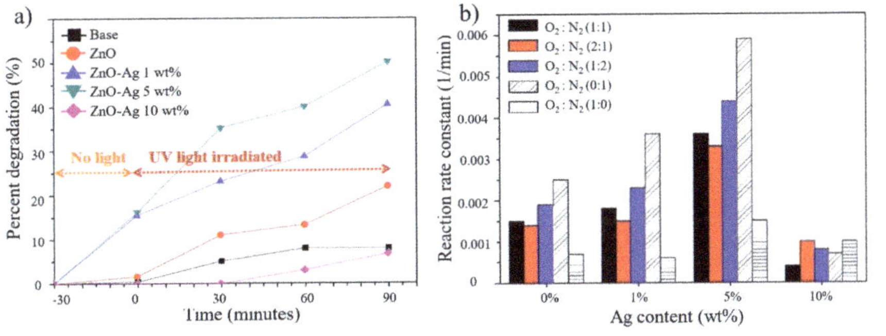

Figure 7. (**a**) Photocatalytic degradation percentage of the pollutants using ZnO-Ag synthesized by pure nitrogen as the carrier gas and (**b**) reaction rate constant of the produced nanoparticles under UV light irradiation.

The absorbance of the prepared samples with the mixed carrier gases and Ag contents were observed (Figure 8). The absorbance can be used to estimate the band gap energy (Figure 9) from $(\alpha h v)^n = A(hv - E_g)$, where α, hv, A, and Eg are the absorption, photon energy, constant, and band gap, respectively [28]. The absorbance measurement showed good response to the shorter wavelength of light, which indicated better performance under UV light irradiation. In addition, the existence of Ag metal nanoparticles was also proven from the absorbance spectra, with Ag metal nanoparticle peaks at wavelengths of 350–400 nm, which is a result from the localized plasmon resonance of Ag metal nanoparticles. The band gap energy of pristine ZnO showed a lower band gap energy of 3 eV compared to the typical band gap energy of ZnO that is 3.37 eV. A slight decrease in band gap energy with the increase in Ag content can be observed, which indicates a good photo response with increase in Ag content.

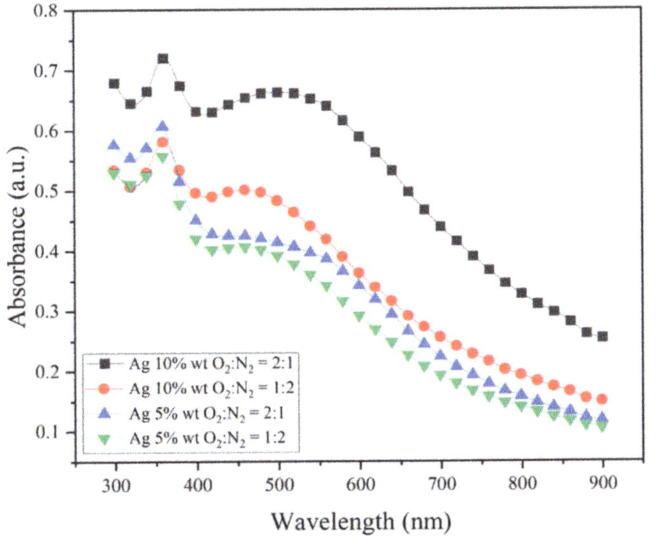

Figure 8. Absorbance spectra of ZnO-Ag nanoparticles with different carrier gas ratios and Ag contents.

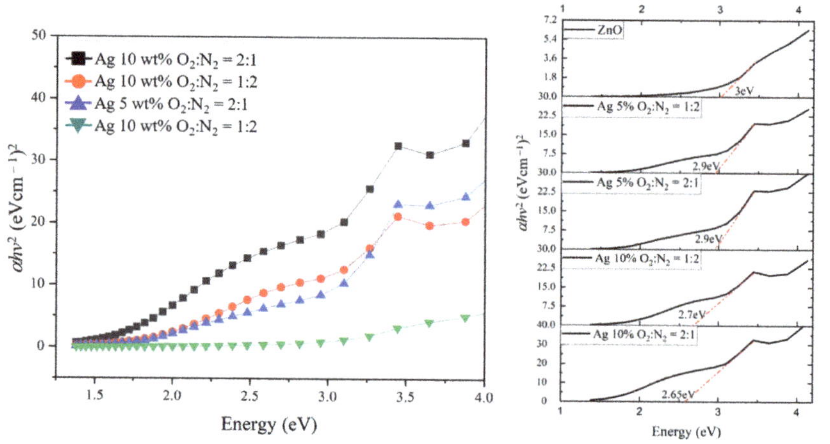

Figure 9. Band gap of ZnO-Ag nanoparticles with different carrier gas ratios and Ag contents.

Despite the higher surface area and lower band gap energy obtained with a Ag content of 10 wt%, the optimum photocatalytic activity is obtained at a Ag content of 5 wt%. This could be attributed to the excessive amount of Ag at a Ag content of 10 wt%, which also acts as charge recombination centers. The negatively charged Ag increases the capturing of holes, which reduces the electron–hole separation and hence decreases the degradation of organic compounds [12,29]. In addition, the increase in foreign material can decrease the photocatalytic activity because of the shielding effect toward the active sites of the photocatalyst [28]. Thus, this photocatalytic degradation is affected by several parameters such as the surface area [2,12], crystallinity [26], crystallite size [6] of ZnO as well as the concentration of the loaded materials [2]. The possible photocatalytic activity of ZnO is described by other research as shown below [8].

$$ZnO + h\upsilon\ (UV) \rightarrow ZnO\ (e_{CB}^{-} + h_{VB}^{+})$$

$$h_{VB}^{+} + OH^{-} \rightarrow {}^{\bullet}OH$$

$$e_{CB}^{-} + O_2 \rightarrow {}^{\bullet}O_2$$

$${}^{\bullet}O^{-} + 2H^{+} \rightarrow 2{}^{\bullet}OH$$

$${}^{\bullet}OH + \text{organic pollutants} \rightarrow 2{}^{\bullet}OH$$

The UV photon energy excites the electrons from the valence band of ZnO, which induces the formation of holes. These holes react with the existing OH^-, forming $^{\bullet}OH$ radicals. Furthermore, electrons can react with oxygen to form $^{\bullet}OH$ radicals. These species are known to have significant roles in the degradation process.

3. Materials and Methods

3.1. Preparation Methods

Nanoparticles were formed using an ultrasonic spray pyrolysis system (Figure 10) following the experimental method given in our previous study [4]. Zinc acetate dehydrate crystals ($Zn(CH_3COO)_2 \cdot 2H_2O$, 99.5%, E. Merck, D-6100 Darmstadt, FR Germany) were used as the precursor to fabricate ZnO nanoparticles. Distilled water was added to the powder to form a zinc acetate aqueous solution (0.1 M). To uniformly dissolve zinc acetate, the solution was ultrasonicated for 30 min. Afterward, the aqueous solution was continuously fed to the nebulizer using a peristaltic pump (Omron, NE-U17, Kyoto, Japan). The precursor was aerosolized, and the produced droplets were continuously carried to the tubular furnace by the carrier gas at a total flow rate of 2 L/min. Various ratios of O_2:N_2 were used in the carrier gas (1:0, 1:2, 1:1, 2:1, and 0:1). The tubular furnace temperature was maintained at 400 °C. Evaporation of the solvent and decomposition of the precursor to form nanoparticles occurred inside the furnace. The produced nanoparticle powder was collected in the electrostatic precipitator at an applied voltage of 40 kV and a temperature of 120 °C, which was maintained to ensure that there was no condensation.

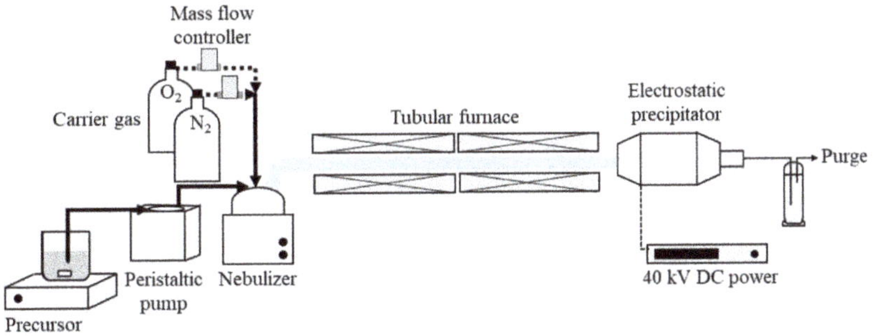

Figure 10. Experimental setup of the nanoparticle synthesis using a spray pyrolysis method.

ZnO-Ag nanocomposites were fabricated by adding silver nitrate ($AgNO_3$, 99.5%, E. Merck, D-6100 Darmstadt, FR Germany) at concentrations of 1, 5, and 10 wt% to the zinc acetate solution. The same method mentioned above was used to fabricate and collect the ZnO-Ag nanocomposite.

3.2. Materials Characterization

The particle morphology was observed using SEM (FlexSEM1000, Hitachi High Technologies, Tokyo, Japan), and further observation of the morphology and elemental analysis was conducted by TEM-EDS (JEM-2010, JEOL, Tokyo, Japan). The crystallinity and phase composition of the nanoparticles were determined by XRD (Philip XPERT MPD, Philips, Almelo, The Netherlands) operated at a 40 kV and 30 mA. The XRD patterns were obtained within a 2θ range of 20–80°. The crystallite size (D) of the nanoparticles was determined by the Scherrer formula shown below:

$$D = \frac{k\,\lambda}{B\,cos\,\theta} \tag{1}$$

where k is a constant (k = 0.9), λ is the X-ray wavelength (λ = 0.154 nm), B is the full width at half maximum of the peak, and θ is the XRD peak angle [6,23]. Furthermore, The SBET of the nanoparticles was determined by a nitrogen gas adsorption device (Quantachrome Instruments, Boynton Beach, FL, USA) based on the BET method.

3.3. Photocatalytic Test

The photocatalytic activity of the nanoparticles was tested for the degradation of organic dye pollutants found in real textile wastewater obtained from UD, ATBM Jufri Hartono, Gresik, East Java. The wastewater was firstly 10× diluted by the as-received distilled water. Next, the diluted pollutant solution (30 mL) was poured inside a beaker glass, in which the nanoparticle powder (10 mg) was added and continuously stirred throughout the photocatalytic test. The sample was placed in a dark chamber for 30 min to obtain the adsorption–desorption equilibrium of the photocatalyst and dye pollutants. The photocatalytic test was conducted for 90 min, in which the absorbance was measured at 30 min intervals using a UV–Vis spectrophotometer. The sample was centrifuged at 5000 rpm for each measurement to separate the powder catalyst from the liquid pollutant. The absorbance of the obtained supernatant was then measured using the UV–Vis spectrophotometer. The concentration of the dye is proportional to the measured intensity [26]. Thus, the reaction rate constant was calculated based on the slope of the dye concentration–intensity curve, i.e., $\ln(C_t/C_o) = kt$ [30], where k, t, C_t, and C_o are the reaction rate constant, reaction time, final concentration, and initial concentration, respectively.

4. Conclusions

Crumpled pristine ZnO and ZnO-Ag nanoparticles with high surface area were fabricated by a one-step ultrasonic spray pyrolysis method at different Ag contents and O_2:N_2 carrier gas ratios. The carrier gas ratio did not significantly affect the morphology, crystallite phase, or primary size of the nanoparticles. However, the SBET and photocatalytic activity of the nanoparticles changed with the increase in the nitrogen ratio. The nanoparticles exhibited good performance, and optimum activity (reaction rate constant = 0.0059 min^{-1}) was observed for the ZnO nanoparticles with a Ag content of 5 wt%, which were fabricated in a nitrogen-rich system.

Author Contributions: Conceptualization, K.K. and S.W.; Methodology, K.K., W.W. and M.H.; Formal analysis, M.H., N.R.P., W.W., M.S. and K.K.; Investigation, K.K., N.R.P. and M.H.; Writing—M.H.; Writing—review and editing, K.K., M.H. and M.S.; Supervision: K.K. and S.W. All authors have read and agreed to the published version of the manuscript.

Funding: Financial support from DRPM Kementrian Pendidikan, Kebudayaan, Riset dan Teknologi Indonesia with contract numbers: 084/E5/PG.02.00.PT/2022 and 1491/PKS/ITS/2022 through "Penelitian Dasar Unggulan Perguruan Tinggi" scheme. This research was also partly supported by Japan Society for the Promotion of Science (KAKENHI grant; grant number JP21K04750).

Data Availability Statement: In this manuscript, our characterizations were SEM, XRD, TEM, BET, and UV Vis. All data have been reported as the images.

Acknowledgments: Our biggest gratitude goes to the persons who supported this research, to Timotius Candra Kusuma, Nungki Widya Savitri, and Syamsul Muarafi Subekhi for the experimental assistance, and to UD. ATBM Jufri Hartano for providing the tested wastewater.

Conflicts of Interest: The authors declare no conflict of interest.

References

1. Paździor, K.; Bilińska, L.; Ledakowicz, S. A review of the existing and emerging technologies in the combination of AOPs and biological processes in industrial textile wastewater treatment. *Chem. Eng. J.* **2019**, *376*, 120597. [CrossRef]
2. Saravanan, R.; Karthikeyan, N.; Gupta, V.; Thirumal, E.; Thangadurai, P.; Narayanan, V.; Stephen, A. ZnO/Ag nanocomposite: An efficient catalyst for degradation studies of textile effluents under visible light. *Mater. Sci. Eng. C* **2013**, *33*, 2235–2244. [CrossRef] [PubMed]
3. Akkari, M.; Aranda, P.; Belver, C.; Bedia, J.; Amara, A.B.H.; Ruiz-Hitzky, E. ZnO/sepiolite heterostructured materials for solar photocatalytic degradation of pharmaceuticals in wastewater. *Appl. Clay Sci.* **2018**, *156*, 104–109. [CrossRef]
4. Kusdianto, K.; Hudandini, M.; Kusuma, T.C.; Widiyastuti, W.; Madhania, S.; Machmudah, S.; Nurtono, T.; Puspitasari, D.; Winardi, S. Photocatalytic degradation of organic waste derived from textile dye by ZnO-Ag nanocomposite synthesized by spray pyrolysis. *AIP Conf. Proc.* **2020**, *2219*, 30001.
5. Ong, C.B.; Ng, L.Y.; Mohammad, A.W. A review of ZnO nanoparticles as solar photocatalysts: Synthesis, mechanisms and applications. *Renew. Sustain. Energy Rev.* **2018**, *81*, 536–551. [CrossRef]
6. El-Bindary, A.A.; El-Marsafy, S.M.; El-Maddah, A.A. Enhancement of the photocatalytic activity of ZnO nanoparticles by silver doping for the degradation of AY99 contaminants. *J. Mol. Struct.* **2019**, *1191*, 76–84. [CrossRef]
7. Kusdianto, K.; Sari, T.D.; Laksono, M.A.; Madhania, S.; Winardi, S. Fabrication and application of ZnO-Ag nanocomposite materials prepared by gas-phase methods. *IOP Conf. Ser. Mater. Sci. Eng.* **2021**, *1053*, 12023. [CrossRef]
8. Fageria, P.; Gangopadhyay, S.; Pande, S. Synthesis of ZnO/Au and ZnO/Ag nanoparticles and their photocatalytic application using UV and visible light. *RSC Adv.* **2014**, *4*, 24962–24972. [CrossRef]
9. Lee, Y.; Fujimoto, T.; Yamanaka, S.; Kuga, Y. Evaluation of photocatalysis of Au supported ZnO prepared by the spray pyrolysis method. *Adv. Powder Technol.* **2021**, *32*, 1619–1626. [CrossRef]
10. Muñoz-Fernandez, L.; Sierra-Fernandez, A.; Milošević, O.; Rabanal, M.E. Solvothermal synthesis of Ag/ZnO and Pt/ZnO nanocomposites and comparison of their photocatalytic behaviors on dyes degradation. *Adv. Powder Technol.* **2016**, *27*, 983–993. [CrossRef]
11. Zhang, Q.; Li, J.; Xu, M. Ag-decorated ZnO-based nanocomposites for visible light-driven photocatalytic degradation: Basic understanding and outlook. *J. Phys. D Appl. Phys.* **2022**, *55*, 483001. [CrossRef]
12. Dermenci, K.B.; Genc, B.; Ebin, B.; Olmez-Hanci, T.; Gürmen, S. Photocatalytic studies of Ag/ZnO nanocomposite particles produced via ultrasonic spray pyrolysis method. *J. Alloys Compd.* **2014**, *586*, 267–273. [CrossRef]
13. Leng, J.; Wang, Z.; Wang, J.; Wu, H.-H.; Yan, G.; Li, X.; Guo, H.; Liu, Y.; Zhang, Q.; Guo, Z. Advances in nanostructures fabricated via spray pyrolysis and their applications in energy storage and conversion. *Chem. Soc. Rev.* **2019**, *48*, 3015–3072. [CrossRef]

14. Emil, E.; Alkan, G.; Gurmen, S.; Rudolf, R.; Jenko, D.; Friedrich, B. Tuning the Morphology of ZnO Nanostructures with the Ultrasonic Spray Pyrolysis Process. *Metals* **2018**, *8*, 569. [CrossRef]
15. Iskandar, F. Nanoparticle processing for optical applications—A review. *Adv. Powder Technol.* **2009**, *20*, 283–292. [CrossRef]
16. Jongthammanurak, S.; Witana, M.; Cheawkul, T.; Thanachayanont, C. The effects of carrier gas and substrate temperature on ZnO films prepared by ultrasonic spray pyrolysis. *Mater. Sci. Semicond. Process.* **2013**, *16*, 625–632. [CrossRef]
17. Wittawat, R.; Rittipun, R.; Jarasfah, M.; Nattaporn, B. Synthesis of ZnO/TiO_2 spherical particles for blue light screening by ultrasonic spray pyrolysis. *Mater. Today Commun.* **2020**, *24*, 101126. [CrossRef]
18. Vorovsky, V.Y.; Kovalenko, A.V.; Kushneryov, A.I.; Khmelenko, O.V. Preparation of zinc oxide nanopowders doped with manganese, which have ferromagnetic properties at room temperature. *Funct. Mater.* **2018**, *25*, 61–66. [CrossRef]
19. Xu, Y.; Zhang, C.; Cheng, Y.; Li, Z.; Cheng, Y.; Feng, Q.; Chen, D.; Zhang, J.; Hao, Y. Influence of Carrier Gases on the Quality of Epitaxial Corundum-Structured α-Ga_2O_3 Films Grown by Mist Chemical Vapor Deposition Method. *Matererials* **2019**, *12*, 3670. [CrossRef]
20. Rahemi Ardekani, S.; Sabour Rouhaghdam, A.; Nazari, M. N-doped ZnO-CuO nanocomposite prepared by one-step ultrasonic spray pyrolysis and its photocatalytic activity. *Chem. Phys. Lett.* **2018**, *705*, 19–22. [CrossRef]
21. El Rouby, W.M.A. Crumpled graphene: Preparation and applications. *RSC Adv.* **2015**, *5*, 66767–66796. [CrossRef]
22. Wang, R.-C.; Tsai, C.-C. Efficient synthesis of ZnO nanoparticles, nanowalls, and nanowires by thermal decomposition of zinc acetate at a low temperature. *Appl. Phys. A* **2009**, *94*, 241–245. [CrossRef]
23. Kusdianto, K.; Hudandini, M.; Jiang, D.; Kubo, M.; Shimada, M. Effect of Heating Rate on the Photocatalytic Activity of Ag–TiO_2 Nanocomposites by One-Step Process via Aerosol Routes. *Catalysts* **2021**, *12*, 17. [CrossRef]
24. Li, J.; Huang, J.-H.; Zhang, Y.-L.; Yang, Y.; Song, W.-J.; Li, X.-M. Effects of rapid thermal annealing in different ambients on structural, electrical, and optical properties of ZnO thin films by sol-gel method. *J. Electroceramics* **2011**, *26*, 84–89. [CrossRef]
25. Yan, Y.; Al-Jassim, M.M.; Wei, S.H. Doping of ZnO by group-IB elements. *Appl. Phys. Lett.* **2006**, *89*, 181912. [CrossRef]
26. Tripathy, N.; Ahmad, R.; Kuk, H.; Lee, D.H.; Hahn, Y.-B.; Khang, G. Rapid methyl orange degradation using porous ZnO spheres photocatalyst. *J. Photochem. Photobiol. B Biol.* **2016**, *161*, 312–317. [CrossRef]
27. Ma, H.; Yue, L.; Yu, C.; Dong, X.; Zhang, X.; Xue, M.; Zhang, X.; Fu, Y. Synthesis, characterization and photocatalytic activity of Cu-doped Zn/ZnO photocatalyst with carbon modification. *J. Mater. Chem.* **2012**, *22*, 23780–23788. [CrossRef]
28. Liu, C.; Mao, S.; Wang, H.; Wu, Y.; Wang, F.; Xia, M.; Chen, Q. Peroxymonosulfate-assisted for facilitating photocatalytic degradation performance of 2D/2D WO_3/BiOBr S-scheme heterojunction. *Chem. Eng. J.* **2022**, *430*, 132806. [CrossRef]
29. Georgekutty, R.; Seery, M.K.; Pillai, S.C. A Highly Efficient Ag-ZnO Photocatalyst: Synthesis, Properties, and Mechanism. *J. Phys. Chem. C* **2008**, *112*, 13563–13570. [CrossRef]
30. Kusdianto, K.; Jiang, D.; Kubo, M.; Shimada, M. Effect of annealing temperature on the photocatalytic activity of Ag–TiO_2 nanocomposite films by one-step gas-phase deposition. *Mater. Res. Bull.* **2018**, *97*, 497–505. [CrossRef]

catalysts

Article

Synthesis and Characterization of SiO_2/TiO_2 as Photocatalyst on Methylene Blue Degradation

Aleksandra Babyszko, Agnieszka Wanag *, Marcin Sadłowski, Ewelina Kusiak-Nejman and Antoni W. Morawski

Faculty of Chemical Technology and Engineering, Department of Inorganic Chemical Technology and Environment Engineering, West Pomeranian University of Technology in Szczecin, Pułaskiego 10, 70-322 Szczecin, Poland
* Correspondence: awanag@zut.edu.pl

Abstract: The paper presents a modification of titanium dioxide with fumed silica. The SiO_2/TiO_2 photocatalysts were obtained by the sol-gel method and then were calcined under an argon atmosphere. Various SiO_2 weights (2–17.2 wt.%) were used in the materials' preparation stage. The obtained samples were characterized using advanced analytical methods, such as FT-IR/DRS infrared spectroscopy, X-ray diffraction, SEM scanning electron microscopy, and UV-Vis/DRS spectroscopy. The BET specific surface area and zeta potential of samples were also measured. Based on the obtained results, it was observed that the modification of titanium dioxide with SiO_2 effectively inhibited the increase in crystallite size of anatase and brookite during calcination and the decrease in specific surface area values. Moreover, the presence of SiO_2 in the nanomaterials contributed to the increase in the size of specific surface area and the change in band gap energy values. The photocatalytic activity was determined based on the decomposition of methylene blue under UV irradiation. Thermal modification in an inert gas atmosphere significantly increased the dye removal rate. It should be noted that all the obtained SiO_2/TiO_2 photocatalysts showed higher activity compared to the starting TiO_2. It was also found that the photocatalytic activity increased along with the increase in SiO_2 content in the sample (up to 14.3 wt.% of SiO_2). The highest activity was recorded for SiO_2(11.1%)/TiO_2_400 and SiO_2(14.3%)/TiO_2_400 samples.

Keywords: photocatalysis; titanium dioxide; fumed silica; methylene blue decomposition

Citation: Babyszko, A.; Wanag, A.; Sadłowski, M.; Kusiak-Nejman, E.; Morawski, A.W. Synthesis and Characterization of SiO_2/TiO_2 as Photocatalyst on Methylene Blue Degradation. *Catalysts* **2022**, *12*, 1372. https://doi.org/10.3390/catal12111372

Academic Editors: Jorge Bedia, Carolina Belver and Gassan Hodaifa

Received: 30 September 2022
Accepted: 3 November 2022
Published: 5 November 2022

1. Introduction

Advanced oxidation processes are promising methods for removing pollutants from the environment. One is heterogeneous photocatalysis, which can purify water and air [1–4]. Due to its properties, titanium dioxide is one of the most commonly used semiconductors in the photocatalytic oxidation of organic compounds. It is characterized by a relatively low price, non-toxicity, high photoactivity, and chemical stability [5]. At the same time, intensive research is being conducted worldwide to increase the efficiency of photocatalytic processes. The studies aim to improve the physicochemical properties of titanium dioxide affecting photocatalytic activity.

There are many methods used to obtain TiO_2, such as hydrothermal, solvothermal, sol-gel, or precipitation [6], but the sol-gel method is effective for controlling the size and morphology of the synthesized particles while obtaining homogeneous materials [7–11]. During the sol-gel process, factors such as the initial composition of the reaction mixture, i.e., the molar ratio of water to the metal precursor, type of solvent, type and concentration of catalyst, temperature, and the pH of the solution affect the rate of hydrolysis and polycondensation [12]. In addition, calcination also significantly influences the physical and chemical properties of the final product [13]. For example, Ciesielczyk et al. [14] proved that calcination causes significant changes in the porous structure parameters. In particular, there was a significant reduction in the surface area and pore volume for the calcined samples.

Photocatalysts with enhanced activity can be obtained by modifying titanium dioxide using various compounds, e.g., metals or their oxides and non-metals [15]. Recently, the silica modification of titanium dioxide is becoming more popular. It has been proven that the addition of SiO_2 increases the surface of the photocatalyst, increasing the adsorption of pollutants [16]. The enhanced adsorption of contaminants on the silica surface, in turn, improves the photocatalytic activity of mixed SiO_2–TiO_2 oxides compared to pure TiO_2 [17,18]. Jimmy et al. [19] and Yu et al. [20] found that the addition of SiO_2 causes the inhibition of TiO_2 crystallization. Moreover, adding SiO_2 increases the amount of water and hydroxyl groups adsorbed on the surface, improving the hydrophilic and photocatalytic properties [21–23].

Several methods of preparation of TiO_2 with silica are presented in the literature. The most commonly used silica precursors are tetraethoxysilane (TEOS), tetramethoxysilane (TMOS), and silica gel [24,25]. The sol-gel method for the synthesis of TiO_2/SiO_2 was used by Fatimah [24]. Titanium(IV) isopropoxide (TTIP) was used as a TiO_2 precursor, while tetramethoxysilane (TMOS) and tetraethoxysilane (TEOS) were used as silica precursors. The molar ratio of titanium dioxide to silicon was 4:9. The obtained nanomaterials were then calcined at 500 °C for 4 h. The photocatalytic activity of new nanomaterials was determined based on the decomposition of methylene blue. It was found that using tetraethoxysilane as a silica precursor results in obtaining material with a larger specific surface area and pore volume, which in turn increases photocatalytic efficiency. Qourzal et al. [25] obtained a TiO_2-SiO_2 photocatalyst by hydrolyzing TTIP in silica gel. The material was calcined at 400 °C for 2 h in the air atmosphere. It was noted that the presence of silica suppresses the anatase to rutile phase transformation. Based on the decomposition of β-naphthol under UV light, they found that the prepared material showed 2.7, 4, and 7.8 times higher photocatalytic activity than commercial TiO_2 P25, TiO_2 PC50, and TiO_2 "Aldrich" photocatalysts, respectively. Enhances photocatalytic activity has been attributed to the high adsorption capacity of the photocatalyst and a large specific surface area. Nandanwar et al. [26] obtained composites by the sol-gel method, using titanium(IV) isopropoxide as TiO_2 precursor, tetraethoxysilane (TEOS) as silica precursor, ethanol as solvent, and hydrochloric acid as a promoter of the hydrolysis reaction. In addition, polyethylene glycol (PEG) and the non-ionic Triton X-100 surfactant were used during the preparation phase. The photocatalytic activity was measured by methylene blue degradation. In the case of the Triton X-100 sample, after only a few minutes of exposure to the UV light, approx. 50% of methylene blue was degraded. These results confirm that TiO_2/SiO_2 with the addition of Triton X-100 shows a higher degradation efficiency than TiO_2/SiO_2 with polyethylene glycol.

This work proposes a new method of titanium dioxide nanomaterial modification with fumed silica. SiO_2/TiO_2 nanomaterials were prepared using the sol-gel method. An essential element of the preparation was introducing the heating stage of the obtained nanomaterials in the argon atmosphere. This represents a kind of novelty concerning the preparation methods because, in the methods described in the literature for obtaining this type of materials, the calcination process was carried out mainly in an air atmosphere. The photocatalytic activity of gained samples was determined based on the methylene blue decomposition under UV irradiation. The effect of specific physicochemical properties on photocatalytic activity was also investigated. To the best of our knowledge, this is the first paper in which fumed silica with average primary particle size of 7–14 nm was used as a silica precursor. According to the manufacturer, the used silica characterizes a relatively large specific surface area (>200 m^2/g). The novelty of the presented study was also the determination of the calcination effect on the photocatalytic and physicochemical properties of TiO_2 nanomaterials modified with fumed silica obtained in an inert gas atmosphere.

2. Results and Discussion

2.1. Characterization of the Photocatalysts

The surface character of the prepared TiO_2 samples was analysed by FT-IR/DRS spectroscopy. In Figure 1A,B, the FT-IR/DR spectra of starting TiO_2, silica-modified TiO_2 and samples calcined in an argon atmosphere are shown. All spectra presented characteristic bands for TiO_2-based nanomaterials. Analysing the obtained spectra, the bands in the range of 3730–2500 cm^{-1} attributed to the stretching mode of O–H group were confirmed [27]. As can be seen from Figure 1A,B, the intensity of the mentioned peaks increased with the increasing amount of silica used for modification. It was related to the presence of Si–OH groups and adsorbed water formed by silica deposition on titanium particles [28]. After the calcination process (see Figure 1B), a decrease in the intensity of this band was observed compared to the unheated materials due to the change in the content of hydroxyl groups on the semiconductor surface [29]. The narrow band located at 1610 cm^{-1} is characteristic of the vibrations of molecular water bending mode [30]. Band located at 3710 cm^{-1} corresponds to the OH groups surface-bonded to titanium [31,32]. At 930 cm^{-1}, an intense band assigned to the O–Ti–O stretching modes was observed, and for photocatalysts modified with silica, the band in question was slightly shifted, which indicates the presence of interaction between titanium and silicon [33]. After silica modification, a new band located at 1160 cm^{-1} attributed to the asymmetrical stretching of Si–O–Si bonds was noticed [34]. It can be noted that the intensity of the upper-mentioned peak increased with the increase in silica amount used for TiO_2 modification. The peak noticed at 1780 cm^{-1} was ascribed to the C=O stretching vibration.

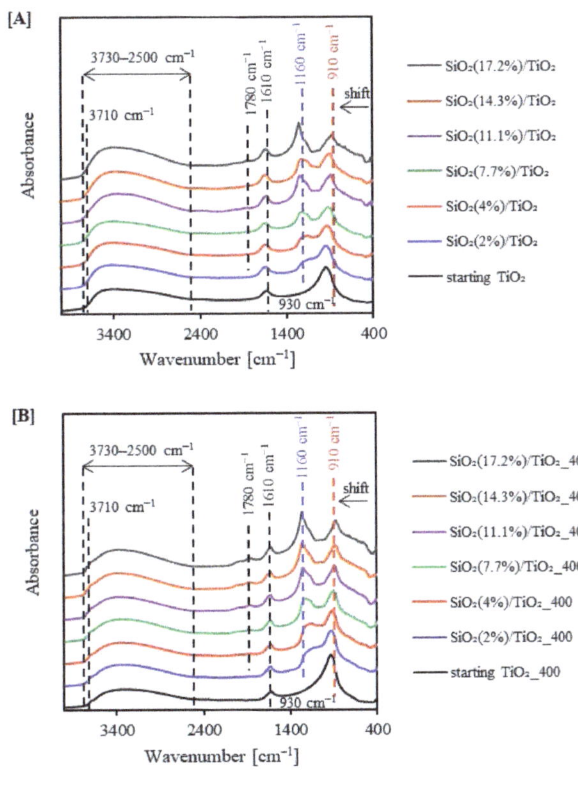

Figure 1. FT-IR/DR spectra of starting TiO_2 and silica-modified TiO_2 prior (**A**) and after heat treatment (**B**).

In Figure 2A,B, the XRD diffraction patterns of the starting TiO_2, silica-modified TiO_2 and samples calcined in an argon atmosphere are shown. The phase composition and average crystallite sizes of the studied samples are compared in Table 1. It should be noted that the presented results refer to the crystalline phase. Based on the analysis of obtained data, the presence of reflections characteristic of anatase and brookite was found. The characteristic reflections (101), (004), (200), (105), (204), (116), and (215), corresponding to the anatase phase, were recorded at 2θ = 25.6, 38.6, 48.8, 55.2, 63.6, 70.1, and 76.3°, respectively (JCPDS 04-002-8296 PDF4+ card). Two reflections (111), (121) characteristic of the brookite phase were recorded at 2θ = 25.6 and 30.3° (JCPDS 04-007-0758 PDF4+ card). The reflection from brookite located at 2θ = 25.6° overlapped with the reflection characteristic of anatase. Therefore, the size of brookite crystallites was calculated based on the reflectance located at 2θ = 30.3°. The presence of the brookite phase was due to the use of hydrochloric acid as a catalyst for the hydrolysis reaction of titanium(IV) isopropoxide [35]. It can be noted that calcination at 400 °C did not contribute to the phase transition of anatase to rutile, and this is a typical phenomenon because the phase transition of anatase to rutile occurs at temperatures above 600 °C [36]. It is also worth noting that the reflections narrowed and sharpened after calcination. This was attributed to eliminating defects at grain boundaries during heating at higher temperatures [37]. According to the data in Table 1, the anatase to brookite ratio remained similar, i.e., 59:41, only when the amount of silica precursor introduced was 11.1wt.% the anatase content started to increase. This can be explained by the fact that the amorphous phase was not included in the presented results. As it is known, silica used for modification was amorphous. Therefore, the observed increase in the anatase phase with increasing SiO_2 content may be overestimated by up to 20–30%. After the calcination process, a decrease in brookite content was observed, which was related to the greater crystallization of anatase than brookite. Only when the amount of silica precursor introduced was 11.1wt.% the brookite content start increasing, which was related to the crystallization process progress. It should also be noted that the calcination process had a significant effect on the increase in average anatase and brookite crystallite size. The average crystallite sizes of anatase for non-calcined materials were 4–5 nm and 8–16 nm for the calcined samples. In contrast, the crystallite sizes of brookite ranged from 2 to 6 nm for non-calcined photocatalysts and 7–9 nm for calcined photocatalysts. This indicates an increase in the crystallinity of samples after calcination. However, comparing the crystallite size of SiO_2/TiO_2 materials after heating with starting TiO_2_400 sample, it can be observed that the crystallite size of anatase and brookite was smaller for silica-modified photocatalysts than for the sample without SiO_2 addition. For example, the size of anatase crystallites for starting TiO_2_400 was 14 nm, while for the SiO_2(17.2%)/TiO_2_400 sample, it was only 8 nm. According to Xu et al. [38] and Lu et al. [39], silica can effectively prevent the growth of TiO_2 crystallites during the calcination process.

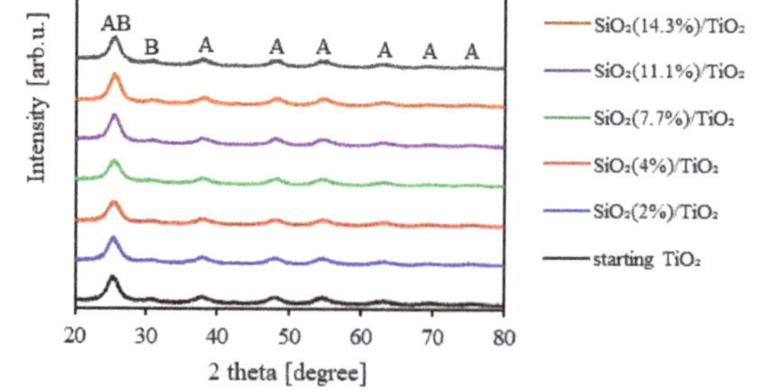

Figure 2. *Cont.*

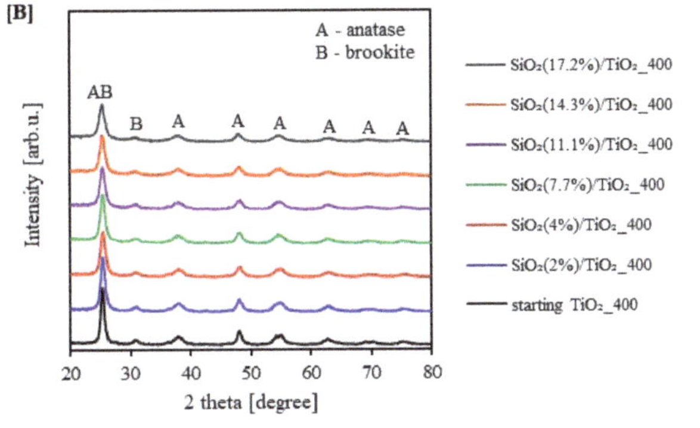

Figure 2. XRD patterns of starting TiO_2 and silica-modified TiO_2 prior (**A**) and after heat treatment (**B**).

Table 1. XRD phase composition, average crystallites size, specific surface area and pore volume distribution of starting TiO_2 and silica-modified photocatalysts.

Sample Code	Phase Composition [%]		Mean Crystallite Size [nm]		S_{BET} [m^2/g]	V_{total} [cm^3/g]	V_{micro} [cm^3/g]	V_{meso} [cm^3/g]
	Anatase	Brookite	Anatase	Brookite				
starting TiO_2	57	43	5 ± 0.2	2 ± 0.2	193	0.109	0.079	0.030
SiO_2(2%)TiO_2	58	42	5 ± 0.2	2 ± 0.2	234	0.159	0.090	0.069
SiO_2(4%)/TiO_2	59	41	5 ± 0.2	4 ± 0.2	221	0.148	0.090	0.058
SiO_2(7.7%)/TiO_2	59	41	5 ± 0.2	4 ± 0.2	218	0.171	0.085	0.171
SiO_2(11.1%)/TiO_2	81	19	5 ± 0.2	5 ± 0.2	208	0.265	0.080	0.185
SiO_2(14.3%)/TiO_2	77	23	5 ± 0.2	6 ± 0.2	207	0.285	0.080	0.205
SiO_2(17.2%)/TiO_2	81	19	4 ± 0.2	6 ± 0.2	228	0.366	0.088	0.278
starting TiO_2_400	63	37	14 ± 0.4	9 ± 0.4	54	0.101	0.020	0.081
SiO_2(2%)/TiO_2_400	66	34	16 ± 0.4	7± 0.4	89	0.177	0.031	0.146
SiO_2(4%)/TiO_2_400	68	32	11 ± 0.4	7 ± 0.4	98	0.160	0.036	0.124
SiO_2(7.7%)/TiO_2_400	66	34	11 ± 0.4	7 ± 0.4	104	0.189	0.039	0.150
SiO_2(11.1%)/TiO_2_400	67	33	10 ± 0.4	7 ± 0.4	121	0.259	0.043	0.216
SiO_2(14.3%)/TiO_2_400	72	28	9 ± 0.4	8 ± 0.4	135	0.313	0.051	0.262
SiO_2(17.2%)/TiO_2_400	72	28	8 ± 0.4	8 ± 0.4	146	0.371	0.056	0.315

The adsorption–desorption N_2 isotherms of starting TiO_2 and the SiO_2/TiO_2 materials are shown in Figures 3 and 4. It can be noted that the samples presented three kinds of isotherm types. Most of the photocatalysts (starting TiO_2, SiO_2(2, 4%)/TiO_2, starting TiO_2_400, SiO_2(2–7.7%)/TiO_2_400) displayed a typical type IV isotherm with an H3 hysteresis loop, which is characteristic of mesoporous materials, according to the IUPAC classification [40]. The H3-type hysteresis loop does not show any adsorption limit in the high relative pressure range and is characteristic of slotted pores [41,42]. The sample (starting TiO_2) showed a type I isotherm characteristic of microporous materials. On the other hand, SiO_2(2%)/TiO_2 and SiO_2(4%)/TiO_2 isotherms in the initial p/p0 range were defined as type I isotherm, and in the intermediate and higher pressure range as type II isotherm. The isotherms of starting TiO_2_400, SiO_2(2%)/TiO_2_400, SiO_2(4%)/TiO_2_400 and SiO_2(7.7%)/TiO_2_400 samples demonstrated an asymmetric and triangular type H2 of the hysteresis loop, what was attributed to blockage of pores/percolation in a narrow range of pore necks [40,43]. For starting TiO_2, the hysteresis loop did not occur.

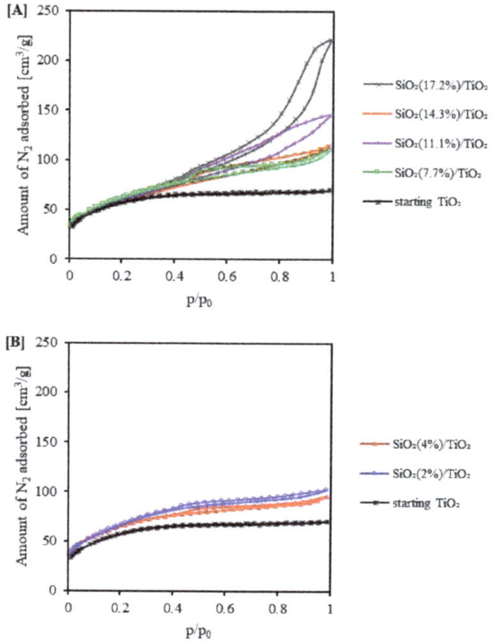

Figure 3. Adsorption–desorption isotherms of: (**A**) starting TiO_2, SiO_2(2%)/TiO_2 and SiO_2(4%)/TiO_2, (**B**) starting TiO_2, SiO_2(7.7%)/TiO_2, SiO_2(11.1%)/TiO_2, SiO_2(14.3%)/TiO_2 and SiO_2(17.2%)/TiO_2.

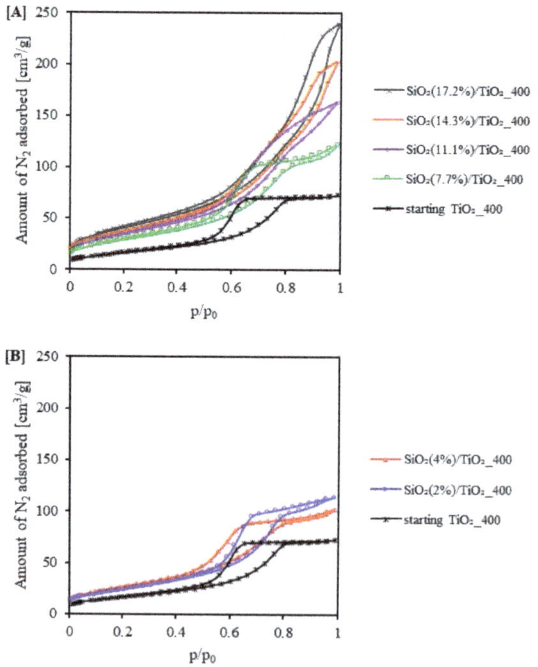

Figure 4. Adsorption–desorption isotherms of: (**A**) starting TiO_2_400, SiO_2(2%)/TiO_2_400 and SiO_2(4%)/TiO_2_400, (**B**) starting TiO_2_400, SiO_2(7.7%)/TiO_2_400, SiO_2(11.1%)/TiO_2_400, SiO_2(14.3%)/TiO_2_400 and SiO_2(17.2%)/TiO_2_400.

Table 1 shows the specific surface area value and the pore size distribution for all the obtained materials. The specific surface area of the starting TiO_2 was 193 m^2/g. The values presented in Table 1 show that the photocatalysts modified with silica were characterized by a higher specific surface area (up to 234 m^2/g) in comparison to the materials obtained the same way but without the addition of SiO_2. The average size of anatase crystallites has not changed. Therefore, the use of silica in the titanium dioxide modification process contributed to the increase in the specific surface area of the photocatalysts. The specific surface area of the silica used for modification was >200 m^2/g [44]. Bao et al. [45] also observed an increase in the surface area for the materials modified with silica. In this case, the specific surface area changed from 15.4 m^2/g for pure TiO_2 to 127.7 m^2/g for silica-modified TiO_2. Table 1 noted that, as the SiO_2 content increased, the total pore volume also increased. This is a normal phenomenon associated with an increase in the size of a specific surface area. For example, the pore volume for the starting TiO_2 was 0.109 cm^3/g, and for SiO_2(17.2%)/TiO_2 was 0.366 cm^3/g. It can be concluded that the SiO_2 modification had a significant effect on the increase in S_{BET} and the total pore volume. After the heat treatment, the specific surface area of the reference photocatalysts decreased from 193 m^2/g for the starting TiO_2 sample to 54 m^2/g for the sample calcined at 400 °C (starting TiO_2_400). Thus, the correlation between calcination and the decrease in the size of the specific surface area is possible to observe. Calcining titanium dioxide causes the sintering and agglomeration of TiO_2 particles, which then results in a decrease in the size of the specific surface area [46]. It should also be noted that the specific surface area did not decrease in the range of materials after calcination. This is related to the relatively large specific surface area of silica and the fact that the addition of silica inhibits the growth of crystallites. The specific surface area of photocatalysts changed from 54 m^2/g for the starting TiO_2_400 to 146 m^2/g for the SiO_2(17.2%)/TiO_2_400 sample.

Scanning electron microscope analysis in the SE (secondary electron) mode was conducted to determine the morphology of obtained photocatalysts and the approximate size of agglomerates/aggregates that form SiO_2/TiO_2 particles. Figure 5A–C shows example SEM images for the starting TiO_2 and the SiO_2(14.3%)/TiO_2_400 sample. The SEM images shown in Figure 5A,B indicate that the particles of starting TiO_2 form large aggregates consisting of finely agglomerated particles and larger clusters at the periphery, forming a matrix of large aggregates. An irregular and indeterminate shape characterized the individual particles. It is also worth noting that the TiO_2 particles were embedded in unreacted TTIP matrix. After the modification process using SiO_2 and calcination in an argon atmosphere, the grains had a more regular spherical shape. An increase in the size of photocatalyst aggregates was also observed after modification (350–500 nm) compared to the starting TiO_2 (150–350 nm). This results from the tendency of silica particles to form larger structures [47]. In addition, silica can effectively prevent the growth of TiO_2 crystallites after calcination, and it is well known that the tendency of particle agglomeration/aggregation is especially observed for small particles [38,39]. Both the starting TiO_2 and SiO_2(14.3%)/TiO_2_400 particles tended to form large aggregates. However, it can be noted that the aggregation ability of the photocatalyst particles increased due to the modification of titanium dioxide with silica.

In Figure 6A,B, the UV-Vis/DR spectra of starting TiO_2, silica-modified TiO_2 and samples calcined in an argon atmosphere are shown. The SiO_2/TiO_2 materials exhibited typical absorption of radiation in the UV area due to the intrinsic absorption of TiO_2 [48]. For the starting TiO_2_400 sample, an absorption peak was observed at 305 nm, which was related to the transition of an electron from the valence band (O 2p) to the conduction band (Ti 3d) [49]. Compared with the reference sample, all silica-modified materials showed a shift in the absorption edge toward lower wavelengths (blue shift). The values of the band gap energy of SiO_2/TiO_2 photocatalysts were higher than the starting TiO_2 because of the blue shift that occurred due to modification. According to the literature [45], the increase in the value of band gap energy can be attributed to the quantum effect. This increase in the value of band gap energy causes a decrease in the energy valence band and an increase in the edge of the conductivity band, which may lead to the slowdown of the electron–hole pair recombination process. In turn, the slowing the electron–hole pair recombination process contributes to the increase in photocatalytic activity.

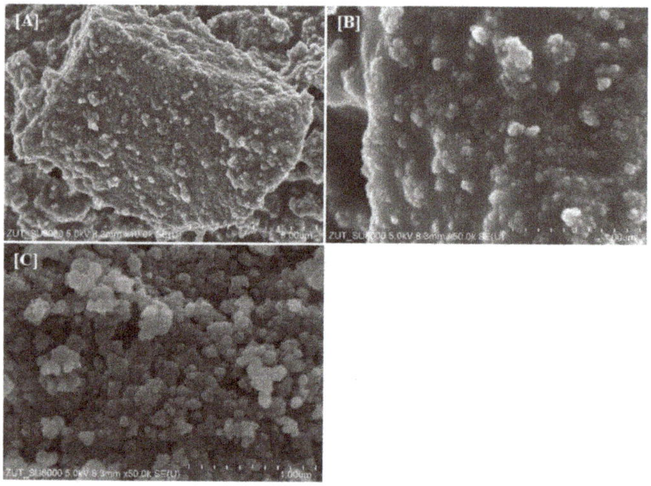

Figure 5. SEM images taken for (**A**,**B**) starting TiO_2 captured at different magnifications and (**C**) SiO_2(14.3%)/TiO_2_400.

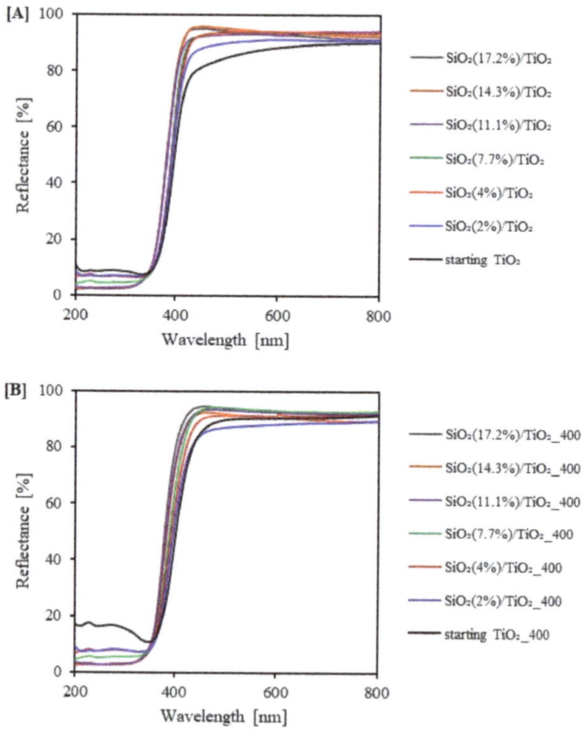

Figure 6. UV-Vis/DR spectra of starting TiO_2 and silica-modified TiO_2 prior (**A**) and after heat treatment (**B**).

According to the data presented in Table 2, silica modification did not change the surface character of the tested materials. After the modification, the surface of the SiO_2/TiO_2

photocatalysts remained positively charged, as did the surface of the reference sample without any SiO_2 content. Silicon-derived bands were observed on the FT-IR spectra of the studied samples. Ferreiry-Neto et al. [50] showed that modification of TiO_2 with silica causes a decrease in the zeta potential. Therefore, the TiO_2 photocatalysts modified with silica showed lower zeta potential values than the starting TiO_2. It is also important to note the decrease in zeta potential for SiO_2/TiO_2 photocatalysts after heating compared to non-calcined materials. Moreover, as the pH increased, the zeta potential value for these samples decreased. The higher the pH, the lower the zeta potential value of tested materials. The SiO_2(17.2%)/TiO_2_400 sample showed the lowest zeta potential (+14.6 mV). Our observations agree with those of Nowacka et al. [51], who found that the zeta potential decreases with increasing pH value, confirming that the zeta potential value strongly depends on pH.

Table 2. The band gap energy and zeta potential values of starting TiO_2 and silica-modified titania photocatalysts.

Sample Code	E_g [eV]	pH	Zeta Potential δ [mV]
starting TiO_2	3.05 ± 0.01	3.1	+38.9
SiO_2(2%)/TiO_2	3.09 ± 0.01	2.8	+35.8
SiO_2(4%)/TiO_2	3.10 ± 0.01	3.2	+35.7
SiO_2(7.7%)/TiO_2	3.12 ± 0.01	3.2	+32.8
SiO_2(11.1%)/TiO_2	3.22 ± 0.01	3.3	+33.6
SiO_2(14.3%)/TiO_2	3.22 ± 0.01	3.4	+31.0
SiO_2(17.2%)/TiO_2	3.23 ± 0.01	3.4	+31.0
starting TiO_2_400	2.98 ± 0.01	4.1	+39.0
SiO_2(2%)/TiO_2_400	3.07 ± 0.01	4.0	+33.2
SiO_2(4%)/TiO_2_400	3.11 ± 0.01	4.2	+30.2
SiO_2(7.7%)/TiO_2_400	3.15 ± 0.01	4.3	+19.5
SiO_2(11.1%)/TiO_2_400	3.23 ± 0.01	4.4	+19.7
SiO_2(14.3%)/TiO_2_400	3.22 ± 0.01	4.4	+17.9
SiO_2(17.2%)/TiO_2_400	3.24 ± 0.01	4.5	+14.6

2.2. Photocatalytic Activity Test

The photocatalytic activity of the starting TiO_2 and the new silica-modified TiO_2 photocatalysts were assessed based on the methylene blue degradation under the influence of UV radiation. The photolysis of dye solution (without the addition of a photocatalyst, see Figure 7) was also investigated. The experiment showed that the MB decomposition due to the photolysis process was negligible (ca. 0.5%). The distribution of methylene blue in the presence of starting TiO_2 and TiO_2 modified with silica before and after heat treatment is shown in Figure 8A,B. In addition, the results of the dye decomposition rate after 5 h of UV (138 W/m^2 UV in the range of 280–400 nm and 167 W/m^2 VIS in the range of 300–2800 nm) irradiation were compared, as shown in Figure 9A,B. It should be noted that all SiO_2/TiO_2 photocatalysts showed higher activity compared to the photocatalyst without the addition of SiO_2. Undoubtedly, the amount of silica used for modification played an important role in the photoactivity of the tested samples. It was observed that the decomposition degree increased with an increase in the SiO_2 content in the sample (up to 14.3 wt.% of SiO_2). SiO_2(11.1%)/TiO_2 and SiO_2(14.3%)/TiO_2 samples demonstrated the highest activity, for which the methylene blue degradation degree was 40.99 and 42.73%, respectively. In comparison, the MB decomposition degree for the starting TiO_2 was 7.19%. The silica modification contributed to an increase in specific surface area and pore volume, as well as a change in the band gap energy value. All SiO_2-modified samples showed a higher specific surface area and pore volume than the starting TiO_2. For example, the specific surface area and pore volume of starting TiO_2 sample were 193 m^2/g and 0.109 cm^3/g, respectively, and those of the SiO_2(11.1%)/TiO_2 sample were 208 m^2/g and 0.265 cm^3/g. In addition, the use of silica in the titanium dioxide modification process contributed to the inhibition of S_{BET} decrease in the range of materials after calcination. For example, the size of specific surface area for starting TiO_2_400 sample was 54 m^2/g, while that for SiO_2(11.1%)/TiO_2_400 and SiO_2(14.3%)/TiO_2_400 was 121 m^2/g and 135 m^2/g, respectively. It is well known that the specific surface area has an important role in increasing photocatalytic activity [52]. The high surface area provides a number of active centers that can adsorb a large number of pollutant molecules [53]. In the case of post-calcination materials, a relationship between the increase in specific surface area and the increase in the applied silica weighting was also noted. The higher the amount of silica in the material composition, the higher the specific surface area. For example, the size of the specific surface area for the sample SiO_2(2%)/TiO_2_400 was 89 m^2/g, and for the sample SiO_2(17.2%)/TiO_2_400, it was already 146 m^2/g. This was related to the fact that silica inhibits the growth of crystallites. Another parameter influencing the improvement of the photocatalytic activity was the change in the band gap energy. The band gap energy for the SiO_2(11.1%)/TiO_2 and SiO_2(14.3%)/TiO_2 (3.22 eV) samples was higher than for the starting TiO_2 (3.05 eV). Moreover, a relationship between the increase in activity and the value of band gap energy was observed. The higher the value of band gap energy, the higher the activity. The highest activity was observed in photocatalysts, whose E_g value was 3.22 eV and 3.23 eV. Bao et al. [45] and Periyat et al. [54] suggested that the increase in band gap energy value slows down the electron–hole pair recombination process. This is since an increase in the band gap energy reduces the energy valence band and increases the edge of the conductivity band. The phase composition was also an important parameter determining the activity. It is important to note that the photocatalysts that showed the highest activity comprised 66–72% anatase and 28–34% brookite. Similar results were also obtained by Allen et al. [55], who studied the effect of the brookite phase on the photoactivity of TiO_2. They showed that photocatalysts mixtures of anatase and brookite show high photocatalytic activity. A photocatalyst obtained by calcination at 400 °C and containing 69% anatase and 31% brookite in its composition removed about 97% of methyl orange after 180 min of irradiation. This was explained by the fact that the conductivity band of the TiO_2 brookite phase is shifted more cathodically by 0.14 eV than that of anatase. This displacement facilitates interfacial electron transfer to molecular oxygen. It should be noted that photocatalysts after calcination showed higher activity compared to non-calcined materials. The higher activity for the samples after calcination was attributed to the larger crystallite size of anatase. The recorded increase in anatase crystallite size indicates the transformation of the amorphous phase present in non-calcined materials. Zhang et al. [56] determined that the photocatalytic activity is the highest when the anatase crystallite size is around 10 nm, which concord with our observations. In our case, the photocatalysts with a crystallite size up to 10 nm showed the highest activity.

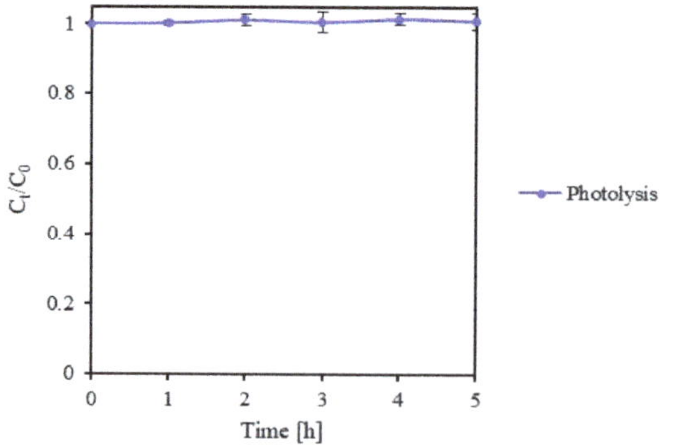

Figure 7. Methylene blue photolysis curve during 5 h of UV light irradiation.

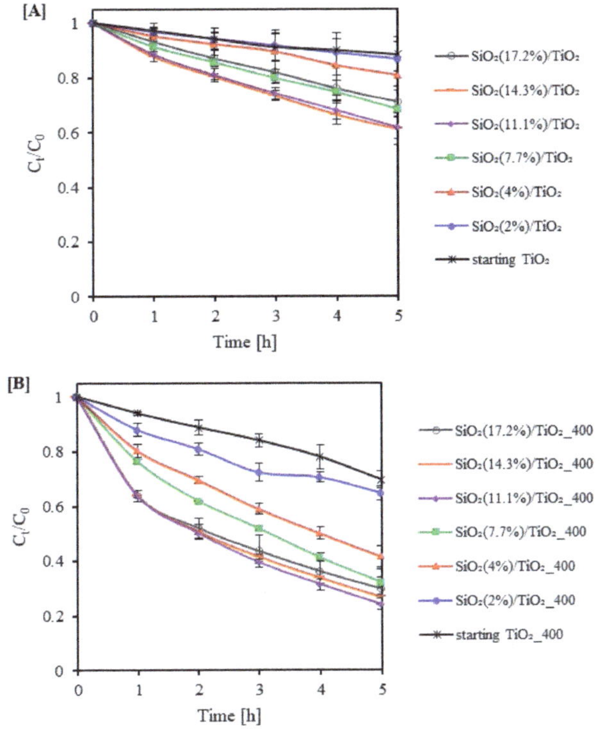

Figure 8. Methylene blue decomposition under UV irradiation of starting TiO_2 and silica-modified TiO_2 prior (**A**) and after calcination at 400 °C (**B**).

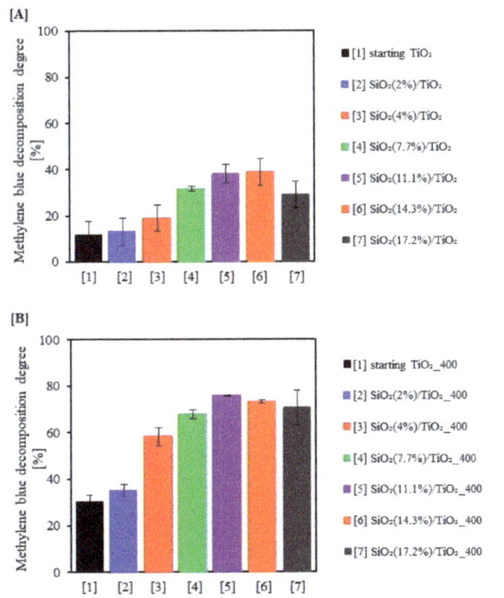

Figure 9. Methylene blue decomposition degree after 5 h of UV light irradiation for starting TiO_2 and silica-modified TiO_2 prior (**A**) and after calcination at 400 °C (**B**).

In order to better understand the photocatalytic degradation of methylene blue, the apparent reaction rate constants were determined. The degradation of methylene blue for the starting TiO_2_400 followed the zero-order model. However, for SiO_2(4%)/TiO_2, SiO_2(7.7%)/TiO_2, SiO_2(17.2%)/TiO_2, SiO_2(4%)/TiO_2_400, SiO_2(7.7%)/TiO_2_400 and SiO_2(11.1%)/TiO_2_400, the dye degradation followed a pseudo-first-order model. For all other samples, the degradation followed a pseudo-second-order model. The linear transformations of zero order, pseudo-first order, and pseudo-second order are shown in Figure 10A–C. It can be seen that, after 3 h of radiation, the points on the graphs started to deviate slightly from the typical linear curve. The observed reduction in reaction rate was due to the formation of intermediates during the dye decomposition. According to the data presented in Table 3, the highest k_1 values of methylene blue decomposition were obtained for the heat-treated photocatalysts. Similar to the pseudo-first-order model, higher k_2 values were recorded for the materials after calcination. It should be noted that the highest rate constant (0.052 L/(min·mg) was observed for SiO_2(14.3%)/TiO_2_400, and this value is 17.3 times higher than for starting TiO_2.

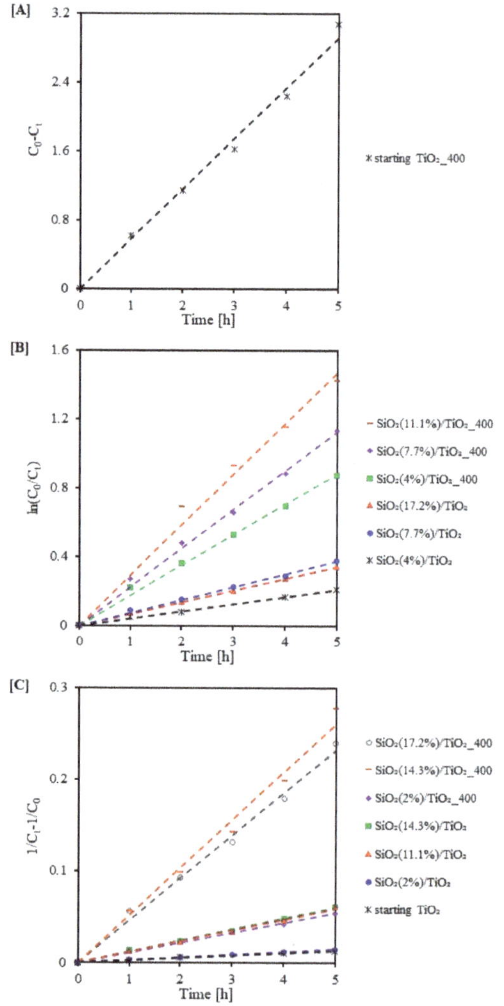

Figure 10. The zero-order plot (**A**), the pseudo-first-order (**B**), and the pseudo-second-order plot (**C**) of methylene blue decomposition.

Table 3. The fitting parameters, zero, pseudo-first, and pseudo-second reaction rate constants for methylene blue decomposition (after 5 h of UV radiation).

Sample Code	k_0 (mg/(L·min))	R^2	Sample Code	k_1 (1/min)	R^2	Sample Code	k_2 (L/(min·mg))	R^2
starting TiO_2_400	0.583	0.99	SiO_2(4%)/TiO_2	0.042	0.99	starting TiO_2	0.003	0.99
			SiO_2(17.2%)/TiO_2	0.068	0.99	SiO_2(2%)/TiO_2	0.003	0.99
			SiO_2(7.7%)/TiO_2	0.075	0.99	SiO_2(2%)/TiO_2_400	0.011	0.99
			SiO_2(4%)/TiO_2_400	0.176	0.99	SiO_2(11.1%)/TiO_2	0.012	0.99
			SiO_2(7.7%)/TiO_2_400	0.225	0.99	SiO_2(14.3%)/TiO_2	0.012	0.99
			SiO_2(11.1%)/TiO_2_400	0.294	0.98	SiO_2(17.2%)/TiO_2_400	0.046	0.99
						SiO_2(14.3%)/TiO_2_400	0.052	0.98

2.3. Photocatalytic Mechanism

The proposed mechanism for methylene blue (MB) degradation in the presence of SiO_2/TiO_2 is shown in Figure 11. When the photocatalyst absorbs a photon with an energy equal or greater than the band gap energy, the activation occurs, which means an electron ejection from the valence band and its transfer to the conduction band with the generation of an electron gap. The active hydroxyl radicals are generated. As an active oxidizing agent, the hydroxyl radicals attack the MB presents at the surface of TiO_2 and decompose it into harmless species, such as carbon dioxide and water. However, it is well known that the high recombination of photogenerated e^--h^+ pairs makes it difficult for the effective photodegradation of pollutants. According to the literature [57,58], connection between TiO_2 and SiO_2 (in our case, confirmed by the DRIFT analysis suggesting the presence of Ti–O–Si band) enhances charge carriers' separation (e^- and h^+) and facilitates the transfer between each other. Thus, the suppression of the electron–hole recombination is limited, causing the enhancement in photoactivity.

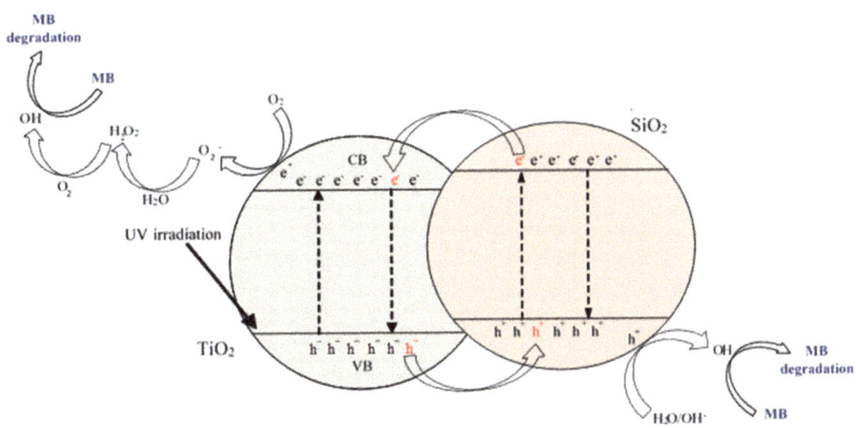

Figure 11. Schematic of the photocatalysis mechanism on SiO_2/TiO_2 surface (based on [57]).

3. Materials and Methods

3.1. Materials and Reagents

The SiO_2/TiO_2 photocatalysts were obtained with the sol-gel method using titanium(IV) isopropoxide TTIP (≥97%, Sigma-Aldrich Co., Saint Louis, MO, USA) as a precursor of TiO_2, fumed silica (≥99.8%, S_{BET} > 200 m^2/g, average particle size 7–14 nm, PlasmaChem GmbH, Germany) as a silica precursor, isopropyl alcohol (pure p.a, Firma Chempur®, Piekary Śląskie, Poland) as a solvent and hydrochloric acid (35–38% pure, Firma Chempur®, Piekary Śląskie, Poland) as a promoter of the hydrolysis reaction. Methylene blue (purity ≥ 82%, Firma Chempur®, Piekary Śląskie, Poland) was used as an organic dye compound in photocatalytic tests.

3.2. Preparation of SiO_2/TiO_2 Photocatalysts

The sol-gel method was used to obtain SiO_2/TiO_2 photocatalysts. Titanium(IV) isopropoxide and fumed silica were chosen as TiO_2 and SiO_2 precursors, respectively. At first, 5 mL of titanium precursor was dropwise added to 15 mL of isopropyl alcohol. Then, 2, 4, 7.7, 11.1, 14.3, or 17.2 wt.% of silica precursor was added. Next, the pH of the obtained solution was adjusted to 2 with hydrochloric acid. Then, the hydrolysis process was started by slowly adding 100 mL of water: isopropyl alcohol mixture (25:75 *v/v*). In the end, the solution was placed in a magnetic stirrer for 1 h and then left for the ageing process taking 24 h. The obtained gel was dried in a muffle furnace at 100 °C for 24 h. Finally, the received material was calcined at 400 °C in a tube furnace under an argon atmosphere (60 mL/min, purity: 5.0, Messer Polska Sp. z o.o., Chorzów, Poland). The obtained photocatalysts were determined as follows: $SiO_2(x\%)/TiO_2\_400$, where x is the weight percentage of SiO_2 in the sample and 400 is the heating temperature. The reference sample, labelled as a starting TiO_2, was also received using the same method but without silica.

3.3. Characterization Methods

Diffuse reflectance DRIFT spectra were recorded using FT-IR-4200 spectrometer (JASCO International Co. Ltd., Tokyo, Japan) equipped with a DiffuseIR accessory (PIKE Technologies, Cottonwood Dr, Fitchburg, WI, USA) and were examined in the range of 4000–400 cm^{-1}. The crystalline structure of received photocatalysts was identified utilizing XRD analysis (Malvern PANalytical B.V., Almelo, The Netherlands), using Cu Kα radiation (λ = 0.154056 nm). The XRD diffractograms were collected in the range of 20–80° on 2θ scale. The PDF-4+ 2014 International Centre for Diffraction Data database (04-0028296 PDF4+ card for anatase and 04-007-0758 PDF4+ card for brookite) was used for the specification of the phase composition. The mean crystallite sizes of the nanomaterials were calculated using the Rietveld method. The surface area (S_{BET}) and pore volume of the tested photocatalysts were calculated from the nitrogen adsorption–desorption measurements at 77 K carried out in QUADRASORB evoTM Gas Sorption analyzer (Anton Paar GmbH, Graz, Austria). All samples were degassed at 100 °C for 16 h under a high vacuum before measurements to preclean the surface of the tested samples. The total pore volume (V_{total}) was calculated from the adsorbed nitrogen after the pore condensation was completed at a relative pressure p/p_0 = 0.99. The volume of micropores (V_{micro}) was determined using the Dubinin–Radushkevich equation. The volume of mesopores (V_{meso}) was determined as the difference between V_{total} and V_{micro}. The surface nature of the studied materials was determined from scanning electron microscopy SEM images. The SU8020 Ultra-High Resolution Field Emission microscope (Hitachi Ltd., Tokyo, Japan) was used for the measurements. The UV-Vis/DR diffuse reflection spectra of the prepared samples were recorded in the range of 200–800 nm using a V-650 UV-Vis spectrophotometer (JASCO International Co., Tokyo, Japan) equipped with an PIV-756 integrating sphere accessory for studying DR spectra. Spectralon (Spectralon® Diffuse Reflectance Material, Labsphere, USA) was used as the standard sample. The values of the band gap energy (E_g) of the studied photocatalysts were calculated with the Kubelk–Munk Equation (1) [59]:

$$F(R) = (1 - R)^2/2R \quad (1)$$

where

F(R)—radiation absorption coefficient;
R—reflectance.

From the above equation, the radiation absorption coefficient was determined. Next, the dependence of energy corresponding to each wavelength was plotted against the square of the inverse of the energy product and the absorption coefficient $(F(R)\cdot E)^{1/2}$.

Finally, a tangent was drawn where the fit is closest to the energy change curve. After transforming the equation of the plotted line, the value of band gap energy was obtained. ZetaSizer NanoSeries ZS (Malvern Panalytical Ltd., Malvern, UK) was used to determine the zeta potential values. The zero-order reaction rate constant can be described as (Equation (2)) [60]:

$$C_0 - C_t = k_0 t \tag{2}$$

The pseudo-first reaction rate constant was determined using the Langmuir–Hinshelwood kinetics model, as shown in Equation (3) [61]:

$$\ln(C_0/C_t) = kKt = k_1 t \tag{3}$$

While the pseudo-second reaction rate constant can be described as (Equation (4)) [62]:

$$1/C_t - 1/C_0 = k_2 t \tag{4}$$

where:

C_0—initial concentration of the methylene blue in solution [mg/L];
C_t—concentration of the methylene blue at time t [mg/L];
k_0—zero-order reaction rate constant [mg/(L·min)];
k_1—pseudo-first reaction rate constant [1/min];
k_2—pseudo-second reaction rate constant [L/(min·mg)];
t—time of irradiation [min];
K—adsorption coefficient of the reactant [L/mg].

3.4. Photocatalytic Activity Test

The photocatalytic activity of the obtained SiO_2/TiO_2 materials was determined using a methylene blue (MB) solution (10 mg/L) as a model water pollutant. During the experiment, a glass beaker containing 0.2 g/L of the tested photocatalyst and 0.5 L of MB solution was placed under an artificial UV-VIS light source comprised of six lamps with the power of 20 W each (Philips) with the radiation intensity of about 138 W/m^2 UV (for the range of 280–400 nm) and 167 W/m^2 VIS (for the range of 300–2800 nm). The emission spectrum of the lamp used is shown in Figure 12. Prior to irradiation, the suspension was stirred in the dark for 1 h to establish the adsorption–desorption equilibrium. After that, the lamps were switched on, and the suspension was irradiated for 5 h while maintaining vigorous stirring. A sample (10 mL) of the suspension was collected every 1 h and then centrifuged to separate the photocatalyst from the MB solution to determine the MB degradation. The methylene blue absorbance was measured by the V-630 UV-Vis spectrometer (Jasco International Co., Tokyo, Japan). The degradation degree of methylene blue was described with C_t/C_0 formula, where C_t stands for the dye absorbance at a given time point and C_0 for the dye absorbance after adsorption.

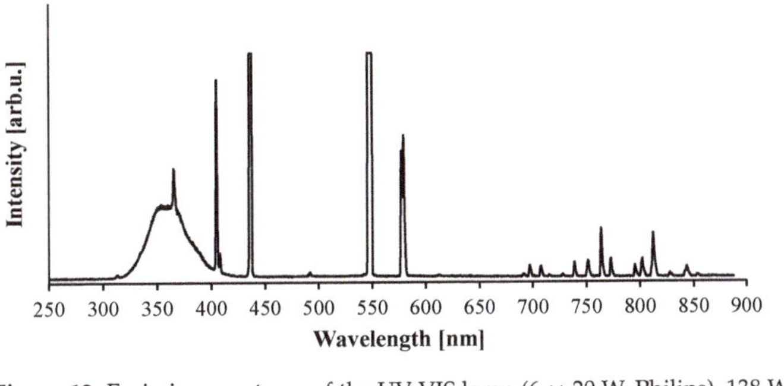

Figure 12. Emission spectrum of the UV-VIS lamp (6 × 20 W, Philips), 138 W/m^2 UV for 280–400 nm and 167 W/m^2 VIS for 300–2800 nm.

4. Conclusions

In conclusion, SiO_2/TiO_2 nanomaterials were prepared by a sol-gel method combined with a calcination process at 400 °C in an inert gas atmosphere. In this paper, fumed silica was used as a silica precursor for the first time. The presence of SiO_2 on the TiO_2 surface was confirmed by FT-IR/DRS infrared spectroscopy. The relationships between the effects of different properties on the photocatalytic activity of the obtained materials were determined. It was found that the modification with SiO_2 increased the specific surface area and total pore volume. In addition, the presence of a modifier contributed to a change in the value of band gap energy and effectively inhibited the growth of crystallites during thermal modification. The photoactivity of the obtained samples was investigated by the degradation of methylene blue under UV irradiation. In general, it was found that the applied modification of titanium dioxide using SiO_2, both without and with the calcination phase, contributed to the increase in photocatalytic activity. All the obtained photocatalysts showed higher activity (up to 75.81%) than the starting TiO_2, which removed only 7.19% of methylene blue. Moreover, the photocatalytic activity increased with increasing SiO_2 content in the sample. Samples containing 11.1 and 14.3 wt.% SiO_2 showed the highest removal of methylene blue.

Author Contributions: Conceptualization: A.B. and A.W.; investigation: A.B., A.W. and M.S.; data curation: A.W., E.K.-N. and A.W.M.; writing—original draft preparation: A.B.; writing—review and editing: A.W., E.K.-N. and A.W.M.; project administration: A.W.M.; funding acquisition: A.W.M. All authors have read and agreed to the published version of the manuscript.

Funding: This work was supported by grant 2017/27/B/ST8/02007 from the National Science Centre, Poland.

Data Availability Statement: The data presented in this study are available upon request from the corresponding author.

Conflicts of Interest: The authors declare no conflict of interest.

References

1. Liu, C.; Mao, S.; Wang, H.; Wu, Y.; Wang, F.; Xia, M.; Chen, Q. Peroxymonosulfate-assisted for facilitating photocatalytic degradation performance of 2D/2D WO_3/BiOBr S-scheme heterojunction. *Chem. Eng. J.* **2022**, *430*, 132806. [CrossRef]
2. Liu, C.; Mao, S.; Shi, M.; Wang, F.; Xia, M.; Chen, Q.; Ju, X. Peroxymonosulfate activation through 2D/2D Z-scheme CoAl-LDH/BiOBr photocatalyst under visible light for ciprofloxacin degradation. *J. Hazard. Mater.* **2021**, *420*, 126613. [CrossRef] [PubMed]
3. Liu, C.; Mao, S.; Shi, M.; Hong, X.; Wang, D.; Wang, F.; Xia, M.; Chen, Q. Enhanced photocatalytic degradation performance of $BiVO_4$/BiOBr through combining Fermi level alteration and oxygen defect engineering. *Chem. Eng. J.* **2022**, *449*, 137757. [CrossRef]
4. Boyjoo, Y.; Sun, H.; Liu, J.; Pareek, V.K.; Wang, S. A review on photocatalysis for air treatment: From catalyst development to reactor design. *Chem. Eng. J.* **2017**, *310*, 537–559. [CrossRef]
5. Reddy, P.V.L.; Kavitha, B.; Reddy, P.A.K.; Kim, K.H. TiO_2-based photocatalytic disinfection of microbes in aqueous media: A review. *Environ. Res.* **2017**, *154*, 296–303. [CrossRef]
6. Nyamukamba, P.; Okoh, O.; Mungondori, H.; Taziwa, R.; Zinya, S. Synthetic methods for titanium dioxide nanoparticles: A review. In *Titanium Dioxide—Material for a Sustainable Environment*; IntechOpen: London, UK, 2018. [CrossRef]
7. Chen, H.S.; Kumar, R.V. Sol-gel TiO_2 in self-organization process: Growth, ripening and sintering. *RSC Adv.* **2012**, *2*, 2294–2301. [CrossRef]
8. Badmus, K.O.; Wewers, F.; Al-Abri, M.; Shahbaaz, M.; Petrik, L.F. Synthesis of Oxygen Deficient TiO_2 for Improved Photocatalytic Efficiency in Solar Radiation. *Catalysts* **2021**, *11*, 904. [CrossRef]
9. Tsebriienko, T.; Popov, A.I. Effect of poly (titanium oxide) on the viscoelastic and thermophysical properties of interpenetrating polymer networks. *Crystals* **2021**, *11*, 794. [CrossRef]
10. Serga, V.; Burve, R.; Krumina, A.; Romanova, M.; Kotomin, E.A.; Popov, A.I. Extraction-pyrolytic method for TiO_2 polymorphs production. *Crystals* **2021**, *11*, 431. [CrossRef]
11. Jiang, Y.; Amal, R. Selective synthesis of TiO_2-based nanoparticles with highly active surface sites for gas-phase photocatalytic oxidation. *Appl. Catal. B Environ.* **2013**, *138*, 260–267. [CrossRef]
12. Danks, A.E.; Hall, S.R.; Schnepp, Z.J.M.H. The evolution of 'sol-gel'chemistry as a technique for materials synthesis. *Mater. Horiz.* **2016**, *3*, 91–112. [CrossRef]

13. Tryba, B.; Tygielska, M.; Colbeau-Justin, C.; Kusiak-Nejman, E.; Kapica-Kozar, J.; Wróbel, R.; Żołnierkiewicz, G.; Guskos, N. Influence of pH of sol-gel solution on phase composition and photocatalytic activity of TiO_2 under UV and visible light. *Mater. Res. Bull.* **2016**, *84*, 152–161. [CrossRef]
14. Ciesielczyk, F.; Przybysz, M.; Zdarta, J.; Piasecki, A.; Paukszta, D.; Jesionowski, T. The sol-gel approach as a method of synthesis of xMgO· ySiO_2 powder with defined physicochemical properties including crystalline structure. *J. Sol-Gel Sci. Technol.* **2014**, *71*, 501–513. [CrossRef]
15. Daghrir, R.; Drogui, P.; Robert, D. Modified TiO_2 for environmental photocatalytic applications: A review. *Ind. Eng. Chem. Res.* **2013**, *52*, 3581–3599. [CrossRef]
16. Pal, A.; Jana, T.K.; Chatterjee, K. Silica supported TiO_2 nanostructures for highly efficient photocatalytic application under visible light irradiation. *Mater. Res. Bull.* **2016**, *76*, 353–357. [CrossRef]
17. Raj, K.; Smith, Y.R.; Subramanian, V.R.; Viswanathan, B. Structural studies of silica modified titania and its photocatalytic activity of 4-chlorophenol oxidation in aqueous medium. *Indian J. Chem.* **2010**, *49*, 867–875.
18. Fu, X.; Clark, L.A.; Yang, Q.; Anderson, M.A. Enhanced photocatalytic performance of titania-based binary metal oxides: TiO_2/SiO_2 and TiO_2/ZrO_2. *Environ. Sci. Technol.* **1996**, *30*, 647–653. [CrossRef]
19. Jimmy, C.Y.; Yu, J.; Ho, W.; Zhao, J. Light-induced super-hydrophilicity and photocatalytic activity of mesoporous TiO_2 thin films. *J. Photochem. Photobiol. A Chem.* **2002**, *148*, 331–339.
20. Yu, J.; Zhao, X.; Jimmy, C.Y.; Zhong, G.; Han, J.; Zhao, Q. The grain size and surface hydroxyl content of super-hydrophilic TiO_2/SiO_2 composite nanometer thin films. *J. Mater. Sci. Lett.* **2001**, *20*, 1745–1748.
21. Chun, H.; Yizhong, W.; Hongxiao, T. Influence of adsorption on the photodegradation of various dyes using surface bond-conjugated TiO_2/SiO_2 photocatalyst. *Appl. Catal. B Environ.* **2001**, *35*, 95–105. [CrossRef]
22. Xie, C.; Xu, Z.; Yang, Q.; Xue, B.; Du, Y.; Zhang, J. Enhanced photocatalytic activity of titania–silica mixed oxide prepared via basic hydrolyzation. *Mater. Sci. Eng.S B* **2004**, *112*, 34–41. [CrossRef]
23. Ding, Z.; Lu, G.Q.; Greenfield, P.F. Role of the crystallite phase of TiO_2 in heterogeneous photocatalysis for phenol oxidation in water. *J. Phys. Chem. B* **2000**, *104*, 4815–4820. [CrossRef]
24. Fatimah, I. Preparation of TiO_2-SiO_2 via sol-gel method: Effect of silica precursor on catalytic and photocatalytic properties. In Proceedings of the 11th Joint Conference on Chemistry in Conjunction with the 4th Regional Biomaterials Scientific Meeting, Purwokerto, Indonesia, 15–16 September 2016. [CrossRef]
25. Qourzal, S.; Barka, N.; Tamimi, M.; Assabbane, A.; Nounah, A.; Ihlal, A.; Ait-Ichou, Y. Sol-gel synthesis of TiO_2–SiO_2 photocatalyst for β-naphthol photodegradation. *Mater. Sci. Eng. C* **2009**, *29*, 1616–1620. [CrossRef]
26. Nandanwar, R.; Singh, P.; Syed, F.F.; Haque, F.Z. Preparation of TiO_2/SiO_2 nanocomposite with non-ionic surfactants via sol-gel process and their photocatalytic study. *Orient. J. Chem.* **2014**, *30*, 1577–1584. [CrossRef]
27. Morawski, A.W.; Kusiak-Nejman, E.; Wanag, A.; Kapica-Kozar, J.; Wróbel, R.J.; Ohtani, B.; Aksienionek, M.; Lipińska, L. Photocatalytic degradation of acetic acid in the presence of visible light-active TiO_2-reduced graphene oxide photocatalysts. *Catal. Today* **2017**, *280*, 108–113. [CrossRef]
28. Resende, S.F.; Nunes, E.H.M.; Houmard, M.; Vasconcelos, W.L. Simple sol-gel process to obtain silica-coated anatase particles with enhanced TiO_2-SiO_2 interfacial area. *J. Colloid Interface Sci.* **2014**, *433*, 211–217. [CrossRef]
29. Górska, P.; Zaleska, A.; Kowalska, E.; Klimczuk, T.; Sobczak, J.W.; Skwarek, E.; Janusz, W.; Hupka, J. TiO_2 photoactivity in vis and UV light: The influence of calcination temperature and surface properties. *Appl. Catal. B Environ.* **2008**, *84*, 440–447. [CrossRef]
30. Maira, A.J.; Coronado, J.M.; Augugliaro, V.; Yeung, K.L.; Conesa, J.C.; Soria, J. Fourier transform infrared study of the performance of nanostructured TiO_2 particles for the photocatalytic oxidation of gaseous toluene. *J. Catal.* **2001**, *202*, 413–420. [CrossRef]
31. Chesalov, Y.A.; Chernobay, G.B.; Andrushkevich, T.V. FTIR study of the surface complexes of β-picoline, 3-pyridine-carbaldehyde and nicotinic acid on sulfated TiO_2 (anatase). *J. Mol. Catal. A Chem.* **2013**, *373*, 96–107. [CrossRef]
32. Yang, G.; Jiang, Z.; Shi, H.; Xiao, T.; Yan, Z. Preparation of highly visible-light active N-doped TiO_2 photocatalyst. *J. Mater. Chem.* **2010**, *20*, 5301–5309. [CrossRef]
33. Araghi, M.A.; Shaban, N.; Bahar, M. Synthesis and characterization of nanocrystalline barium strontium titanate powder by a modified sol-gel processing. *Mater. Sci.-Pol.* **2016**, *34*, 63–68. [CrossRef]
34. Li, Z.; Hou, B.; Xu, Y.; Wu, D.; Sun, Y.; Hu, W.; Deng, F. Comparative study of sol–gel-hydrothermal and sol-gel synthesis of titania–silica composite nanoparticles. *J. Solid State Chem.* **2005**, *178*, 1395–1405. [CrossRef]
35. Andrade-Guel, M.; Díaz-Jiménez, L.; Cortés-Hernández, D.; Cabello-Alvarado, C.; Ávila-Orta, C.; Bartolo-Pérez, P.; Gamero-Melo, P. Microwave assisted sol–gel synthesis of titanium dioxide using hydrochloric and acetic acid as catalysts. *Boletín Soc. Española Cerámica Vidr.* **2019**, *58*, 171–177. [CrossRef]
36. Pelaez, M.; Nolan, N.T.; Pillai, S.C.; Seery, M.K.; Falaras, P.; Kontos, A.G.; Dunlop, P.S.M.; Hamilton, J.W.J.; Byrne, J.A.; O'Shea, K.; et al. A review on the visible light active titanium dioxide photocatalysts for environmental applications. *Appl. Catal. B Environ.* **2012**, *125*, 331–349. [CrossRef]
37. Kim, M.G.; Kang, J.M.; Lee, J.E.; Kim, K.S.; Kim, K.H.; Cho, M.; Lee, S.G. Effects of calcination temperature on the phase composition, photocatalytic degradation, and virucidal activities of TiO_2 nanoparticles. *ACS Omega* **2021**, *6*, 10668–10678. [CrossRef] [PubMed]
38. Xu, G.; Zheng, Z.; Wu, Y.; Feng, N. Effect of silica on the microstructure and photocatalytic properties of titania. *Ceram. Int.* **2009**, *35*, 1–5. [CrossRef]

39. Lu, Z.; Jiang, X.; Zhou, B.; Wu, X.; Lu, L. Study of effect annealing temperature on the structure, morphology and photocatalytic activity of Si doped TiO_2 thin films deposited by electron beam evaporation. *Appl. Surf. Sci.* **2011**, *257*, 10715–10720.
40. Sing, K.S.W. Reporting physisorption data for gas/solid systems with special reference to the determination of surface area and porosity (Provisional). *Pure Appl. Chem.* **1982**, *54*, 2201–2218. [CrossRef]
41. Sing, K.S.; Williams, R.T. Physisorption hysteresis loops and the characterization of nanoporous materials. *Adsorpt. Sci. Technol.* **2004**, *22*, 773–782.
42. Kutarov, V.V.; Tarasevich, Y.I.; Aksenenko, E.V.; Ivanova, Z.G. Adsorption hysteresis for a slit-like pore model. *Russ. J. Phys. Chem. A* **2011**, *85*, 1222–1227. [CrossRef]
43. Leofanti, G.; Padovan, M.; Tozzola, G.; Venturelli, B.J.C.T. Surface area and pore texture of catalysts. *Catal. Today* **1998**, *41*, 207–219. [CrossRef]
44. SiO#2—Krzemionka Koloidalna, Nanoproszek, Hydrofilowy. Available online: https://3d-nano.com/pl/catalogue/sio2-krzemionka-koloidalna-nanoproszek-hydrofilowy/ (accessed on 22 September 2022).
45. Bao, N.; Wei, Z.; Ma, Z.; Liu, F.; Yin, G. Si-doped mesoporous TiO_2 continuous fibers: Preparation by centrifugal spinning and photocatalytic properties. *J. Hazard. Mater.* **2010**, *174*, 129–136. [CrossRef] [PubMed]
46. Nyamukamba, P.; Tichagwa, L.; Greyling, C. The influence of carbon doping on TiO_2 nanoparticle size, surface area, anatase to rutile phase transformation and photocatalytic activity. *Mater. Sci. Forum* **2012**, *712*, 49–63. [CrossRef]
47. Jesionowski, T.; Żurawska, J.; Krysztafkiewicz, A.; Pokora, M.; Waszak, D.; Tylus, W. Physicochemical and morphological properties of hydrated silicas precipitated following alkoxysilane surface modification. *Appl. Surf. Sci.* **2003**, *205*, 212–224. [CrossRef]
48. Tan, L.L.; Ong, W.J.; Chai, S.P.; Mohamed, A.R. Reduced graphene oxide-TiO_2 nanocomposite as a promising visible-light-active photocatalyst for the conversion of carbon dioxide. *Nanoscale Res. Lett.* **2013**, *8*, 465. [CrossRef]
49. Nilchi, A.; Janitabar-Darzi, S.; Mahjoub, A.R.; Rasouli-Garmarodi, S. New TiO_2/SiO_2 nanocomposites—Phase transformations and photocatalytic studies. *Colloids Surf. A: Physicochem. Eng. Asp.* **2010**, *361*, 25–30. [CrossRef]
50. Ferreira-Neto, E.P.; Ullah, S.; Simões, M.B.; Perissinotto, A.P.; de Vicente, F.S.; Noeske, P.L.M.; Ribeiro, S.J.L.; Rodrigues-Filho, U.P. Solvent-controlled deposition of titania on silica spheres for the preparation of SiO_2@ TiO_2 core@ shell nanoparticles with enhanced photocatalytic activity. *Colloids Surf. A Physicochem. Eng. Asp.* **2019**, *570*, 293–305. [CrossRef]
51. Nowacka, M.; Ambrożewicz, D.; Jesionowski, T. TiO_2-SiO_2/Ph-POSS functional hybrids: Preparation and characterisation. *J. Nanomater.* **2013**, *2013*, 1–10. [CrossRef]
52. Carp, O.; Huisman, C.L.; Reller, A. Photoinduced reactivity of titanium dioxide. *Prog. Solid State Chem.* **2004**, *32*, 33–177. [CrossRef]
53. Alvaro, M.; Aprile, C.; Benitez, M.; Carbonell, E.; García, H. Photocatalytic activity of structured mesoporous TiO_2 materials. *J. Phys. Chem. B* **2006**, *110*, 6661–6665. [CrossRef]
54. Periyat, P.; Baiju, K.V.; Mukundan, P.; Pillai, P.K.; Warrier, K.G.K. High temperature stable mesoporous anatase TiO_2 photocatalyst achieved by silica addition. *Appl. Catal. A Gen.* **2008**, *349*, 13–19. [CrossRef]
55. Allen, N.S.; Mahdjoub, N.; Vishnyakov, V.; Kelly, P.J.; Kriek, R.J. The effect of crystalline phase (anatase, brookite and rutile) and size on the photocatalytic activity of calcined polymorphic titanium dioxide (TiO_2). *Polym. Degrad. Stab.* **2018**, *150*, 31–36. [CrossRef]
56. Zhang, Z.; Wang, C.C.; Zakaria, R.; Ying, J.Y. Role of particle size in nanocrystalline TiO_2-based photocatalysts. *J. Phys. Chem. B* **1998**, *102*, 10871–10878. [CrossRef]
57. Nabih, S.; Shalan, A.E.; Serea, E.S.A.; Goda, M.A.; Sanad, M.F. Photocatalytic performance of TiO_2@ SiO_2 nanocomposites for the treatment of different organic dyes. *J. Mater. Sci. Mater. Electron.* **2019**, *30*, 9623–9633. [CrossRef]
58. Gholami, T.; Bazarganipour, M.; Salavati-Niasari, M.; Bagheri, S. Photocatalytic degradation of methylene blue on TiO_2@ SiO_2 core/shell nanoparticles: Synthesis and characterization. *J. Mater. Sci. Mater. Electron.* **2015**, *26*, 6170–6177. [CrossRef]
59. Chen, D.; Zou, L.; Li, S.; Zheng, F. Nanospherical like reduced graphene oxide decorated TiO_2 nanoparticles: An advanced catalyst for the hydrogen evolution reaction. *Sci. Rep.* **2016**, *6*, 20335. [CrossRef]
60. Abdullah, M.A.; Chong, F.K. Dual-effects of adsorption and photodegradation of methylene blue by tungsten-loaded titanium dioxide. *Chem. Eng. J.* **2010**, *158*, 418–425.
61. Sahoo, C.; Gupta, A.K.; Sasidharan Pillai, I.M. Photocatalytic degradation of methylene blue dye from aqueous solution using silver ion-doped TiO_2 and its application to the degradation of real textile wastewater. *J. Environ. Sci. Health* **2012**, *47*, 1428–1438. [CrossRef]
62. Khan, F.; Wahab, R.; Hagar, M.; Alnoman, R.; Rashid, M. Nanotransition materials (NTMs): Photocatalysis, validated high effective sorbent models study for organic dye degradation and precise mathematical data's at standardized level. *Nanomaterials* **2018**, *8*, 134. [CrossRef]

catalysts

Article

Fluoride-Doped TiO_2 Photocatalyst with Enhanced Activity for Stable Pollutant Degradation

Aida M. Díez [1,2,*], Iván Núñez [1], Marta Pazos [2], M. Ángeles Sanromán [2] and Yury V. Kolen'ko [1]

1 Nanochemistry Research Group, International Iberian Nanotechnology Laboratory, Avenida Mestre José Veiga, 4715-330 Braga, Portugal
2 CINTECX, Grupo de Bioingeniería y Procesos Sostenibles, Departamento de Ingeniería Química, Campus Lagoas-Marcosende, Universidade de Vigo, 36310 Vigo, Spain
* Correspondence: adiez@uvigo.es

Abstract: Fluoride-doped TiO_2 (F-TiO_2) was synthesized by an efficient and simple one-step synthesis and successfully used for the UV-photo-degradation of the toxic and stable pollutants methylene blue (MB) and bisphenol A (BPA). Initially, the synthesized catalyst was characterized and compared to untreated TiO_2 (P25 Degussa) by different physical–chemical analyses such as XRD, band gap calculation, SEM, EDS, FITR, ECSA, or EIS. F-TiO_2 defeated commercial TiO_2, and almost complete pollutant removal was achieved within 30 min. The energy consumption was reduced as a result of the suitable reactor set-up, which reduced light scattering, and by the application of a long-pulse radiation procedure, where the lamp was switched off during periods where the radical degradation continued. This enhanced the overall photocatalysis process performance. Under these conditions, 80% of MB removal was attained within 15 min radiation with an energy consumption of only 0.070 Wh min^{-1}, demonstrating a much better efficiency when compared to previously reported data. The catalyst was reusable, and its performance can be improved by the addition of H_2O_2. The results were validated by BPA degradation and the treatment of real wastewaters with both pollutants. The results were so encouraging that a scale-up reactor has been proposed for future studies.

Keywords: catalyst synthesis; fluoride-doped titanium dioxide; pulse radiation; process evaluation; photocatalyst characterization

Citation: Díez, A.M.; Núñez, I.; Pazos, M.; Sanromán, M.Á.; Kolen'ko, Y.V. Fluoride-Doped TiO_2 Photocatalyst with Enhanced Activity for Stable Pollutant Degradation. *Catalysts* **2022**, *12*, 1190. https://doi.org/10.3390/catal12101190

Academic Editors: Jorge Bedia and Carolina Belver

Received: 13 September 2022
Accepted: 2 October 2022
Published: 7 October 2022

1. Introduction

One of the main global problems is the limited access to uncontaminated water, particularly in developing countries. Among other alternatives, a promising remediation method is photocatalysis. This process is based on the action of a photocatalyst that, after being irradiated with a specific wavelength radiation, has the capacity of generating active species that decompose pollutants [1,2].

The most used photocatalyst is TiO_2, a well-known semiconductor that is not toxic, insoluble, cheap, and very selective on its absorption spectrum [3,4]. This last property makes TiO_2 very active in the UV range. TiO_2 commercialized from Degussa (so-labelled P25) has shown the best performance. Nevertheless, there is still room for improvement with the aim of using this process in real scenarios. Among the parameters to be optimized, there is the widening of the required radiation wavelength of activation and the inhibition of the electron-hole (e^-/h^+) recombination [5].

Indeed, the e^-/h^+ generated on the semiconductor are the responsible species for, respectively, the reduction and oxidation of pollutants and O_2 or H_2O. These cause the generation of oxidant radical species such as HO· or HO_2·. TiO_2 has a high e^-/h^+ recombination rate which diminishes its degradation performance [1,4]. Consequently, several TiO_2 modifications have been tried in order to capture the e^-; hence, the recombination rate would be reduced, providing more time for the reactions of free radical production to happen [2,3].

However, new photocatalysts usually require long and expensive synthesis processes. For instance, Zhan et al. [2] reported the application of a V/Mo-TiO_2 catalyst, the synthesis of which involved several steps and calcination during 4 h at 500 °C. They attained almost complete MB degradation after 1 h. In other studies, the catalyst contained expensive elements such as in the Bi/Ag-TiO_2 catalyst [6]. Similarly, the WO_3/$KNbO_3$ photocatalyst was used for attaining BPA and MB degradation within 20 min by using a 375 W UV lamp. Thus, lamp consumption and the catalyst selection increases the complexity and price of the process [7]. In fact, other authors have pointed out the necessity of more environmentally friendly synthesis processes [8].

We hereby propose a more realistic photocatalyst, where the synthesis process is simple, inexpensive, and environmentally friendly based on the single addition of an element in what we have labelled as one-step synthesis. In this context, several studies stand out, in which catalysts were synthesized by TiO_2 doped with graphene [9] or doped with electronegative compounds such as N, P, S, or halogens, reducing the band gap and shifting the absorption edges favoring photocatalysis [3].

Nevertheless, recent studies have demonstrated that the well-known methylene blue (MB) is still being treated by the usage of complex catalysts. For instance, Bibi et al. [10] required the addition of reduced GO and Fe_3O_4 to TiO_2 on different steps so they arrived at 99% degradation of 3 mg/L of MB effluent within 70 min. Likewise, other authors [11] made a composite Ni/TiO_2/zeolite by several calcination steps and even with this catalyst, they needed long treatment times (2 h) for complete MB removal. On the other hand, when the synthesis process is simpler, the MB photodegradation requires the usage of high intensity lamps [12]. In this study, the addition of fluorine to the TiO_2 surface, in order to avoid or retard e^-/h^+ recombination is reported. This catalyst is based on a previous paper about gas treatment [13] and it may also be useful for effluent treatment. Other authors have essayed the synthesis of F-TiO_2 throughout more complex processes, attaining good efficiencies on MB removal [14].

This one-step synthesized F-TiO_2 catalyst would be, for the first time, assessed for liquid effluent treatment, using two stable model pollutants, namely MB and bisphenol A (BPA). Several authors have carried out the procedure of complementing MB degradation study with a BPA plasticizer [7,15]. This procedure favors the validation of photo-based processes for the degradation of aromatic stable compounds, regardless of being colorful (MB) or colorless (BPA). Moreover, these compounds cause environmental problems due to their stability, toxicity, and increasing usage [15]. Sadly, traditional treatment alternatives such as biodegradation are not efficient for BPA and MB removal due to their stability and toxicity to microorganisms [15]. Therefore, photocatalysis has been proposed as an alternative, although the up-to-date proposed catalysts are complex and have been synthesized by environmentally unfriendly methods. For instance, the glass fibers of Fe-TiO_2–carbon quantum dots [4] or nanodiamond-TiO_2 [8] have been used for the elimination of these pollutants.

Another bottleneck in the photocatalysis processes is related to efficiency problems due to both (i) light scattering with solution depth [16] and (ii) high energy consumption (EC) related to the requirement for UVA lamps, as it is the most useful radiation for TiO_2-based photocatalysis [9].

This research attempts, for the first time, the usage of the one-step-synthesized F-TiO_2 on liquid stream photocatalysis. The scarce studied long-pulse procedure with the usage of a UVA-LED lamp was proposed in order to reduce EC. The H_2O_2 addition was also considered in order to promote higher degradation rates [17]. What is more, a novel reactor design based on the reduction in solution depth and the increase in radiated surface area was utilized, reducing light scattering. Considering this used set-up, a scale-up reactor is proposed.

2. Results

2.1. Preliminary Tests

2.1.1. Rector Set-up

To begin with, and taking into consideration the influence of the radiated area and light scattering on photo-based processes [16], the degradation of a 20 mg/L MB solution was carried out in both an open-wide reactor and a narrow-open reactor using TiO_2 (800 mg/L) as the model photocatalyst (Figure S1). By increasing the solution depth and reducing the radiated area, the MB degradation was reduced by 65% and the pseudo-first kinetic rate of the degradation was reduced by 10 units, that is from 0.024 min^{-1} (R^2 = 0.9970) to 0.0027 min^{-1} (R^2 = 0.9906).

In order to standardize the results, the EC per mg of MB was calculated following Equation (1), where L is the lamp power (W), t is the time the lamp is switched on (h), and m is the mass of the eliminated MB (mg).

$$EC\ (Wh\ mg^{-1}) = L \cdot t/m \quad (1)$$

Indeed, on the open-wide reactor, the EC was 0.154 Wh/mg, whereas the EC on the narrow-open reactor was 0.521 Wh/mg.

2.1.2. Catalyst Doping Dosage

Then, the variation of catalyst performance depending on F dosage was evaluated (Figure S2). No significant differences were found when the content of TiO_2:F was varied from 1:0 to 1:2. The ratio of TiO_2:F 1:1 was selected to continue the experiences due to slightly quicker degradation kinetics. Smaller quantities resulted in a degradation profile equal to raw TiO_2. Thus, the F-TiO_2 catalyst with a ratio 1:1 was used thereafter, and it was named F-TiO_2.

2.1.3. Catalyst Concentration

The study of the catalyst dosage was studied with the optimal F-TiO_2. Figure S3 depicts both the degradation profile in time and the pseudo-first order kinetic constants. As can be seen, increasing the catalyst dosage favored the degradation performance and rate until 1600 mg/L of F-TiO_2 was used. Nevertheless, slight differences were found between 800 and 1600 mg/L and, what is more, the latter exhibited 10% of adsorption, which made the actual degradation 85%, instead of 97% in the case of using 800 mg/L of F-TiO_2. A total of 3200 mg/L caused 17% initial adsorption and a detriment on overall MB removal after 30 min. Consequently, 800 mg/L of F-TiO_2 was selected as the optimal dosage and used thereafter.

2.2. Characterization

The characterization of F-TiO_2 was carried out and it was compared to commercial Degussa P25 TiO_2.

2.2.1. XRD

Crystalline structures of TiO_2 and F-TiO_2 were analyzed throughout the XRD analysis. The spectra of both samples are shown in Figure S4. Typical anatase and rutile phases confirmed the stability of the TiO_2 sample after F doping. F content was not detected on XRD.

2.2.2. Raman

Raman spectra show the characteristics peaks of anatase TiO_2 (Figure 1), namely those placed at 148.5 [Eg (1)], 397.5 [B1g], 516.15 [A1g], and 635.2 [Eg (2)] cm^{-1} [8]. Slight differences between TiO_2 and F-TiO_2 were detected (Figure 1B) which were related to successful F doping [18].

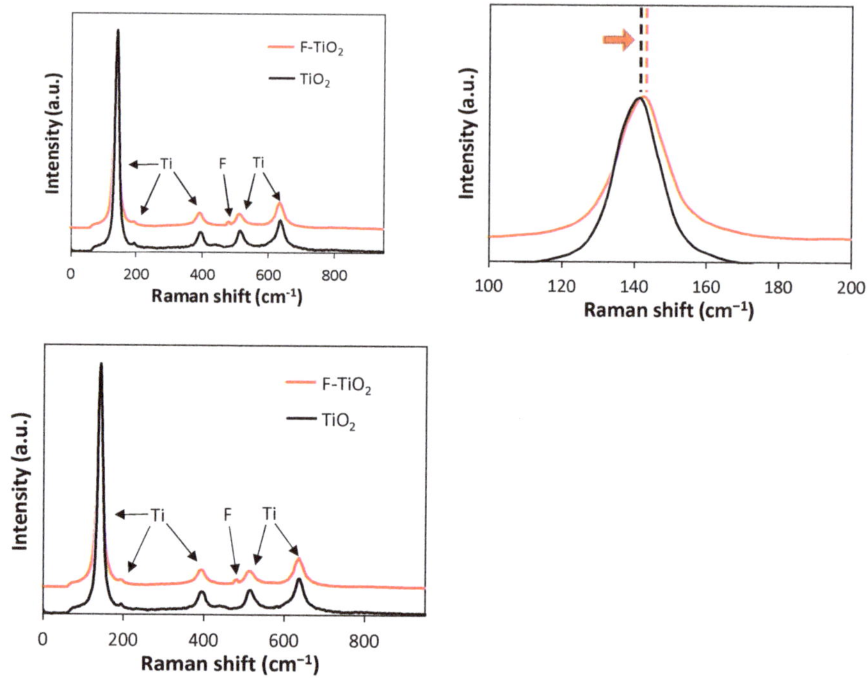

(**A**) (**B**)

Figure 1. Raman spectra of F-TiO_2 and TiO_2 catalysts (**A**), zoom in on the Raman spectra in the range 100–200 cm^{-1} (**B**).

2.2.3. SEM, EDS and ICP

The SEM image (Figure S5) demonstrates that the particle size was in the μm order due to the agglomeration of the particles caused by the synthesis process. A small C quantity was detected maybe due to the remaining IPA from the synthesis process. The EDS results demonstrate the successful F doping where 8% of F was detected. These data were corroborated by ICP measurements after acid digestion of the catalyst. The ICP results showed that each 0.25 mg of sample contained 0.11 mg of Ti (that would be 0.184 mg of TiO_2) and 0.015 mg of F.

2.2.4. UV-Vis Spectra

Spectrophotometric measurements were taken in order to evaluate the differences in the UV-Vis responses of both the TiO_2 and F-TiO_2 catalysts. The UV-Vis response was used for calculating the Tauc plots (Figure S6) following Equation (2), where v is the frequency, A is the proportional constant, n is 4 due to an indirect transition, and E_g is the band energy [19]. The TiO_2 band gap resulted in 3.2 eV, demonstrating the high percentage of the anatase phase on commercial TiO_2 (the rutile phase had a band gap of 3 eV) [3]. On the other hand, F-TiO_2 provided a band gap of 1.5 eV, demonstrating that the addition of F narrowed the band gap of TiO_2.

$$h\upsilon \cdot \alpha = (Ah\upsilon \cdot E_g)^{n/2} \tag{2}$$

2.2.5. Electrochemical Measurements

The electrochemical performance of materials has been demonstrated to be related to their photo-catalytic performance, as it is indicative of electron mobility (Electrochemical Impedance Spectroscopy: EIS) [20] and active surface area (ECSA) [19].

EIS

EIS results are provided in Figure 2, where the equivalent circuits are depicted. The series resistance represents the electrode [21]; this is why the values were practically constant (14.9 and 12.4 Ω), as in both cases Ni-Foam was the supporting working electrode. The parallel resistance stands for the layer resistance of the material [21], and in this case F-TiO_2 provided a much smaller value than TiO_2 (1.77 vs. 12.5 kΩ), demonstrating the electron mobility was much higher on F-TiO_2 [20], which may be related to better photocatalytic activity.

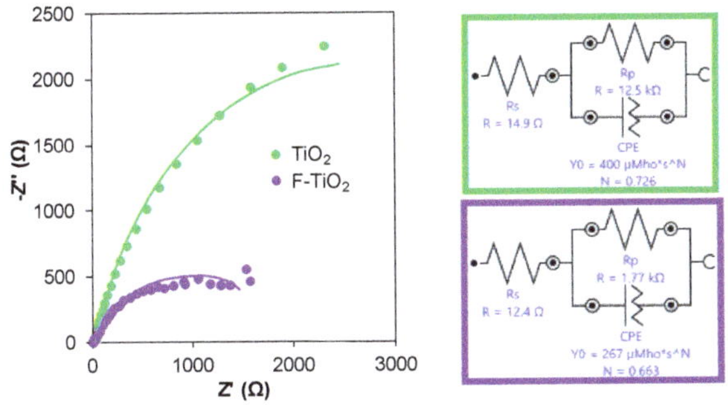

Figure 2. EIS measurement of TiO_2 (green) and F-TiO_2 (purple) with the fitting of the equivalent circuits.

ECSA

ECSA is also related to catalytic activity [22]. After the representation of CVs regarding the scan rate (Figure S7) and following Equation (3), CDL was 0.3 and 0.2 uF cm^{-2} for F-TiO_2 and TiO_2, respectively. That means the ECSA values were 7.5 cm^2 and 5 cm, respectively. Therefore, the surface area useful for electrochemical reactions was 30% higher in the case of F-TiO_2.

2.3. *Pollutant Abatement*

Subsequently, the degradation of the model pollutant MB was carried out, and different parameters were considered.

2.3.1. Catalyst Evaluation

The comparison of synthesized F-TiO_2 with commercial TiO_2 on the performance of MB degradation was accomplished on the wide-open reactor (Figure S2). The results manifest the suitability of the reactor set-up and the efficacy of TiO_2 to work as UV-photocatalyst, as in both cases more than 80% degradation was attained within half an hour. Moreover, the higher performance of F-TiO_2 was remarkable; after the one-step synthesis, it provided an extra 25% MB elimination. Indeed, the pseudo-first order kinetic values also denote a much higher efficiency of F-TiO_2 vs. TiO_2 being, respectively, 0.1367 min^{-1} and 0.0220 min^{-1}. The usage of this one-step synthesized photocatalyst also favored the efficiency of the process as the EC was 0.152 and 0.121 Wh mg^{-1} for TiO_2 and F-TiO_2, respectively, demonstrating the efficacy of the latter.

The high efficiency of the F-TiO_2 catalyst was not only demonstrated by its comparison to the well-performant TiO_2, but also to previous recent data (Table 1). The high complexity of the synthesis processes is remarkable.

Table 1. Last reported data on MB and BPA UV-photocatalysis degradation.

MB (mg/L)	Catalyst (mg/L)	Lamp (nm, w)	Time (min)	Degradation (%)	EC (W·h/mg)	Reference
20	F-TiO_2 (800)	360–365 nm, 4.8 W	15	78.84	0.070	This study
20	F-TiO_2 (800)	360–365 nm, 4.8 W	30	96.24	0.121	This study
73.57	Graphene-TiO_2 (50)	360 nm, 17 W	480	87	2.12	[9]
10	F-TiO_2 (200)	300 W	20	88	11.25	[14]
5	GO-TiO_2 (200)	500 W	60	92	108.7	[1]
3.2	V/Mo-TiO_2 (1000)	365 nm, 8 W	60	86.7	2.88	[2]
10	Zeolite/Ni-TiO_2 (50)	UV, 16 W	120	100	3.20	[11]
10	TiO_2 nanoparticles (60)	UV, 300 W, >420 nm	24	99	12.12	[12]
10	$WO_3/KNbO_3$ (3000)	365 nm, 375 W	40	98	24.3	[7]
3	rGO-Fe_3O_4/TiO_2 (150)	UV, 500 W	70	99	196.4	[10]
BPA (mg/L)	**Catalyst (mg/L)**	**Lamp (nm, w)**	**Time (min)**	**Degradation (%)**	**EC (W·h/mg)**	**Reference**
20	F-TiO_2 (800)	360–365 nm, 4.8 W	30	98	0.122	This study
10	$WO_3/KNbO_3$ (3000)	365 nm, 375 W	240	49	300	[7]
10	Nanodiamond-TiO_2 (80) Glass Fibers	UV, 20 W	100	96	3.47	[8]
20	Fe-TiO_2-carbon quantum dots (n.r.)	$\lambda < 472$ nm, 55 W	60	100	2.75	[4]
20	ZnO-Graphene oxide	254 nm, 15 W	60	99.5	0.75	[23]

n.r. = not reported.

2.3.2. Long-Pulse Radiation Procedure

Taking into consideration the fact that photocatalysis degradation is a radical attack, we considered the poorly studied possibility of treating the effluent by alternating radiated and dark periods. That is, we proposed the usage of the lamp in the long-pulse radiation procedure, so the lamp consumption could be reduced for future usages. In this procedure, the lamp was switched on from 0 to 5 min treatment time and from 10 to 20 min of treatment time. As it can be seen in Figure 3, after 10 and 30 min the detriment on MB degradation was only, 20 and 10%, respectively, whereas the lamp usage was reduced to a half. This means that EC was reduced, roughly, to a half (Figure 3). Consequently, the EC for MB degradation within 30 min was 0.070 Wh mg^{-1} with the long-pulse radiation procedure.

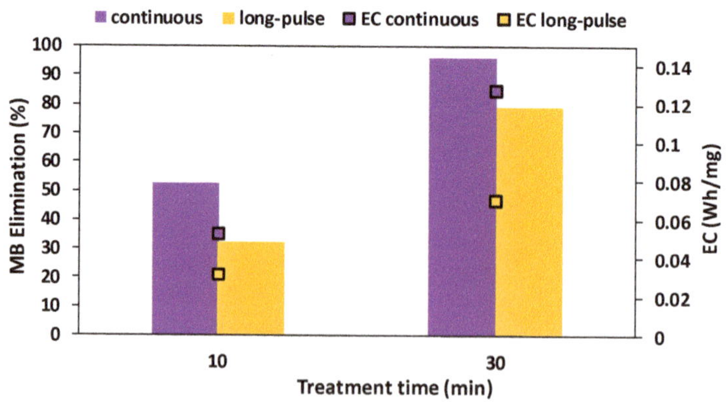

Figure 3. MB elimination (20 mg/L) at a catalyst concentration of 800 mg/L. Continuous radiation (purple), and long-pulse radiation (yellowish).

2.3.3. Reusability

Evaluating the reusability of the catalyst is mandatory for future applications [19] and thus, it was evaluated. For practical purposes, the experiences were tested with continuous radiation. First of all, the reutilization of F-TiO_2 led to a 17% detriment on the third cycle (Figure 4A) which is within typical values [19]. Samples were filtered by 0.22 μm and measured by ICP, where F and Ti content was below detection limits, demonstrating that the doping process is not only simple and economic, but also efficient, leading to a stable F-TiO_2 photocatalyst.

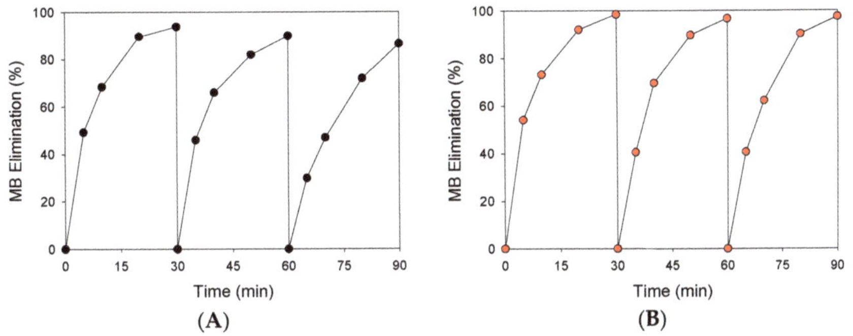

Figure 4. MB elimination (20 mg/L each cycle), using the same photocatalyst in three cycles (800 mg/L) (**A**). MB elimination (40 mg/L each cycle), using the same photocatalyst in three cycles (800 mg/L), and adding 0.3 mg/ml of H_2O_2 (**B**).

Nevertheless, the addition of H_2O_2 as an extra oxidant has been reported to favor photocatalysis [17] and hence, the addition of 0.3 mg/mL of H_2O_2 in each cycle was evaluated (Figure 4B). Indeed, the process was so favorable that 40 mg/L of MB was used in order to have enough room for improvement as well as a better quantification (20 mg/L was eliminated too quickly). One can compare Figure 4A,B and the detriment detected on the degradation of 20 mg/L of MB without H_2O_2 (Figure 4A) was not detected when H_2O_2 was added, even though the initial MB concentration was 40 mg/L (Figure 4B). In any case, the MB concentration effect is studied in next section in order to compare the results with these data.

2.3.4. Pollutant Concentration

The effect of MB concentration was then studied (Figure 5). When the initial MB concentration was set at 40 mg/L, the degradation plunged to 50% (Figure 5A). Concentration differences were more significant during the first 10 min of radiation (Figure 5B) where one can notice that the MB degradation was practically constant in the range 5–20 mg/L, whereas the performance abruptly decreased at 40 mg/L. With 0.3 mg/L of H_2O_2, this detriment did not take place and, what is more, the catalyst could be reused for up to three cycles (Figure 4B).

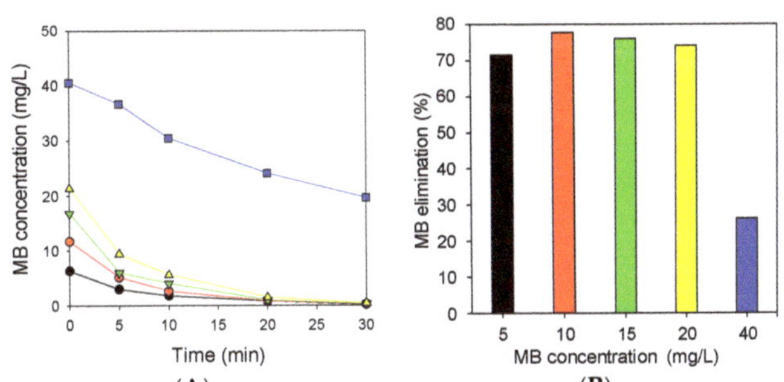

Figure 5. MB degradation under UV radiation depending on the initial MB concentration (**A**). MB degradation after 10 min treatment (**B**).

2.3.5. Treatment of More Complex Effluents

Then, the photocatalyst and reactor set-up suitability for the treatment of polluted effluents were corroborated by treating a solution with the complex polymer BPA as well as real wastewaters with both MB and BPA (Figure 6). This procedure allows the suitability of this process with colorful and colorless effluents to be confirmed. Both pollutants suffer photocatalytic degradation, which is started by the radical attack, so hydroxylation occurs in areas close to the aromatic ring, which are electronegative. Then, the scission of the ring takes place, liberating some carbonaceous chains which end up on simple carboxylic acids [24,25].

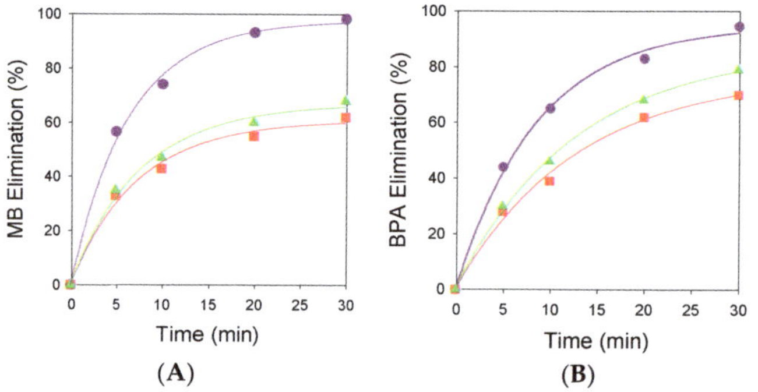

Figure 6. MB (**A**) and BPA (**B**) elimination (20 mg/L) at a catalyst concentration of 800 mg/L with pseudo-first order kinetics adjustment (line) in wastewater that just received a primary treatment (red squares), that received a secondary treatment (green triangles), and Milli-Q water (purple circles).

Regarding MB degradation (Figure 6A) the efficiency was reduced (30%) when using real primary-treated wastewater, contrasting with the results using the distilled Milli-Q matrix. Moreover, when using secondary wastewater, the result was slightly better due to the reduction of the organic matter due to the biological degradation (Table S1). The total organic carbon (TOC) of the solution was 13 mg/L when MB was dissolved in Milli-Q water and 86% mineralization was attained within 30 min of photocatalysis. At this point, the remaining carbon was related to carboxylic acids, which reduced the solution pH from 5.8 to 5.4. On the other hand, the presence of a more complex wastewater matrix increased the initial TOC to 91 and 54 mg/L for, the primary and secondary-treated wastewaters, respectively. After the photodegradation process, TOC was reduced by 66% and 71%, respectively, and the pH was reduced by 0.2 units in both cases. This demonstrates there acid intermediates were also generated although the pH drop was smaller due to the buffer behavior of the salt content (both samples had high conductivity: Table S1). Moreover, the remaining TOC was related to the MB molecule, which was not completely removed due to the degradation of the organic matter already present on the real wastewaters.

Regarding the pseudo-first order kinetic adjustment (Table S2), the slope was around 0.13 min^{-1} for MB treatment on real wastewaters due to the increased complexity of the effluent, as the constant rate was 0.137 min^{-1} when treating MB in Milli-Q water.

Figure 6B demonstrates the suitability of the degradation process for different stable and toxic pollutants, as even BPA was almost completely removed within 30 min. In this case, the treatment of BPA in a primary or secondary wastewater also caused a detriment of around 20% on the performance of the photocatalysis process. In this case, mineralization on Milli-Q water achieved 83% in 30 min and the pH was reduced from 6 to 5.4, which may have been due to the generation of more carboxylic acids, as the BPA-Milli-Q solution had an initial TOC of 16 mg/L. In the case of treating real wastewaters, mineralization dropped to 69% and 76%, for, primary and secondary-treated wastewaters, respectively. This demonstrates that BPA degradation was easier than in the case of MB, due to the fact BPA and its by-products do not cause light scattering. In this case, the pH dropped by 0.4 and 0.3 units, respectively, demonstrating the generation of carboxylic acids [25]. Further studies need to be completed on the scale-up study of this process.

3. Discussion

In order to understand the efficiency of the proposed process, specifically in terms of photocatalyst activity and reactor set-up, some previous studies depicted on Table 1 were used as references and are commented on throughout this section.

3.1. Preliminary Tests

3.1.1. Reactor Set-up

The preliminary test for the MB degradation with TiO_2 photocatalysis was completed. With the TiO_2 catalyst, the comparison between the performance of the wide and the narrow-open reactors (Figure S1) demonstrated the degradation was improved when (i) the solution depth was smaller and (ii) when the radiation area was bigger (open-wide reactor: Figure 7) due to a better activation of the photocatalyst and a reduction in the light scattering. This reactor is notably thinner than the typical cylindrical reactor and moreover, it has the radiation source placed at 3 cm from the solution. Jafari et al. [4] also placed their lamp at 3 cm for BPA degradation; however, their reactor was 5 cm in diameter and 10 cm in height, producing an elevated light scattering, which may cause a lower efficiency than the hereby attained value (2.75 vs. 1.22 Wh/mg, respectively). In fact, they required twice the time for achieving the same degradation of 20 mg/L of BPA (Table 1).

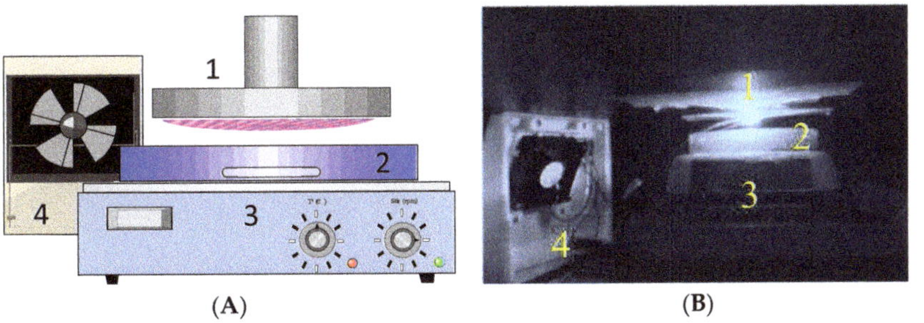

Figure 7. Reactor set-up, schema (**A**), and real (**B**): UV LED (1), crystal open-wide-reactor (2), magnetic stirrer (3), and refrigerator fan (4).

The EC attained in this study was smaller than most of previous references (Table 1). This may indicate, apart from a proper photocatalyst selection, the suitability of the reactor set-up. Indeed, several authors have reported the positive effect of reducing the treated effluent depth when working with photocatalysis systems [26]. Hence, Shiraishi et al. [16] concluded that the UV intensity received on the photocatalyst decreases as liquid depth increases. Nevertheless, they highlighted that the negative effect of solution depth can be overcome with high mixing. In fact, a smaller solution depth reduces light scattering [6]. In our study, the good results may have been fomented by a small solution depth, an elevated radiated area, and a high mixing rate. For instance, Zhang et al. [2] attained an extremely high EC (Table 1). These authors may also have used an elevated catalyst concentration (1000 mg/L) of the complex V/Mo-TiO_2 catalyst. This might have produced light scattering, making the catalyst activation difficult. In fact, these authors used a reactor set-up with ian nappropriate photo-activation, as the lamp was placed at 8 cm instead of the 3 cm fixed in this study. Moreover, the sample was placed in a beaker which undoubtedly has a smaller width than the open-wide reactor hereby proposed (Figure 7). These authors may have attained higher efficiencies by the utilization of a wide-open reactor and higher stirring rates.

Another example was brought by Niu et al. [12] who placed the lamp 15 cm above of the sample, increasing the EC to 12.12 Wh mg^{-1}. Both results could have been promising if the lamp was placed closer to the sample, as the intensity of the lamp is reduced exponentially with distance from the source [27,28]. Indeed, the catalytic performance is related to photo-current intensity, as a greater intensity engenders more e^-/h^+ pairs due to the generation of easier electron transitions [12]. Considering the results attained in this research study, a new reactor set-up is proposed (Figure S8), based on the proximity of the radiation source from the bulb solution and the reduction in solution depth.

3.1.2. Catalyst Doping Dosage

Different F dosages were evaluated (Figure S2), namely, TiO_2:F ratios of 1:0.5, 1:1 and 1:2 as other authors noticed differences related to the performance of UV photocatalysis related to the doping dosage [23,29]. Considering the slight differences, the most efficient photocatalyst to work with was TiO_2:F 1:1 which was further characterized and compared to commercial TiO_2. The exact amount of F dopant was evaluated by subsequent ICP and EDS.

3.1.3. Catalyst Concentration

The effect of photocatalyst dosage was in concordance with previous authors. Thus, a lack of catalyst amount prevents the availability of active sites to be activated by irradiation, whereas an excess of photocatalyst causes the agglomeration of the particles and light scattering [10,30]. This was clearly noticed when testing 3,200 mg/L of F-TiO_2 (Figure S3) where the performance was similar to that of 800 mg/L until it reached 20 min of treatment time. Indeed, after 20 min the performance was even reduced which may be consistent with the excessive generation of oxidants which react with themselves due to a lack of MB particles (as the initial concentration was diminished due to degradation) [31]. Likewise, Bibi et al. [10] demonstrated that MB removal increased from 60 to 100% when switching the rGO-Fe_3O_4/TiO_2 catalyst concentration from 50 to 150 mg/L. Oppositely, increasing their photocatalyst dosage to 200 mg/L led to a 15% detriment in relation to MB elimination. In our study, slight differences were found between the 800 and 1600 mg/L catalyst dosage, although considering the economic aspects and the fact that 1600 mg/L overestimated the degradation performance due to adsorption, 800 mg/L was selected as the preferred concentration.

3.2. *Characterization*

3.2.1. XRD

XRD spectra (Figure S4) showed the characteristics peaks of anastase TiO_2 at 25.2°, 37.9°, 47.8°, 54.3°, and 62.7° θ, which correspond to the planes (101), (004), (200), (105), and (204), which is the most photo-active phase of TiO_2 [6]. Both TiO_2 and F-TiO_2 samples provided a similar XRD profile. This was caused by the fact the ionic radii of F and O are practically the same (0.133 and 0.132 nm) [32]. Moreover, the analogous XRD profiles of both TiO_2 and F-TiO_2 demonstrated that the synthesis process at low temperatures and atmospheric pressure is considerate of the TiO_2 crystalline structure.

3.2.2. Raman

Raman spectra of TiO_2 and F-TiO_2 (Figure 1) demonstrated that TiO_2 was chemically stable in the synthesis process as the spectra were mainly constant. However, F addition caused a defect introduction at 490 cm^{-1} which can be related to F content [33]. The characteristic peak of Ti at 148.5 cm^{-1} suffered a shift and a peak broadening (Figure 1B) due to the formation of F-Ti-O [18].

3.2.3. SEM and EDS

The particle agglomeration (Figure S5) on the F-TiO_2 catalyst can be positive for the photocatalysis performance [8]. EDS provided the experimental weight ratio Ti:F of 8.25:1 which would be an atomic ratio of 3.3:1. The ICP results showed the actual atomic ratio of

Ti:F was 2.8:1. These slight differences can be explained by the superficial character of the EDS measurement.

3.2.4. UV-Vis Spectra

The fluoride doping caused an increase in the solar spectrum [3] as can be seen in Figure S6, where the band gap of F-TiO_2 was 1.5 eV vs. the 3.2 of commercial TiO_2. This was caused by the addition of F, which adds defects that not only reduce the band gap but also act as active centers [2]. Consequently, visible radiation absorption is favored as well as UV absorption [14]. In fact, several authors have demonstrated the dual UV and visible activity of their synthesized catalysts such as V/Mo-TiO_2 [2] or TiO_2-SiO_2 [34]. This dual activity needs to be corroborated in future studies with F-TiO_2.

By doping TiO_2, electron trapping is favored and thus e^-/h^+ recombination is diminished [3] which is related to a better photocatalysis performance. This electron trapping may be favored by electronegative compounds such as F.

3.2.5. Electrochemical Measurements

EIS

The smaller radius arc of the F-TiO_2 sample when compared to TiO_2 (Figure 2) is related to a better charge transfer and to a lower $^-/h^+$ recombination [20,35]. The linear part of the Nyquist plot (0–200 Ω of real Z) is proportional to the limited diffusion processes [36]. Consequently, the fact that the slope is less blatant for F-TiO_2 indicates that this catalyst favors a better diffusion when compared to commercial TiO_2. Indeed, the wider semicircle on the TiO_2 catalyst indicates the electrons reach the semiconductor/interface where the photocatalysis occurs with more difficulty, and thus these electrons could recombine with holes before reacting with pollutants or oxygen molecules [20]. This high charge separation and conductivity of F-TiO_2 indicates it would act as a better catalyst [36]. Indeed, a lower resistance, which is related to a faster migration of electrons and holes, has been associated with a higher generation of radicals throughout the degradation process. This lower resistance enhances the photocatalytic activity so the pollutant can be degraded [12]. Other authors have demonstrated a better charge separation efficiency by the attainment of smaller arc radius [37].

ECSA

Active surface area controls the activity of the photocatalysts due to the interaction between the pollutant and photocatalyst. Indeed, a higher ECSA implies an increased adsorption of the target compound due to the presence of a higher surface area as well as the presence of more active sites [19]. Thus, higher ECSA values favor photodegradation processes [22,38]. In fact, the higher ECSA of F-TiO_2 (Figure S7) demonstrated that this catalyst favors light capture, carrier separation, and transfer ability [22].

3.3. Pollutant Abatement

3.3.1. Catalyst Evaluation

The performance enhancement can be explained by the reduction in the e^-/h^+ recombination caused by the substitution of F with O atoms in TiO_2, forming a Ti-F bond. These Ti-F bonds have a higher e^- attraction and the e^-/h^+ recombination is more difficult [3]. This is probably due to the strong F electronegativity. This provides extra time for the electrons to react to produce radicals involved in the pollutant's degradation [39].

These EC results were considerably smaller than the previously reported data (Table 1). Other authors have used inefficient photocatalyst and reactor set-ups which they have tried to overcome by the utilization of elevated consumption lamps. For intance, Badvi and Javanbakht [11] were far from our one-step-synthesis of the catalyst; indeed, they needed calcination steps (which increase the operational costs) and several sol-gel reactions. However, 2 h was necessary for complete MB degradation and the efficiency was 23 and 41 times lower than in the case of using our F-TiO_2 process under continuous radiation or the long-pulse procedure, respectively. It is remarkable that these authors used 1000 mg/L of the initial catalyst concentration which may have had a detrimental effect (Section 3.1.3).

In general, the appropriate photocatalyst selection, reactor, and lamp configuration decreased the EC. This is discussed in the last section (Section 3.3.6).

3.3.2. Long-Pulse Radiation Procedure

This novel set-up was extremely efficient (Figure 3); indeed, the F-TiO_2 system required 0.125 Wh mg^{-1} for the degradation of MB, which would mean a cost of 2×10^{-5} EUR/mg^{-1} using the actual energy cost in Spain. Considering the long-pulse procedure, the EC was reduced to 0.070 Wh mg^{-1} (1.1×10^{-5} EUR/mg^{-1}) demonstrating the suitability of this procedure on radical-based processes. With this lower EC, the application costs of this process would be proportionally lower. In fact, the results of radiation on switch on/off mode were in concordance with some studies where the switching off of the lamps did not cause a direct elimination of the current response, demonstrating the presence of e^- and h^+ [28,37].

The suitability of the reactor set-up, the efficiency of the photocatalyst, and the possibility of working with radiation on long-pulses increased the efficiency of the process when compared with recent studies (Table 1).

For instance, Gao et al. [14] reduced both the initial concentration and the treatment time to 20 min. Nevertheless, the EC was extremely high due to the usage of a high consumption lamp (300 W). These authors could have reduced EC roughly to a half by using the long-pulse procedure and probably by selecting a higher efficient photocatalyst.

3.3.3. Reusability

F-TiO_2 showed a high reusability performance (Figure 4) due to the F doping which reduces the transformation from anatase to rutile phases, making it more stable throughout the degradation process [3]. The slight reduction in performance meant the generated by-products did not block significantly the active sites of the catalyst [2]. Moreover, considering no Ti nor F were detected after three degradation cycles, one can conclude the F-TiO_2 synthesis provides a stable and well-performant catalyst.

Nonetheless, the reusability was assessed with the presence of H_2O_2, and consequently, the degradation performance was kept constant during three batches even at double MB initial concentration. This was due to the fact the F-TiO_2 catalyzes the decomposition of H_2O_2 into HO· radicals [11] and thus, the presence of an additional amount of radicals make the process more performant. Moreover, H_2O_2 acts as electron acceptor, avoiding e^-/h^+ recombination [11]. This result confirms that H_2O_2 is a powerful substance that can prolong the life of a catalyst and can burst this process.

3.3.4. Pollutant Concentration

Initially, the influence of MB concentration was tested so the process was validated to work with concentration variations. The process was stable at around 85% degradation in 30 min at concentrations between 10 and 20 mg/L (Figure 5). Nevertheless, reducing the concentration to 5 mg/L caused an MB degradation of 75%. This can be explained by the fact that too few pollutant molecules make the oxidant species react between themselves, losing effectivity [31]. Indeed, other authors reported a lower degradation of MB when its initial concentration was reduced [40,41]. On the other hand, increasing the concentration to 40 mg/L reduced MB elimination to 52.4%. Other authors such as Chakinala et al. [6] reported this behavior. Indeed, they suffered a reduction in dye degradation of around 30% after increasing its concentration from 10 to 45 mg/L.

In the case of Kurniawan et al. [1], 30% of MB degradation detriment was attained when switching the initial concentration from 5 to 25 mg/L. This was caused by (i) the radiation diffraction due to the effluent darkness, impeding photocatalyst activation [1,6], (ii) the clogging of the photocatalyst surface hindering F-TiO_2 activation [11], (iii) the excessive amount of molecules which, after being in radical state due to the photocatalysis process, react with themselves instead than with oxidants, stopping chain radical reactions, and (iv) the higher amount of generated by-products which compete with the target pollutant for the active sites of the photocatalyst [4].

In all cases, the process goes slower as time passes, for instance from treatment time 20 min, the rate is much slower. This is explained by the fact that small by-products generated due to MB degradation act as radical scavengers [6]. Acosta-Esparza et al. [9] required up to 8 h to treat a quite concentrated MB effluent, and consequently, the EC increased (Table 1). Considering our data, reducing the initial concentration could have reduced the treatment time.

3.3.5. Treatment of More Complex Effluents

As was reported on Figure 6, MB and BPA degradation was achieved within 30 min. This defeats previous studies (Table 1) and demonstrates the efficiency of the process, as for instance, BPA natural degradation takes 90 years [4].

Nevertheless, the slower BPA degradation rate at the beginning of the process when compared to MB degradation is significant (Figure 6 and Table S2). This could be due to the non-polar character of BPA which makes both aqueous solubilization and oxygen radical attack more difficult. Nevertheless, as treatment time passed, degradation rose due to the simpler structure of BPA when compared to MB (graphical abstract), which meant that almost complete degradation in both cases was achieved within 30 min.

The degradation of both MB and BPA was essayed on real wastewaters after primary and secondary treatment. In this case, both pollutants suffered a reduction in degradation performance due to the higher complexity of these matrixes (Table S1). Nevertheless, the results were promising and considering the detriment was less than 40% in MB and less than 30% in BPA treatment, it can be concluded that this process copes better with matrix variations than those offered in other studies. This adaptability might be due to the optimized reactor set-up and the selection of an active photocatalyst.

For instance, Jafari et al. [4] suffered a 40% detriment on 20 mg/L BPA degradation when switching from distilled water to real wastewater. Likewise, Murcia et al. [42] experienced a detriment of more than 60%, as MB was completely degraded in 2 h by TiO_2 when using distilled water as a matrix and in 5 h when treating a real sample collected from a handicraft factory. This may have been caused by the increased turbidity and the presence of other organic and inorganic content which consume the oxidant species available on the system and poison the photocatalyst surface due to adsorption into it [4,42].

Considering our results (Section 2.3.3), the addition of 0.3 mg/mL of H_2O_2 could favor the photocatalyst performance even on these complex cases, and consequently further research should focus on the addition of this oxidant and its concentration optimization. For instance, when H_2O_2 was added to the 20 mg/L MB photocatalytic degradation, mineralization increased from 86% to 94% when Milli-Q water was used as a working matrix. Consequently, the results demonstrate that this process could be applied as a post-treatment for real wastewaters, although H_2O_2 addition or longer treatment times may be necessary [4].

3.3.6. Comparison with Previous Studies

The reported results highlight the high efficiency of the set degradation process. Namely, the high performance of these data can be explained by the proper photocatalyst selection and reactor set-up, as well as the consideration of the long-pulse radiation or H_2O_2 addition.

As it can be seen in Table 1, the synthesized F-TiO_2 photocatalyst performed better than others reported in the literature, or at least, more efficiently. Indeed, other authors reached quite high MB degradation rates [1,14,34]. Nevertheless, high energy expenditure lamps (order of hundreds of watts) or long treatment times were required. For example, the EC attained by Kurniawan et al. [1] was extremely elevated due to the high power lamp utilized (500 W) and the fact that in this case the treated effluent had only 5 mg/L of

MB concentration, a fact than could also have caused a slight reduction in the degradation performance (Figure 5). Likewise, Acosta-Esparza et al. [9] required 8 h for almost 90% MB degradation. This could be explained by the high initial MB concentration used by those authors (Table 1). Indeed, even in our case, increasing the MB concentration from 20 to 40 mg/L led to an MB degradation switch from 88 to 52.4% in 30 min (Figure 5).

Moreover, it should be noted that it is extremely important to stop the photocatalysis process at the optimal treatment time in order to have a more efficient output. Thus, adsorption and biological or other oxidation processes (Fenton, ozonation, electrolysis, etc.) could be coupled as final polishing [43]. For instance, Gao et al. [14] achieved quite a low efficiency due to the usage of an extremely powerful lamp. However, if they had stopped the treatment time after 10 min (80% of MB degradation), the EC would have been reduced to a half (6.25 Wh/mg). Another example was brought by Badvi and Javanbakht [11] who required 2 h for complete MB degradation. Nevertheless, approximately 90% was attained in 40 min. By stopping the photodegradation process at that point, the energy expenditure would drop from having an EC of 3.2 to 1.19 Wh/mg. Those alternatives may be suitable as it has been reported that the by-products above 80% degradation are mainly simple carboxylic acids which are bio-degradable [24].

The attained EC values are related to the future application costs of these processes, and consequently, it seems the proposed catalyst and set-up could be scaled-up in further studies (Figure S8). In fact, Pham et al. [44] scaled-up the photodegradation of 20 mg/L of MB with H_2O_2 and even though their lamp consumed much more (25 W) than the used here (4.8 W), they established an energy cost of 8.41 EUR/m^3 which is competitive with other treatment processes. Our process may reduce those costs considering our degradation rate of 0.1367 min^{-1} defeated the degradation rate of Pham et al. [44] (0.0105 min^{-1}). Indeed, the energy costs for the proposed system of 50 mL with the long-pulse procedure radiation would be 9.6 EUR/m^3 which, as it was reported in Section 3.3.3, could be improved with the addition of H_2O_2 [44].

As was described in the previous section, under the worst case scenario the performance can be enhanced by the addition of H_2O_2 (Figure 4B). With that, EC is controlled, as when the treatment time increases, the EC consumption increases more dramatically if the pollutant is being degraded slowly. As an example, in this study the EC for the degradation of 20 mg/L MB was 0.093 Wh/mg after 20 min, whereas at 30 min it was 0.125 Wh/mg (and the MB degradation only switched from 82 to 94%). In Figure 3 one can see how the efficiency of the process was reduced with time, which demonstrates it is not worth increasing the treatment time; thus, adding H_2O_2 may be a good alternative (Figure 4B).

The good performance attained by Garg et al. is remarkable [23]. Those authors attained a small EC due to the usage of a low-consumption UVC lamp and an increase in the irradiated surface area. Nevertheless, the effect of the adsorption on their ZnO-graphene oxide catalyst should be considered, as it adsorbed around 20% of BPA. This may make the reutilization of the catalyst difficult, which was not assessed.

In general, the proposed set-up and photocatalyst are suitable for stable and threatening pollutants. Consequently, more studies should be carried out in order to complete this study, i.e., the optimization of H_2O_2 dosage and treatment time, in order to scale-up the process (Figure S8), fixing properly the residence time.

4. Materials and Methods

4.1. Reagents

Titanium dioxide (TiO_2) Aeroxide P25 was bought from Acros Organics (Madrid, Spain); nitric acid (HNO_3), 70%, from Fischer Chemical (Madrid, Spain); sodium fluoride (NaF) >99%, from Sigma-Aldrich; and isopropyl alcohol (IPA) (C_3H_8O), from Honeywell (Madrid, Spain), as well as the ethanol (EtOH) (>99.8%). Nafion was acquired from Sigma-Aldrich (Madrid, Spain). MB (>82%) was purchased from Sigma-Aldrich. The water was supplied by a Milli-Q system (Advantage A10, from Merck, Darmstadt, Germany).

4.2. Catalyst Synthesis

The synthesis was adapted from Shayegan and co-workers [13]. Briefly, 3 g of NaF was added to 230 ml of IPA, and placed on a magnetic stirrer (approx. 700 rpm) for 2 h at room temperature. Next, 0.2 M nitric acid was added until pH ≈ 3.5. After that, TiO_2 was added to the solution at different TiO_2:F ratios of 1:0.5, 1:1, and 1:2. The stirring was continued for another 3 h. During that time, nitric acid should be added to keep pH ≈ 4, if needed. When the time was finished, the obtained substance was centrifuged and washed several times, until it acquired a homogeneous color. It was dried in an oven at 100 °C for at least 12 h so F-TiO_2 with a molar ratio (Ti:F 1:1) catalyst was attained. Different TiO_2:F contents were essayed and for that, instead of 3 g of NaF, 1.5 or 6 g was used so F-TiO_2 was attained with a ratio of Ti:F of1:0.5 and 1:2, respectively.

4.3. Reactor Set-Up

The photocatalytic performance was evaluated on a novel open-wide reactor which was cylindrical with a diameter of 8 cm and a height of 3 cm. It promoted photocatalyst activation by the lamp placed 3 cm above (Figure 7) and reduced light scattering due to solution depth. Indeed, another narrow-open reactor (4 cm diameter, 10 cm height) was also evaluated with comparative aims. In both cases, magnetic sitting at 800 rpm was set. Under this stirring and in dark conditions, 30 min was left to attain the adsorption equilibrium which may take place between the catalyst and pollutant. The lamp used was a low-consumption UV-LED lamp (365 nm, 4.8 W, 78 CMF-AR-A03, from Seoul Viosys, Gyeonggi-do, Republic of Korea) and it had a small fan in order to keep the temperature constant at 22 ± 2 °C, so evaporation of the sample was avoided.

During the long-pulse radiation procedure, the lamp was switched on from 0 to 5 and from 10 to 20 min treatment time, whereas from 5 to 10 and 20 to 30 min the reaction continued in dark conditions. That meant 15 min radiation time out of 30 min of overall treatment time.

Unless stated differently, 50 mL was used and the catalyst concentration was set at 800 mg/L and the initial MB or BPA concentration was fixed at 20 mg/L, values found in real landfill leachates [4]. All experiences were performed using Milli-Q water as the working matrix (Sartorious, Göttingen, Germany). However, validation experiences were performed with real wastewater, kindly donated by WTP (wastewater treatment plant) of Guillarei, Tui, Galicia. This WTP remediated the municipal effluents by primary treatment (after filtration) and subsequent secondary treatment (after biological remediation). The characteristics of those effluents are provided in the supplementary material (Table S1). All experiences were performed in duplicate, and the deviation was found to be smaller than 4.3%. Consequently, the graphs show the average values.

4.4. Characterization

4.4.1. X-ray Diffraction (XRD)

Room-temperature XRD was carried out on a Rigaku diffractometer with an X PERT PRO MRD (Pananalytical, MA, USA) using Cu-Kα radiation (λ = 1.54187 Å). Each pattern was recorded in the 2θ range from 20° to 80° with a step of 0.03° and a total collection time of 2 h. The analysis of XRD patterns was carried out with Higscore software (Pananalytical, MA, USA).

4.4.2. Raman

Raman spectra were collected using the Jobin Yvon HR800UV spectrometer (Horiba, Kyoto, Japan) equipped with Ar laser.

4.4.3. SEM, EDS, and ICP

SEM and EDS analyses were completed using the microscope JSM-6700 F (JEOL, Tokyo, Japan) (CACTI). Inductively coupled plasma (ICP) coupled to optical emission spectrometry (ICP-OES Optima 4300, PekinElmer, MA, USA) (CACTI) was used for measuring leaching after the experiments and the initial photocatalyst composition. For that, acid digestion of 0.25 mg of the catalyst was carried out by adding 5 mL HNO_3 7 M at 120 °C for 2 h in a 100 mL autoclave.

4.4.4. UV-Vis

UV-Vis spectra of the samples for band gap calculation (Tauc plot [2]) were measured in a 1% EtOH solution on the spectrophotometer (V-2550, Shimadzu, Kyoto, Japan). This equipment was also used for MB removal monitorization with time.

4.4.5. Electrochemical Measurements

Electrochemical Impedance Spectroscopy (EIS) and Electrochemically Active Surface Area (ECSA) were measured using the Autolab PGSTAT302N of Metrohm (Herisau, Switzerland). A three-electrode system was used, with Pt as the counter electrode, the calomel electrode as the reference, and Ni-foam (1 cm^2) as the working electrode where the catalysts were impregnated. A total of 3 mg of the catalyst was dissolved in 630 µL of a mixture Nafion:EtOH (1:20). The experiences were performed in a 0.5 M Na_2SO_4 solution. EIS was measured on a frequency range of 10^5–0.1 Hz with a sinusoidal perturbation of 10 mV. ECSA was measured in a range of 0.1 V at scan rates between 10 and 200 mV s^{-1}. Then, a linear trend was attained by plotting scan rate vs. the difference in current density between anodic and cathodic sweeps divided by two. The slope provided after linear fitting of this graph provided geometric double-layer capacitance (CDL). ECSA was calculated following Equation (3), where Cs is the specific capacitance and has a value of 40 µF cm^{-2} for experiences on a 0.5 M H_2SO_4 media [45].

$$ECSA = CDL/Cs \tag{3}$$

5. Conclusions

The F-TiO_2 catalyst was easily synthesized in a one-step synthesis. This catalyst was characterized by several means (XRD, FTIR, EDS, band gap, EIS, and ECSA calculations) which anticipated the good performance (F-TiO_2 showed a high electron mobility, a small band gap, and an elevated surface area). Indeed, the results demonstrated that the photocatalytic activity was better than the previously reported data, even better than the commercial well performant TiO_2 from Degussa. The wide-open reactor was selected based on a preliminary test which highlighted the importance of reducing the solution depth and increasing the radiation surface. The effect of pollutant and photocatalyst concentration, the alternative of working in a long-pulse radiation procedure, and the effect of the working matrix was evaluated. The reactor set-up and the synthesized catalyst made possible the efficient treatment of methylene blue and bisphenol A even when being in real wastewater matrixes. H_2O_2 addition is presented as an alternative to improve the degradation and reusability process. Thus, given that photocatalysis is a clean and sustainable technology, the fact the photocatalyst is reusable and the process can be used for real wastewater treatment, this process could be studied at higher scale in order to open a path for real applications.

Supplementary Materials: The following supporting information can be downloaded at: https://www.mdpi.com/article/10.3390/catal12101190/s1, Figure S1: MB degradation (20 mg/L) with 800 mg/L of TiO2 under UV radiation on the open-wide reactor (black circles) or the narrow-open reactor (red triangles). The lines represent the pseudo-first order kinetic adjustment; Figure S2: MB degradation under UV radiation depending on the selected ratio Ti:F between 1:0.5 (triangles), 1:1 (circles) and 1:2 (squares) (A) and pseudo-first order kinetic constants (B); Figure S3: MB degradation under UV radiation depending on the catalyst concentration. 200 mg/L (circles), 400 mg/L (triangles), 800 mg/L (squares), 1600 mg/L (rhombus) and 3200 mg/L (crosses) where the lines are the pseudo-first kinetic adjustment and B) depicts the pseudo-first order kinetic constants; Figure S4: TiO_2 and F-TiO_2 XRD with their peaks assignment; Figure S5: SEM image of F-TiO_2 photocatalyst (A) with EDS analysis (B); Figure S6: UV-Vis spectra of TiO_2 (A) and F-TiO_2 (B) photo-catalyst with band gap calculation with Tau equation (embed graphs); Figure S7: CDL measurement of TiO_2 (A) and F-TiO_2 (B) by measuring CVs at different scan rates and plotting the j difference vs the scan rate (embedded figures); Figure S8. Proposed scaled-up reactor; Table S1: Real wastewaters used as working matrix for MC degradation; Table S2: Pseudo-first order kinetic constants for the MB and BPA degradation under UV radiation with 800 mg/L of F-TiO_2.

Author Contributions: Conceptualization, A.M.D.; Methodology, A.M.D.; Software, A.M.D. and I.N.; Validation, A.M.D.; Formal Analysis, A.M.D.; Investigation, A.M.D. and I.N.; Resources, M.P., M.Á.S. and Y.V.K.; Data Curation, A.M.D. and I.N.; Writing—Original Draft Preparation, A.M.D.; Writing—Review and Editing, M.P., M.Á.S. and Y.V.K.; Visualization, A.M.D.; Supervision, M.P., M.Á.S. and Y.V.K.; Project Administration, M.P., M.Á.S. and Y.V.K.; Funding Acquisition, A.M.D., M.P., M.Á.S. and Y.V.K. All authors have read and agreed to the published version of the manuscript.

Funding: This research was funded by Xunta de Galicia grant numbers ED481B 2019/091 and ED431C 2021-43. Additionally, the research has been provided with the financial support under the BiodivRestore ERA-Net COFUND programme, Project PCI2022-132941 funded by MCIN/AEI/10.13039/501100011033 and European Union Next Generation EU/PRTR and PID2020-113667GB-I00 funded by MCIN/AEI/10.13039/501100011033.

Data Availability Statement: Data is contained within the article and Supplementary Materials.

Conflicts of Interest: The authors declare no conflict of interest.

References

1. Kurniawan, T.A.; Mengting, Z.; Fu, D.; Yeap, S.K.; Othman, M.H.D.; Avtar, R.; Ouyang, T. Functionalizing TiO_2 with graphene oxide for enhancing photocatalytic degradation of methylene blue (MB) in contaminated wastewater. *J. Environ. Manag.* **2020**, *270*, 110871–110879. [CrossRef] [PubMed]
2. Zhang, X.; Chen, W.; Bahmanrokh, G.; Kumar, V.; Ho, N.; Koshy, P.; Sorrell, C.C. Synthesis of V- and Mo-doped/codoped TiO_2 powders for photocatalytic degradation of methylene blue. *Nano-Struct. Nano-Objects* **2020**, *24*, 100557–100568. [CrossRef]
3. D'Souza, L.P.; Prakash, R.M.; Balakrishna, R.G. Chapter 3 Anion-Modified Photocatalysts. In *Photocatalytic Systems by Design*; Sakar, M., Balakrishna, R.G., Do, T., Eds.; Elsevier: Amsterdam, The Netherlands, 2021; pp. 55–83.
4. Jafari, A.J.; Kalantary, R.R.; Esrafili, A.; Moslemzadeh, M. Photo-catalytic degradation of bisphenol-a from aqueous solutions using GF/Fe-TiO_2-CQD hybrid composite. *J. Environ. Health Sci. Eng.* **2021**, *19*, 837–849. [CrossRef]
5. Humayun, M.; Raziq, F.; Khan, A.; Luo, W. Modification strategies of TiO_2 for potential applications in photocatalysis: A critical review. *Green Chem. Lett. Rev.* **2018**, *11*, 86–102. [CrossRef]
6. Chakinala, N.; Gogate, R.R.; Chakinala, G. Highly efficient bi-metallic bismuth-silver doped TiO_2 photocatalyst for dye degradation. *Korean J. Chem. Eng.* **2021**, *38*, 2468–2478. [CrossRef]
7. Zheng, X.; Han, H.; Ye, X.; Meng, S.; Zhao, S.; Wang, X.; Chen, S. Fabrication of Z-scheme WO_3/$KNbO_3$ photocatalyst with enhanced separation of charge carriers. *Chem. Res. Chin. Univ.* **2020**, *36*, 901–907. [CrossRef]
8. Hunge, Y.M.; Yadav, A.A.; Khan, S.; Takagi, K.; Suzuki, N.; Teshima, K.; Terashima, C.; Fujishima, A. Photocatalytic degradation of bisphenol A using titanium dioxide@ nanodiamond composites under UV light illumination. *J. Colloid. Interface Sci.* **2021**, *582*, 1058–1066. [CrossRef]
9. Acosta-Esparza, M.A.; Rivera, L.P.; Pérez-Centeno, A.; Zamudio-Ojeda, A.; González, D.R.; Chávez-Chávez, A.; Santana-Aranda, M.A.; Santos-Cruz, J.; Quiñones-Galván, J. UV and Visible light photodegradation of methylene blue with graphene decorated titanium dioxide. *Mater. Res. Express.* **2020**, *7*, 35504–35512. [CrossRef]
10. Bibi, S.; Ahmad, A.; Anjum, M.A.R.; Haleem, A.; Siddiq, M.; Shah, S.S.; Kahtani, A.A. Photocatalytic degradation of malachite green and methylene blue over reduced graphene oxide (rGO) based metal oxides (rGO-Fe_3O_4/TiO_2) nanocomposite under UV-visible light irradiation. *J. Environ. Chem. Eng.* **2021**, *9*, 105580–105591. [CrossRef]
11. Badvi, K.; Javanbakht, V. Enhanced photocatalytic degradation of dye contaminants with TiO_2 immobilized on ZSM-5 zeolite modified with nickel nanoparticles. *J. Clean Prod.* **2021**, *280*, 124518–124531. [CrossRef]
12. Niu, L.; Zhao, X.; Tang, Z.; Lv, H.; Wu, F.; Wang, X.; Zhao, T.; Wang, J.; Wu, A.; Giesy, J. Difference in performance and mechanism for methylene blue when TiO_2 nanoparticles are converted to nanotubes. *J. Clean Prod.* **2021**, *297*, 126498–126509. [CrossRef]
13. Shayegan, Z.; Lee, C.; Haghighat, F. Effect of surface fluorination of P25-TiO_2 coated on nickel substrate for photocatalytic oxidation of methyl ethyl ketone in indoor environments. *J. Environ. Chem. Eng.* **2019**, *7*, 103390–103401. [CrossRef]
14. Gao, Q.; Si, F.; Zhang, S.; Fang, Y.; Chen, X.; Yang, S. Hydrogenated F-doped TiO_2 for photocatalytic hydrogen evolution and pollutant degradation. *Int. J. Hydrogen Energy* **2019**, *44*, 8011–8019. [CrossRef]
15. Rani, M.; Keshu Shanker, U. Efficient degradation of organic pollutants by novel titanium dioxide coupled bismuth oxide nanocomposite: Green synthesis, kinetics and photoactivity. *J. Environ. Manag.* **2021**, *300*, 113777–113791. [CrossRef] [PubMed]
16. Shiraishi, F.; Ueno, M.; Chand, R.; Shibata, Y.; Luitel, H.N. Effect of silanization of titanium dioxide on photocatalytic decomposition of 2,4-dinitropheonol under irradiation with artificial UV light and sunlight. *J. Chem. Technol. Biotechnol.* **2014**, *89*, 81–87. [CrossRef]

17. Habib, I.Y.; Burhan, J.; Jaladi, F.; Lim, C.M.; Usman, A.; Kumara, N.T.R.N.; Tsang, S.C.E.; Mahadi, A.H. Effect of Cr doping in CeO2 nanostructures on photocatalysis and H_2O_2 assisted methylene blue dye degradation. *Catal. Today* **2020**, *375*, 506–513. [CrossRef]
18. Diao, W.; Xu, J.; Rao, X.; Zhang, Y. Facile Synthesis of Fluorine Doped Rutile TiO2 Nanorod Arrays for Photocatalytic Removal of Formaldehyde. *Catal. Lett.* **2022**, *152*, 1029–1039. [CrossRef]
19. Ahmad, N.; Sultana, S.; Sabir, S.; Khan, M.Z. Exploring the visible light driven photocatalysis by reduced graphene oxide supported Ppy/CdS nanocomposites for the degradation of organic pollutants. *J. Photochem. Photobiol. A* **2020**, *386*, 112129–112142. [CrossRef]
20. Ângelo, J.; Magalhães, P.; Andrade, L.; Mendes, A. Characterization of TiO_2-based semiconductors for photocatalysis by electrochemical impedance spectroscopy. *Appl. Surf. Sci.* **2016**, *387*, 183–189. [CrossRef]
21. Díez, A.M.; Valencia, H.E.; Meledina, M.; Mayer, J.; Kolen'ko, Y.V. Photocatalytic-fenton process under simulated solar radiation promoted by a suitable catalyst selection. *Catalysts* **2021**, *11*, 885. [CrossRef]
22. Lian, X.; Chen, S.; He, F.; Dong, S.; Liu, E.; Li, H.; Xu, K. Photocatalytic degradation of ammonium dinitramide over novel S-scheme g-C3N4/BiOBr heterostructure nanosheets. *Sep. Purif. Technol.* **2022**, *286*, 120449–120459. [CrossRef]
23. Garg, R.; Gupta, R.; Bansal, A. Degradation mechanism, reaction pathways and kinetics for the mineralization of Bisphenol A using hybrid ZnO/graphene oxide nano-catalysts. *Korean J. Chem. Eng.* **2021**, *38*, 485–497. [CrossRef]
24. Arias, M.C.; Aguilar, C.; Piza, M.; Zarazua, E.; Anguebes, F.; Cordova, V. Removal of the Methylene Blue Dye (MB) with Catalysts of Au-TiO_2: Kinetic and Degradation Pathway. *Mod. Res. Catal.* **2021**, *10*, 1–14. [CrossRef]
25. Brahmi, C.; Benltifa, M.; Ghali, M.; Dumur, F.; Simonnet-Jégat, C.; Monnier, V.; Morlet-Savary, F.; Bousselmi, L.; Lalevée, J. Polyoxometalates/polymer composites for the photodegradation of bisphenol-A. *J. Appl. Polym. Sci.* **2021**, *138*, 50864–50876. [CrossRef]
26. Pupo Nogueira, R.F.; Trovó, A.G.; Modé, D.F. Solar photodegradation of dichloroacetic acid and 2,4-dichlorophenol using an enhanced photo-Fenton process. *Chemosphere* **2002**, *48*, 385–391. [CrossRef]
27. Bangun, J.; Adesina, A.A. The photodegradation kinetics of aqueous sodium oxalate solution using TiO_2 catalyst. *Appl. Catal. A Gen.* **1998**, *175*, 221–235. [CrossRef]
28. Sun, X.; Xu, K.; Chatzitakis, A.; Norby, T. Photocatalytic generation of gas phase reactive oxygen species from adsorbed water: Remote action and electrochemical detection. *J. Environ. Chem. Eng.* **2021**, *9*, 104809–104818. [CrossRef]
29. Pascariu, P.; Cojocaru, C.; Samoila, P.; Airinei, A.; Olaru, N.; Rusu, D.; Rosca, I.; Suchea, M. Photocatalytic and antimicrobial activity of electrospun ZnO:Ag nanostructures. *J. Alloy. Compd.* **2020**, *834*, 155144–155155. [CrossRef]
30. Arunagiri, C. *Enhanced Visible Light Photocatalytic Degradation of Fe-Doped ZnO Nanoparticles For Organic Dyes*; Research Square: Newcastle, UK, 2021.
31. Peng, H.; Xu, L.; Zhang, W.; Liu, F.; Lu, X.; Lu, W.; Danish, M.; Lin, K. Different kinds of persulfate activation with base for the oxidation and mechanism of BDE209 in a spiked soil system. *Sci. Total Environ.* **2017**, *574*, 307–313. [CrossRef] [PubMed]
32. Subalakshmi, K.; Senthilselvan, J. Effect of fluorine-doped TiO_2 photoanode on electron transport, recombination dynamics and improved DSSC efficiency. *Sol. Energy* **2018**, *171*, 914–928. [CrossRef]
33. Noh, K.; Oh, H.; Bo-Ra, K.; Jung, S.; Kang, W.; Sun-Jae, K. Photoelectrochemical Properties of Fe_2O_3 Supported on TiO2-Based Thin Films Converted from Self-Assembled Hydrogen Titanate Nanotube Powders. *J. Nanomater.* **2012**, *2012*, 1–6. [CrossRef]
34. Mahanta, U.; Khandelwal, M.; Deshpande, A.S. TiO_2@SiO_2 nanoparticles for methylene blue removal and photocatalytic degradation under natural sunlight and low-power UV light. *Appl. Surf. Sci.* **2022**, *576*, 151745–151756. [CrossRef]
35. Pant, B.; Ojha, G.P.; Kuk, Y.; Kwon, O.H.; Park, Y.W.; Park, M. Synthesis and Characterization of ZnO-TiO_2/Carbon Fiber Composite with Enhanced Photocatalytic Properties. *Nanomaterials* **2020**, *10*, 1960. [CrossRef] [PubMed]
36. Mahanthappa, M.; Kottam, N.; Yellappa, S. Enhanced photocatalytic degradation of methylene blue dye using CuSCdS nanocomposite under visible light irradiation. *Appl. Surf. Sci.* **2019**, *475*, 828–838. [CrossRef]
37. Liu, J.; Shi, H.; Sans, C.; Sun, L.; Yuan, X.; Pan, F.; Xia, D. Insights into the photocatalytic ozonation over Ag_2O-ZnO@g-C3N4 composite: Cooperative structure, degradation performance, and synergistic mechanisms. *J. Environ. Chem. Eng.* **2022**, *10*, 107285–107298. [CrossRef]
38. Sultana, S.; Ahmad, N.; Ahmad, E.; Sabir, S.; Khan, M.Z. Electrochemical synthesis of novel aluminium oxyhydroxide-decorated MnO_2/chitosan nanocomposite with efficient photocatalytic and antibacterial activity. *Nanotechnol. Environ. Eng.* **2020**, *5*, 20. [CrossRef]
39. Peng, X.; Zhang, Y.; Liu, Y. Fabrication of a novel high photocatalytic Ag/Ag_3PO_4/P25 (TiO_2) heterojunction catalyst for reducing electron-hole pair recombination and improving photo-corrosion. *Mater. Res. Express.* **2019**, *6*, 65515–65525. [CrossRef]
40. Arifin, M.N.; Rezaul Karim, K.M.; Abdullah, H.; Khan, M.R. Synthesis of titania doped copper ferrite photocatalyst and its photoactivity towards methylene blue degradation under visible light irradiation. *Bull. Chem. React. Eng. Catal.* **2019**, *14*, 219–227. [CrossRef]
41. Chen, H.; Luo, S.; Lei, X.; Liu, H.; Liu, Y.; Xu, X.; Jiang, Z.; Li, X. Synthesis and photocatalytic performance of nano-CeO_2 by a PVP-assisted microwave interface method for organic dye degradation. *Ionics* **2020**, *26*, 5829–5839. [CrossRef]
42. Murcia, J.J.; Cely, Á.C.; Rojas, H.A.; Hidalgo, M.C.; Navío, J.A. Fluorinated and platinized titania as effective materials in the photocatalytic treatment of dyestuffs and stained wastewater coming from handicrafts factories. *Catalysts* **2019**, *9*, 179. [CrossRef]

43. Titchou, F.E.; Zazou, H.; Afanga, H.; El Gaayda, J.; Ait Akbour, R.; Nidheesh, P.V.; Hamdani, M. Removal of organic pollutants from wastewater by advanced oxidation processes and its combination with membrane processes. *Chem. Eng. Process.-Process Intensif.* **2021**, *169*, 108631–108653. [CrossRef]
44. Pham, D.C.; Cao, T.M.D.; Nguyen, M.C.; Nguyen, T.D.; Nguyen, V.H.; Bui, V.H.; Nguyen, T.T.T. Integrating Photocatalysis and Microfiltration for Methylene Blue Degradation: Kinetics and Cost Estimation. *Chem. Eng. Technol.* **2022**, *45*, 1748–1758. [CrossRef]
45. Coleman, N.; Lovander, M.D.; Leddy, J.; Gillan, E.G. Phosphorus-Rich Metal Phosphides: Direct and Tin Flux-Assisted Synthesis and Evaluation as Hydrogen Evolution Electrocatalysts. *Inorg. Chem.* **2019**, *58*, 5013–5024. [CrossRef] [PubMed]

catalysts

Article

Zinc–Acetate–Amine Complexes as Precursors to ZnO and the Effect of the Amine on Nanoparticle Morphology, Size, and Photocatalytic Activity

Jerry D. Harris [1,*], Emily A. Wade [1], Emmaline G. Ellison [1], Cecelia C. Pena [1], Stephen C. Bryant [1], Nicholas L. McKibben [1], Allison J. Christy [1], Kevin O. Laughlin [1], Ashley E. Harris [1], Kenrik V. Goettsche [1], Chad E. Larson [1], Seth M. Hubbard [2], Jonathan E. Cowen [3], Josh Eixenberger [4], David Estrada [4] and Jennifer R. Chase [5]

1 Department of Chemistry, Northwest Nazarene University, Nampa, ID 83686, USA
2 Department of Physics, Rochester Institute of Technology, Rochester, NY 14623, USA
3 Swagelok Center for the Surface Analysis, Case Western Reserve University, Cleveland, OH 44106, USA
4 Micron School of Materials Science and Engineering, Boise State University, Boise, ID 83725, USA
5 Department of Biology, Northwest Nazarene University, Nampa, ID 83686, USA
* Correspondence: jdharris@nnu.edu; Tel.: +1-208-467-8883

Abstract: Zinc oxide is an environmentally friendly and readily synthesized semiconductor with many industrial applications. ZnO powders were prepared by alkali precipitation using different $[Zn(acetate)_2(amine)_x]$ compounds to alter the particle size and aspect ratio. Slow precipitations from 95 °C solutions produced micron-scale particles with morphologies of hexagonal plates, rods, and needles, depending on the precursor used. Powders prepared at 65 °C with rapid precipitation yielded particles with minimal morphology differences, but particle size was dependent on the precursor used. The smallest particles were produced using precursors that yielded crystals with low aspect ratios during high-temperature synthesis. Particles produced during rapid synthesis had sizes ranging from 21–45 nm. The materials were characterized by scanning electron microscopy, transmission electron microscopy, X-ray diffraction, thermogravimetric analysis, BET, and diffuse reflectance. The materials prepared using precursors with less-volatile amines were found to retain more organic material than ZnO produced using precursors with more volatile amines. The amount of organic material associated with the nanoparticles influenced the photocatalytic activity of the ZnO, with powders containing less organic material producing faster rate constants for the decolorizing of malachite green solutions under ultraviolet illumination, independent of particle size. $[Zn(acetate)_2(hydrazine)_2]$ produced ZnO with the fastest rate constant and was recycled five times for dye degradation studies that revealed minimal to no reduction in catalytic efficiency.

Keywords: nanoparticles; photocatalyst; thermal analysis; mass spectrometry; surface area; synthesis; alkali precipitation; diffuse reflectance

Citation: Harris, J.D.; Wade, E.A.; Ellison, E.G.; Pena, C.C.; Bryant, S.C.; McKibben, N.L.; Christy, A.J.; Laughlin, K.O.; Harris, A.E.; Goettsche, K.V.; et al. Zinc–Acetate–Amine Complexes as Precursors to ZnO and the Effect of the Amine on Nanoparticle Morphology, Size, and Photocatalytic Activity. *Catalysts* **2022**, *12*, 1099. https://doi.org/10.3390/catal12101099

Academic Editors: Jorge Bedia and Carolina Belver

Received: 19 August 2022
Accepted: 19 September 2022
Published: 23 September 2022

1. Introduction

Zinc oxide is a II-VI-wide bandgap semiconductor that has many industrial applications [1]. It is relatively inexpensive, easy to prepare, has attractive electronic properties, and is an environmentally friendly alternative to other semiconductors, such as CdS and ZnSe [2]. Like other materials, the physicochemical properties of ZnO nanoparticles show various enhancements compared to bulk ZnO. These enhancements include, but are not limited to, increased optical properties and sensor sensitivity [3,4], increased reactivity toward cancer cells and bacteria [5,6], and increased catalytic activity [7]. Both particle size and shape influence these enhancements. Micro- and nanoparticles of ZnO have been prepared in a wide range of shapes and sizes that have included nanorods [8,9], nanowires [8,10], nanotubes [11,12], nanosheets [13,14], plates [15,16], and multi-segmented structures [17–19].

The richness of the size and morphologies obtained is dependent on the synthesis method (sol-gel, precipitation, hydrothermal, etc.), other molecules and ions present during synthesis [20–23], solvents used [15,24], rate of precipitation [19], and annealing time and temperature [25,26].

Air-stable molecules used to prepare ZnO include zinc acetate, zinc chloride, zinc nitrate, and zinc acetylacetonate [3,27–35]. All of these compounds are water-soluble, enabling a greener synthesis, and their physicochemical properties can be modified by adding other ligands. Amines as ligands have been utilized during ZnO synthesis to act as solution stabilizers, modify decomposition temperatures and solubility, used as N-doping sources, and used as capping ligands on nanoparticles, but the precursor molecules are seldom isolated and characterized [29,31,36–38].

In our previous work, we isolated, characterized and utilized [Zn(acetate)$_2$(amine)$_2$] single-source precursors to prepare ZnO films and powders [39]. We demonstrated that the size and morphology of ZnO particles prepared by alkali precipitation could be modified using different amines and changing the precipitation rate. However, several precursors lost solubility during the work, upon isolation and subsequent storage under ambient conditions and in an inert nitrogen-filled glove glovebox. For this work, to minimize solubility issues, the [Zn(acetate)$_2$(amine)$_x$] precursors were generated in solution, and the ZnO nanoparticles were prepared by alkali precipitation without the additional steps of isolation and redissolution. Maintaining the precursors in solution simplifies the synthesis of ZnO, decreases the synthesis time, and allows more amines to be used in the synthesis while still using fully characterized precursors.

The synthesis and characterization of several of the materials made using [Zn(acetate)$_2$ (amine)$_x$] compounds are presented here. The presence of the different amines influences the physicochemical properties (particle size and shape, surface area, etc.) of the precipitated particles. To explore how modifying these properties affects ZnO's photocatalytic properties, the nanomaterials were used to photocatalytically decolorize solutions of the triarylmethane dye malachite green. Malachite green is used in many industrial applications, including dyeing textiles, plastics, and paper. It is also used as a fungicide and parasiticide in aquaculture applications, but its use in many countries has been banned because of its toxicity to animals and humans [40,41]. Therefore, it is desirable to develop low-cost, effective processes for removing malachite green from industrial effluents and the environment using eco-friendly chemicals such as ZnO nanoparticles. Many have studied ZnO as a potential photocatalyst for water treatment [42,43]. Here, we demonstrate how the physicochemical properties of ZnO nanoparticles can be modified using [Zn(acetate)$_2$(amine)$_x$] precursors and how these modifications impact the material's ability to degrade an organic dye. Using these materials with the simplified synthetic process may provide an effective solution to treating wastewater. Additionally, ZnO composites with chitosan, graphene, graphene oxide, and carbon nanotubes have recently demonstrated excellent photocatalytic activity for degrading organic dyes [44–48]. One of the benefits of using ZnO composites as the photocatalyst is that they minimize the photocorrosion that can occur to ZnO in solution.

However, few authors who report ZnO photocatalysis ever characterize their ZnO precursors. Many use simple Zn^{2+} salts as their zinc source, while others use zinc-containing molecules or zinc ions with other ions in solution, but seldom do they report the characterization of their zinc oxide precursors [44–46,49–60]. In contrast, here we report the synthesis, NMR characterization, and CHN elemental analysis of the precursors. We have previously reported the crystal structure of [Zn(acetate)(ethylenediamine)] [39]. Characterizing the precursors allows for the correlation of precursor composition with the ZnO particle size and shape, the kind and mount of organic ligands associated with the ZnO, and the material's photocatalytic performance. Additionally, the materials reported here have similar photocatalytic performance to the ZnO composite materials.

2. Results and Discussion

2.1. Materials Characterization

Several [Zn(acetate)$_2$(amine)$_x$] compounds were synthesized and characterized. The compounds were isolated from solution and characterized by carbon, hydrogen, and nitrogen elemental analysis and NMR to verify that the precursors had the correct stoichiometry. For the amines utilized extensively in this work (Tris, 2-thiazolamine, hydrazine, and ethylenediamine), the analysis results are consistent with tetrahedrally coordinated zinc compounds, with both the amines and acetates binding at two coordination sites around the zinc. Tris, 2-thiazolamine, and hydrazine yield compounds that contain two amines bound to the zinc, whereas the compound prepared using ethylenediamine has only one of the bidentate ligands bound to the zinc. All precursor compounds are soluble and were characterized by NMR, except the one prepared with hydrazine. The NMR experiments and elemental analysis are consistent with the expected stoichiometry. The crystal structure of [Zn(acetate)$_2$(ethylenediamine)] was reported in our previous work [39].

The compounds were maintained in solution and then used to prepare ZnO nanoparticles by alkali precipitation. For this method, a zinc ion in solution is precipitated by adding an alkali hydroxide. The precipitate is then isolated, washed, dried, and annealed or calcined to convert any $Zn(OH)_2$ to ZnO and remove any solvent still associated with the product [61]. Zinc is an amphoteric element and, without coordinating ligands, can precipitate as $Zn(OH)_2$ at pHs near neutral. At higher pHs, ZnO can be formed from hydrolysis and condensation, leading to quasi-spherical particles if there is no preferential growth along the c-axis [62,63]. Mechanisms for the growth of ZnO nanoparticles from Zn(acetate)$_2$ in the presence of amines have been proposed by several investigators, and all utilize $Zn(OH)_2$ and Zn_4O(acetate)$_6$ intermediates [38,64,65].

Ligands in solution during particle growth and precipitation can be used to modify the shape and size of the ZnO particles. Meagley and Garcia modified the aspect ratio of ZnO particles using carboxylate ligands of different denticity [22]. In their work, no added ligands and monodentate ligands yielded long rods, bidentate ligands yielded short rods, and tridentate ligands yielded plates. The [Zn(acetate)$_2$(amine)$_x$] molecules in this work do not follow this trend during alkali precipitation. Hydrazine and 2-thiazolamine yield rods, ethylenediamine yields long needles, and Tris yields plates when grown at 95 °C in aqueous solutions (Figure 1). Particle sizes are 700 nm × 1 µm hexagonal plates for Tris, 700 nm × 200 nm rods for 2-thiazolamine, 8 µm × 2 µm rods for hydrazine, and 5 µm × 500 nm needles for ethylenediamine when prepared in water and with the slow addition of NaOH over 30 min. If zinc acetate is used without any added amine, round rods are formed. When using [Zn(cyclohexyl)$_2$(hexadecylamine)] to prepare ZnO, Zheng et al. found temperature to be the primary factor controlling morphology, with spherical nanoparticles produced at temperatures over 60 °C and rods produced at temperatures less than 60 °C [66]. Zhang et al. used zinc acetate solutions containing different long-chained amines (oleylamine, dioctylamine, hexadecylamine, and dodecylamine) to modify particle morphology and produced ZnO nanoparticles with shapes ranging from prisms to wires to rods [29]. Mechanisms for the growth of nanoparticles from precursors have been proposed but have been dependent on the synthetic protocol and required optimization for a particular morphology from a specific precursor [67].

Since particle size is likely driven by nucleation kinetics, lower temperatures and faster precipitation rates were expected to yield smaller particles because of the limited nucleation time during rapid precipitation. To achieve this, the solvent was changed from 100% water to a 50/50 mixture of methanol/water to prevent early precipitation before NaOH addition, the synthesis temperature was lowered to 65 °C, and the NaOH was added rapidly. A schematic of the rapid synthesis protocol is provided in Figure 2. For these experiments, we utilized the same amine precursors that produced micron-sized particles and the various morphologies seen in Figure 1, (Tris = plates, 2-thiazolamine = short rods, hydrazine = long rods, and ethylenediamine = long needles). Under these synthesis conditions, the size of the particles decreases to 20–45 nm, depending on which precursor was used. Unlike the previous high temperature, slow NaOH addition route, the morphology differences were minimal with the low temperature, rapid NaOH addition route (Figure 3). Close inspection reveals hexagonal morphology is still observed in the nanoparticles. In addition to temperature and precipitation rate, the amine on the precursor also influenced the particle size. Precursors that produce large decreases in the particle aspect ratio in the slow precipitation method also cause the greatest size reduction during rapid precipitation. For example, [Zn(acetate)$_2$(Tris)$_2$] yielded plates with a 0.7 to 1 (L to W) aspect ratio with slow precipitation when prepared at 95 °C and under the rapid precipitation protocol yielded the smallest particles (20 nm). [Zn(acetate)$_2$(ethylenediamine)] produced crystals with a 10 to 1 aspect ratio during slow precipitation at 95 °C and during rapid precipitation yielded the largest particles (45 nm). [Zn(acetate)$_2$(2-thiazolamine)$_2$] and [Zn(acetate)$_2$(hydrazine)$_2$] produced particles with intermediate aspect ratios of 3.1 to 1 and 4 to 1, respectively, during slow precipitation at 95 °C and yielded comparably sized particles during the rapid synthesis (34 nm and 27 nm, respectively).

All ZnO powders were found to be phase pure wurtzite and highly crystalline when characterized by X-ray powder diffraction (Figure 4). The average particle size was also calculated using the Scherrer equation and diffraction data from the (100), (002), and (101) peaks. Calculated particle sizes ranged from 15–19 nm (Table 1). The Scherrer equation calculations indicate that the particles observed in the TEM are aggregates of small crystallites. The particle sizes are consistent with what others have observed for precursor grown ZnO prepared by precipitation and sol-gel synthesis, with particle sizes ranging from 10–165 nm [63,68].

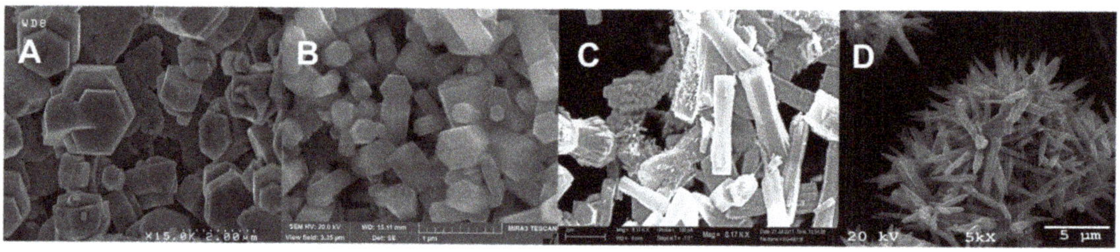

Figure 1. SEM images of ZnO powders prepared from [Zn(acetate)$_2$(amine)$_x$] precursors at 95 °C in an aqueous solution. The precursors used were (**A**) [Zn(acetate)$_2$(Tris)$_2$], (**B**) [Zn(acetate)$_2$(2-thiazolamine)$_2$], (**C**) [Zn(acetate)$_2$(hydrazine)$_2$], and (**D**) [Zn(acetate)$_2$(ethylenediamine)].

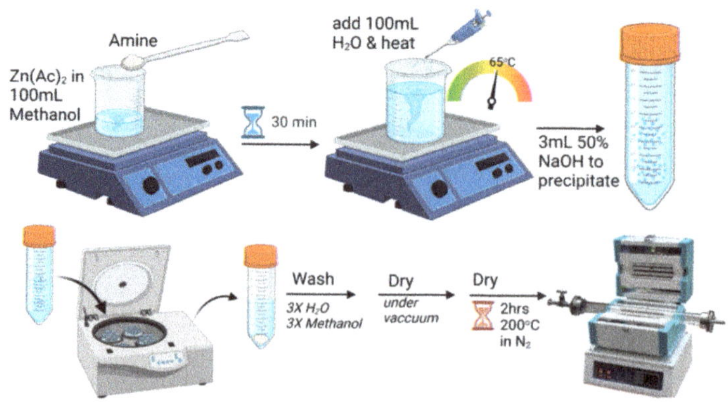

Figure 2. A schematic for the synthesis of ZnO from [Zn(acetate)$_2$(amine)$_x$] precursors with rapid addition of NaOH to precipitate the product. This figure was created with BioRender.com.

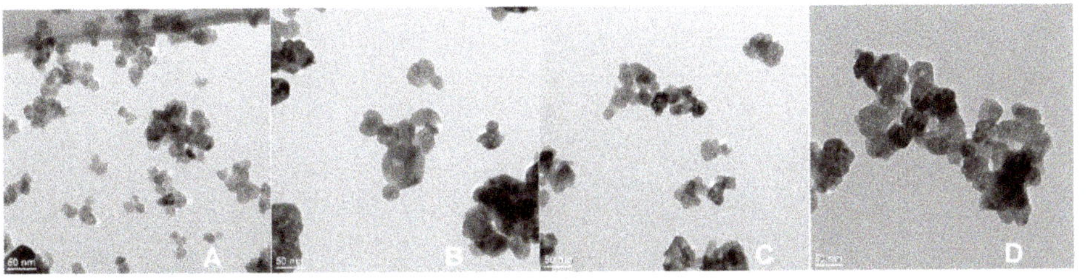

Figure 3. TEM images of ZnO nanoparticles prepared from [Zn(acetate)$_2$(amine)$_x$] precursors in a water/methanol mixture with rapid NaOH addition at 65 °C. Average particle size is given for each material. The precursors used were (**A**) [Zn(acetate)$_2$(Tris)$_2$] (20 ± 4 nm), (**B**) [Zn(acetate)$_2$(2-thiazolamine)$_2$] (34 ± 9 nm), (**C**) [Zn(acetate)$_2$(hydrazine)$_2$] (27 ± 7 nm), and (**D**) [Zn(acetate)$_2$(ethylenediamine)] (45 ± 7 nm).

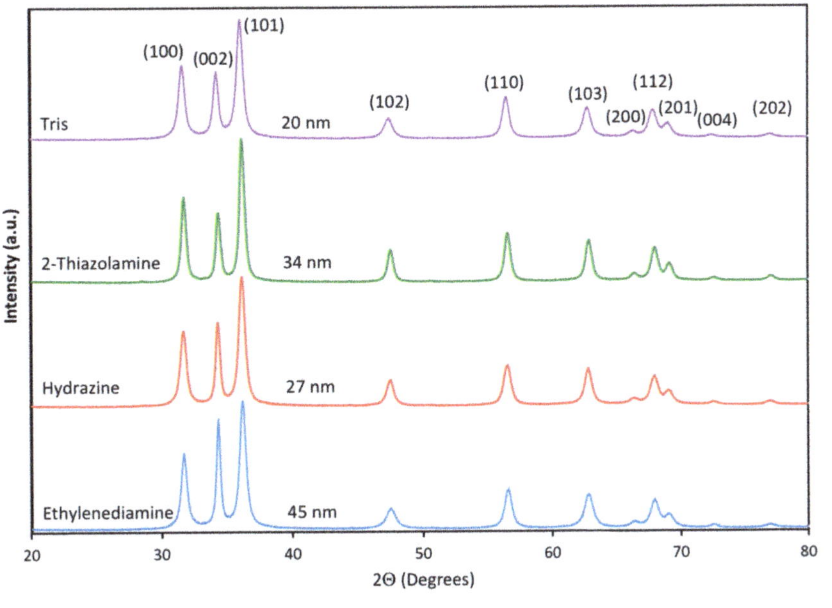

Figure 4. X-ray powder diffraction data for ZnO nanomaterials prepared using [Zn(acetate)$_2$(amine)$_x$] precursors by alkali precipitation in water/methanol solutions at 65 °C. The labels are the amine on the precursor during synthesis. The average particle size given is that observed with the TEM measurements.

Table 1. Size and surface area of the ZnO nanoparticles produced using the different [Zn(acetate)$_2$(amine)$_x$] precursors during rapid precipitation in methanol/water solutions. Sizes are average sizes measured with the TEM, and calculated with the Scherrer and surface area equations. Standard deviations for measurements are provided in parentheses.

Amine	Particle Size (nm) (TEM)	Particle Size (nm) (Scherrer)	Particle Size (nm) (Surface Area)	Surface Area (m^2/g)
Tris	20(4)	15(1)	24.5	43.7
Hydrazine	27(7)	17(4)	30.0	35.6
2-Thiazolamine	34(9)	19(1)	30.5	35.1
Ethylenediamine	45(7)	18(4)	66.4	16.1

The surface area of the nanoparticles was obtained using the BET method from nitrogen adsorption/desorption isotherms to determine how the different precursors used during synthesis influence the surface area of the materials (Table 1). As expected, ZnO with smaller particle sizes produced higher surface areas. Surface areas ranged from 16.1 m^2/g for 45 nm ZnO particles produced using ethylenediamine to 43.7 m^2/g for 20 nm ZnO particles prepared using Tris. These surface areas are consistent with results obtained by others [68,69]. Absorption/desorption isotherms for each compound are provided in the supplemental material (Figures S1–S4). Particle size and surface area are related by Equation (1), allowing the surface area results to be used as another means of determining particle size [70]. The equation assumes that all particles are solid spheres of uniform size with a smooth surface, which is not strictly true given the hexagonal shape of these nanoparticles. However, the equation does provide another means of determining the particle size for each material. The average particle size, D_{BET} (nm), is determined using the theoretical density of the material, ρ, (ZnO = 5.61 g/cm^3), and the specific surface area obtained from the BET measurements, S_{BET} (m^2/g). Equation (1) calculates particle sizes of 24.5 nm for the nanoparticles made using the precursor [Zn(acetate)$_2$(Tris)$_2$], 30.5 nm with that prepared using [Zn(acetate)$_2$(2-thiazolamine)$_2$], 30.0 nm with [Zn(acetate)$_2$(hydrazine)$_2$], and 66.4 nm for the particles made using [Zn(acetate)$_2$(ethylenediamine)]. These calculated values are consistent with the particle sizes observed with TEM measurements.

$$D_{BET} = {}^{6000}/_{\rho \cdot S_{BET}} \tag{1}$$

Thermogravimetric analysis—mass spectrometry (TGA-MS) was also used to characterize the materials prepared by rapid precipitation. The TGA data revealed that ZnO prepared with lower boiling amines resulted in less mass loss when heated to 1000 °C (Figure 5). Since all the materials were annealed at 200 °C as part of the synthesis, the observed weight loss under 200 °C in the TGA was assumed to be from loss of adsorbed atmospheric moisture. All of the materials were observed to have some degree of hygroscopicity when removed from the annealing furnace and weighed for yield. The mass loss above 200 °C was assumed to be from decomposition and volatilization of capping ligands (acetate and amines) attached to the zinc oxide nanoparticles. The materials prepared using Tris, 2-thiazolamine, hydrazine, and ethylenediamine lose 2.2%, 1.8%, 1.6%, and 1.5% of their mass, respectively, when heated from 200 °C to 1000 °C in the TGA. Hydrazine and ethylenediamine have boiling points less than the 200 °C annealing temperature (114 °C and 116 °C, respectively), whereas 2-thiazolamine and Tris have boiling points higher than the annealing temperature (216 °C and 219 °C, respectively). Consistent with the material retaining some of the amines, ZnO prepared using Tris and 2-thiazolamine lose more mass above 200 °C when heated to 1000 °C than hydrazine and ethylenediamine.

To better understand the thermal decomposition results, to confirm the identity of the materials lost during heating, and to help quantify the number and kinds of ligands attached to the nanoparticles, all materials were characterized by TGA-MS. Additionally, nanoparticles were made using $Zn(NO_3)_2$ with the added amines and Zn(acetate)$_2$ with no added amines and were characterized by TGA-MS to help understand how the reactants affect the final materials. Individual TGA-MS data for each ZnO material are provided in the supplemental material (Figures S5–S13). For these nanomaterials, the majority of the mass loss below 200 °C comes from the loss of H_2O, which starts to be released at 30–50 °C, has a maximum around 100 °C, and trails off by 200 °C. Additionally, all materials made using amines, except 2-thiazolamine, release CO_2 concurrent with the low temperature release of H_2O. Materials prepared using Zn(acetate)$_2$ and hydrazine or Tris also release NO_2 at the same time as their low temperature release of H_2O and CO_2. These results are consistent with ZnO previously being shown to absorb CO_2 and NO_2 and ZnO's use in sensors for CO_2 and NO_2 detection [71,72].

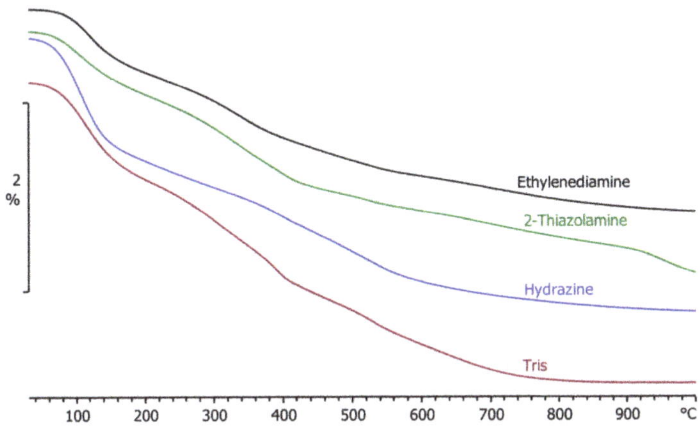

Figure 5. TGA curves showing the percent weight loss during heating for ZnO powders prepared using four different precursors. The amine label by the curve corresponds to the amine used during synthesis of the ZnO.

To help determine how much acetate is still attached to each nanomaterial, ZnO was prepared using Zn(acetate)$_2$ with no added amine. The TGA data for this material show a pronounced weight loss starting around 250 °C, from the loss of water, carbon dioxide, and the acetate ligand (m/z = 58) (Figure S9). The m/z = 58 was monitored because that mass to charge ratio was found to have the highest intensity during the TGA-MS characterization of solid Zn(acetate)$_2$. The acetate signal peaks at 300 °C and is gone by 360 °C. However, the signals for water and CO_2 continue to increase until about 400 °C, indicating that the remaining acetate ligands start to be converted to CO_2 and water above 300 °C and are entirely burned above 360 °C. All remaining hydrocarbons are converted to CO_2 and H_2O as the temperature ramps to 1000 °C, yielding a 2.7% mass loss during the heating from 200 °C to 1000 °C.

The TGA-MS data for zinc oxide prepared using zinc acetate with no added amines provide insight into the composition of the nanoparticles prepared using Zn(acetate)$_2$ with added amines. No acetate signal is observed in the mass spectrometer for any samples prepared using zinc acetate and amines, indicating that minimal acetate is still attached to the material. It has been proposed that the amine can nucleophilic attack the acetate carbonyl and cause the elimination of the amide during zinc oxide synthesis [29]. This would explain why acetate ions are not retained in some of the materials. The lack of acetate on these materials gives them a lower mass loss than the ZnO made from Zn(acetate)$_2$ and no amines. To help confirm the lack of acetate ions, ZnO nanomaterials were also prepared using zinc nitrate and each amine, and the materials were analyzed by TGA-MS (Figures S10–S13). The ZnO samples made using ethylenediamine and both zinc sources have similar weight losses (to two significant digits) from 200 °C to 1000 °C. Although the sample prepared using Zn(acetate)$_2$ has a higher CO_2 loss above 600 °C than the sample using zinc nitrate, this sample likely contains minimal acetate since the acetate signal is missing in the MS. Neither of these samples had a mass spectrometer signal for NO or NO_2, or any other mass associated with loss of the diamine, indicating that it was lost either during precipitation or the 200 °C annealing step.

In comparing the TGA-MS data for the materials prepared using hydrazine and the two zinc sources, the sample prepared using the nitrate had a larger mass loss than the sample prepared using zinc acetate, with much of the mass lost as water between 200 °C and 600 °C. Since this is well above the vaporization temperature of water, this is likely from hydroxides on the particle surface decomposing to water and forming the surface oxide [61,73]. The sample prepared using zinc acetate does show a small CO_2 loss with a peak near 390 °C, but likely contains only minimal acetate ions, given the lack of observed acetate signal in the MS. These samples also lose tiny amounts of NO and NO_2 at the same temperature as the CO_2 loss, indicating that most of the hydrazine was lost during the synthesis or annealing steps.

Zinc oxide samples made using zinc acetate and the two less volatile amines show minimal low temperature (<200 °C) loss of CO_2 in the TGA-MS, indicating that the ZnO surface of these particles might be partially covered with amines or acetate ligands and not readily accessible for absorbing CO_2. In the samples made using zinc acetate and Tris, CO_2 and NO_2 start to be lost a little above 200 °C, and the shape of the CO_2 signal resembles that of the CO_2 signal for the zinc oxide prepared using zinc acetate with no added amines. Both of the samples made with Tris continue to produce CO_2 and H_2O to at least 800 °C, consistent with them burning off attached hydrocarbons. The sample made using Tris and zinc acetate loses 2.2% total mass during heating from 200 °C to 1000 °C compared to the sample made with zinc nitrate losing 1.9% mass during the same temperature interval. This greater weight loss is consistent with the sample made using zinc acetate retaining some of the acetate on the particles.

The materials prepared using 2-thiazolamine produce similar TGA signals, with each having a high temperature mass loss which starts at 850 °C for the sample made with zinc nitrate and 920 °C for the material made using zinc acetate. The product made with zinc acetate also has a higher mass loss of 1.8%, compared to 1.7% for the sample made using zinc nitrate. The mass spectrometry data for the two products poorly resemble each other. For the product made with $Zn(NO_3)_2$, the signal from CO_2 peaks at 310 °C, with a long trailing tail that plateaus from 380–410 °C, compared to the sample made using $Zn(acetate)_2$, which peaks at 390 °C, dies away quickly and then climbs to a small peak at 920 °C. The data are consistent with both materials retaining some of the amine and continuing to burn it off all the way to 1000 °C. Additionally, the material made using zinc acetate likely retains some of the acetate ligands, given the small increase in mass loss over that of the sample made using zinc nitrate.

In an attempt to quantify the number of ligands associated with each ZnO prepared from zinc acetate, each material was characterized by CHN elemental analysis. However, only ZnO prepared using Tris and 2-thiazolamine had carbon percentages above the 0.5% reporting limit (Tris: C = 1.10% and 2-thiazolamine: C = 0.74%). Results for hydrogen and nitrogen were below the reporting limit for all samples, which could have been used to quantify the amount of amine bound to each ZnO. Using the carbon percentages obtained for the materials made from $[Zn(acetate)_2(Tris)_2]$ and $[Zn(acetate)_2(2\text{-thiazolamine})_2]$, and assuming a composition of $ZnO(acetate)_x(amine)_x$ to determine stoichiometry, yields formulas of $ZnO(acetate)_{0.0128}(Tris)_{0.0128}$ (one set of ligands for every 78 ZnO formula units) and $ZnO(acetate)_{0.0103}(2\text{-thiazolamine})_{0.0103}$ (one set of ligands for every 97 ZnO formula units). Assuming the observed mass loss in the TGA from 200–1000 °C for these materials is from loss of attached ligands, the 2.2% mass loss for the material made using Tris would yield a calculated stoichiometry of $ZnO(acetate)_{0.0102}(Tris)_{0.0102}$. The 1.8% mass loss for 2-thiazolamine would yield $ZnO(acetate)_{0.0096}(2\text{-thiazolamine})_{0.0096}$. These stoichiometries are in reasonably good agreement with the CHN data, considering both materials were still losing mass at the 1000 °C TGA cutoff temperature. Performing the same calculations for the material made with ethylenediamine and a 1.5% TGA mass loss would yield $ZnO(acetate)_{0.014}(ethylenediamine)_{0.007}$, based on the stoichiometry of the ligands on the precursor carrying over to the final composition of the ZnO. This stoichiometry is equivalent to a 0.6% carbon content and should have exceeded the elemental analysis detection limit. The fact that carbon was not detected in the CHN analysis supports the hypothesis that very little of the acetate was incorporated into the final material, or that most of the diamine was lost during the 200 °C annealing step, and that the extra weight loss observed in the TGA is from an undetermined number of hydroxides being lost as water during heating. Likewise, for the material prepared using hydrazine to have a 1.6% TGA weight loss would require a final stoichiometry of $ZnO(acetate)_{0.0146}(hydrazine)_{0.0146}$ and would correspond to a 0.4% carbon content and a 0.5% nitrogen content. Although the carbon content is below the CHN detection limit, the nitrogen content is at the detection limit. It could have been determined, further supporting the hypothesis that much of the weight loss for this material is from the loss of water from hydroxides. Attempts were also made to quantify the number of ligands bound to each ZnO using FT-IR and X-ray photoelectron spectroscopy. With such small quantities, vibrations associated with the functional groups on the ligands were not visible above the ZnO background in the IR spectra (Figure S14). Neither nitrogen nor sulfur was observed in the XPS spectrum for any of the materials, consistent with them containing a very limited number of ligands.

To better understand how the synthesis conditions influence the optical properties of the materials, the optical bandgap for each ZnO nanomaterial prepared by the rapid precipitation method was determined by ultraviolet-visible diffuse reflectance spectroscopy. The data were transformed using the Kubelka–Munk equation (Equation (2)) [74,75], where $R_\infty = R_{sample}/R_{standard}$ and is the measured diffuse reflectance of the sample, $F(R_\infty)$ is the Kubelka–Munk function or the so-called remission, $h\nu$ is the photon energy, C_2 is a proportionality constant, and E_g is the optical band gap of the powdered sample. The band gap is obtained from a plot of $[F(R_\infty)\,h\nu]^2$ against $h\nu$, where a linear fit through the steepest part of the data intersects the photon energy axis (x-axis) is taken as the value for E_g.

$$[F(R_\infty)\,h\nu]^2 = C_2\,(h\nu - E_g) \tag{2}$$

Optical bandgaps for the materials were found to be 3.31(1) eV for ZnO prepared using hydrazine, 3.33(1) eV for ZnO made with ethylenediamine, 3.34(2) eV for Tris, and 3.35(4) eV for ZnO made using 2-thiazolamine (Figure 6). These values are within three standard deviations of each other, compare well with what others have measured for ZnO nanoparticles [53,74], and are consistent with the 3.3 eV optical bandgap of bulk ZnO [76]. Changes in semiconductor bandgaps are generally attributed to reduction in crystallite size, the number of structural defects, and the overall crystallinity of the material [77,78]. Although the different amines can be used to modify the size and shape of ZnO particles and lead to different amounts of organic material attached to the nanoparticles, the crystallite size in the nanoparticles are very similar, as was observed with the Scherrer calculations, yielding similar bandgaps for all of the materials.

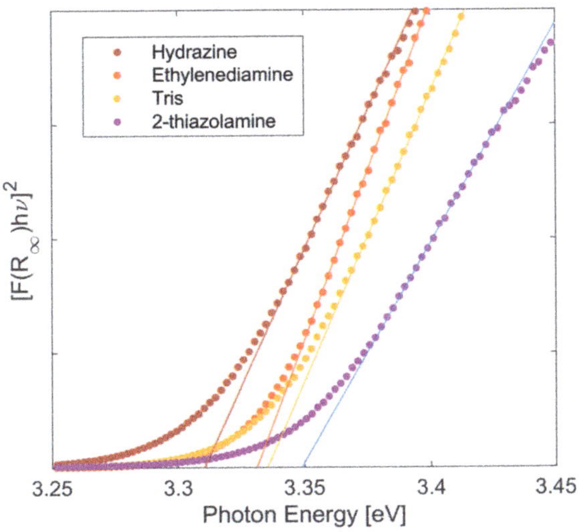

Figure 6. Kubelka–Munk transformed diffuse reflectance spectra of annealed ZnO powders prepared by rapid precipitation. The amine in the legend corresponds to the amine used during the synthesis of each material.

2.2. Photocatalytic Activity

Because of ZnO's attractive physical and chemical properties, it has been extensively studied as a photocatalyst [79]. To explore how the synthesis conditions impact the material's chemical reactivity, the ZnO prepared by rapid precipitation was used to photocatalyze the decolorization of the organic dye malachite green under ultraviolet (UVA) radiation. The mechanism for the photodegradation of dyes and other organic compounds by semiconductors has previously been described by many investigators [52,53,79–88]. Several catalysts have already been used to decolorize malachite green including, but not limited to, ZnO [53,81], NiS [82], TiO_2 [83], $Ag/TiO_2/Nb_2O_5$ composites [84], BiOBr composites [85], ZnO on activated carbon [86], Mn-doped BiOCl [87], and a Pd/WO_3 mixture [88]. Under UV illumination in aqueous environments, ZnO generates reactive oxygen species responsible for the reactions that decolorize and degrade the dye [89–91]. In the reaction mechanism, ZnO absorbs a photon of UV light and generates an electron-hole pair on the particle surface (Equation (3)), with the electron being promoted to the conduction band (e_{cb}^-) and leaving the hole in the valence band (h_{vb}^+). The electrons react with surface adsorbed oxygen to form the superoxide radical ion $O_2^{\bullet -}$ (Equation (4)). The holes react with adsorbed water molecules to form hydroxide radicals ($HO^\bullet$) and protons (Equation (5)), or in basic solutions, the holes can react with hydroxide molecules to produce additional hydroxide radicals (Equation (6)). The $O_2^{\bullet -}$ can react with protons to form $HO_2^\bullet$ (Equation (7)), which can further react to form hydrogen peroxide and O_2 (Equation (8)). The peroxide then decomposes to generate more hydroxide radicals (Equation (9)). The hydroxide radicals generated by the mechanism are responsible for degrading the dye and converting the organic molecules to colorless intermediate products and, theoretically, carbon dioxide and water (Equation (10)). Saad et al. have proposed a degradation pathway for malachite green [44], and Yong and co-workers have identified 40 different intermediate products using liquid chromatography–mass spectrometry to analyze malachite green dye solutions during photodegradation [91].

$$ZnO + h\nu \rightarrow e_{cb}^{-} + h_{vb}^{+} \tag{3}$$

$$O_2 + e_{cb}^{-} \rightarrow O_2^{\bullet -} \tag{4}$$

$$h^+ + H_2O \rightarrow HO^\bullet + H^+ \tag{5}$$

$$h^+ + HO^- \rightarrow HO^\bullet \tag{6}$$

$$O_2^{\bullet -} + H^+ \rightarrow HO^\bullet{}_2 \tag{7}$$

$$2\,HO^\bullet{}_2 \rightarrow H_2O_2 + O_2 \tag{8}$$

$$H_2O_2 \rightarrow 2HO^\bullet \tag{9}$$

$$HO^\bullet + \text{dye} \rightarrow \text{intermediates} \rightarrow CO_2 + H_2O \tag{10}$$

The decolorization of the dye proceeds by way of pseudo first-order rate kinetics. The reaction can be studied by measuring the dye's absorbance (A) in solution over time (Equation (11)). The observed rate constants (k_{obs}) include the degradation, and any physical adsorption of the dye onto the catalyst. The rate equation can be rearranged to give the rate constant as a positive slope (Equation (12)). In the equations, the initial absorbance of the dye in solution is A*.

$$\ln A = -k_{obs}t + \ln A^* \tag{11}$$

$$\ln(A^*/A) = k_{obs}t \tag{12}$$

This work found that different sizes and morphologies of the ZnO particles influence the material's ability to catalyze the photochemical reaction. In general smaller particles yielded faster decomposition rates than larger particles, given their larger surface area for particle sizes on the order of hundreds of nanometers to several micrometers. Additionally, rod and needle morphologies, with high surface area-to-volume ratios, also appear to increase photocatalytic effectiveness [92]. Wang et al. observed that 50 nm ZnO particles degraded methyl orange with the highest efficiency under UV light [93], and others observed that ZnO nanoparticles smaller than 50 nm were faster at degrading methylene blue than particles larger than 50 nm [49]. For this study, when ZnO particle size decreases to that produced by the rapid synthesis, the observed first-order rate constant (k_{obs}) for the photodecomposition appears to be driven by multiple factors, including the quantity of ligands associated with the nanoparticles, as well as particle size and surface area (Figures 7 and 8 and Table 2). Zinc oxide prepared using hydrazine had the fastest rate constant, k_{obs} = 0.038(6) min^{-1}, and degraded the dye more than twice as fast as the ZnO prepared using 2-thiazolamine, with k_{obs} = 0.019(2) min^{-1}, even though the particles have similar size and surface area. ZnO prepared using ethylenediamine, k_{obs} = 0.024(1) min^{-1}, and Tris, k_{obs} = 0.021(1) min^{-1}, yielded intermediate decolorization rates, even though the material made using ethylenediamine had the largest particles and smallest surface area, whereas the material made using Tris had the smallest particles and largest surface area. A control was also prepared using $Zn(NO_3)_2$ with no added amine and used to decolor the dye solution and had a k_{obs} = 0.015(1) min^{-1}. Additionally, it was observed that in the absence of a catalyst, the dye degrades under the UVA light with a k_{obs} = 0.005(2) min^{-1}. The rate constants measured for the catalysts are consistent with what others have measured for the photodegradation of malachite green using ZnO [53–56,58,59]. Table 2 compares the photodegradation results for malachite green using zinc oxide prepared from several precursors.

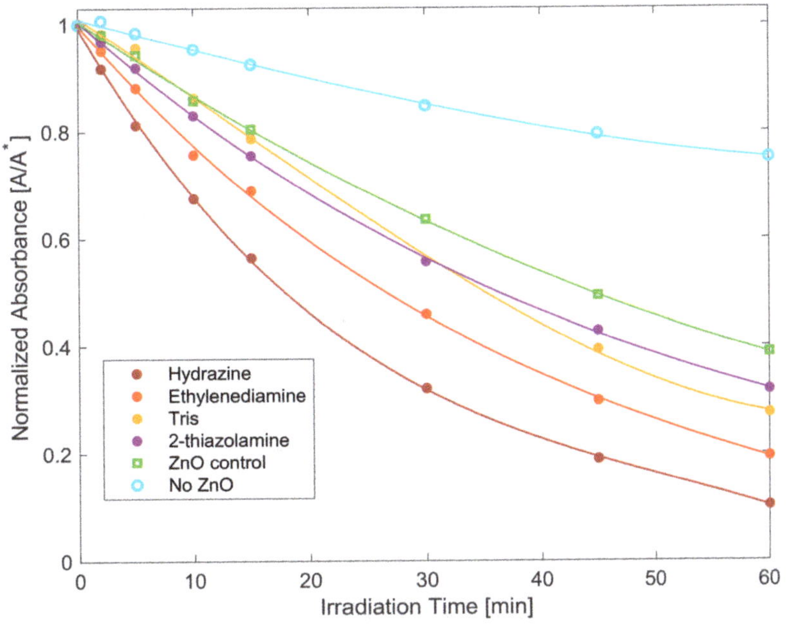

Figure 7. Absorption data for the decolorization of malachite green solutions using different zinc oxide photocatalysts prepared by rapid precipitation. The amines used to make the $[Zn(acetate)_2(amine)_x]$ precursors are identified by the labels in the legend. The ZnO control was prepared using $Zn(NO_3)_2$ without added amines. "No ZnO" corresponds to the decolorization of malachite green with no catalyst in solution under UVA.

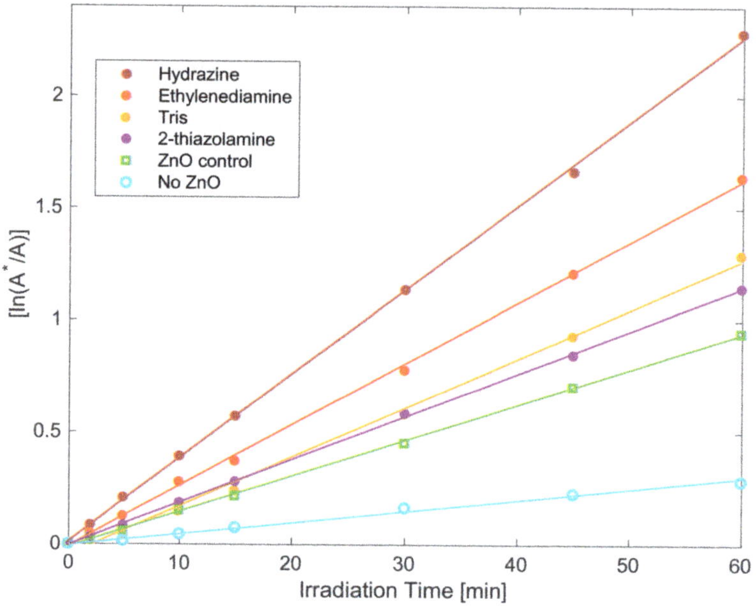

Figure 8. Absorption data of malachite green solutions during photodegradation by ZnO prepared by rapid precipitation. The amines used to make the $[Zn(acetate)_2(amine)_x]$ precursors are identified by the labels in the legend. R^2 values for all of the best fit lines exceed 0.997.

Table 2. Photodegradation of malachite green using zinc oxide made from different precursors.

Precursor	Dye Conc. (g/L)	Catalyst Conc. (g/L)	k_{obs} (min^{-1})	Ref
$ZnCl_2$	0.01	0.2	0.010	[58]
$Zn(acetate)_2$	0.2	0.2	0.014	[53]
$Zn(acetate)_2 + NH_3$	0.02	0.05	0.009	[94]
$Zn(acetate)_2$ + TEA	0.01	0.12	0.0046	[56]
$Zn(NO_3)_2$ + PEG	0.2	0.2	0.023	[53]
$Zn(NO_3)_2$ + EDTA	.0093	0.2	0.058	[55]
$[Zn(acetate)_2(hydrazine)_2]$	0.027	1.67	0.038	this work
$[Zn(acetate)_2(ethylenediamine)]$	0.027	1.67	0.024	this work
$[Zn(acetate)_2(Tris)_2]$	0.027	1.67	0.021	this work
$[Zn(acetate)_2(2\text{-thiazolamine})_2]$	0.027	1.67	0.019	this work

TEA = triethylamine, PEG = polyethylene glycol, EDTA = ethylenediaminetetraacetic acid.

The measurements were also obtained in the dark to determine the contribution of absorption on the decolorization constants. The $k_{obs,dark}$ values for absorption in the dark were nearly identical for all of the materials, with the $k_{obs,dark}$ = 0.007(1) min^{-1} for ZnO prepared from hydrazine, ethylenediamine, and 2-thizolamine and 0.008(1) min^{-1} for the ZnO prepared using Tris. Absorption onto the catalyst accounts for approximately one-third of the observed decolorization rate for the materials prepared using ethylenediamine, Tris, and 2-thiazolamine, but less than 20% of the k_{obs} for the material prepared using hydrazine.

As noted above in the discussion of the TGA results, ZnO produced using hydrazine and ethylenediamine retained the lowest amount of organic material and liberated the largest quantities of CO_2 at temperatures under 200 °C, indicating that their surfaces are more readily accessible for absorbing CO_2 than the materials prepared using the less volatile amines. This result likely contributes to their improved catalysis with a surface readily available in solution for water and oxygen sorption, which is necessary for the catalyst mechanism described above.

Additionally, much of the weight loss during heating from the sample made with hydrazine was in the form of water, likely from the decomposition of hydroxides. In contrast, the other materials all liberate more CO_2 from bound hydrocarbons, allowing the hydrazine-derived material to have a more accessible surface and contributing to it being the fastest catalyst. The Tris and 2-thiazolamine-derived materials have hydrocarbons bound to them which impedes the ability of oxygen and water to reach the ZnO surface, slowing the degradation mechanism and kinetics. Even though both Tris and 2-thiazolamine have higher measured surface areas than the material prepared using ethylenediamine, their rates for decolorization are slower. Although the Tris-made ZnO contains more organic material than the 2-thiazolamine-made material, the Tris-made ZnO produced faster decolorization kinetics. The Tris-made material released more CO_2 at temperatures below 200 °C than the ZnO made using 2-thiazolamine during TGA-MS measurements, suggesting that the Tris-made ZnO has a more accessible surface to CO_2 in the air and perhaps H_2O and O_2 in solution. The Tris-made ZnO also contained smaller particle sizes with a higher surface area than the ZnO prepared using 2-thiazolamine. The authors speculate that these attributes contribute to the ZnO made using Tris degrading the dye faster than the ZnO made using 2-thiazolamine.

2.3. Catalyst Reusability

The ZnO made using hydrazine was evaluated for its reusability, as it had the highest rate constant in the study. A single sample of the material was used to decolorize five samples of the dye solution with no significant loss in catalytic efficiency (Figure 9). For these experiments, 0.100 g of catalyst was added to 50 mL of a 28.9 μM dye solution and allowed to stir for two hours under UVA illumination. At the end of each experiment, the nanomaterial was recovered by centrifugation and then added to the next dye solution. The percent dye decolorization ranged from 96–97% for all five samples, indicating very good recyclability of the material.

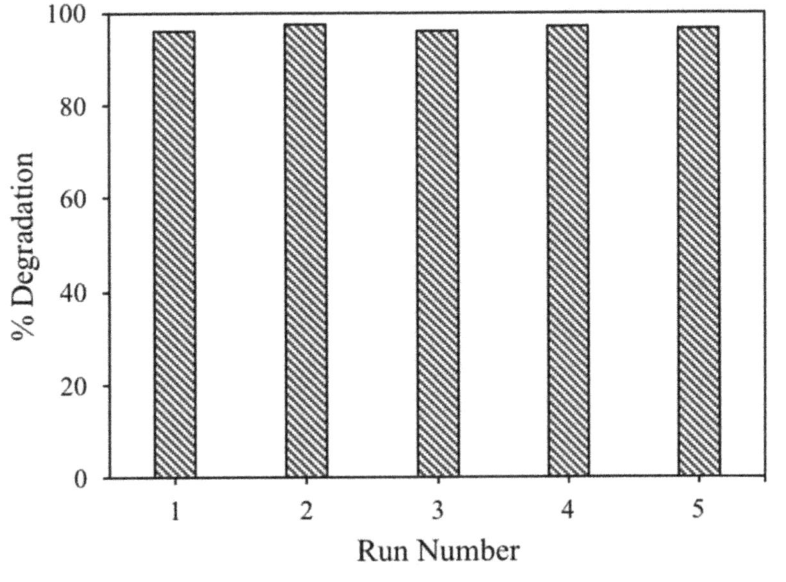

Figure 9. Reusability study for ZnO prepared using [Zn(acetate)$_2$(hydrazine)$_2$] under UVA illumination for two hours. A single sample of ZnO decolored five malachite green dye samples with minimal efficiency loss.

3. Materials and Methods

3.1. Techniques and Materials

Zinc acetate dihydrate (ACS reagent grade, Fisher Chemicals, Hampton, NH, USA) was used as received. Amines were either used as received (ACS reagent grade, Sigma-Aldrich Chemicals, St. Louis, MO, USA) or distilled under nitrogen before use. All synthesis was done in containers open to the atmosphere. Solvents were either ACS or HPLC grade and used as received. Ultrapure water for solutions and synthesis was obtained using a Millipore Milli-Q system.

3.2. ZnO Precursor Isolation and Characterization

Each ZnO precursor was isolated from solution for characterization by ^{1}H NMR and CHN elemental analysis to verify precursor composition. The synthesis and characterization of [Zn(acetate)$_2$(Tris)$_2$] and [Zn(acetate)$_2$(ethylenediamine)] were reported in our previous publication [39].

For the synthesis of [Zn(acetate)$_2$(2-thiazolamine)$_2$], zinc acetate dihydrate (1.0311 g, 47 mmo) and 2-thiazolamine (0.9412 g, 94 mmol) were dissolved in a mixed solution of anhydrous ethanol (20 mL) and methanol (20 mL) under ambient conditions. The solution was stirred for one hour and then was evaporated to dryness under vacuum. Yield: 1.6863 g (93.5%). Soluble in dimethylsulfoxide, and methanol. Elem. Anal. for $ZnC_{10}H_{14}N_4O_4S_2$: C = 31.30; H = 3.68; N = 14.60%. Found: C = 31.44; H = 3.65; N = 14.58%. ^{1}H NMR (CD_3OD): δ 2.015 (s, 6H, $OOCCH_3$); 6.599 (s, 2H, $NH_2CSCH_2CH_2N$), 6.943 (s, 2H, $NH_2CSCH_2CH_2N$).

For the synthesis of [Zn(acetate)$_2$(hydrazine)$_2$], zinc acetate dihydrate (1.0314 g, 47 mmol) and hydrazine (0.5160 g) were dissolved in a mixed solution of anhydrous ethanol (20 mL) and water (20 mL) under ambient conditions. The solution was stirred for one hour and then was evaporated to dryness under vacuum. Yield: 0.9751 g (83.8%). Insoluble in all solvents. Elem. Anal. for $ZnC_4H_{14}N_4O_4$: C = 19.41; H = 5.70; N = 22.63%. Found: C = 19.37; H = 5.67; N = 22.52%.

3.3. ZnO Particle Synthesis

For a typical nanoparticle synthesis, zinc acetate dihydrate (5.3938 g, 24.6 mmol) was dissolved in 100 mL of methanol with stirring. An amine was then added (49.1 mmol for monodentate amines and 25.4 mmol for bidentate amines). The solution was allowed to stir for 30 min at room temperature, 100 mL of water was added, the solution was heated to 65 °C and rapidly titrated with 3 mL of an aqueous 50% NaOH solution to precipitate ZnO. Upon cooling, the reaction solutions were centrifuged to isolate the product, rinsed three times with water and three times with methanol, dried under vacuum, and then dried at 200 °C for two hours under a nitrogen atmosphere. The amines used in the synthesis included hydrazine, ethylenediamine, 2-thiazolamine, and tris(hydroxymethyl)aminomethane (Tris). The product prepared using hydrazine partially precipitates before the NaOH addition. Typical ZnO yield: 1.8089 g, 90%.

3.4. Powder X-ray Diffraction

Powder X-ray diffraction (XRD) was utilized to confirm the crystallographic structure of the materials and estimate their particle size using the Scherrer equation [95]. All XRD measurements were performed on a Rigaku 2200 D/Max X-ray diffractometer. The diffractometer was equipped with a copper sealed tube anode utilizing Cu Kα radiation (λ = 1.5418 Å). Specimens were scanned from 20° to 95° 2Θ with a step size of 0.02° and a dwell time of 1 s. Phase composition was determined by comparison to the International Centre for Diffraction Data (ICDD) patterns.

3.5. Scanning Electron Microscopy

Field emission scanning electron microscopy (FESEM) (Hitachi, S-4500, Tokyo, Japan) was used to determine size and morphology of all ZnO samples. SEM micrographs were acquired and processed using Quartz PCI Version 8 image processing software.

Accelerating voltages of 7 kV to 20 kV were used for imaging, with magnification ranging from 1000X to 200,000X, depending on particle size and morphology. The ZnO powders were deposited onto wet silver paint applied to the aluminum sample stage to minimize charging and loose particles in the FESEM vacuum chamber. The paint was allowed to dry, and then compressed air was applied to the sample to remove any loose particulates.

3.6. Transmission Electron Microscopy

Powders prepared by rapid addition of NaOH were characterized by Transmission Electron Microscopy (TEM). Sample preparation was done by dispersing 4 mg of the ZnO nanoparticles into 4 mL of methanol and ultrasonicating for five minutes. Immediately after ultrasonication, one drop of the ZnO nanoparticle suspension was placed on an ultrathin holey carbon grid (Ted Pella, product # 01824, Redding, CA, USA). Analysis of the ZnO nanoparticles was carried out using a Zeiss Libra 200 transmission electron microscope, and micrographs were acquired using a Gatan Orius CCD digital camera. Analytical TEM (ATEM) was employed by running scanning transmission electron microscopy (STEM) and performing X-ray energy dispersive spectroscopy (XEDS) to confirm the elemental composition of the nanoparticles. STEM images were acquired with a high angle annular dark field detector (HAADF). The Zeiss Libra 200 is equipped with a Noran XEDS system, with a Li drifted Si detector. All image acquisitions and XEDS measurements were carried out with an acceleration voltage of 200 kV and an emission current of 304 μA. XEDS measurements were made with a beam spot size of 20 nm. Average particle sizes and standard deviations were determined from measurements of 50–60 particles for each material.

3.7. Ultraviolet–Visible Diffuse Reflectance Spectroscopy

The optical band gap of the ZnO nanoparticles was determined at room temperature by diffuse reflectance, using an Agilent Cary 100 spectrophotometer equipped with an internal diffuse reflectance attachment and a powder sample cell. Samples were scanned from 300 to 500 nm. Data were transformed using the Kubelka–Munk equation [74,75]. The band gap was determined using the R software package grofit from the intersection of the tangent at the maximum slope of the smoothed cubic spline fit to the x-axis [96,97].

3.8. Surface Area Measurements

The surface area of the powders was measured with a surface area and pore size analyzer (Quantachrome Instruments, NOVA 2200 e, Boynton Beach, FL, USA). Approximately 0.3 gm of each powder was weighed in a bulbous glass vial, subsequently vacuum degassed at 200 °C for 12 h, and then backfilled with helium. Adsorption and desorption curves were then obtained at liquid-nitrogen temperatures using N_2 as the adsorbate for P/Po ranging from 0.05 to 0.99. An automatic multipoint Brunauer, Emmett, and Teller (BET) model was fitted to the adsorption curve from P/Po ranging from 0.05 to 0.3 and was used to calculate the surface area. Before analyzing the ZnO samples, a certified standard (105.57 ± 4.41 m^2/g) was measured to be 107.54 m^2/g.

3.9. Thermal Analysis

All ZnO samples were analyzed by thermogravimetric analysis using a Mettler Toledo TGA/DSC1 thermal analyzer connected to a Pfieffer Vacuum ThermoStar mass spectrometer. Samples were heated with ramp rates ranging from 5–50 °C/min from 30 °C to 1000 °C in alumina crucibles in a dry air atmosphere. Blanks for each crucible were measured before each sample. Samples sizes of 10–20 mg were routinely used. Samples sizes of 50–60 mg were used when characterizing the volatile compounds produced during heating by mass spectrometry.

3.10. Photocatalytic Activity

The photocatalytic activity of the different ZnO nanoparticles was evaluated by photocatalytic decolorization of malachite green oxalate dye solutions (28.9 μM) under ultraviolet

(UV) illumination. For these experiments, ZnO nanoparticles (0.1 g) were magnetically stirred in 10 mL of water for 10 min to disperse the nanoparticles in solution, then 50 mL of the dye solution was added and irradiated with UVA illumination (two 13W Damar 26289A F13TT/BLB bulbs). The intensity of the light at the surface of the dye was measured as approximately 3.4 W/m^2 using a Vernier UVA Sensor. Aliquots of the dye solution were removed at specific time intervals (0, 2, 5, 10, 15, 30, 45, and 60 min), centrifuged for one minute at 12,000 rpm to isolate the ZnO, and the absorbance of the solution was measured at 617 nm in quartz cuvettes using an Agilent Cary 100 spectrophotometer. Experiments in the dark were performed in the absence of both UVA and ambient light. For consistency, all experiments were performed in the same 600 mL beaker using the same two-inch Teflon-coated stir bar to stir the solutions. Average kinetic rate constants with standard deviations were determined from seven to ten experiments using three different sample preparations with the same amine. All data were manipulated using Microsoft Excel.

4. Conclusions

We have demonstrated that [Zn(acetate)$_2$(amine)$_x$] compounds can be used to prepare wurtzite zinc oxide nanoparticles by alkali precipitation. The size and aspect ratio (length to width) of ZnO particles can be modified using different amines to produce needle-, rod- or plate-shaped particles. Nucleation kinetics can further control particle size, with larger particles prepared using slow precipitation rates and higher solution temperatures and fast precipitation rates and lower solution temperatures producing nanoparticles. Amines that resulted in the smallest aspect ratios during higher temperatures and slow precipitation rates yielded the smallest particles during rapid precipitation, and amines that yielded particles with large aspect ratios during slow precipitation produced the largest nanoparticles during rapid precipitation, even though all produced mostly spherical, hexagonal particles during rapid precipitation. The surface area of the nanomaterials was found to depend only on particle size, with the material with the smallest particles yielding the highest surface area. The optical bandgap of the materials was not altered by particle size or the amines bound to the precursor compounds, which is consistent with the materials having similar crystallite sizes, regardless of particle size. The different amines on the precursors alter the number of organic ligands associated with each material. In general, ZnO made using precursors containing less-volatile amines retain more organic material than those prepared using more-volatile amines. The presence of the organic material on the nanoparticles impacts the photocatalytic activity of the materials, with the ZnO containing the lowest amount of associated hydrocarbons yielding the highest rate constants for the decolorization of malachite green, regardless of particle size. [Zn(acetate)$_2$(hydrazine)$_2$] produced zinc oxide with the least organic material bound to the nanoparticles and degraded the dye the fastest. In recycling experiments, zinc oxide made using [Zn(acetate)$_2$(hydrazine)$_2$] degraded five malachite green solutions with minimal loss of efficiency.

Supplementary Materials: The following supporting information can be downloaded at: https://www.mdpi.com/article/10.3390/catal12101099/s1, Figure S1: Absorption, desorption isotherms for BET surface area measurements for ZnO prepared using Tris. Figure S2: Absorption, desorption isotherms for BET surface area measurements for ZnO prepared using 2-thiazolamine. Figure S3: Absorption, desorption isotherms for BET surface area measurements for ZnO prepared using hydrazine. Figure S4: Absorption, desorption isotherms for BET surface area measurements for ZnO prepared using ethylenediamine. Figure S5: Thermogravimetric analysis (TGA) and mass spectrometry (MS) curves for the heating of ZnO prepared using [Zn(acetate)$_2$(ethylenediamine)]. Figure S6: Thermogravimetric analysis (TGA) and mass spectrometry (MS) curves for the heating of ZnO prepared using [Zn(acetate)$_2$(hydrazine)$_2$]. Figure S7: Thermogravimetric analysis (TGA) and mass spectrometry (MS) curves for the heating of ZnO prepared using [Zn(acetate)$_2$(2-thiazolamine)$_2$]. Figure S8: Thermogravimetric analysis (TGA) and mass spectrometry (MS) curves for the heating of ZnO prepared using [Zn(acetate)$_2$(Tris)$_2$]. Figure S9: Thermogravimetric analysis (TGA) and mass spectrometry (MS) curves for the heating of ZnO prepared using Zn(acetate)$_2$ with no added amine. Figure S10: Thermogravimetric analysis (TGA) and mass spectrometry (MS) curves for the heating of ZnO prepared using Zn(nitrate)$_2$ and ethylenediamine. Figure S11: Thermogravimetric analysis (TGA) and mass spectrometry (MS) curves for the heating of ZnO prepared using Zn(nitrate)$_2$ and 2-thiazolamine. Figure S12: Thermogravimetric analysis (TGA) and mass spectrometry (MS) curves for the heating of ZnO prepared using Zn(nitrate)$_2$ and hydrazine. Figure S13: Thermogravimetric analysis (TGA) and mass spectrometry (MS) curves for the heating of ZnO prepared using Zn(nitrate)$_2$ and Tris. Figure S14: Infrared spectra of ZnO prepared using [Zn(acetate)$_2$(Tris)$_2$], [Zn(acetate)$_2$(2-thiazolamine)$_2$], [Zn(acetate)$_2$(hydrazine)$_2$] and [Zn(acetate)$_2$(ethylenediamine)].

Author Contributions: Conceptualization, J.D.H.; methodology, J.D.H., C.C.P. and N.L.M.; software, J.D.H., C.C.P., N.L.M. and J.R.C.; validation, J.D.H., C.C.P. and J.R.C.; formal analysis, J.D.H., C.C.P. and J.R.C.; investigation, J.D.H., E.A.W., E.G.E., C.C.P., S.C.B., N.L.M., A.J.C., K.O.L., A.E.H., K.V.G., C.E.L., S.M.H., J.E., D.E. and J.E.C.; resources, J.D.H.; data curation, J.D.H.; writing—original draft preparation, J.D.H., S.C.B., K.O.L. and J.R.C.; writing—review and editing, J.D.H., K.V.G., J.E., D.E. and N.L.M. and J.R.C.; visualization, J.D.H., N.L.M. and J.R.C.; supervision, J.D.H.; project administration, J.D.H.; funding acquisition, J.D.H. All authors have read and agreed to the published version of the manuscript.

Funding: This research was funded by the National Science Foundation through grant DMR 0840265, an Institutional Development Award (IDeA) from the National Institute of General Medical Sciences of the National Institutes of Health under grant P20 GM103408, NASA Idaho Space Grant Consortium, M.J. Murdock Charitable Trust, and Northwest Nazarene University's Science, Math Associates.

Acknowledgments: We extend our thanks to several faculty members and graduate students at Boise State University's Department of Materials Science and Engineering. We thank William Knowlton, Jason Brotherton, and Pamela Walker for SEM characterization of several zinc oxide samples, Darryl Butt and Joe Croteau for collecting the BET measurements, and Kyle Nogales for NMR measurements.

Conflicts of Interest: The authors declare no conflict of interest.

References

1. Janotti, A.; van de Walle, C.G. Fundamentals of zinc oxide as a semiconductor. *Rep. Prog. Phys.* **2009**, *72*, 126501. [CrossRef]
2. Özgür, Ü.; Alivov, Y.I.; Liu, C.; Teke, A.; Reshchikov, M.A.; Doğan, S.; Avrutin, V.; Cho, S.-J.; Morko, H. A comprehensive review of ZnO materials and devices. *J. Appl. Phys.* **2005**, *98*, 041301. [CrossRef]
3. Zang, Z.; Tang, X. Enhanced fluorescence imaging performance of hydrophobic colloidal ZnO nanoparticles by a facile method. *J. Alloys Compd.* **2015**, *619*, 98–101. [CrossRef]
4. Liu, S.; Yu, B.; Zhang, H.; Fei, T.; Zhang, T. Enhancing NO_2 gas sensing performances at room temperature based on reduced graphene oxide-ZnO nanoparticles hybrids. *Sens. Actuators B Chem.* **2014**, *202*, 272–278. [CrossRef]
5. Padmavathy, N.; Vijayaraghavan, R. Enhanced bioactivity of ZnO nanoparticles—An antimicrobial study. *Sci. Technol. Adv. Mater.* **2008**, *9*, 035004. [CrossRef]
6. Premanathan, M.; Karthikeyan, K.; Jeyasubramanian, K.; Manivannan, G. Selective toxicity of ZnO nanoparticles toward Gram-positive bacteria and cancer cells by apoptosis through lipid peroxidation. *Nanomed. Nanotechnol. Biol. Med.* **2011**, *7*, 184–192. [CrossRef]
7. Babitha, N.; Priya, L.S.; Christy, S.R.; Manikandan, A.; Dinesh, A.; Durka, M.; Arunadevi, S. Enhanced Antibacterial Activity and Photo-Catalytic Properties of ZnO Nanoparticles: Pedalium Murex Plant Extract-Assisted Synthesis. *J. Nanosci. Nanotechnol.* **2019**, *19*, 2888–2894. [CrossRef]
8. Zhou, H.; Li, Z. Synthesis of nanowires, nanorods and nanoparticles of ZnO through modulating the ratio of water to methanol by using a mild and simple solution method. *Mater. Chem. Phys.* **2005**, *89*, 326–331. [CrossRef]
9. Bhushan, B.; Murty, B.; Mondal, K. A new approach for synthesis of ZnO nanorod flowerets and subsequent pure free-standing ZnO nanorods. *Adv. Powder Technol.* **2019**, *30*, 30–41. [CrossRef]
10. Cossuet, T.; Roussel, H.; Chauveau, J.-M.; Chaix-Pluchery, O.; Thomassin, J.-L.; Appert, E.; Consonni, V. Well-ordered ZnO nanowires with controllable inclination on semipolar ZnO surfaces by chemical bath deposition. *Nanotechnology* **2018**, *29*, 475601. [CrossRef]
11. Liu, B.; Zeng, H.C. Direct growth of enclosed ZnO nanotubes. *Nano Res.* **2009**, *2*, 201–209. [CrossRef]
12. Choi, K.-S.; Chang, S.-P. Effect of structure morphologies on hydrogen gas sensing by ZnO nanotubes. *Mater. Lett.* **2018**, *230*, 48–52. [CrossRef]
13. Mina, S.-K.; Manea, R.S.; Joob, O.-S.; Ganesha, T.; Choc, B.W.; Han, S.-H. Upright-standing ZnO nano-sheets growth using wet chemistry. *Curr. Appl. Phys.* **2009**, *9*, 492–495. [CrossRef]
14. Wang, M.; Luo, Q.; Hussain, S.; Liu, G.; Qiao, G.; Kim, E.J. Sharply-precipitated spherical assembly of ZnO nanosheets for low temperature H2S gas sensing performances. *Mater. Sci. Semicond. Process.* **2019**, *100*, 283–289. [CrossRef]

15. Li, Y.; Liu, C.-S. Hydro/solvo-thermal synthesis of ZnO crystallite with particular morphology. *Trans. Nonferrous Metal Soc. China* **2009**, *19*, 399–403. [CrossRef]
16. Li, P.; Zhu, S.; Hu, H.; Guo, L.; He, T. Influence of defects in porous ZnO nanoplates on CO_2 photoreduction. *Catal. Today* **2019**, *335*, 300–305. [CrossRef]
17. Zhang, L.; Jeem, M.; Okamoto, K.; Watanabe, S. Photochemistry and the role of light during the submerged photosynthesis of zinc oxide nanorods. *Sci. Rep.* **2018**, *8*, 177. [CrossRef]
18. Xu, Z.; Ben, Y.; Chen, Z.; Qi, F. Facile synthesis of snowflake-like ZnO nanostructures at low temperature and their super catalytic activity for the ozone decomposition. *Mater. Res. Bull.* **2013**, *48*, 1725–1727. [CrossRef]
19. Sigoli, F.A.; Davolos, M.R.; Jafelicci, M., Jr. Morphological evolution of zinc oxide originating from zinc hydroxide carbonate. *J. Alloys Compd.* **1997**, *262–263*, 292–295. [CrossRef]
20. Wang, L.; Muhammed, M. Synthesis of zinc oxide nanoparticles with controlled morphology. *J. Mater. Chem.* **1999**, *9*, 2871–2878. [CrossRef]
21. Liu, Y.; Liu, Z.; Wang, G. Synthesis and characterization of ZnO nanorods. *J. Cryst. Growth* **2003**, *252*, 213–218. [CrossRef]
22. Meagley, K.L.; Garcia, S.P. Chemical Control of Crystal Growth with Multidentate Carboxylate Ligands: Effect of Ligand Denticity on Zinc Oxide Crystal Shape. *Cryst. Growth Des.* **2012**, *12*, 707–713. [CrossRef]
23. Anders, C.B.; Eixenberger, J.E.; Franco, N.A.; Hermann, R.J.; Rainey, K.D.; Chess, J.J.; Punnoose, A.; Wingett, D.G. ZnO nano-particle preparation route influences surface reactivity, dissolution and cytotoxicity. *Environ. Sci. Nano* **2018**, *5*, 572–588. [CrossRef]
24. Marin, O.; González, V.; Tirado, M.; Comedi, D. Effects of methanol on morphology and photoluminescence in solvothermal grown ZnO powders and ZnO on Si. *Mater. Lett.* **2019**, *251*, 41–44. [CrossRef]
25. Andrés-Vergés, M.; Martínez-Gallego, M. Spherical and rod-like zinc oxide microcrystals: Morphological characterization and microstructural evolution with temperature. *J. Mater. Sci.* **1992**, *27*, 3756–3762. [CrossRef]
26. Noack, V.; Eychmüller, A. Annealing of Nanometer-Sized Zinc Oxide Particles. *Chem. Mater.* **2002**, *14*, 1411–1417. [CrossRef]
27. Ohyama, M.; Kouzuka, H.; Yoko, T. Sol-Gel preparation of ZnO films with extremely preferred orientation along (002) plane from zinc acetate solution. *Thin Solid Films* **1997**, *306*, 78–85. [CrossRef]
28. Yang, Y.; Chen, H.; Zhao, B.; Bao, X. Size control of ZnO nanoparticles via thermal decomposition of zinc acetate coated on organic additives. *J. Cryst. Growth* **2004**, *263*, 447–453. [CrossRef]
29. Zhang, Z.; Lu, M.; Xu, H.; Chin, W.-S. Shape-Controlled Synthesis of Zinc Oxide: A Simple Method for the Preparation of Metal Oxide Nanocrystals in Non-aqueous Medium. *Chem. Eur. J.* **2006**, *13*, 632–638. [CrossRef]
30. Bacaksiz, E.; Parlak, M.; Tomakin, M.; Özçelik, A.; Karakız, M.; Altunbaş, M. The effects of zinc nitrate, zinc acetate and zinc chloride precursors on investigation of structural and optical properties of ZnO thin films. *J. Alloy. Compd.* **2008**, *466*, 447–450. [CrossRef]
31. Ebrahimifard, R.; Abdizadeh, H.; Golobostanfard, M.R. Controlling the extremely preferred orientation texturing of sol–gel derived ZnO thin films with sol and heat treatment parameters. *J. Sol-Gel Sci. Technol.* **2020**, *93*, 28–35. [CrossRef]
32. Ravichandran, K.; Begum, N.J.; Snega, S.; Sakthivel, B. Properties of Sprayed Aluminum-Doped Zinc Oxide Films—A Review. *Mater. Manuf. Process.* **2016**, *31*, 1411–1423. [CrossRef]
33. Eensalu, J.S.; Krunks, M.; Gromyko, I.; Katerski, A.; Mere, A. A comparative study on physical properties of Al-doped zinc oxide thin films deposited from zinc acetate and zinc acetylacetonate by spray pyrolysis. *Energetika* **2017**, *63*, 46–55. [CrossRef]
34. Lanje, A.S.; Sharma, S.J.; Ningthoujam, R.S.; Ahn, J.-S.; Pode, R.B. Low temperature dielectric studies of zinc oxide (ZnO) na-noparticles prepared by precipitation method. *Adv. Powder Technol.* **2013**, *24*, 331–335. [CrossRef]
35. Li, P.; Wei, Y.; Liu, H.; Wang, X.-K. Growth of well-defined ZnO microparticles with additives from aqueous solution. *J. Solid State Chem.* **2005**, *178*, 855–860. [CrossRef]
36. Cao, Y.; Miao, L.; Tanemura, S.; Tanemura, M.; Kuno, Y.; Hayashi, Y. Low resistivity p-ZnO films fabricated by sol-gel spin coating. *Appl. Phys. Lett.* **2006**, *88*, 251116. [CrossRef]
37. Kittilstved, K.R.; Norberg, N.S.; Gamelin, D.R. Chemical manipulation of high-TC ferromagnetism in ZnO diluted magnetic semiconductors. *Phys. Rev. Lett.* **2005**, *94*, 147209. [CrossRef]
38. Vajargah, P.H.; Abdizadeh, H.; Ebrahimifard, R.; Golobostanfard, M.R. Sol-gel derived ZnO thin films: Effect of ami-no-additives. *Appl. Surf. Sci. B* **2013**, *285*, 732–743. [CrossRef]
39. Hyslop, J.S.; Boydstun, A.R.; Fereday, T.R.; Rusch, J.R.; Strunk, J.L.; Wall, C.T.; Pena, C.C.; McKibben, N.L.; Harris, J.D.; Thurber, A.; et al. Synthesis and characterization of [Zn(acetate)2(amine)] compounds (x = 1 or 2) and their use as precursors to ZnO. *Mater. Sci. Semicond. Process.* **2015**, *38*, 278–289. [CrossRef]
40. Culp, S.J.; Beland, F.A. Malachite Green: A Toxicological Review. *J. Am. Coll. Toxicol.* **1996**, *15*, 219–238. [CrossRef]
41. Srivastava, S.; Sinha, R.; Roy, D. Toxicological effects of malachite green. *Aquat. Toxicol.* **2004**, *66*, 319–329. [CrossRef]
42. Lee, K.M.; Lai, C.W.; Ngai, K.S.; Juan, J.C. Recent developments of zinc oxide based photocatalyst in water treatment technology: A review. *Water Res.* **2016**, *88*, 428–448. [CrossRef]
43. Oliveira, A.G.; Andrade, J.D.L.; Montanha, M.C.; Lima, S.M.; Andrade, L.H.D.C.; Hechenleitner, A.A.W.; Pineda, E.A.G.; de Oliveira, D.M.F. Decontamination and disinfection of wastewater by photocatalysis under UV/visible light using nano-catalysts based on Ca-doped ZnO. *J. Environ. Manag.* **2019**, *240*, 485–493. [CrossRef] [PubMed]

44. Saad, A.M.; Abukhadra, M.R.; Ahmed, S.A.-K.; Elzanaty, A.M.; Mady, A.H.; Betiha, M.A.; Shim, J.-J.; Rabie, A.M. Photocatalytic degradation of malachite green dye using chitosan supported ZnO and Ce–ZnO nano-flowers under visible light. *J. Environ. Manag.* **2020**, *258*, 110043. [CrossRef] [PubMed]
45. Tayyebi, A.; Outokesh, M.; Tayebi, M.; Shafikhani, A.; Şengör, S.S. ZnO quantum dots-graphene composites: Formation mecha-nism and enhanced photocatalytic activity for degradation of methyl orange dye. *J. Alloys Compd.* **2016**, *663*, 738–749. [CrossRef]
46. Tayyebi, A.; Soltani, T.; Lee, B.-K.; Outokesh, M.; Tayebi, M. Novel Visible Light Photocatalytic and Photoelectrochemical (PEC) Activity of Carbon-doped Zinc Oxide/Reduced Graphene Oxide: Supercritical Methanol Synthesis with Enhanced Photocor-rosion Suppression. *J. Alloys Compd.* **2017**, *723*, 1001–1010. [CrossRef]
47. Tayebi, M.; Tayyebi, A.; Masoumi, Z.; Lee, B.-K. Photocorrosion suppression and photoelectrochemical (PEC) enhancement of ZnO via hybridization with graphene nanosheets. *Appl. Surf. Sci.* **2020**, *502*, 144189. [CrossRef]
48. Elias; Uddin, N.; Saha, J.; Hossain, A.; Sarker, D.; Akter, S.; Siddiquey, I.; Uddin, J. A Highly Efficient and Stable Photocatalyst; N-Doped ZnO/CNT Composite Thin Film Synthesized via Simple Sol-Gel Drop Coating Method. *Molecules* **2021**, *26*, 1470. [CrossRef]
49. Shen, W.; Li, Z.; Wang, H.; Liu, Y.; Guo, Q.; Zhang, Y. Photocatalytic degradation for methylene blue using zinc oxide prepared by codeposition and sol-gel methods. *J. Hazard. Mater.* **2008**, *152*, 172–175. [CrossRef]
50. Shifu, C.; Wei, Z.; Sujuan, Z.; Wei, L. Preparation, characterization and photocatalytic activity of N-containing ZnO powder. *Chem. Eng. J.* **2009**, *148*, 263–269. [CrossRef]
51. Yang, Y.; Li, Y.; Zhu, L.; He, H.; Hu, L.; Huang, J.; Hu, F.; He, B.; Ye, Z. Shape control of colloidal Mn doped ZnO nanocrystals and their visible light photocatalytic properties. *Nanoscale* **2013**, *5*, 10461–10471. [CrossRef] [PubMed]
52. Giahi, M.; Farajpour, G.; Taghavi, H.; Shokri, S. Preparation of Photocatalytic ZnO Nanoparticles and Application in Photo-chemical Degradation of Betamethasone Sodium Phosphate Using Taguchi Approach. *Russ. J. Phys. Chem. A* **2014**, *88*, 1241–1247. [CrossRef]
53. Saikia, L.; Bhuyan, D.; Saikia, M.; Malakar, B.; Dutta, D.K.; Sengupta, P. Photocatalytic performance of ZnO nanomaterials for self sensitized degradation of malachite green dye under solar light. *Appl. Catal. A Gen.* **2015**, *490*, 42–49. [CrossRef]
54. Khezami, L.; Taha, K.K.; Ghiloufi, I.; Mir, L.E. Adsorption and photocatalytic degradation of malachite green by vanadium doped zinc oxide nanoparticles. *Water Sci. Technol.* **2015**, *73*, 881–889. [CrossRef]
55. Meena, S.; Vaya, D.; Das, B.K. Photocatalytic degradation of Malachite Green dye by modified ZnO nanomaterial. *Bull. Mater. Sci.* **2016**, *39*, 1735–1743. [CrossRef]
56. Kaneva, N.; Bojinova, A.; Papazova, K. Photocatalytic degradation of Reactive Black 5 and Malachite Green with ZnO and lanthanum doped nanoparticles. *J. Phys. Conf. Ser.* **2016**, *682*, 12022. [CrossRef]
57. Horzum, N.; Hilal, M.E.; Isık, T. Enhanced bactericidal and photocatalytic activities of ZnO nanostructures by changing the cooling route. *New J. Chem.* **2018**, *42*, 11831–11838. [CrossRef]
58. Babajani, N.; Jamshidi, S. Investigation of photocatalytic malachite green degradation by iridium doped zinc oxide nanoparti-cles: Application of response surface methodology. *J. Alloys Compd.* **2019**, *782*, 533–544. [CrossRef]
59. Rabie, A.M.; Abukhadra, M.R.; Rady, A.M.; Ahmed, S.A.; Labena, A.; Mohamed, H.S.H.; Betiha, M.A.; Shim, J.-J. Instantaneous photocatalytic degradation of malachite green dye under visible light using novel green Co–ZnO/algae composites. *Res. Chem. Intermed.* **2020**, *46*, 1955–1973. [CrossRef]
60. Vitiello, G.; Iervolino, G.; Imparato, C.; Rea, I.; Borbone, F.; de Stefano, L.; Aronne, A.; Vaiano, V. F-doped ZnO nano- and meso-crystals with enhanced photocatalytic activity in diclofenac degradation. *Sci. Total Environ.* **2021**, *762*, 143066. [CrossRef]
61. Moezzi, A.; Cortie, M.; McDonagh, A. Transformation of zinc hydroxide chloride monohydrate to crystalline zinc oxide. *Dalton Trans.* **2016**, *45*, 7385–7390. [CrossRef] [PubMed]
62. Dutta, D. Optimization of process parameters and its effect on particle size and morphology of ZnO nanoparticle synthesized by sol–gel method. *J. Sol-Gel Sci. Technol.* **2016**, *77*, 48–56. [CrossRef]
63. Harun, K.; Hussain, F.; Purwanto, A.; Sahraoui, B.; Zawadzka, A.; Mohamad, A.A. Sol–gel synthesized ZnO for optoelectronics applications: A characterization review. *Mater. Res. Express* **2017**, *4*, 122001. [CrossRef]
64. Spanhel, L. Colloidal ZnO nanostructures and functional coatings: A survey. *J. Sol-Gel Sci. Technol.* **2006**, *39*, 7–24. [CrossRef]
65. Briois, V.; Giorgetti, C.; Baudelet, F.; Blanchandin, S.; Tokumoto, M.S.; Pulcinelli, S.H.; Santilli, C.V. Dynamical Study of ZnO Nanocrystal and Zn-HDS Layered Basic Zinc Acetate Formation from Sol−Gel Route. *J. Phys. Chem. C* **2007**, *111*, 3253–3258. [CrossRef]
66. Zheng, Z.; Butynska, R.; Serrano, C.V.; Marty, J.-D.; Mingotaud, C.; Kahn, M.L. One-Step Synthesis of Hybrid Liquid-Crystal ZnO Nanoparticles: Existence of a Critical Temperature Associated with the Anisotropy of the Nanoparticles. *Chem. Eur. J.* **2016**, *22*, 15614–15618. [CrossRef]
67. Jia, G.; Xu, S.; Wang, A. Emerging strategies for the synthesis of monodisperse colloidal semiconductor quantum rods. *J. Mater. Chem. C* **2015**, *3*, 8284–8293. [CrossRef]
68. Herrera-Rivera, R.; de la Olvera, M.L.; Maldonado, A. Synthesis of ZnO Nanopowders by the Homogeneous Precipitation Method: Use of Taguchi's Method for Analyzing the Effect of Different Variables. *J. Nanomater.* **2017**, *2017*, 4595384. [CrossRef]

69. Alnarabiji, M.S.; Yahya, N.; Hamed, Y.; Ardakani, S.E.M.; Azizi, K.; Klemeš, J.J.; Abdullah, B.; Tasfy, S.F.H.; Hamid, S.B.A.; Nashed, O. Scalable bio-friendly method for production of homogeneous metal oxide nanoparticles using green bovine skin gel-atin. *J. Cleaner Prod.* **2017**, *162*, 186–194. [CrossRef]
70. Li, J.-G.; Ikegami, T.; Wang, Y.; Mori, T. 10-mol%-Gd2O3-Doped CeO_2 Solid Solutions via Carbonate Coprecipitation: A Com-parative Study. *J. Am. Ceram. Soc.* **2003**, *86*, 915–921. [CrossRef]
71. Kanaparthi, S.; Singh, S.G. Chemiresistive Sensor Based on Zinc Oxide Nanoflakes for CO_2 Detection. *ACS Appl. Nano Mater.* **2019**, *2*, 700–706. [CrossRef]
72. Quy, C.T.; Thai, N.X.; Hoa, N.D.; Le, D.T.T.; Hung, C.M.; Duy, N.V.; Hieu, N.V. C_2H_5OH and NO_2 sensing properties of ZnO nanostructures: Correlation between crystal size, defect level and sensing performance. *RSC Adv.* **2018**, *8*, 5629–5639. [CrossRef] [PubMed]
73. Shaporev, A.S.; Ivanov, V.K.; Baranchikov, A.E.; Polezhaeva, O.S.; Tret'Yakov, Y.D. ZnO formation under hydrothermal conditions from zinc hydroxide compounds with various chemical histories. *Russ. J. Inorg. Chem.* **2007**, *52*, 1811–1816. [CrossRef]
74. Morales, A.E.; Mora, E.S.; Pal, U. Use of diffuse reflectance spectroscopy for optical characterization of un-supported nanostructures. *Rev. Mex. Fís. S* **2007**, *53*, 18–22.
75. Kubelka, P.; Munk, F. Ein Beitrag zur Optik der Farbanstriche. *Z. Tech. Phys.* **1931**, *12*, 593–601.
76. Srikant, V.; Clarke, D.R. On the optical band gap of zinc oxide. *J. Appl. Phys.* **1998**, *83*, 5447. [CrossRef]
77. Ramelan, A.H.; Wahyuningsih, S.; Munawaroh, H.; Narayan, R. ZnO wide bandgap semiconductors preparation for optoelectronic devices. *IOP Conf. Ser. Mater. Sci. Eng.* **2017**, *176*, 012008. [CrossRef]
78. Eixenberger, J.E.; Anders, C.B.; Wada, K.; Reddy, K.M.; Brown, R.J.; Moreno-Ramirez, J.; Weltner, A.E.; Karthik, C.; Tenne, D.A.; Fologea, D.; et al. Defect Engineering of ZnO Nanoparticles for Bioimaging Applications. *ACS Appl. Mater. Interfaces* **2019**, *11*, 24933–24944. [CrossRef] [PubMed]
79. Theerthagiri, J.; Salla, S.; Senthil, R.A.; Nithyadharseni, P.; Madankumar, A.; Arunachalam, P.; Maiyalagan, T.; Kim, H.-S. A review on ZnO nanostructured materials: Energy, environmental and biological applications. *Nanotechnology* **2019**, *30*, 392001. [CrossRef]
80. Rauf, M.A.; Ashraf, S.S. Fundamental principles and application of heterogeneous photocatalytic degradation of dyes in solu-tion. *Chem. Eng. J.* **2009**, *151*, 10–18. [CrossRef]
81. Rabieh, S.; Bagheri, M.; Heydari, M.; Badiei, E. Microwave assisted synthesis of ZnO nanoparticles in ionic liquid [Bmim]cl and their photocatalytic investigation. *Mater. Sci. Semicond. Process.* **2014**, *26*, 244–250. [CrossRef]
82. Prasad, K.S.; Prajapati, S.; Selvaraj, K. Efficient sorption and photocatalytic degradation of malachite green dye onto NiS na-noparticles prepared using novel green approach. *Korean J. Chem. Eng.* **2015**, *32*, 1986–1992. [CrossRef]
83. Jo, W.-K.; Park, G.T.; Tayade, R.J. Synergetic effect of adsorption on degradation of malachite green dye under blue LED irra-diation using spiral-shaped photocatalytic reactor. *J. Chem. Technol. Biotechnol.* **2015**, *90*, 2280–2289. [CrossRef]
84. Goswami, T.; Reddy, K.M.; Bheemaraju, A. Silver Nanocluster Anchored TiO_2/Nb_2O_5 Hybrid Nanocomposite as Highly Efficient and Selective Visible-Light Sensitive Photocatalyst. *ChemistrySelect* **2019**, *4*, 6790–6799. [CrossRef]
85. Nethaji, S.; Tamilarasan, G.; Neehar, P.; Sivasamy, A. Visible light photocatalytic activities of BiOBr-activated carbon (derived from waste polyurethane) composites by hydrothermal process. *J. Environ. Chem. Eng.* **2018**, *6*, 3735–3744.
86. Ghaedi, M.; Ansari, A.; Habibi, M.H.; Asghari, A.R. Removal of malachite green from aqueous solution by zinc oxide nanopar-ticle loaded on activated carbon: Kinetics and isotherm study. *J. Ind. Eng. Chem.* **2014**, *20*, 17–28. [CrossRef]
87. Pare, B.; Sarwan, B.; Jonnalagadda, S.B. Photocatalytic mineralization study of malachite green on the surface of Mn-doped BiOCl activated by visible light under ambient condition. *Appl. Surf. Sci.* **2011**, *258*, 247–253. [CrossRef]
88. Liu, Y.; Ohko, Y.; Zhang, R.; Yang, Y.; Zhang, Z. Degradation of malachite green on Pd/WO3 photocatalysts under simulated solar light. *J. Hazard. Mater.* **2010**, *184*, 386–391. [CrossRef]
89. Qi, K.; Cheng, B.; Yu, J.; Ho, W. Review on the improvement of the photocatalytic and antibacterial activities of ZnO. *J. Alloy. Compd.* **2017**, *727*, 792–820. [CrossRef]
90. Ancona, A.; Dumontel, B.; Garino, N.; Demarco, B.; Chatzitheodoridou, D.; Fazzini, W.; Engelke, H.; Cauda, V. Lipid-Coated Zinc Oxide Nanoparticles as Innovative ROS-Generators for Photodynamic Therapy in Cancer Cells. *Nanomaterials* **2018**, *8*, 143. [CrossRef]
91. Yong, L.; Zhanqi, G.; Yuefei, J.; Xiaobin, H.; Cheng, S.; Shaogui, Y.; Lianhong, W.; Qingeng, W.; Die, F. Photodegradation of mal-achite green under simulated and natural irradiation: Kinetics, products, and pathways. *J. Hazard. Mater.* **2015**, *285*, 127–136. [CrossRef]
92. Hepp, A.F.; Bailey, S.G.; McNatt, J.S.; Chandrashekhar, M.V.S.; Harris, J.D.; Rusch, A.W.; Nogales, K.A.; Goettsche, K.V.; Hanson, W.; Amos, D.; et al. Novel Materials, Processing and Device Technologies for Space Exploration with Potential Dual-Use Applications. In *NASA Technical Memorandum*; NASA/TM—2015-218866; National Aeronautics and Space Administration: Brook park, OH, USA, 2015.
93. Wang, H.; Xie, C.; Zhang, W.; Cai, S.; Yang, Z.; Gui, Y. Comparison of dye degradation efficiency using ZnO powders with various size scales. *J. Hazard. Mater.* **2007**, *141*, 645–652. [CrossRef]
94. Sharma, S.; Mehta, S.K.; Kansal, S.K. N doped ZnO/C-dots nanoflowers as visible light driven photocatalyst for the degradation of malachite green dye in aqueous phase. *J. Alloys Compd.* **2017**, *699*, 323–333. [CrossRef]
95. Langford, J.I.; Wilson, A.J.C. Scherrer after sixty years: A survey and some new results in the determination of crystallite size. *J. Appl. Crystallogr.* **1978**, *11*, 102–113. [CrossRef]

96. Team R. *Core. R: A Language and Environment for Statistical Computing*, The R Foundation for Statistical Computing: Vienna, Austria, 2013. Available online: http://www.R-project.org/.
97. Kahm, M.; Hasenbrink, G.; Lichtenberg-Fraté, H.; Ludwig, J.; Kschischo, M. grofit: Fitting Biological Growth Curves with R. *J. Stat. Softw.* **2010**, *33*, 1–21. [CrossRef]

catalysts MDPI

Article

Disinfection and Photocatalytic Degradation of Organic Contaminants Using Visible Light-Activated GCN/Ag_2CrO_4 Nanocomposites

Olufemi Oluseun Akintunde [1], Linlong Yu [1], Jinguang Hu [2], Md Golam Kibria [2], Casey R. J. Hubert [3], Samuel Pogosian [4] and Gopal Achari [1,*]

1 Department of Civil Engineering, University of Calgary, ENE 231, 2500 University Drive NW, Calgary, AB T2N 1N4, Canada
2 Department of Chemical and Petroleum Engineering, University of Calgary, ENB 202, 2500 University Drive NW, Calgary, AB T2N 1N4, Canada
3 Department of Biological Sciences, University of Calgary, Biological Sciences 186, 2500 University Drive NW, Calgary, AB T2N 1N4, Canada
4 Nemalux Industrial, 1018 72 Ave NE, Calgary, AB T2E 8V9, Canada
* Correspondence: gachari@ucalgary.ca

Abstract: Visible-light-driven photocatalysts have gained increasing attention in the past few decades in treating emerging contaminants in water and wastewater. In this work, the photocatalytic activity of the coupled graphitic carbon nitride (GCN) and silver chromate (Ag_2CrO_4), herein denoted as GCN/Ag_2CrO_4, nanocomposites was evaluated for degrading organic pollutants and inactivating microorganisms under visible light irradiation using a royal blue light-emitting diode (LED). The organic pollutants studied were 2,4-dichlorophenoxyacetic acid (2,4-D) and methyl chlorophenoxy propionic acid (MCPP or Mecoprop-P) present in Killex^R, a commercially available herbicide, bovine serum albumin (BSA) protein, and SARS-CoV-2 spike protein. The disinfection experiments were conducted on wastewater secondary effluent. The results showed that over 85% degradation was achieved for both 2,4-D and Mecoprop-P in 120 min while 100% of BSA protein and 77.5% of SARS-CoV-2 protein were degraded in 20 min and 30 min, respectively. Additionally, GCN/Ag_2CrO_4 nanocomposites led to over one log reduction of cellular *ATP* (*cATP*), total coliforms, and *E. coli* in wastewater treatment plant (WWTP) secondary effluent after 60 min of royal blue LED irradiation. It was observed that the degradation performance of a photocatalyst under light irradiation is contaminant-specific. The binding affinity of the released metal ions from GCN/Ag_2CrO_4 with protein and *ATP* functional groups was responsible for the degradation of proteins and the reduction of *cATP*, while the generated ROS was responsible for the disinfection of total coliforms and *E. coli*. Overall, the results indicate that GCN/Ag_2CrO_4 nanocomposite is a promising photocatalyst in degrading organic pollutants and disinfecting microorganisms under visible light irradiation within a reasonable time.

Keywords: GCN/Ag_2CrO_4; photocatalysis; organic pollutants; microorganisms; wastewater; royal blue LED

Citation: Akintunde, O.O.; Yu, L.; Hu, J.; Kibria, M.G.; Hubert, C.R.J.; Pogosian, S.; Achari, G. Disinfection and Photocatalytic Degradation of Organic Contaminants Using Visible Light-Activated GCN/Ag_2CrO_4 Nanocomposites. *Catalysts* **2022**, *12*, 943. https://doi.org/10.3390/catal12090943

Academic Editors: Jorge Bedia and Carolina Belver

Received: 6 August 2022
Accepted: 21 August 2022
Published: 25 August 2022

1. Introduction

Semiconductor photocatalysis is fast emerging as the advanced oxidation process to address environmental pollution and global warming such as organic pollutant removal, disinfection of microorganisms, hydrogen evolution, and CO_2 reduction [1–5]. In semiconductor photocatalysis, photoexcited electrons and holes are generated when the energy absorbed from light is greater than or equal to the bandgap energy of the photocatalysts. These electrons react with the dissolved oxygen in the solution while the holes react with water molecules to produce superoxide radical (O_2^-) and hydroxyl radical (•OH), respectively. These radicals are referred to as reaction oxidative species (ROS) and they can oxidize organic pollutants, finally producing mineralized products such as CO_2 and H_2O. They can also be used for disinfection purposes. Titanium dioxide (TiO_2), a metal-based semiconductor, is the most widely studied photocatalyst, but it suffers from limited applications such as wide bandgap (3.2 eV), low visible light utilization, and high recombination rate of electron-hole pair [6–9]. Therefore, various semiconductor photocatalysts with narrow bandgap and visible-light activity were developed such as Bi_2WO_6 [10], CuO [11], and CdS [12]. To this end, modified and doped TiO_2 [13], photosensitizers [14], as well as the use of low-energy systems such as light-emitting diodes [15,16] and "adsorb and shuttle" wherein adsorbent along with a photocatalyst [17] have also been investigated.

Graphitic carbon nitride (g-C_3N_4 or GCN) is a metal-free photocatalyst, with a bandgap of 2.7 eV that has gained a lot of attention in photocatalytic water splitting [18], photocatalytic degradation of organic pollutants [19], and disinfection of microorganisms [20] under visible light irradiation [21,22]. However, the major drawback of bulk GCN is low surface area, high rate of electron-hole pair recombination, and low utilization of visible light up to only 460 nm (blue region) which limits its photocatalytic activity [23].

It is well known that silver-based photocatalysts such as silver halides [24,25], Ag_3PO_4 [26], and Ag_2CrO_4 [27] are potential catalysts because of their excellent light sensitivity and photocatalytic activity under visible light irradiation. Of note is Ag_2CrO_4 with its high visible-light absorption efficiency, and electronic and crystal structure [28]. However, Ag_2CrO_4, like other silver-based photocatalysts, has similar drawbacks which are aggregated particle size and ease of photocorrosion leading to poor stability and limited photocatalytic performance.

Coupling of GCN and the silver-based photocatalysts to form heterojunctions can address the individual shortcomings of GCN and silver-based photocatalysts by improving charge separation, reducing the rate of recombination of electron-hole pair, improving bandgap optical properties, and enhancing visible light absorption and utilization of GCN, as well as reducing the aggregation of silver-based photocatalysts. For example, GCN/Ag [23], GCN/AgBr [29], GCN/Ag_3PO_4 [30], and GCN/Ag_2CrO_4 [31] showed higher degradation of methylene blue (MB), *E. coli* disinfection, 4-chlorophenol (4-CP) degradation, and methyl orange (MO) degradation, respectively, under visible light when compared to GCN.

The purpose of this study is to investigate synthesized GCN/Ag_2CrO_4 to photocatalytically degrade organic pollutants as well as disinfect microorganisms. Killex^R (2,4-D and Mecoprop-P), BSA, and SARS-CoV-2 spike protein were selected to study while WWTP secondary effluent was used to assess photocatalytic disinfection. The addition of g_2CrO_4 to GCN to form the GCN/Ag_2CrO_4 heterojunction enhances visible-light activity. Ag_2CrO_4 being a strong oxidizing agent serves as electron trap by accepting photogenerated electrons from GCN thereby reducing electron-hole pair recombination rate and improving charge separation. The narrow band gap of Ag_2CrO_4 (1.80 eV) indicates strong absorption in the visible-light region which could enhance its photocatalytic activity. Both 2,4-D and certain microbial constituents in wastewater such as *E. coli* have known negative health and environmental impacts [32,33]. Moreover, 2,4-D and Mecoprop-P are common constituents of herbicides that can be found on the surface and in groundwater because of their use in lawns and agricultural lands [34]. A study in Alberta, Canada showed that 2,4-D is one of the most common pollutants in water and precipitate samples [35]. Mecoprop-P has been detected as a minor pollutant in U.S. and European soils, as well as in ground and surface waters, and it is the most often found herbicide in drinking waters [36,37]. Bovine serum albumin protein (BSA) is a commonly used standard for protein analysis selected for studying the interaction between proteins and other chemicals. SARS-CoV-2, the Coronavirus that causes COVID-19, has resulted in a worldwide pandemic with over 280 million infections and caused over 5 million deaths globally [38]. This viral disease continues to ravage the world with severe or mild variants. COVID-19 burden in communities has been monitored by testing wastewater to provide an early warning of a community spread [39]. Sources of coliforms can be humans, livestock, wildlife, and even pets. *E. coli* is considered a potential pathogen and its presence in the water body can indicate the presence of other waterborne pathogens such as bacteria, viruses, and parasites [40].

Photocatalytic degradation of 2,4-D and Killex using TiO_2 has been widely reported [41–43]. However, only a few studies focused on the photocatalytic degradation of 2,4-D using GCN. The degradation of 2,4-D using GCN-based composite under natural sunlight showed over 90% removal after 330 min [44]. About 100% degradation of a very small concentration (50 μg/L) of Mecoprop-P was achieved in 7 min using a GCN-based catalyst under simulated solar irradiation [45]. Heydari et al. degraded Killex^R (2,4-D and Mecoprop-P) using TiO_2 under natural sunlight exposure with 97% degradation of 2,4-D after 15 days and 100% degradation of Mecoprop-P after 22 days [46]. The enhancement of the photocatalytic activity of GCN in degrading methylene blue (MB) was achieved with its combination with Co-NiS and sulfur to form 1D/2D Co-NiS/S-g-C_3N_4 heterojunction with 98% degradation in 32 min [47]. $ZnFe_2O_4$/S-g-C_3N_4 heterojunction was synthesized with varying wt% of S-g-C_3N_4. The $ZnFe_2O_4$/S-g-C_3N_4 (50%) showed the maximum MB degradation under solar irradiation in 150 min [48]. The interaction between BSA and nanoparticles and the structural change to BSA have been investigated with nanoparticles such as MnO_2 [49], CuS [50], and silver nanoparticles (AgNP) [51,52] at different concentrations and compositions. The results show increasing interactions and structural changes to BSA with increasing concentration of the nanoparticles. The disinfection of *E. coli* using GCN-based composite under a xenon lamp as a solar simulator showed complete inactivation (6.5 log) after 60 min [29]. In the published literature, the focus was on either natural sunlight or solar simulator as broadband light source for the photocatalytic degradation and disinfection. To the best of our knowledge, the combination of a narrowband low-energy visible light source (in this case royal blue LED with a peak wavelength of 460 nm) and GCN-based nanocomposites has not been reported on degrading herbicides and protein and disinfecting WWTP secondary effluent. The benefits of narrowband light are its consistency, directionality (narrow and focused beam), and narrow frequency range [53]. The photons in narrowband light move at the same wavelength (consistency) and in the same direction (directionality) resulting in high light intensity.

This work aims to address the knowledge gaps in evaluating the photocatalytic performance of GCN/Ag_2CrO_4 nanocomposite under a narrowband low-energy visible light source (royal blue LED) in degrading different organic pollutants such as Killex^R (2,4-D and MCPP), BSA protein, SARS-CoV-2 spike protein, and disinfecting WWTP secondary effluent. The reaction kinetics of 2,4-D, Mecoprop-P, and BSA protein were also investigated. The interaction between BSA and the photocatalysts, and the structural changes to BSA were studied using a UV-Vis spectrophotometer; log reduction of cellular *ATP* (*cATP*), total coliforms, and *E. coli* in WWTP secondary effluent was used to determine the efficiency of the catalyst in wastewater disinfection.

2. Results and Discussions

2.1. Photocatalytic Degradation of Killex^R

The dark adsorption of 2,4-D and Mecoprop-P on the synthesized photocatalyst are presented in Figure 1. It is obvious that GCN alone did not adsorb and thus there was no decrease in the concentration of 2,4-D or Mecoprop-P during the dark experiment. Equilibrium was reached within 30 min for Ag_2CrO_4 and GCN/Ag_2CrO_4. A total of 8.06% of 2,4-D was adsorbed on Ag_2CrO_4 while 9.32% of Mecoprop-P was adsorbed on Ag_2CrO_4. A total of 6.24% of 2,4-D was adsorbed on GCN/Ag_2CrO_4 while 15.32% of Mecoprop-P was adsorbed on GCN/Ag_2CrO_4. The photocatalytic degradation of GCN/Ag_2CrO_4 on the 2,4-D and Mecoprop-P in Killex^R was investigated and the results are presented in Figure 2. In the control experiment, 2,4-D and Mecoprop-P did not decrease with irradiation by royal blue LED, indicating that the royal blue light (peak wavelength = 460 nm) did not degrade 2,4-D or Mecoprop-P. Using GCN, 16.85% of 2,4-D and 10.06% of Mecoprop-P degraded after 120 min. Ag_2CrO_4 remains largely unchanged under royal blue LED irradiation when compared to the dark experiment results. It was 7.05% for 2,4-D and 9.19% for Mecoprop-

P. This is because silver-based photocatalysts are prone to photolysis and cannot stand long-run duration of photocatalytic degradation [30], and the low surface area of Ag_2CrO_4 (1.32 m^2/g) [21] available for the adsorption of 2,4-D and Mecoprop-P and the eventual photocatalytic activity. Silver-based particles have been reported to aggregate [54] while GCN in the GCN/Ag_2CrO_4 nanocomposite serves to prevent the aggregation of Ag_2CrO_4. GCN/Ag_2CrO_4 showed the highest percent degradation for both 2,4-D and Mecoprop-P at 86.32% and 86.39% in 120 min.

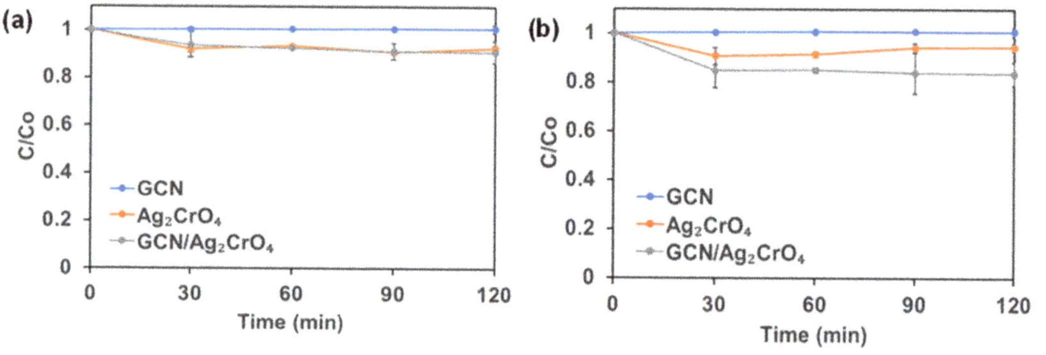

Figure 1. The dark experiment of Killex^R: (**a**) 2,4-D and (**b**) Mecoprop-P.

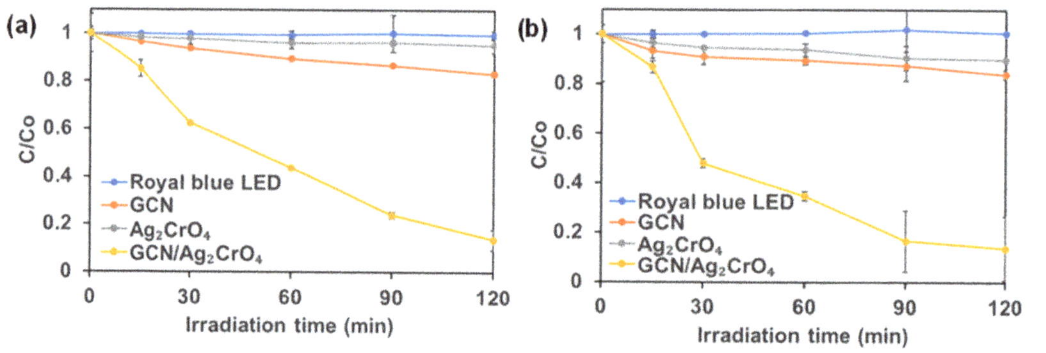

Figure 2. Photocatalytic degradation (royal blue LED irradiation) of Killex^R: (**a**) 2,4-D and (**b**) Mecoprop-P.

The first-order rate was calculated to evaluate the performance of the different photocatalysts. The rate constant (k) values are summarized in Table 1. For 2,4-D, the k values of GCN and GCN/Ag_2CrO_4 were 0.16×10^{-2} min^{-1} and 1.59×10^{-2} min^{-1} respectively. The highest k value for degradation using GCN/Ag_2CrO_4 was 9.9 times that with GCN alone. For Mecoprop-P, the k values of GCN and GCN/Ag_2CrO_4 were 0.11×10^{-2} min^{-1} and 1.79×10^{-2} min^{-1} respectively. The highest k value for the photocatalyst GCN/Ag_2CrO_4 was 1.79×10^{-2} min^{-1} which was 16.3 times that of GCN. This result shows that the deposition of Ag_2CrO_4 on GCN enhanced its photocatalytic activity by serving as an electron trap for photogenerated electrons in GCN.

Table 1. Summary of first-order photocatalytic degradation rate constants.

Experimental Condition		GCN	Ag_2CrO_4	GCN/Ag_2CrO_4
2,4-D	Rate constant, k ($\times 10^{-2}$) min^{-1}	0.16	1.20	1.59
	R^2	0.9891	0.6227	0.9959
Mecoprop-P	Rate constant, k ($\times 10^{-2}$) min^{-1}	0.11	1.26	1.79
	R^2	0.8167	0.6377	0.9874

In our previous publication, we reported that the photocatalytic degradation of 4-CP by GCN/Ag_2CrO_4 was better than GCN [21], while in this study (Figure 2a,b) we find a better performance was observed for GCN than Ag_2CrO_4 in the photocatalytic degradation of 2,4-D and Mecoprop-P. This indicates that the photocatalytic degradation performance of Ag_2CrO_4 is contaminant-specific. This can be attributed to the bond energies existing in the contaminants such as C-C, C=C, O-H, C=O, and C-Cl. Bond energy or bond dissociation energy is the measure of bond strength, and it is an indication of the easiness to break a chemical bond. The bond energies of bonds existing in 4-CP, 2,4-D, and Mecoprop-P are shown in Table 2.

Table 2. Bond energies at 273K adapted from Ref. [55].

Bond	Bond Energy (KJ/mol)
C-Cl	327
C-C	346
C-O	358
C-H	411
O-H	459
C=C	602
C=O	749

There are fewer bonds in 4-CP than 2,4-D and Mecoprop-P that need to be broken by ROS generated from Ag_2CrO_4. In 4-CP, apart from the aromatic ring, there are C-Cl, C-O, and O-H corresponding to 327 KJ/mol, 358 KJ/mol, and 459 KJ/mol, while 2,4-D and Mecoprop-P have all the bonds listed in Table 2 in addition to the aromatic ring. This means with 2,4-D and Mecoprop-P having more bonds and higher bond energies, they will be more difficult to break by the ROS generated from Ag_2CrO_4 and hence the observed low photocatalytic performance compared to 4-CP. The low surface area of Ag_2CrO_4, in addition to the bond energies of the contaminants, also would have contributed to the low performance. With a specific surface area of 1.32 m^2/g [21], it means that less absorption of 2,4-D and Mecoprop-P would take place for sufficient photocatalytic activity compared to single contaminant absorption of 4-CP. Therefore, the combination of the low surface area of Ag_2CrO_4 for the absorption of the two contaminants and the several bonds and higher bond energies of 2,4-D and Mecoprop-P contributed to the observed low performance of Ag_2CrO_4.

In the photocatalytic degradation of pollutants, the molecular structures could determine their susceptibility to ROS attack [45]. The compounds 2,4-D and Mecoprop-P are both chlorophenoxy herbicides (phenoxycarboxylic acid herbicides). These similar structures could explain their similar degradation rates under visible light. In the photocatalytic degradation of 2,4-D, the aromatic ring of 2,4-D is first hydroxylated to the main by-product, 2,4 dichlorophenol (2,4-DCP), followed by other by-products such as chlorohydroquinone, 4-chloropyrocatechol, 2,4-dichloropyrocatechol, and chlorobenzoquinone [56,57] as shown in Equations (1)–(5), where AR is the aromatic ring.

$$AR - Cl_2 - OCH_2 - COOH + {}^{\cdot}OH \rightarrow AR - Cl_2 - OH + CH_2 - OH - COOH \quad (1)$$

$$AR - Cl_2 - OH + OH \rightarrow AR - Cl - (OH)_2 + Cl^- \quad (2)$$

$$AR - Cl - (OH)_2 + OH \rightarrow AR - Cl - (OH)_2^{'} \quad (3)$$

$$AR - Cl - (OH)_2^{'} + OH + Cl^- \rightarrow AR - Cl_2 - (OH)_2 + OH^- \quad (4)$$

$$AR - Cl_2 - (OH)_2 + OH \rightarrow C_6H_3ClO_2 \quad (5)$$

In addition, in the photocatalytic degradation of Mecoprop-P, the aromatic ring is hydroxylated to the main by-product, 4-chloro-o-cresol, followed by other by-products such as 2-methyl hydroquinone, and 2-methyl-p-benzoquinone [37] as proposed in Equations (6)–(8). The main reactive species involved in photocatalytic degradation is the •OH radical as reported for 2,4-D [44,58–60] and Mecoprop-P [45,61]. Table 3 shows the photocatalytic degradation of 2,4-D and Mecoprop-P by different photocatalysts.

$$AR - CH_3Cl - OC_2H_4 - COOH + OH \rightarrow AR - CH_3Cl - OH + C_2H_4 - OH - COOH \quad (6)$$

$$AR - CH_3Cl - OH + OH \rightarrow AR - CH_3 - (OH)_2 + Cl^- \quad (7)$$

$$AR - CH_3 - (OH)_2 + OH \rightarrow C_6H_3ClCH_3 \quad (8)$$

Table 3. Comparison of 2,4-D and Mecoprop-P degradation with the results of previously published articles.

S/N	Photocatalyst	Visible Light Source	Organic Pollutant	Concentration (mg/L)	Degradation Time (min)	Degradation Efficiency (%)	Ref.
1	TiO_2	Natural sunlight	2,4-D and Mecoprop-P	49.4 and 27.3	1,296,000 and 1,900,800	97 and 100	[46]
2	MoO_3/g-C_3N_4	Natural sunlight	2,4-D	50	300	99	[44]
3	GCN	Xenon lamp (200–800 nm, 300 W)	Mecoprop-P	0.05	7	100	[45]
4	GCN/Ag_2CrO_4	Royal blue LED (460 nm)	2,4-D and Mecoprop-P	12.1 and 6.7	120 and 120	86.32 and 86.39	Current study

2.2. Photocatalytic Degradation of BSA Protein and SARS-CoV-2 Spike Protein

The interaction between the synthesized photocatalysts and BSA in the absence of light (dark experiment) was investigated and the results are shown in Figure 3a. It was observed that the concentration of BSA did not change when GCN was present in the solution, indicating that the GCN had a negligible impact in the dark experiment. However, it was observed that the concentration of BSA decreased significantly by 93.1% in 30 min in the presence of GCN/Ag_2CrO_4. The photocatalytic degradation results of BSA are presented in Figure 3b. It shows that royal blue LED irradiation alone and GCN/royal blue LED irradiation could not degrade BSA. The royal blue LED with peak wavelength of 460 nm and low energy (2.7 eV) cannot alone degrade BSA with absorption peak of 280 nm and high energy (4.4 eV) indicating that BSA requires high energy radiation greater than 4.4 eV for degradation to take place. The GCN/royal blue LED could not degrade BSA because of the high electron-hole pair recombination rate of GCN owing to its medium band gap (2.7 eV). However, more than 99.9% of BSA was degraded in 20 min in the presence of GCN/Ag_2CrO_4 under royal blue LED irradiation. The first-order rate constant (k) values of GCN and GCN/Ag_2CrO_4 were calculated, and they are 0.002×10^{-1} min^{-1} and 3.64×10^{-1} min^{-1} respectively.

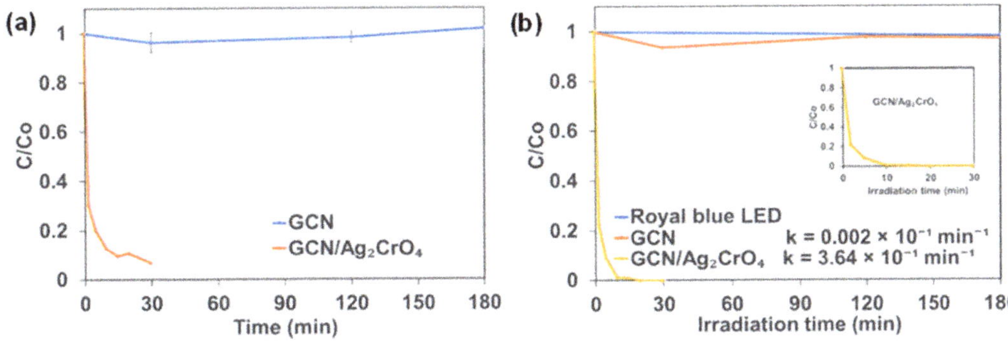

Figure 3. Change in BSA concentration with time: (**a**) Dark experiment and (**b**) Royal blue LED irradiation.

In a recently published work, the photocatalytic degradation mechanism of protein using silver nanoparticles (AgNPs) was reported by investigating the effects of the following 3-step phenomenon; release of silver ions, generation of reactive oxidative species (ROS), and the light-induced protein oxidation through the bind-and-damage model [62]. It was postulated and concluded by asserting an alternate mechanism, which is that light-induced protein oxidation is responsible for the degradation rather than the popular notions that the mechanisms of Ag^+ release from AgNPs and the generation of ROS are the driving forces for protein degradation and bacterial death. For the bind-and-damage model of the light-induced protein oxidation to take place, the AgNPs released the absorbed visible light energy in the form of emission. This emission of light is then quenched by a protein which indicates that the absorbed light energy was transferred to the protein, causing structural damage to the proteins.

The assertions were applied to this work for verification by carrying out UV-vis absorption spectra measurements of BSA for GCN and GCN/Ag_2CrO_4 with and without royal blue LED irradiation to further understand the degradation profile and the change in concentration with time. UV-vis, as an analytical method, can be used in absorption spectra measurement in understanding structural changes to protein and the interaction between protein and nanoparticles [49,50,63,64]. The observation was that the phenomena: release of Ag^+ from the Ag_2CrO_4, generation of ROS through photocatalytic reaction, and the bind-and-damage model of the light-induced protein oxidation, have varying degrees of contribution to the degradation of protein in the presence of GCN/Ag_2CrO_4 with and without blue LED irradiation. The roles these phenomena play in either protein degradation or bacterial death will depend on the type of photocatalyst, light intensity, radiation spectrum, and energy of the light source. In this work, two ions (Ag^+ and CrO_4^{2-}) were released against Ag^+ of AgNPs since GCN/Ag_2CrO_4 composite was used, indicating the possibility of the impact of CrO_4^{2-} in the degradation mechanism as an electron scavenger during photocatalytic activity. The UV-vis absorption spectrum in Figure 4 shows the effect of GCN and GCN/Ag_2CrO_4 on BSA with and without royal blue LED irradiation. At time zero (Figure 4a–d), BSA shows the typical absorption peaks at a lower wavelength of 200 nm and a higher wavelength of 280 nm. The absorption peak at 200 nm is attributed to the BSA's polypeptide backbone and the absorption peak at 280 nm is due to its aromatic amino acids (tryptophan, tyrosine, and phenylalanine). In Figure 4a,c, the GCN without and with royal blue LED irradiation revealed similar absorption spectra. In Figure 4a, without light (dark), the BSA peak did not change because the conditions required for protein damage which are light-induced protein oxidation and ROS generation are not present since photocatalytic activity did not take place. It also indicates no interaction between the BSA and GCN, hence no structural change to the protein. In Figure 4c, under royal blue LED irradiation, the bind-and-damage model of the light-induced protein oxidation and the generated ROS did not have an impact on the degradation of the protein. This suggests that the emitted light from the absorbed light energy by the GCN was neither sufficient nor transferred to the protein for degradation, and the generated ROS was not sufficient to cause degradation because of the GCN's poor charge transfer and fast electron-hole recombination rate characteristics. In the presence of GCN/Ag_2CrO_4 (Figure 4b,d), an absorption peak at 374 nm was observed. This peak is attributed to the presence of silver chromate in the BSA slurry. Silver compounds are known to have absorption peaks between 320 nm and 410 nm [51,52,65]. In Figure 4b,d, the GCN/Ag_2CrO_4 without and with royal blue LED irradiation show similar spectral patterns. In Figure 4b, without light (dark), Ag^+ and CrO_4^{2-} were released by Ag_2CrO_4 in the GCN/Ag_2CrO_4 nanocomposite. The interactions of the Ag^+ with BSA resulted in higher BSA absorption peaks compared to GCN without light (dark). In Figure 4d, under royal blue LED irradiation the bind-and-damage model light-induced protein oxidation and the ROS generation did not significantly contribute to the protein degradation since the peaks without light and with royal blue LED are similar. This is evident in the change in BSA concentration when exposed to GCN/Ag_2CrO_4 with and without royal blue LED irradiation (Figure 3a,b), where the degradation is similar at 90% and 99% in 15 min respectively. Therefore, we can conclude the following: (1) the Ag^+ released from GCN/Ag_2CrO_4 was largely responsible for BSA protein degradation with and without royal blue LED irradiation, and (2) the bind-and-damage model of light-induced protein oxidation and the generation of ROS did not significantly contribute to photocatalytic degradation of BSA protein, indicating that the light irradiation had less impact in the photocatalytic process. The less impact of light irradiation is expressed in the proximities of the degradation kinetics of GCN/Ag_2CrO_4 with royal blue irradiation (3.64×10^{-1} min^{-1}) which is less than two times without light, that is dark (1.86×10^{-1} min^{-1}). The reason for the dominance of these silver ions in the degradation of BSA protein could be attributed to their binding interactions with BSA. Previous studies show that silver ions and silver nanoparticles interact with protein functional groups, such as carboxylic groups (COOH), thiol groups (SH), and amino groups (NH) of protein, to form bonds resulting in protein inactivation [51,66–69]. Zaher et al. investigated the interaction between Ag^+ and AgNPs with BSA protein and reported that Ag^+ has stronger binding and quenching efficiency towards BSA than AgNPs which is evidenced in a higher binding constant of Ag^+ [70]. BSA is a polymer of about 583 amino acids joined together by peptide linkage between two tryptophan residues. Each amino acid has two pH-sensitive functional groups, COOH and NH. The proposed binding mechanism was due to the electrostatic attraction between the ionized COO^- and the positive charge of Ag^+ at physiological pH (pH = 7.4). This was confirmed in the reported UV-vis spectra showing the peak intensity increasing with rising Ag^+ concentration. This correlates with our UV-vis spectra results for GCN/Ag_2CrO_4 with and without royal blue LED irradiation where the BSA peak intensity increases indicating a rise in Ag^+ concentration with time in the reacting medium and therefore the increasing interaction between the COO^- and Ag^+ electrostatically. This interaction leads to denaturing, unfolding, aggregation, and the eventual inactivation of the protein. Jaiswar et al. investigated the binding interaction between protein and chromate ions and reported structural changes to protein, indicating successful interactions [71].

In Figure 5, SARS-CoV-2 (Coronavirus) spike protein result is presented. It shows that GCN/Ag_2CrO_4 also can degrade the spike protein under royal blue LED. A photocatalytic degradation efficiency of 77.5% was achieved in 30 min. The lower degradation efficiency observed for SARS-CoV-2 spike protein (SP) can be attributed to its heavier molecular weight and structure. SARS-CoV-2 SP is 135 kDa and 1213 amino residues which is twice the size of BSA protein which is 66 kDa and 583 amino acid residues.

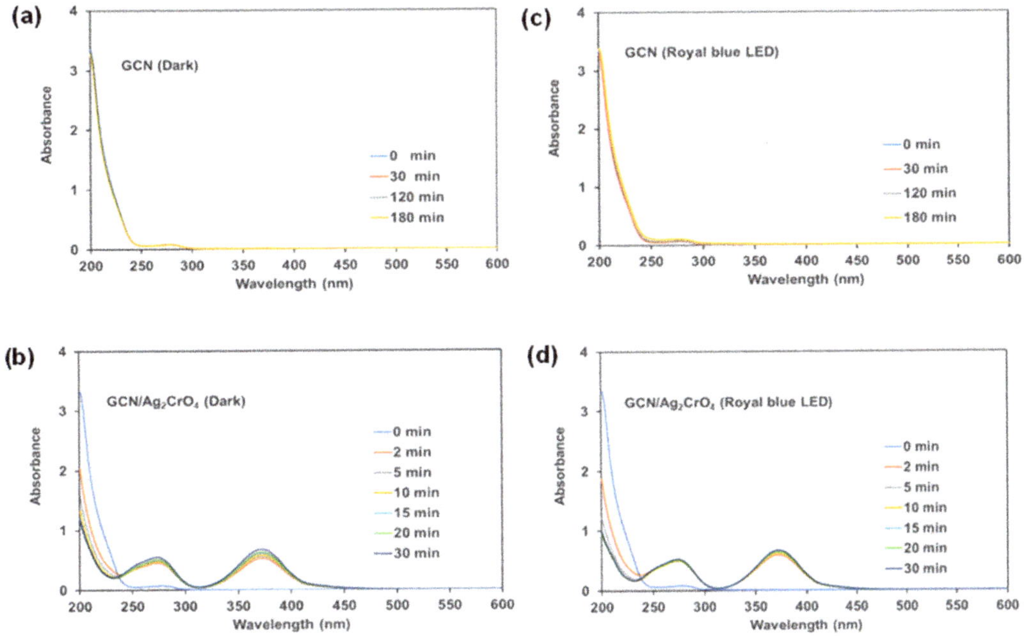

Figure 4. UV-vis absorption spectra of BSA: Dark experiment (**a**,**b**) and Royal blue LED irradiation (**c**,**d**).

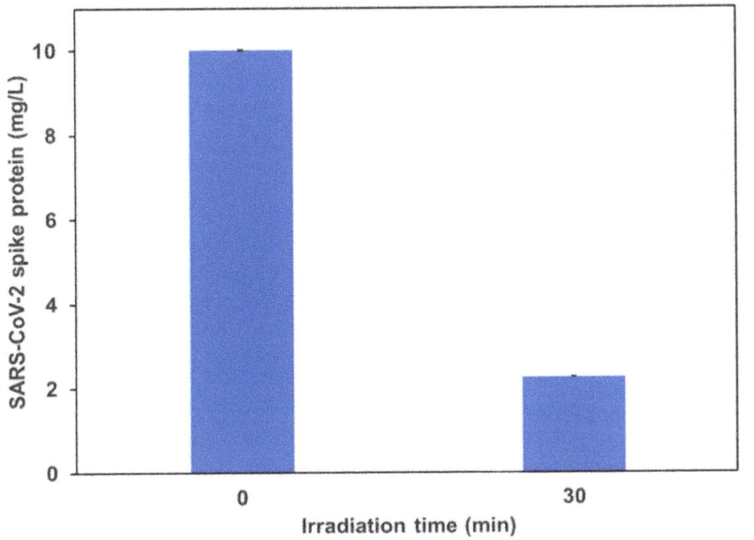

Figure 5. Photocatalytic degradation of SARS-CoV-2 (Coronavirus) spike protein using GCN/Ag_2CrO_4. Under royal blue LED irradiation.

2.3. Photocatalytic Disinfection of Wastewater Secondary Effluent

Results on the disinfection of secondary effluent from a municipal wastewater treatment plant by GCN/Ag_2CrO_4 are presented in Figures 6 and 7. The log reduction of cellular Adenosine Triphosphate (*cATP*) is presented in Figure 6, while the log reduction of total coliforms and *E. coli* is presented in Figure 7. As shown in Figure 6, around 0.2 log reduction of *cATP* was observed after 60 min of royal blue LED irradiation or after 60 min of dark (control experiment). The presence of GCN/Ag_2CrO_4 in the dark or royal blue LED irradiation both increased the log reduction of *cATP* in wastewater. Moreover, 1.2 log reduction was observed for *cATP* when wastewater was mixed with GCN/Ag_2CrO_4 in the dark for 60 min. An even higher log reduction (>1.7) was observed in the presence of GCN/Ag_2CrO_4 under royal blue LED irradiation for 60 min.

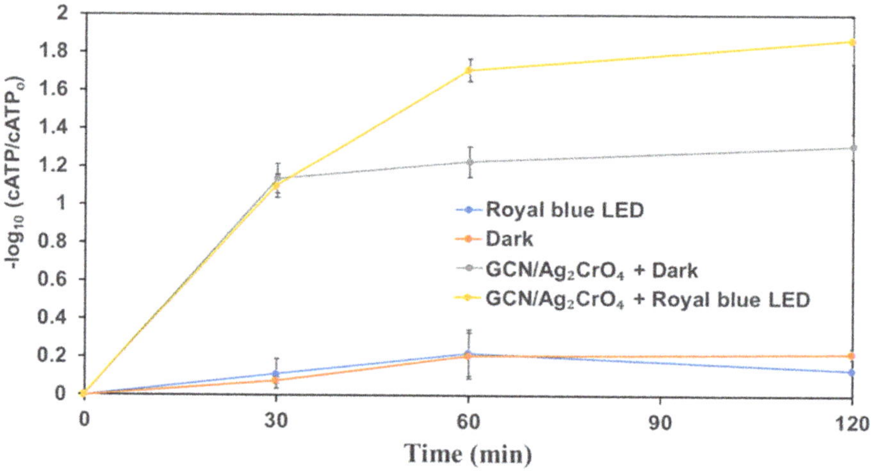

Figure 6. The change of *ATP* in wastewater secondary effluent in the presence of GCN/Ag_2CrO_4 with and without royal blue LED irradiation.

In the control experiments of royal blue LED and GCN (dark experiment), the total coliforms and *E. coli* did not decrease during the experimental duration of 60 min. This indicates that the royal blue LED irradiation and GCN (dark experiment) did not impact the total coliforms and *E. coli* counts in the wastewater. The dark control experiment of GCN/Ag_2CrO_4 and the use of GCN under royal blue LED both led to lower log reduction (<1) in 60 min for total coliforms and *E. coli*. However, the use of GCN/Ag_2CrO_4 under royal blue LED irradiation led to higher log reduction (>1) in 60 min with 2.43 log reduction of total coliforms and 1.48 log reduction of *E. coli*. This result indicates that the disinfection rate of *E. coli* was slower than that of the total coliforms.

The interaction of silver compounds with the cell membrane of the bacteria is one of the important mechanisms of silver compound toxicity [54]. This can occur both in the dark and visible light irradiation. Over one log reduction was observed for *ATP*, total coliforms, and *E. coli* in 60 min during the photocatalytic degradation process. This indicates GCN/Ag_2CrO_4 can potentially photocatalytically inactivate, under royal blue LED irradiation, over 98% (1.7 log) of the microbial activity (*ATP*) in 60 min which includes inactivation of over 99.6% (2.43 log) of total coliforms.

The bind-and-damage model of protein oxidation, release of ions from the photocatalyst, and the ROS generation processes of protein degradation described in Section 3.2 can be adopted for this disinfection process since it was used for protein degradation and bacterial disinfection by Shi et al. [62]. In Figure 6, the GCN/Ag_2CrO_4 with and without royal blue LED irradiation showed a significant impact on the reduction of *cATP* of microorganisms. Without light (GCN/Ag_2CrO_4 + Dark), the release of Ag^+ and CrO_4^{2-} contributed to the reduction of *cATP* by 1.2 log in 60 min. However, with irradiation (GCN/Ag_2CrO_4 + Royal blue LED), the ROS generation in addition to Ag^+ and CrO_4^{2-} released resulted in >1.7 log reduction of *cATP* in 60 min. With the marginal difference of >0.5 log reduction of *cATP*, it means the release of the two ions contributed far more than

ROS generation in the overall reduction of *cATP*. In Figure 7a, the GCN/Ag_2CrO_4 with and without royal blue LED irradiation showed a different response to the disinfection of total coliforms. Without light (GCN/Ag_2CrO_4 + Dark), the release of Ag^+ and CrO_4^{2-} resulted in a 0.3 log reduction of total coliforms in 60 min while the irradiation (GCN/Ag_2CrO_4 + Royal blue LED), the ROS generation and the light-induced protein oxidation in addition to the released two ions resulted in 2.4 log reduction of total coliforms in 60 min. The log reduction difference >2 log is largely due to the contributions of the generated ROS and the light-induced protein oxidation. In Figure 7b, the GCN/Ag_2CrO_4, with and without royal blue LED irradiation showed a different response to *E. coli* disinfection. Without light (GCN/Ag_2CrO_4 + dark), the Ag^+ and CrO_4^{2-} released resulted in a 0.1 log reduction of *E. coli* in 60 min while with irradiation (GCN/Ag_2CrO_4 + Royal blue LED), the generated ROS and light-induced protein oxidation in addition to the released ions resulted in 1.5 log reduction of *E. coli* in 60 min. The log reduction difference >1 log is largely due to the contributions of the ROS generated and the light-induced protein oxidation.

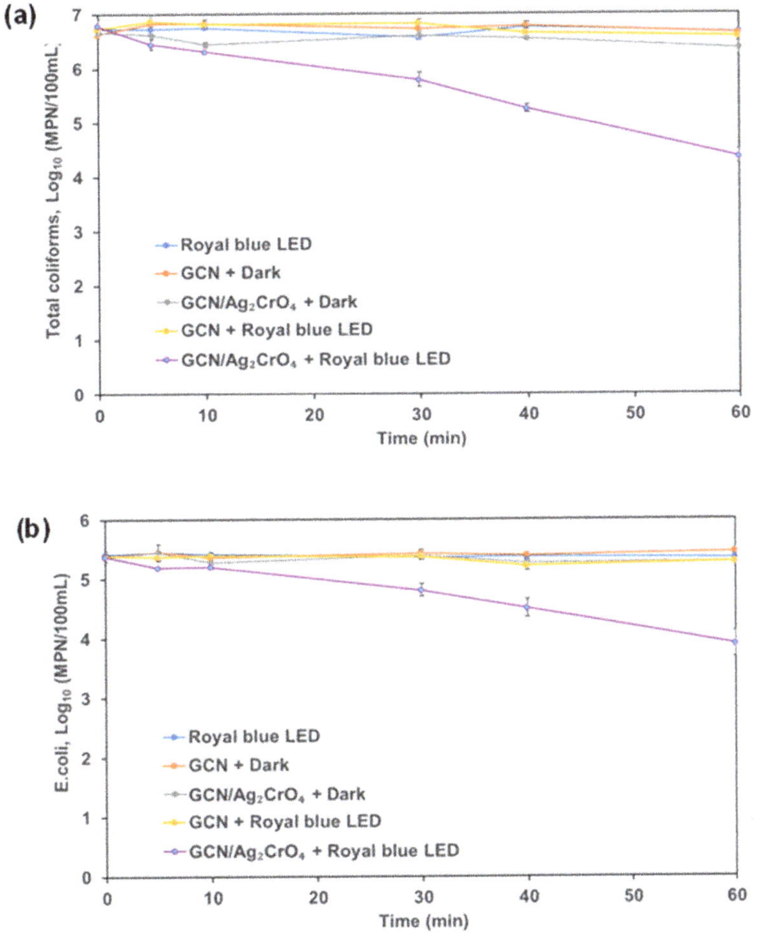

Figure 7. Coliforms in wastewater secondary effluent: (**a**) total coliforms and (**b**) *E. coli*.

Zhang et al. investigated the impact of light irradiation on the reduction of Ag^+ to AgNPs by bacteria and reported that visible lights (monochromatic light at wavelengths 415 nm and 600 nm) accelerated the reduction of Ag^+ to AgNPs by bacteria and almost no

AgNPs were formed in the dark [72]. Furthermore, the visible light irradiation reversibly can excite the AgNPs for their surface plasmon resonance (SPR) and accelerate the electrons from the bacteria to the adjacent Ag^+. It reported that the bacteria can reduce Ag^+ and the process can be accelerated by light irradiation. Shi et al. focused on the bactericidal effect of AgNPs and identified a new antibacterial mechanism different from the ROS-induced photocatalyst [62]. AgNPs can penetrate the cell wall and membrane and gather in the cytosol and bind directly with the cytosol protein leading to the inducement of protein aggregation, an indication of protein degradation, caused by the bind-and-damage model of the light-induced protein oxidation. Deng et al. investigated the effects of silver ions, chromium VI, Cr(VI), and ROS on the photocatalytic disinfection mechanism of *E. coli* [29]. The results show that the combined effects of the release of silver ion and the generation of ROS of holes (h^+) and hydroxyl (·OH) radicals played important roles in the photocatalytic disinfection process. The silver ions interact with the thiol (-SH) functional group causing cell damage and the generated ROS could inactivate bacterial cells by interacting with the cell wall and destabilizing the membrane causing leakage of potassium ions (K^+). The interaction of Ag^+ with bacterial cell membranes can also lead to lower *ATP* readings caused by the leakage of potassium ions [54]. The addition of hexavalent chromium to the photocatalytic system enhanced the photocatalytic disinfection kinetics. This is because Cr(VI) can behave as an electron scavenger, directly reduces electron-hole recombination, enhances the amount of h^+ available for disinfection, and ultimately leads to better disinfection performance. Table 4 shows the photocatalytic disinfection of microorganisms by different photocatalysts.

Table 4. Comparison of microorganim disinfection with the results of previously published articles.

S/N	Photocatalyst	Visible Light Source	Contaminant	Disinfection Time (min)	Concentration (CFU/mL)	Disinfection Efficciency (Log Reduction)	Ref.
1	BiOBr	Xenon lamp (1000 W/m^2, 300 W)	*E. coli*	24	10^7	7	[73]
2	Ag/BiOI	Iodide lamp (400 W)	*E. coli*	20	10^7	7	[74]
3	Ag/ZnO/GCN	Xenon lamp (300 W)	*E. coli*	120	10^7	7	[75]
4	GCN/Ag_2CrO_4	Royal blue LED (460 nm)	Total coliforms and *E. coli*	60 and 60	$10^{6.7}$ and $10^{5.5}$	2.43 and 1.48	Current study

3. Materials and Methods

3.1. Chemicals

Killex^R, a known herbicide was purchased from a local store in Calgary, Alberta. Urea-CH_4N_2O (≥99.5%), silver nitrate-$AgNO_3$ (≥99%), potassium chromate-K_2CrO_4 (≥99%), 1–10 phenanthroline-$C_{12}H_8N_2$, iron III chloride-$FeCl_3$, ammonium iron II sulfate hexahydrate (Mohr's salt, ≥98%)—$(NH_4)_2Fe(O_4)_2(H_2O)_6$ and Bovine Serum Albumin (BSA) protein (hydrolyzed powder, crystallized ≥98.0%) were purchased from Sigma-Aldrich Co. (St. Louis, MO, USA). Luria broth, potassium trioxalatoferrate III trihydrate-$K_3Fe(C_2O_4)_3.3H_2O$ (98%) was purchased from Alfar Aesar (St. Louis, MO, USA). Sodium acetate-$C_2H_3NaO_2$ (>99%) was purchased from Brady (Markham, ON, Canada). BSA Bradford assay protein test kit was purchased from BIORAD (Hercules, CA, USA). SARS-CoV-2 spike protein (SP), S-ECD (>95%) was purchased from Thermofisher Scientific (Ottawa, ON, Canada). *ATP* test kit was purchased from LuminUltra Technologies Ltd. (Fredericton, NB, Canada) while, total coliforms/*E. coli* Colilert test kit was purchased from IDEXX laboratories inc. (Westbrook, ME, USA). Secondary treated wastewater effluent was provided by Advancing Canadian Water

Assets—ACWA (Calgary, AB, Canada), a full-scale testing facility embedded in a wastewater treatment plant in Calgary, Alberta.

3.2. Preparation of GCN, Ag_2CrO_4, and GCN/Ag_2CrO_4 Nanocomposite

The GCN, Ag_2CrO_4, and GCN/Ag_2CrO_4 were synthesized based on a procedure described in detail elsewhere [21]. The GCN was synthesized using the direct pyrolysis method by heating urea at 550 °C for 4 h in a muffle furnace. The GCN/Ag_2CrO_4 composite was synthesized using the dark-induced/In-situ-deposition method by first sonicating one gram of GCN slurry. A known quantity of $AgNO_3$ and 4.83 g/L K_2CrO_4 were added to the GCN slurry under magnetic stirring. The mixture was then magnetically stirred in the dark for 7 h and after that, it was filtered using a 0.45 µm filter and washed about five times with water and alcohol. The residue was left to dry at room temperature and thereafter labeled as GCN/Ag_2CrO_4. The characterizations of these catalysts have been reported elsewhere [21].

3.3. Experimental Set-Up

3.3.1. Photocatalytic Degradation of a Commercial Herbicide

Killex^R is a commercially-available herbicide that contains 95 g/L of 2,4-dichlorophen oxyacetic acid (2,4-D), 52.52 g/L of methyl phenoxy propionic acid (Mecoprop-P or MCPP), and 9 g/L of Dicamba. A quantity of 0.8 mL of Killex^R was dissolved in 79.2 mL of distilled water (DI) which gave an 80 mL solution, resulting in the solution containing 12.1 mg/L of 2,4-D, 6.7 mg/L of Mecoprop-P, and 1.2 mg/L of Dicamba. In this experiment, 80 mg of the photocatalyst was dispersed in an 80 mL solution containing Killex^R in a 100 mL quartz beaker. The intensity of the royal blue LED (peak wavelength = 460 nm) entering the 80 mL solution was estimated to be 11.80×10^{17} photons/s using chemical ferrioxalate actinometry [43,76,77]. The same photoreactor was used for all experiments mentioned in this study. Before LED irradiation, the water/catalysts mixture was stirred magnetically in the dark for 30 min to attain adsorption-desorption equilibrium. Quantities of 5 mL of the samples were collected at different irradiation intervals and filtered using a 0.45 µm syringe filter (PTFE, Chromatographic Specialties Inc., Brockville, ON, Canada) to remove the photocatalysts. Dark experiments with catalysts and light experiments without catalysts were also conducted as controls. HPLC was used to evaluate the concentration change of 2,4-D and Mecoprop-P.

3.3.2. Photocatalytic Degradation of BSA Protein and SARS-CoV-2 Spike Protein

In each experiment, 80 mg of the photocatalyst was dispersed in 80 mL of 100 mg/L BSA solution. During royal blue LED irradiation, 6 mL of sample was collected at different irradiation intervals and filtered using a 0.45 µm syringe filter to remove the photocatalysts. The performance of GCN/Ag_2CrO_4 was also investigated on SARS-CoV-2 spike protein (SP) under royal blue LED. In the experiment, 40 mg of the GCN/Ag_2CrO_4 was dispersed in a 40 mL of 10 mg/L SARS-CoV-2 SP, and 3 mL of the sample was collected at a different irradiation interval and filtered using a 0.45 µm syringe filter to remove the photocatalysts. Bradford assay procedure was used to determine the concentration change for the BSA protein and SARS-CoV-2 SP [78]. UV-Vis analysis was conducted for visible light and dark experiments to study the binding interaction between BSA and GCN/Ag_2CrO_4.

3.3.3. Photocatalytic Disinfection of Wastewater Secondary Effluent

Two sets of disinfection experiments were conducted. The first one was focused on the change of *cATP* level on the collected wastewater during the disinfection experiment, while the second one was aimed at quantifying the log reduction of total coliforms and *E. coli* by the catalyst under royal blue LED irradiation. For the *cATP* experiment, 80 mg of the photocatalyst was dispersed in 80 mL of wastewater secondary effluent and then irradiated with royal blue LED under continuous magnetic stirring. A total of 2.5 mL of the sample was taken at a different time interval and its *cATP* level was quantified using the LuminUltra test procedure for assay analysis [79]. Control experiments were carried out in the dark with the photocatalyst and under the light without the photocatalyst. Before the start of the disinfection experiment, the initial concentrations of total coliforms and *E. coli* were calculated using the tray count method [80]. A stock solution of 1 L secondary treated wastewater effluent containing 50 mg of Luria broth nutrient was first prepared in a 1 L jar and then incubated at 35 °C ± 0.5 °C for 24 h. Then 1 mL of the 80 mL collected from the stock solution was serially diluted 40 times as follows: in the first mixture, 1 mL was added to 3 mL of deionized (DI) water. After mixing, 1 mL from the first mixture was added to 9 mL of DI water to give 40 times serial dilution, this gives the second mixture. Finally, using the tray count method, 1 mL of the second mixture was added to the mixture of 100 mL DI water and colilert reagent, giving the final dilution factor of 4000. The final mixed sample was poured in a quantitray, sealed with an IDEXX sealer, and incubated at 35 °C ± 0.5 °C for 24 h. The tray, which has 49 large wells and 48 small wells, was used to estimate the most probable number (MPN) of coliforms and *E. coli* in a 100 mL wastewater sample. The actual MPN/100 mL of total coliforms or *E. coli* present in the sample was calculated by multiplying the MPN by the dilution factor of 4000. The average initial concentrations of total coliforms and *E. coli* were calculated to be 5.2×10^6 MPN/100 mL and 0.25×10^6 MPN/100 mL respectively. Disinfection experiments were conducted and MPNs of total coliforms and *E. coli* were estimated and concentrations at different time intervals were analyzed and calculated using the tray count method. Control experiments were also performed in the dark with photocatalysts and under the light without photocatalysts.

3.4. Chemical Analysis

3.4.1. Herbicide Analysis

The concentrations of 2,4-D and Mecoprop-P were determined using high-performance liquid chromatography (HPLC; LC-2040C 3D Shimadzu Corporation, Kyoto, Japan) with UV absorbance detection at $\lambda = 280$ nm. A kinetex 2.6 µm PFP column by Phenomenex (Torrance, CA, USA) was used to separate the target compounds and other interferences. The mobile phases were 0.1% phosphoric acid in HPLC grade water and 0.1% phosphoric acid in acetonitrile at a 50:50 mixture with a flow rate of 1 mL/min. The measurements were conducted in duplicate, and the detection limit was around 0.1 mg/L.

3.4.2. Protein Analysis

The concentration of BSA protein was determined using the Bradford protein assay procedure [78]. This procedure is based on the binding of Coomassie brilliant blue dye to protein. Briefly, the diluted dye was prepared by mixing concentrated Coomasie dye (80 mL) and deionized water (DI) of 240 mL and then filtered with 5 µm using vacuum filtration to remove dye particles. From the BSA stock solution (2000 mg/L), five standard dilute solutions were prepared (100 mg/L, 50 mg/L, 25 mg/L, 12.5 mg/L and 6.25 mg/L). Then, 0.8 mL of each standard and photocatalytic degraded BSA samples were collected into the vial using a pipette and 3.2 mL of diluted dye was added. The mixture was put on a rotary shaker and left to incubate for at least 5 min but not more than 60 min at room temperature. This incubation period allows for colour change of the Coomasie dye from red to blue. This blue colour is an indication of the binding of the dye to the protein, that is detected at 595 nm. The spectrophotometer (UV-2600 Shimadzu Corp., Kyoto, Japan) was set at 595 nm to take the absorbance readings of BSA standards and the degraded samples using a disposable cuvette. The concentration of the degraded samples was calculated based on the known concentrations of the BSA standard.

This Bradford protein assay procedure was also used to determine the concentration of the SARS-CoV-2 SP. Five standard solutions of BSA of known concentrations (10 mg/L, 5 mg/L, 2.5 mg/L, and 0.625 mg/L) were prepared. Using the microassay plate, 800 µL of each standard and degraded SARS-CoV-2 SP sample was put in a vial and 200 µL of dye concentrate was added to have the protein-dye mixture. The same incubation period as described earlier was used for the spike protein and after that, the absorbance of the SP-dye complex was read at 595 nm using a spectrophotometer (UV-2600 Shimadzu Corp., Kyoto, Japan).

The UV-vis as an analytical technique was used to study the BSA and photocatalyst interaction and the structural changes to BSA. The UV-vis absorption spectra measurements were carried out for the dark as well as royal blue irradiation samples.

3.4.3. ATP Analysis

The concentration of the microorganisms' *cATP* was determined using the LuminUltra test kit and assay procedure [80]. A total of 5 mL of the water sample was passed through a 0.45 μm syringe filter. The syringe filter containing microorganisms was re-attached to a clean syringe with the plunger removed, and then 1 mL of UltraLyse 7 was pipette into the syringe. The filtration was done slowly, and the sample was collected in a 9 mL UltraLute (dilution) tube. The purpose of the filtration using UltraLyse 7 is to be able to dissolve and remove the bound *ATP* (*ATP* in living microorganisms after disinfection) for analysis. A quantity of 100 μL of UltraLute (dilution) solution was pipette into a clean test tube and then 100 μL Luminase was added. The mixture was shaken about five times, and then the reading was taken within 10 s using a luminometer. The cellular *ATP* is calculated using Equation (9) and the log reduction of the normalized concentration is evaluated using Equation (10):

$$cATP\left(pg\frac{ATP}{mL}\right) = \frac{RLU_{cATP}}{RLU_{ATP1}} \times 10,000\frac{pg\ ATP}{V_{sample}(mL)} \tag{9}$$

$$Log\ reduction = -\log\left(\frac{cATP_t}{cATP_o}\right) \tag{10}$$

where *RLU* is the unit of measurement as it is the relative light unit, *pg* is picogram, V_{sample} is the volume of the sample collected during the photocatalytic experiment, *ATP* is the calibrated *ATP RLU* value for the wastewater before the photocatalytic experiment, and $cATP_o$ and $cATP_t$ are the cellular *ATP* of the microorganisms at time, $t = 0$ and $t = t$ min. Control experiments were conducted in the dark to determine the adsorption performance of the photocatalysts in disinfecting microorganisms and the corresponding estimation of *ATP* levels.

3.4.4. Total Coliform/*E. coli* Analysis

One mL sample was collected at different irradiation intervals and serially diluted using deionized water. This was aimed to (1) reduce the amount of catalysts and other water matrices present in the final tray so as to minimize their interferences on total coliform and *E. coli* measurement; (2) to ensure the concentration of total coliform and *E. coli* fall within the measurement range of the Colilert tray count procedure. The total coliforms and *E. coli* concentrations were determined using the tray count method [80] as described in Section 3.3.3. The total coliform and *E. coli* disinfection efficiency were expressed in log form as shown in Equation (11):

$$Log\ reduction = \log C_o - \log C_t \tag{11}$$

where C_o and C_t are the concentrations of total coliforms or *E. coli* before and after time *t* min, expressed as MPN/100 mL.

4. Conclusions

A visible-light active photocatalyst, GCN/Ag_2CrO_4 nanocomposite was synthesized, and its performance was investigated in the degradation of organic pollutants (2,4-D and Mecoprop-P) in Killex^R, BSA protein, and SARS-CoV-2 spike protein) and the disinfection of WWTP secondary effluent, using a narrowband low-energy visible light source (royal blue LED at 460 nm). The results from this study highlight the following:

- Similar degradation rates and percentage removal visible light irradiation were observed for 2,4-D and Mecoprop-P.
- The degradation performance of a photocatalyst is dependent on the available surface area for absorption for photocatalytic activity and the bond energies of the chemical bonds that exist in the contaminant.
- Released Ag^+ and CrO_4^{2-} from the GCN/Ag_2CrO_4 nanocomposite were largely responsible for the BSA degradation and *cATP* reduction.
- Generated ROS were largely responsible for the disinfection of total coliforms and *E. coli*.
- The lower percentage degradation (70%) of SARS-CoV-2 spike protein can be attributed to its large molecular weight and structure compared to BSA protein.
- The pathways for degradation of proteins and the disinfection of microorganisms depend on any of the following: (1) binding affinity of the released metal ions from GCN/Ag_2CrO_4 with the functional groups of proteins and *ATP*, (2) amount of ROS generated, and (3) the quantum of emission light from the photocatalyst that is absorbed and quenched by the proteins because of the light-induced protein oxidation.
- Significant disinfection results observed in the reduction of microorganisms' *cATP*, total coliforms, and *E. coli* with GCN/Ag_2CrO_4 are also due to the reduction of the released Ag^+ to AgNPs by bacteria which was accelerated by light irradiation.
- Over one log reduction of microorganisms' *cATP*, total coliforms, and *E. coli* in wastewater secondary effluent was achieved in 60 min.
- The combination of GCN/Ag_2CrO_4 and the narrowband low-energy light (royal blue LED at peak wavelength = 460 nm) holds promise in that it can be used in the wastewater treatment plant to degrade multiple recalcitrant organic pollutants and inactivate microorganisms.

Author Contributions: Conceptualization, O.O.A., L.Y. and G.A.; methodology, O.O.A.; experimental design, O.O.A. and L.Y.; formal analysis, O.O.A. and L.Y.; investigation, O.O.A.; resources, G.A.; writing—original draft presentation, O.O.A.; writing—review & editing, L.Y., J.H., M.G.K., C.R.J.H., S.P. and G.A.; supervision, G.A.; project administration, L.Y.; funding acquisition, G.A. All authors have read and agreed to the published version of the manuscript.

Funding: This work was carried out with the support and funding of the Natural Sciences and Engineering Research Council of Canada (NSERC), account number 10029431 and Alberta Innovates under the CASBE program, account number 10035007.

Data Availability Statement: The data presented in this study are available on request from the corresponding author.

Acknowledgments: The authors gratefully acknowledge the funding support provided by Natural Sciences and Engineering Research of Canada (NSERC) and Alberta Innovates under the CASBE program. We thank Advancing Canadian Water Assets (ACWA) for providing the WWTP secondary effluent.

Conflicts of Interest: The authors declare no conflict of interest.

References

1. Jiang, Y.; Liu, Z.; Zeng, G.; Liu, Y.; Shao, B.; Li, Z.; Liu, Y.; Zhang, W.; He, Q. Polyaniline-Based Adsorbents for Removal of Hexavalent Chromium from Aqueous Solution: A Mini Review. *Environ. Sci. Pollut. Res.* **2018**, *25*, 6158–6174. [CrossRef] [PubMed]
2. Liu, Y.; Liu, Z.; Zeng, G.; Chen, M.; Jiang, Y.; Shao, B.; Li, Z.; Liu, Y. Effect of Surfactants on the Interaction of Phenol with Laccase: Molecular Docking and Molecular Dynamics Simulation Studies. *J. Hazard. Mater.* **2018**, *357*, 10–18. [CrossRef]
3. Liu, Y.; Liu, Z.; Huang, D.; Cheng, M.; Zeng, G.; Lai, C.; Zhang, C.; Zhou, C.; Wang, W.; Jiang, D.; et al. Metal or Metal-Containing Nanoparticle@MOF Nanocomposites as a Promising Type of Photocatalyst. *Coord. Chem. Rev.* **2019**, *388*, 63–78. [CrossRef]
4. Ahmadian-Fard-Fini, S.; Ghanbari, D.; Salavati-Niasari, M. Photoluminescence Carbon Dot as a Sensor for Detecting of Pseudomonas Aeruginosa Bacteria: Hydrothermal Synthesis of Magnetic Hollow $NiFe_2O_4$-Carbon Dots Nanocomposite Material. *Compos. Part B Eng.* **2019**, *161*, 564–577. [CrossRef]

5. Mamba, G.; Mishra, A.K. Graphitic Carbon Nitride ($g-C_3N_4$) Nanocomposites: A New and Exciting Generation of Visible Light Driven Photocatalysts for Environmental Pollution Remediation. *Appl. Catal. B Environ.* **2016**, *198*, 347–377. [CrossRef]
6. Kashale, A.A.; Rasal, A.S.; Kamble, G.P.; Ingole, V.H.; Dwivedi, P.K.; Rajoba, S.J.; Jadhav, L.D.; Ling, Y.-C.; Chang, J.-Y.; Ghule, A.V. Biosynthesized Co-Doped TiO_2 Nanoparticles Based Anode for Lithium-Ion Battery Application and Investigating the Influence of Dopant Concentrations on Its Performance. *Compos. Part B Eng.* **2019**, *167*, 44–50. [CrossRef]
7. Liu, Z.; Shao, B.; Zeng, G.; Chen, M.; Li, Z.; Liu, Y.; Jiang, Y.; Zhong, H.; Liu, Y.; Yan, M. Effects of Rhamnolipids on the Removal of 2,4,2,4-Tetrabrominated Biphenyl Ether (BDE-47) by Phanerochaete Chrysosporium Analyzed with a Combined Approach of Experiments and Molecular Docking. *Chemosphere* **2018**, *210*, 922–930. [CrossRef]
8. Zhang, C.; Li, Y.; Shuai, D.; Shen, Y.; Xiong, W.; Wang, L. Graphitic Carbon Nitride ($g-C_3N_4$)-Based Photocatalysts for Water Disinfection and Microbial Control: A Review. *Chemosphere* **2019**, *214*, 462–479. [CrossRef]
9. Zhang, T.; Jiang, B.; Huang, Y. UV-Curable Photosensitive Silicone Resins Based on a Novel Polymerizable Photoinitiator and GO-Modified TiO_2 Nanoparticles. *Compos. Part B Eng.* **2018**, *140*, 214–222. [CrossRef]
10. Fu, H.; Pan, C.; Yao, W.; Zhu, Y. Visible-Light-Induced Degradation of Rhodamine B by Nanosized Bi_2WO_6. *J. Phys. Chem. B* **2005**, *109*, 22432–22439. [CrossRef]
11. Liu, S.; Tian, J.; Wang, L.; Luo, Y.; Sun, X. One-Pot Synthesis of CuO Nanoflower-Decorated Reduced Graphene Oxide and Its Application to Photocatalytic Degradation of Dyes. *Catal. Sci. Technol.* **2012**, *2*, 339–344. [CrossRef]
12. Bao, N.; Shen, L.; Takata, T.; Domen, K. Self-Templated Synthesis of Nanoporous CdS Nanostructures for Highly Efficient Photocatalytic Hydrogen Production under Visible Light. *Chem. Mater.* **2008**, *20*, 110–117. [CrossRef]
13. Izadifard, M.; Achari, G.; Langford, C. Application of Photocatalysts and LED Light Sources in Drinking Water Treatment. *Catalysts* **2013**, *3*, 726–743. [CrossRef]
14. Izadifard, M.; Langford, C.H.; Achari, G. Photocatalytic Dechlorination of PCB 138 Using Leuco-Methylene Blue and Visible Light; Reaction Conditions and Mechanisms. *J. Hazard. Mater.* **2010**, *181*, 393–398. [CrossRef]
15. Ghosh, J.P.; Langford, C.H.; Achari, G. Characterization of an LED Based Photoreactor to Degrade 4-Chlorophenol in an Aqueous Medium Using Coumarin (C-343) Sensitized TiO_2. *J. Phys. Chem. A* **2008**, *112*, 10310–10314. [CrossRef]
16. Malkhasian, A.Y.S.; Izadifard, M.; Achari, G.; Langford, C.H. Photocatalytic Degradation of Agricultural Antibiotics Using a UV-LED Light Source. *J. Environ. Sci. Health Part B* **2014**, *49*, 35–40. [CrossRef]
17. Langford, C.; Izadifard, M.; Radwan, E.; Achari, G. Some Observations on the Development of Superior Photocatalytic Systems for Application to Water Purification by the "Adsorb and Shuttle" or the Interphase Charge Transfer Mechanisms. *Molecules* **2014**, *19*, 19557–19572. [CrossRef]
18. Shi, L.; Liang, L.; Ma, J.; Wang, F.; Sun, J. Remarkably Enhanced Photocatalytic Activity of Ordered Mesoporous Carbon/$g-C_3N_4$ Composite Photocatalysts under Visible Light. *Dalton Trans.* **2014**, *43*, 7236–7244. [CrossRef]
19. Tanveer, M.; Cao, C.; Ali, Z.; Aslam, I.; Idrees, F.; Khan, W.S.; But, F.K.; Tahir, M.; Mahmood, N. Template Free Synthesis of CuS Nanosheet-Based Hierarchical Microspheres: An Efficient Natural Light Driven Photocatalyst. *CrystEngComm* **2014**, *16*, 5290. [CrossRef]
20. Adhikari, S.P.; Pant, H.R.; Kim, J.H.; Kim, H.J.; Park, C.H.; Kim, C.S. One Pot Synthesis and Characterization of Ag-ZnO/$g-C_3N_4$ Photocatalyst with Improved Photoactivity and Antibacterial Properties. *Colloids Surf. A Physicochem. Eng. Asp.* **2015**, *482*, 477–484. [CrossRef]
21. Akintunde, O.O.; Yu, L.; Hu, J.; Kibria, M.G.; Achari, G. Visible-Light Driven Photocatalytic Degradation of 4-Chlorophenol Using Graphitic Carbon Nitride-Based Nanocomposites. *Catalysts* **2022**, *12*, 281. [CrossRef]
22. Jourshabani, M.; Dominic, J.A.; Achari, G.; Shariatinia, Z. Synergetic Photocatalytic Ozonation Using Modified Graphitic Carbon Nitride for Treatment of Emerging Contaminants under UVC, UVA and Visible Irradiation. *Chem. Eng. Sci.* **2019**, *209*, 115181. [CrossRef]
23. Faisal, M.; Ismail, A.A.; Harraz, F.A.; Al-Sayari, S.A.; El-Toni, A.M.; Al-Assiri, M.S. Synthesis of Highly Dispersed Silver Doped $G-C_3N_4$ Nanocomposites with Enhanced Visible-Light Photocatalytic Activity. *Mater. Des.* **2016**, *98*, 223–230. [CrossRef]
24. Wang, D.; Duan, Y.; Luo, Q.; Li, X.; Bao, L. Visible Light Photocatalytic Activities of Plasmonic Ag/AgBr Particles Synthesized by a Double Jet Method. *Desalination* **2011**, *270*, 174–180. [CrossRef]
25. Wang, P.; Huang, B.; Qin, X.; Zhang, X.; Dai, Y.; Wei, J.; Whangbo, M.-H. Ag@AgCl: A Highly Efficient and Stable Photocatalyst Active under Visible Light. *Angew. Chem. Int. Ed.* **2008**, *47*, 7931–7933. [CrossRef]
26. Bi, Y.; Ouyang, S.; Umezawa, N.; Cao, J.; Ye, J. Facet Effect of Single-Crystalline Ag_3PO_4 Sub-Microcrystals on Photocatalytic Properties. *J. Am. Chem. Soc.* **2011**, *133*, 6490–6492. [CrossRef]
27. Liu, Y.; Yu, H.; Cai, M.; Sun, J. Microwave Hydrothermal Synthesis of Ag_2CrO_4 Photocatalyst for Fast Degradation of PCP-Na under Visible Light Irradiation. *Catal. Commun.* **2012**, *26*, 63–67. [CrossRef]
28. Xu, D.; Cao, S.; Zhang, J.; Cheng, B.; Yu, J. Effects of the Preparation Method on the Structure and the Visible-Light Photocatalytic Activity of Ag_2CrO_4. *Beilstein J. Nanotechnol.* **2014**, *5*, 658–666. [CrossRef]
29. Deng, J.; Liang, J.; Li, M.; Tong, M. Enhanced Visible-Light-Driven Photocatalytic Bacteria Disinfection by $g-C_3N_4$-AgBr. *Colloids Surf. B Biointerfaces* **2017**, *152*, 49–57. [CrossRef]
30. Ren, Y.; Zhao, Q.; Li, X.; Xiong, W.; Tade, M.; Liu, L. 2D Porous Graphitic C_3N_4 Nanosheets/Ag_3PO_4 Nanocomposites for Enhanced Visible-Light Photocatalytic Degradation of 4-Chlorophenol. *J. Nanopart. Res.* **2014**, *16*, 2532. [CrossRef]

31. Liu, Z.; Jiang, Y.; Liu, X.; Zeng, G.; Shao, B.; Liu, Y.; Liu, Y.; Zhang, W.; Zhang, W.; Yan, M.; et al. Silver Chromate Modified Sulfur Doped Graphitic Carbon Nitride Microrod Composites with Enhanced Visible-Light Photoactivity towards Organic Pollutants Degradation. *Compos. Part B Eng.* **2019**, *173*, 106918. [CrossRef]
32. Canada, H. Guidelines for Canadian Drinking Water Quality—Summary Table. Available online: https://www.canada.ca/en/health-canada/services/environmental-workplace-health/reports-publications/water-quality/guidelines-canadian-drinking-water-quality-summary-table.html (accessed on 30 June 2021).
33. 2,4-D Technical Fact Sheet. Available online: http://npic.orst.edu/factsheets/archive/2,4-DTech.html (accessed on 5 March 2022).
34. Canada, H. Re-Evaluation Note REV2016-08, Special Review of 2,4-D: Proposed Decision for Consultation. Available online: https://www.canada.ca/en/health-canada/services/consumer-product-safety/pesticides-pest-management/public/consultations/re-evaluation-note/2016/special-review-2-4-d/document.html (accessed on 2 April 2022).
35. Anderson, A.; Byrtus, G.; Thompson, J.; Humphries, D.; Hill, B.; Bilyk, M. Baseline Pesticide Data for Semi-Permanent Wetlands in the Aspen Parkland of Alberta—Open Government. Available online: https://open.alberta.ca/publications/0778524426 (accessed on 2 April 2022).
36. Gray, N.F. *Drinking Water Quality*, 2nd ed.; Cambridge University Press: Cambridge, UK, 2008; p. 538.
37. Flox, C.; Garrido, J.A.; Rodríguez, R.M.; Cabot, P.-L.; Centellas, F.; Arias, C.; Brillas, E. Mineralization of Herbicide Mecoprop by Photoelectro-Fenton with UVA and Solar Light. *Catal. Today* **2007**, *129*, 29–36. [CrossRef]
38. WHO Coronavirus (COVID-19) Dashboard. Available online: https://covid19.who.int (accessed on 31 December 2021).
39. CDC National Wastewater Surveillance System. Available online: https://www.cdc.gov/healthywater/surveillance/wastewater-surveillance/wastewater-surveillance.html (accessed on 5 April 2022).
40. EPA. *Wastewater Technology Fact Sheet: Bacterial Source Tracking*; United States Environmental Protection Agency: Washington, DC, USA, 2002; 12p.
41. Yu, L.; Achari, G.; Langford, C.H. Design and Evaluation of a Novel Light-Emitting Diode Photocatalytic Reactor for Water Treatment. *J. Environ. Eng.* **2018**, *144*, 04018014. [CrossRef]
42. Radwan, E.K.; Yu, L.; Achari, G.; Langford, C.H. Photocatalytic Ozonation of Pesticides in a Fixed Bed Flow through UVA-LED Photoreactor. *Environ. Sci. Pollut. Res.* **2016**, *23*, 21313–21318. [CrossRef]
43. Yu, L.; Achari, G.; Langford, C.H. LED-Based Photocatalytic Treatment of Pesticides and Chlorophenols. *J. Environ. Eng.* **2013**, *139*, 1146–1151. [CrossRef]
44. Alenazi, D.A.K.; Aslam, M.; Chandrasekaran, S.; Soomro, M.T.; Ali, S.; Danish, E.Y.; Ismail, I.M.I.; Hameed, A. Facile Fabrication of MoO_3/g-C_3N_4 p-n Junction for Boosted Photocatalytic Elimination of 2,4-D under Natural Sunlight Exposure. *J. Environ. Chem. Eng.* **2021**, *9*, 106304. [CrossRef]
45. Zhong, J.; Jiang, H.; Wang, Z.; Yu, Z.; Wang, L.; Mueller, J.F.; Guo, J. Efficient Photocatalytic Destruction of Recalcitrant Micropollutants Using Graphitic Carbon Nitride under Simulated Sunlight Irradiation. *Environ. Sci. Ecotechnol.* **2021**, *5*, 100079. [CrossRef]
46. Heydari, G.; Langford, C.H.; Achari, G. Passive Solar Photocatalytic Treatment of Emerging Contaminants in Water: A Field Study. *Catalysts* **2019**, *9*, 1045. [CrossRef]
47. Abubshait, S.A.; Iqbal, S.; Abubshait, H.A.; AlObaid, A.A.; Al-Muhimeed, T.I.; Abd-Rabboh, H.S.M.; Bahadur, A.; Li, W. Effective Heterointerface Combination of 1D/2D Co-NiS/S-g-C_3N_4 Heterojunction for Boosting Spatial Charge Separation with Enhanced Photocatalytic Degradation of Organic Pollutants and Disinfection of Pathogens. *Colloids Surf. A Physicochem. Eng. Asp.* **2021**, *628*, 127390. [CrossRef]
48. Iqbal, S.; Amjad, A.; Javed, M.; Alfakeer, M.; Mushtaq, M.; Rabea, S.; Elkaeed, E.B.; Pashameah, R.A.; Alzahrani, E.; Farouk, A.-E. Boosted Spatial Charge Carrier Separation of Binary $ZnFe_2O_4$/S-g-C_3N_4 Heterojunction for Visible-Light-Driven Photocatalytic Activity and Antimicrobial Performance. *Front. Chem.* **2022**, *10*, 975355. [CrossRef]
49. Baral, A.; Satish, L.; Das, D.P.; Sahoo, H.; Ghosh, M.K. Construing the Interactions between MnO_2 Nanoparticle and Bovine Serum Albumin: Insight into the Structure and Stability of a Protein–Nanoparticle Complex. *New J. Chem.* **2017**, *41*, 8130–8139. [CrossRef]
50. Huang, P.; Li, Z.; Hu, H.; Cui, D. Synthesis and Characterization of Bovine Serum Albumin-Conjugated Copper Sulfide Nanocomposites. *J. Nanomater.* **2010**, *2010*, 641545. [CrossRef]
51. Ballottin, D.; Fulaz, S.; Souza, M.L.; Corio, P.; Rodrigues, A.G.; Souza, A.O.; Gaspari, P.M.; Gomes, A.F.; Gozzo, F.; Tasic, L. Elucidating Protein Involvement in the Stabilization of the Biogenic Silver Nanoparticles. *Nanoscale Res. Lett.* **2016**, *11*, 313. [CrossRef] [PubMed]
52. Banerjee, V.; Das, K.P. Interaction of Silver Nanoparticles with Proteins: A Characteristic Protein Concentration Dependent Profile of SPR Signal. *Colloids Surf. B Biointerfaces* **2013**, *111*, 71–79. [CrossRef] [PubMed]
53. What Is the Difference Between Ordinary Light and Laser Light? Available online: https://www.aydinlatma.org/en/what-is-the-difference-between-ordinary-light-and-laser-light.html (accessed on 7 May 2022).
54. Kędziora, A.; Speruda, M.; Krzyżewska, E.; Rybka, J.; Łukowiak, A.; Bugla-Płoskońska, G. Similarities and Differences between Silver Ions and Silver in Nanoforms as Antibacterial Agents. *Int. J. Mol. Sci.* **2018**, *19*, 444. [CrossRef]
55. Djebbar, K.; Zertal, A.; Sehili, T. Photocatalytic Degradation of 2,4-Dichlorophenoxyacetic Acid and 4-Chloro-2-Methylphenoxyacetic Acid in Water by Using TiO_2. *Environ. Technol.* **2006**, *27*, 1191–1197. [CrossRef]

56. Helz, G.R.; Zepp, R.G.; Crosby, D.G. (Eds.) *Aquatic and Surface Photochemistry*; Lewis Publishers: Boca Raton, FL, USA, 1994; ISBN 978-0-87371-871-4.
57. Kumar, A.; Sharma, G.; Naushad, M.; Al-Muhtaseb, A.H.; Kumar, A.; Hira, I.; Ahamad, T.; Ghfar, A.A.; Stadler, F.J. Visible Photodegradation of Ibuprofen and 2,4-D in Simulated Waste Water Using Sustainable Metal Free-Hybrids Based on Carbon Nitride and Biochar. *J. Environ. Manag.* **2019**, *231*, 1164–1175. [CrossRef]
58. Ejeta, S.Y.; Imae, T. Photodegradation of Pollutant Pesticide by Oxidized Graphitic Carbon Nitride Catalysts. *J. Photochem. Photobiol. A Chem.* **2021**, *404*, 112955. [CrossRef]
59. Gu, J.; Chen, H.; Jiang, F.; Wang, X.; Li, L. All-Solid-State Z-Scheme Co_9S_8/Graphitic Carbon Nitride Photocatalysts for Simultaneous Reduction of Cr(VI) and Oxidation of 2,4-Dichlorophenoxyacetic Acid under Simulated Solar Irradiation. *Chem. Eng. J.* **2019**, *360*, 1188–1198. [CrossRef]
60. Zhong, J.; Ahmed, Y.; Carvalho, G.; Wang, Z.; Wang, L.; Mueller, J.F.; Guo, J. Simultaneous Removal of Micropollutants, Antibiotic Resistant Bacteria, and Antibiotic Resistance Genes Using Graphitic Carbon Nitride under Simulated Solar Irradiation. *Chem. Eng. J.* **2021**, *433*, 133839. [CrossRef]
61. General Chemistry. 2022. 1417.
62. Shi, T.; Wei, Q.; Wang, Z.; Zhang, G.; Sun, X.; He, Q.-Y. Photocatalytic Protein Damage by Silver Nanoparticles Circumvents Bacterial Stress Response and Multidrug Resistance. *MSphere* **2019**, *4*, e00175-19. [CrossRef] [PubMed]
63. Suryawanshi, V.D.; Walekar, L.S.; Gore, A.H.; Anbhule, P.V.; Kolekar, G.B. Spectroscopic Analysis on the Binding Interaction of Biologically Active Pyrimidine Derivative with Bovine Serum Albumin. *J. Pharm. Anal.* **2016**, *6*, 56–63. [CrossRef]
64. Jaiswal, V.D.; Dongre, P.M. Biophysical Interactions between Silver Nanoparticle-Albumin Interface and Curcumin. *J. Pharm. Anal.* **2020**, *10*, 164–177. [CrossRef] [PubMed]
65. Ershov, V.; Tarasova, N.; Ershov, B. Evolution of Electronic State and Properties of Silver Nanoparticles during Their Formation in Aqueous Solution. *Int. J. Mol. Sci.* **2021**, *22*, 10673. [CrossRef] [PubMed]
66. Klueh, U.; Wagner, V.; Kelly, S.; Johnson, A.; Bryers, J.D. Efficacy of Silver-Coated Fabric to Prevent Bacterial Colonization and Subsequent Device-Based Biofilm Formation. *J. Biomed. Mater. Res.* **2000**, *53*, 621–631. [CrossRef]
67. Dakal, T.C.; Kumar, A.; Majumdar, R.S.; Yadav, V. Mechanistic Basis of Antimicrobial Actions of Silver Nanoparticles. *Front. Microbiol.* **2016**, *7*, 1831. [CrossRef]
68. Rai, M.K.; Deshmukh, S.D.; Ingle, A.P.; Gade, A.K. Silver Nanoparticles: The Powerful Nanoweapon against Multidrug-Resistant Bacteria: Activity of Silver Nanoparticles against MDR Bacteria. *J. Appl. Microbiol.* **2012**, *112*, 841–852. [CrossRef]
69. Franci, G.; Falanga, A.; Galdiero, S.; Palomba, L.; Rai, M.; Morelli, G.; Galdiero, M. Silver Nanoparticles as Potential Antibacterial Agents. *Molecules* **2015**, *20*, 8856–8874. [CrossRef]
70. Zaheer, Z.; Kosa, S.A.; Akram, M. Interactions of Ag^+ Ions and Ag-Nanoparticles with Protein. A Comparative and Multi Spectroscopic Investigation. *J. Mol. Liq.* **2021**, *335*, 116226. [CrossRef]
71. Jaiswar, A.; Varshney, D.; Adholeya, A.; Prasad, P. Do Environmentally Induced DNA Variations Mediate Adaptation in Aspergillus Flavus Exposed to Chromium Stress in Tannery Sludge? *BMC Genomics* **2018**, *19*, 868. [CrossRef] [PubMed]
72. Zhang, X.; Yang, C.-W.; Yu, H.-Q.; Sheng, G.-P. Light-Induced Reduction of Silver Ions to Silver Nanoparticles in Aquatic Environments by Microbial Extracellular Polymeric Substances (EPS). *Water Res.* **2016**, *106*, 242–248. [CrossRef] [PubMed]
73. Feng, T.; Liang, J.; Ma, Z.; Li, M.; Tong, M. Bactericidal Activity and Mechanisms of BiOBr-AgBr under Both Dark and Visible Light Irradiation Conditions. *Colloids Surf. B Biointerfaces* **2018**, *167*, 275–283. [CrossRef]
74. Zhu, L.; He, C.; Huang, Y.; Chen, Z.; Xia, D.; Su, M.; Xiong, Y.; Li, S.; Shu, D. Enhanced Photocatalytic Disinfection of *E. coli* 8099 Using Ag/BiOI Composite under Visible Light Irradiation. *Sep. Purif. Technol.* **2012**, *91*, 59–66. [CrossRef]
75. Ma, S.; Zhan, S.; Xia, Y.; Wang, P.; Hou, Q.; Zhou, Q. Enhanced Photocatalytic Bactericidal Performance and Mechanism with Novel Ag/ZnO/g-C_3N_4 Composite under Visible Light. *Catal. Today* **2019**, *330*, 179–188. [CrossRef]
76. Hatchard, C.G.; Parker, C.A.; Bowen, E.J. A New Sensitive Chemical Actinometer—II. Potassium Ferrioxalate as a Standard Chemical Actinometer. *Proc. R. Soc. Lond. Ser. A. Math. Phys. Sci.* **1956**, *235*, 518–536. [CrossRef]
77. Watanabe, K.; Matsuda, S.; Cuevas, C.A.; Saiz-Lopez, A.; Yabushita, A.; Nakano, Y. Experimental Determination of the Photooxidation of Aqueous I—As a Source of Atmospheric I2. *ACS Earth Space Chem.* **2019**, *3*, 669–679. [CrossRef]
78. Quick StartTM Bradford Protein Assay. Available online: https://www.bio-rad.com/en-ca/product/quick-start-bradford-protein-assay?ID=5ec149ee-0cd1-468b-8651-a2fe9de6944d (accessed on 19 June 2022).
79. Whalen, P. *Cellular Adenosine Triphosphate (CATP) Assay Procedure*; LuminUltra Technologies Ltd.: Fredericton, NB, Canada, 2012.
80. Colilert—IDEXX, US. Available online: https://www.idexx.com/en/water/water-products-services/colilert/ (accessed on 19 June 2022).

catalysts

MDPI

Article

Preparation of TiO_2-CNT-Ag Ternary Composite Film with Enhanced Photocatalytic Activity via Plasma-Enhanced Chemical Vapor Deposition

Jianghua Lang [1], Kazuma Takahashi [1], Masaru Kubo [2] and Manabu Shimada [2,*]

[1] Department of Chemical Engineering, Graduate School of Engineering, Hiroshima University, 1-4-1, Kagamiyama, Higashi-Hiroshima 739-8527, Hiroshima, Japan; d196006@hiroshima-u.ac.jp (J.L.); ka-864-@ezweb.ne.jp (K.T.)

[2] Chemical Engineering Program, Graduate School of Advanced Science and Engineering, Hiroshima University, 1-4-1, Kagamiyama, Higashi-Hiroshima 739-8527, Hiroshima, Japan; mkubo@hiroshima-u.ac.jp

* Correspondence: smd@hiroshima-u.ac.jp

Abstract: In this study, a TiO_2-CNT-Ag ternary composite film was successfully synthesized using the plasma-enhanced chemical vapor deposition method by simultaneously feeding a carbon nanotube (CNT)/Ag suspension and titanium tetraisopropoxide gas. The prepared TiO_2-CNT-Ag film was characterized by scanning electron microscopy, transmission electron microscopy, energy-dispersive X-ray spectroscopy, X-ray photoelectron spectroscopy, X-ray diffraction, and ultraviolet-visible spectroscopy. Moreover, the Ag/Ti ratio of the film was confirmed using an inductivity-coupled plasma optical emission spectrometer. The performance of the TiO_2-composite film for the degradation of rhodamine 6G under simulated solar light irradiation was evaluated. The rate constant of the prepared TiO_2-CNT-Ag for rhodamine 6G degradation was approximately 1.8 times greater than that of prepared TiO_2. This result indicates that the addition of CNT and Ag significantly improved the photocatalytic activity of the prepared films.

Keywords: thin film deposition; nanoparticles; aerosols; photocatalysis

Citation: Lang, J.; Takahashi, K.; Kubo, M.; Shimada, M. Preparation of TiO_2-CNT-Ag Ternary Composite Film with Enhanced Photocatalytic Activity via Plasma-Enhanced Chemical Vapor Deposition. *Catalysts* **2022**, *12*, 508. https://doi.org/10.3390/catal12050508

Academic Editor: Jorge Bedia

Received: 28 March 2022
Accepted: 26 April 2022
Published: 30 April 2022

1. Introduction

In recent years, photocatalysis has attracted considerable research attention owing to its wide range of applications, including the construction of organic compounds [1], pollution alleviation [2] and energy storage [3]. However, many photocatalytic materials exhibit drawbacks that hinder their application, necessitating the development of heterojunction materials [4]. Various heterojunction materials have demonstrated advantages properties such as fast charge separation [5], favorable electron transfer [6], hindering of electron-hole recombination [7], and high surface area [8]. TiO_2 is a notable material because of its chemical stability, nontoxicity, and high photocatalytic activity [9,10]. However, its photocatalytic activity is limited by the following factors: First, due to the large band gap (3.2 eV) of TiO_2 (anatase phase), its photocatalytic activity can be initiated only under ultraviolet light irradiation (3–5% of the total solar spectrum) [11,12]. Second, the fast recombination of electron–hole pairs results in low efficiency for photocatalysis [13]. Hence, TiO_2-based heterojunction materials have attracted considerable attention for their enhanced photocatalytic activity.

Researchers have successfully fabricated TiO_2 composites by combining TiO_2 with Au [14], Ag [15], N [16], carbon nanotubes (CNTs) [17], etc. Among these materials, CNTs are promising material owing to their high electrical and thermal conductivity and large surface area [18]. Zhou et al. [19] synthesized a composite of TiO_2/single-walled carbon nanotubes (SWCNTs) by a simple solvothermal technique. At the TiO_2–SWCNT interface, the Fermi level was positively transferred, resulting in electron flow from TiO_2

(higher Fermi level) to the CNTs (lower Fermi level) to align the Fermi energy levels. Thus, the activated electrons were stabilized on the SWCNTs, and the recombination of electron–hole pairs was mitigated. Another study has reposited the existence of excess charge between the metal and semiconductor [20]. Owing to the surface plasmon resonance of Ag, TiO_2-Ag can also be excited under visible light.

Recently, TiO_2/CNT/Ag ternary composites were reported to exhibit excellent performance in dye degradation. Wang et al. [21] reported a ternary composite of Ag-CNT/TiO_2 generated by the sol–gel and photoreduction methods. The ternary structure was achieved by decorating CNTs with both TiO_2 nanoparticles and Ag nanoparticles. The photocatalytic activity of Ag-CNT/TiO_2 was approximately twice that of P25. Koo et al. [22] reported an Ag-TiO_2-CNT composite fabricated by photochemical reduction. The Ag doped TiO_2 nanoparticles were uniformly distributed on the CNT surface. The performance of Ag-TiO_2-CNT in the degradation of methylene blue was enhanced compared to that of Ag-TiO_2. Notably, CNTs are always used as the support material for coating with TiO_2 and Ag nanoparticles. Few researchers have focused on the preparation of TiO_2-based composite films, utilizing Ag, TiO_2, and CNTs. A catalyst film is sometimes more advantageous than a nanosized catalyst powder because it addresses several challenges. For example, this strategy (1) eliminates the need for catalyst separation and filtration, (2) enables application in continuous flow systems, and (3) reduces agglomeration at various catalyst loadings [23]. Various methods have been explored to prepare TiO_2-based composite films, such as the sol–gel [24], photoreduction [25], pulsed laser deposition [26], liquid-phase deposition [27], chemical vapor deposition [28], and plasma-enhanced chemical vapor deposition (PECVD) methods [10]. PECVD is regarded as the most versatile method for the relatively low-cost fabrication of composite films with a uniform morphology and good step coverage. Recently, we successfully prepared TiO_2-Ag binary films using this technique, where $AgNO_3$ aqueous and titanium tetraisopropoxide (TTIP) vapors were supplied simultaneously [15]. The obtained composite film exhibited a higher photocatalytic activity than TiO_2. Through addition of CNTs, we presume that ternary films can exhibit a higher photocatalytic activity than binary films.

In the available literature on ternary composites, a binary composite was always synthesized first. Recently, CNT-Ag composites have been prepared [29,30], which offers a potential route to generate TiO_2-CNT-Ag ternary composites. Nevertheless, we assume that it is possible to simultaneously supply different materials without preparing a binary composite. Furthermore, compared to our work on the preparation of TiO_2-Ag using liquid raw materials, we developed a different process to deposit composite films. We fabricated composite films using solid materials. We have also prepared TiO_2 films embedded with SiO_2 nanoparticles by PECVD [31], where the SiO_2 solid material and TTIP vapors were supplied simultaneously. Thus, we have demonstrated the feasibility of generating TiO_2-CNT-Ag ternary composite films by supplying CNTs and Ag solid mixtures.

In this study, TiO_2-CNT-Ag ternary composite films were prepared using PECVD. During preparation, solid suspensions of the CNT/Ag mixture were sprayed with a nozzle and heated to obtain an aerosol of CNTs and Ag nanoparticles. The aerosol and TTIP vapors were simultaneously fed into a plasma reactor, where the TiO_2-CNT-Ag films were prepared through a simple operation. The ternary structure of the composite films was confirmed by characterizing the morphology and optical properties and through elemental analysis. The photocatalytic activity of the films was evaluated by degrading rhodamine 6G under simulated solar light irradiation.

2. Results and Discussion

2.1. Film Characterization

The TiO_2, TiO_2-CNT, TiO_2-Ag, and TiO_2-CNT-Ag films were generated by supplying only TTIP, a TTIP and 0.25 wt% CNT suspension, a TTIP and 0.1 wt% Ag suspension, and TTIP and CNT/Ag mixed suspensions (0.25 wt%/0.1 wt%). The corresponding products are denoted as T, T-C, T-A, and T-C-A0, respectively. The morphology of the

CNTs is shown in Figure 1a. Figure 1b,c shows the scanning electron microscopy (SEM) images of the prepared T-C films containing cracks. The cracks sometimes appear on the periphery of the annealed film surface, which could be attributed to the heat treatment. The CNTs are exposed, confirming their successful incorporation into the film. Further, the morphologies of the prepared T, T-C, T-A, and T-C-A films are shown in Figure 2. Figure 2a–d show the cross-sectional morphologies of the films, which are apparently dense with a quasi-uniform thickness of approximately 1 μm. The addition of CNTs and Ag does not affect the film thickness. Compared to the T films, the CNTs are attached to TiO_2 on the surfaces of the T-C and T-C-A0 films (Figure 2b,d). Only a few CNTs are observed, which can be attributed to the effective dispersion of the CNT suspension during deposition. However, the presence of Ag nanoparticles is not confirmed in the T-A and T-C-A0 (Figure 2c,d) films. The cross-sectional images of the TiO_2 film show that it consists of large columnar grains. This leads to the formation of a relative regular surface, as shown in Figure 2e. However, with the supply of CNTs and Ag particles, some parts of the film surface become irregular, like the strip structure in Figure 2f, large aggregate particles in Figure 2g, and both phenomena in Figure 2e. The use of additives may lead to the growth of a core-shell structure, which results in the deposition of an irregular structure. In addition, we performed element mapping of the irregular structure that appeared in the other samples prepared under the same conditions (Figure 3). The results confirmed the presence and distribution of Ti, O, Ag, and C. The aggregated particles consisted of Ag and TiO_2, and the strip structures consisted of CNT and TiO_2. Thus, the results, confirmed that TiO_2 was embedded with CNTs and Ag nanoparticles. Furthermore, the influence of different concentrations of the CNT/Ag mixed suspensions, such as 0.25 wt%/0.05 wt%, 0.25 wt%/0.01 wt%, 0.01 wt%/0.1 wt%, and 0.5 wt%/0.1 wt% concentrations, on the T-C-A films surface was investigated. The products were denoted as T-C-A1, T-C-A2, T-C-A3, and T-C-A4, respectively. The SEM images of these films are shown in Figure 4. From the surface structure observations, the amount of CNTs appears to be strongly influenced by the concentration of the mixed suspension. However, considering that more CNTs were embedded into the films (Figure 1), it was difficult to define the amount of CNTs based on the SEM images. In addition, it was challenging to define the morphology of the Ag particles in all the T-C-A films, which may be due to the existence of the embedded structure and the small size of Ag particles.

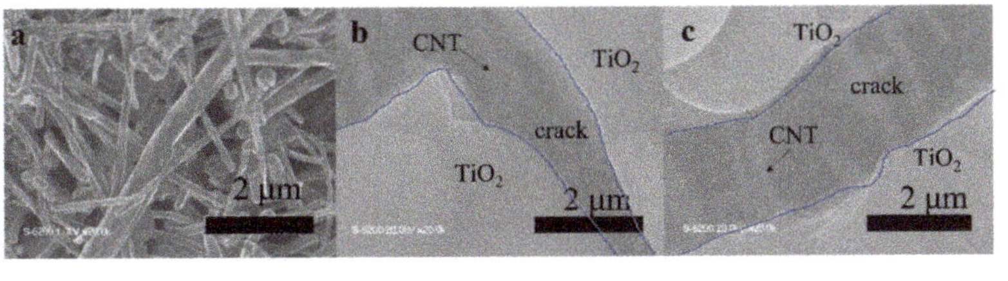

Figure 1. Scanning electron microscopy (SEM) images of: (**a**) CNTs and (**b**) and (**c**) prepared T-C film containing cracks.

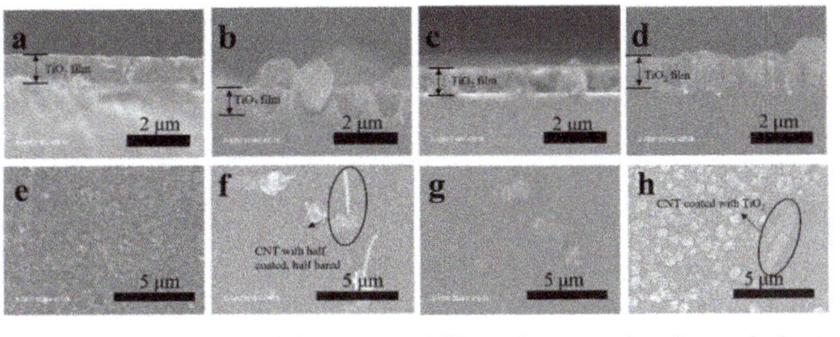

Figure 2. SEM images of the prepared films. Cross-sectional morphologies of (**a**) T, (**b**) T-C, (**c**) T-A, and (**d**) T-C-A0. Surface morphologies of (**e**) T, (**f**) T-C, (**g**) T-A, and (**h**) T-C-A0.

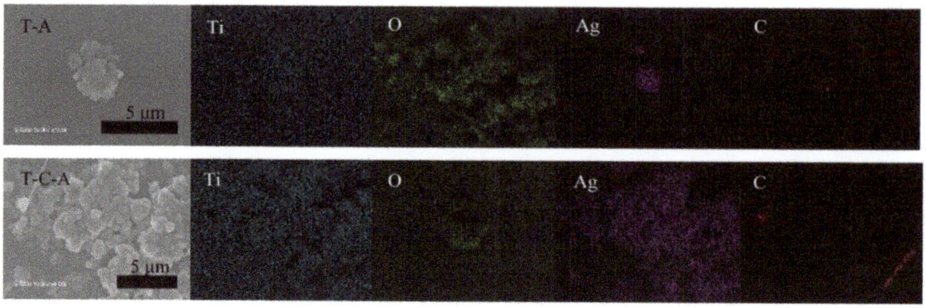

Figure 3. Elemental mapping of T-A and T-C-A.

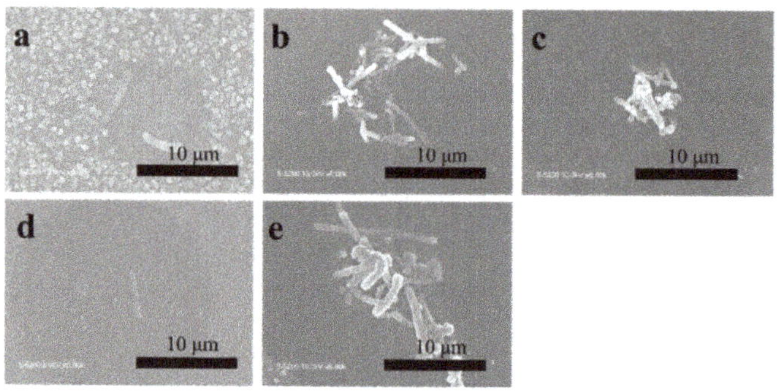

Figure 4. SEM images of T-C-A films deposited using different concentrations of CNT/Ag mixed suspensions: (**a**) T-C-A0 (0.25 wt%/0.1 wt%), (**b**) T-C-A1 (0.25 wt%/0.05 wt%), (**c**) T-C-A2 (0.25 wt%/0.01 wt%), (**d**) T-C-A3 (0.01 wt%/0.1 wt%), and (**e**) T-C-A4 (0.5 wt%/0.1 wt%).

Next, transmission electron microscopy energy dispersive X-ray (TEM-EDS) spectroscopy was performed to confirm the presence of Ag in the prepared T-C-A0 film, which was selected as a representative sample. The observed sample was scratched off from the films and collected by a micro grid, therefore, the original structure was not maintained. Figure 5a depicts the TiO_2 and Ag particles appeared to be attached to the CNT surface. However, the CNTs were embedded in the TiO_2 film, as shown in Figure 1. Figure 5b shows a magnified version of the area in the yellow square in Figure 5a, confirming that the TiO_2 and Ag nanoparticles are aggregated. Further, the EDS analysis of the area in the red square in Figure 5b proves the presence of Ag in the film.

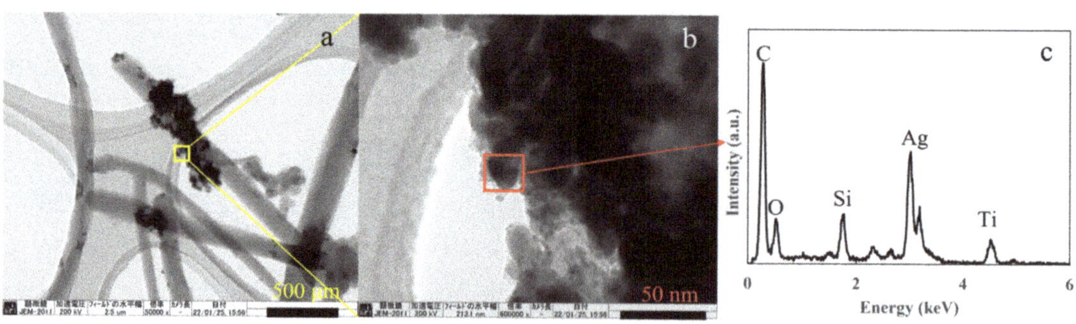

Figure 5. Transmission electron microscopy (TEM) images of the prepared T-C-A0 film: (**a**) CNTs with Ag and TiO_2 particles attached, (**b**) magnified version of the area in the yellow square in (**a**), and (**c**) energy dispersive X-ray spectroscopy (EDS) results corresponding to the red square in (**b**).

Next, X-ray photoelectron spectroscopy (XPS) measurements were performed to examine the composite spices and states in the T-C-A0 film, and the results are presented in Figure 6. The peaks corresponding to Ti 2P, Ag 3d, and O 1s are observed in the wide spectrum of this film (Figure 6a). Figure 6b illustrates the typical Ti 2p spectrum, with Ti $2p_{3/2}$ at 459.0 eV and Ti $2p_{1/2}$ at 464.7 eV. Figure 6c provides evidence of the presence of metallic Ag, with Ag $3d_{5/2}$ at 368.2 eV and Ag $3d_{3/2}$ at 374.1 eV. In addition, the two decomposed peaks from the O 1s profile correspond to the lattice oxygen in TiO_2 (530.5 eV) and absorbed H_2O (532.1 eV), respectively.

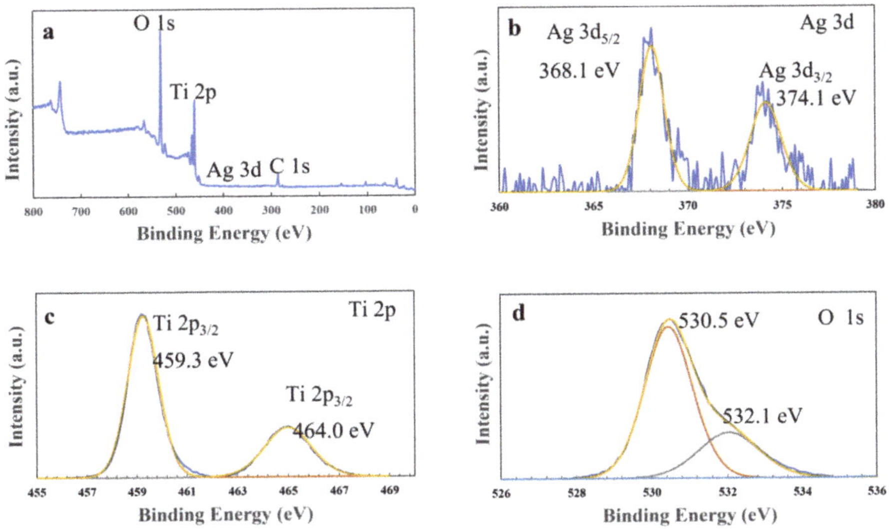

Figure 6. X-ray photoelectron spectra of the T-C-A0 film: (**a**) survey, (**b**) Ag 3d, (**c**) Ti 2p, and (**d**) O 1s spectra.

Figure 7a shows the X-ray diffraction (XRD) patterns of the T, T-C, T-A, and T-C-A0 films. The diffraction peaks marked with "A" correspond to the (101), (004), (112), (200), and (211) planes of anatase, indicating that the developed products existed in an anatase state. There is no peak corresponding to Ag nanoparticles in the patterns of the T-C-A0 and T-A films, which may be because of the low content of Ag nanoparticles in these films. Therefore, the major characteristic peak of Ag at 38.2° (111) overlaps with the characteristic peaks of TiO_2 [22]. The same phenomenon occurs for the CNTs. The diffraction peak of CNTs at 25.9° is attributed to anatase TiO_2. This peak shielded the CNTs at 26.1°, which is not observed in the XRD patterns of the T-C and T-C-A0 samples, suggesting that the film contained a low amount of CNTs compared to TiO_2 [32].

Generally, the abovementioned phase transformation is initiated by annealing at ~600 °C. However, we obtained the only anatase when the samples were annealed at 600 °C. When the annealing temperature was increased to 700 °C, films with an anatase–rutile mixed phase were obtained. The XRD analysis of the films annealed at different temperature are presented in Figure 7b, JCPDS cards of Ag, SiO_2, anatase and rutile were used as the reference. Noticeably, only the anatase phase is evident for the T-C-A0 film annealed at 600 °C, and the peak of the rutile phase appears upon annealing at 700 °C. The phase transfer occurs between 600 and 700 °C. This result agrees well with previous findings [33].

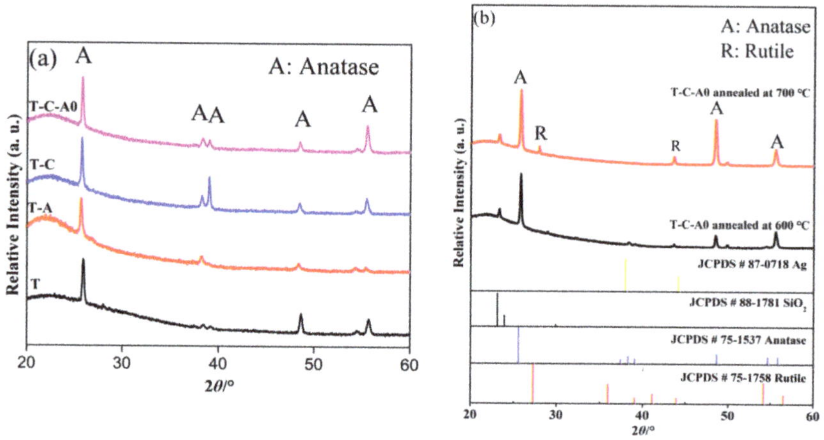

Figure 7. X-ray diffraction (XRD) patterns of (**a**) T and T-composite films and (**b**) T-C-A0 films annealed at different temperatures.

Figure 8a shows the absorption spectra of the T, T-C, T-A, and T-C-A0 films. For all of the films, the absorption increased sharply below 400 nm, probably due to the intrinsic band gap absorption of TiO_2 [34]. Compared with the T film, the absorption thresholds of the T-C, T-A, and T-C-A films extend to the longer wavelength range. This behavior can be attributed to the photosensitizing effect of CNTs and the local surface plasmon resonance (LSPR) effect of the Ag particles [35]. Furthermore, the energy level was presumably aligned in the composite film by the introduction of CNTs and Ag, which can be attributed to the charge transmission among the TiO_2, Ag, and CNTs [36]. The band gap energy can be estimated by employing the Tauc relation: $\alpha hv = B\,(hv - Eg)^{1/2}$, where α is the absorption coefficient, hv is the photon energy, Eg is the optical band gap, and B is the absorption constant for direct transitions. Figure 8b shows the plot of $(\alpha hv)^2$ versus photoenergy (hv), where the intercept of the tangent to the plot corresponds to the bandgap. The bandgap of the T film is 3.47 eV; however, it decreases to 3.39 eV for the T-C film, 3.31 eV for the T-A film, and 3.22 eV for the T-C-A film.

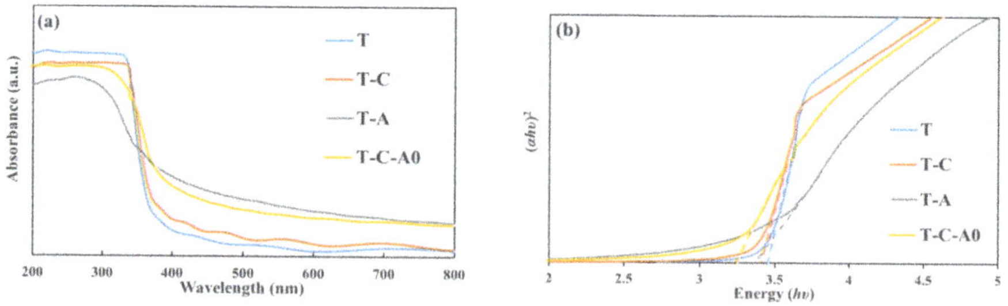

Figure 8. Optical properties of the T and T-composite film: (**a**) absorbance spectra and (**b**) $(\alpha hv)^2$ versus photon energy (hv) plot of the prepared films.

The ratio of Ag/Ti in the film was confirmed by inductivity coupled plasma optical emission spectrometry (ICP-OES) analysis. The results are presented in Table 1. The units of CNT/Ag (wt%/wt%) suspensions refer to the raw materials (relative values) and correspond to the conditions when deposit films were deposited. Ag/Ti (mg/mg) refer to the results of the ICP-OES analysis and corresponds to the content of the film (absolute values (mass)). Ag nanoparticles were present in all of the T-C-A films. However, the Ag/Ti ratio decreases with decreasing Ag concentration. This result indicates that the Ag content in the films was influenced by the Ag concentration.

Table 1. Elemental analysis of the T-C-A film using inductivity coupled plasma optical emission spectrometry.

Samples	Concentration of CNT/Ag Suspensions (wt%/wt%)	Ag/Ti (mg/mg)
T-C-A0	0.25/0.1	0.0182
T-C-A1	0.25/0.05	0.0139
T-C-A2	0.25/0.01	0.0118

2.2. Photocatalytic Activity of the Films

Figure 9 illustrates the photocatalytic degradation of aqueous rhodamine 6G by different films under simulated solar light. The films used in this section were annealed at 600 °C. The plot of C_t/C_0 versus the specified irradiation time (t) for different films is presented in Figure 9a. Furthermore, we prepared the ln (C_t/C_0) versus t plot (Figure 9b) to obtain the corresponding rate constant k from the slope of the fitting curve (Figure 9c). The photocatalytic activity of the film under simulated solar light decreased in the following order: T-C-A0 > T-A > T-C > T. Compared with the T film deposited by PECVD, the composite films performed better. Furthermore, the k of the prepared T-C-A0 film is approximately 1.8 times higher than that of the prepared T film.

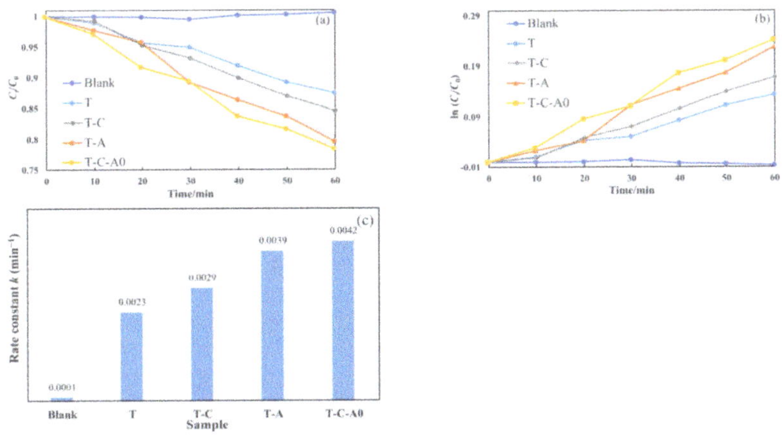

Figure 9. Photocatalytic activity for the degradation of rhodamine 6G by different films under simulated solar light: (**a**) degradation with respect to time, (**b**) ln (C_t/C_0) versus to time plot and (**c**) rate constants degradation for various samples.

In addition, Figure 10a presents the plot of C_t/C_0 as a function of irradiation time for the T-C-A0 films annealed at different temperatures. We also plotted ln (C_t/C_0) against t (Figure 10b) to obtain the corresponding rate constant k from the slope of the fitting curve (Figure 10c). The efficiency of the T-C-A0 films annealed at 700 °C for rhodamine 6G degradation is higher than that of the T-C-A0 films annealed at 600 °C, which can be attributed to the phase transfer of the crystalline phase at different annealing temperatures.

To the best of our knowledge, other researchers have also estimated the photocatalytic activity based on the rate constant. Soltanieh et al. prepared a TiO_2/Ag^0@MWCNTs composite with rate constant of 0.07675 min^{-1} [37], which is higher than that of their prepared TiO_2/Ag binary structure. Wang et al. illustrated an Ag-CNT/TiO_2 composite with a rate constant 0.03052 min^{-1} [21], which is higher than that of their prepared CNT/TiO_2 binary structure. Numerous factors affect the photocatalytic activity, such as the concentration of dyes, power of the irradiation light, and morphology of the catalyst. Thus, it is difficult to compare the activities of the catalysts prepared in different studies. Nevertheless, we proved that the fabricated ternary composites show a higher rate constant than that of fabricated binary composite (TiO_2-CNT and TiO_2-Ag). Therefore, the developed process has significant potential to improve the performance of photocatalysts.

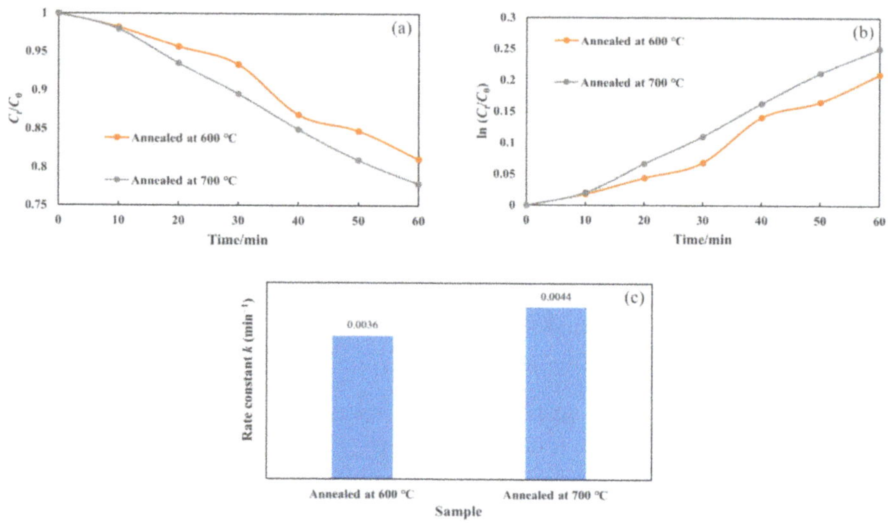

Figure 10. Photocatalytic degradation of rhodamine 6G by T-C-A films annealed at different temperatures: (**a**) degradation with respect to time, (**b**) ln (C_t/C_0) versus to time plot, and (**c**) corresponding rate constants of degradation for samples.

These results confirm that the photocatalytic activity of TiO_2 is significantly enhanced by embedding CNTs and Ag particles into the TiO_2 film. In addition, the ternary composites (T-C-A) show higher photocatalytic activity than binary structure films (T-C and T-A). Thus, we believe that the photoreaction system in this study could not be simply defined as two binary systems. However, we could not determine the detailed structure of this system during photoreaction. Moreover, the photocatalytic mechanisms are so complex that defining them only by film structure is insufficient. Hence, we assume the possible mechanism for the ternary system during photoreaction. In general, electrons are generated by the light irradiation of Ag nanoparticles due to the LSPR effect. These photoexcited electrons can be transferred to the conduction band of TiO_2 through the Schottky barrier formed at the Ag–TiO_2 interface [18]. Simultaneously, TiO_2 can act as a photosensitizer, where the valence band electrons can be excited to a new sub-bandgap state that is formed by the TiO_2-CNT-Ag ternary structure [38]. In addition, CNTs act as electron acceptors, and the electrons on TiO_2 and Ag can also be transferred to the CNTs, which reduces electron-hole recombination. Thus, charge carriers contribute to the formation of active species (O^{2-}, ·OH), thereby enhancing the photocatalytic activity [39]. We propose that TiO_2 acts as the main photosensitizer under solar light irradiation due to the special structure of the ternary films and that the activated electrons can easily be transferred to the CNTs and Ag, reducing the recombination of electron–hole pairs, as shown in Figure 11.

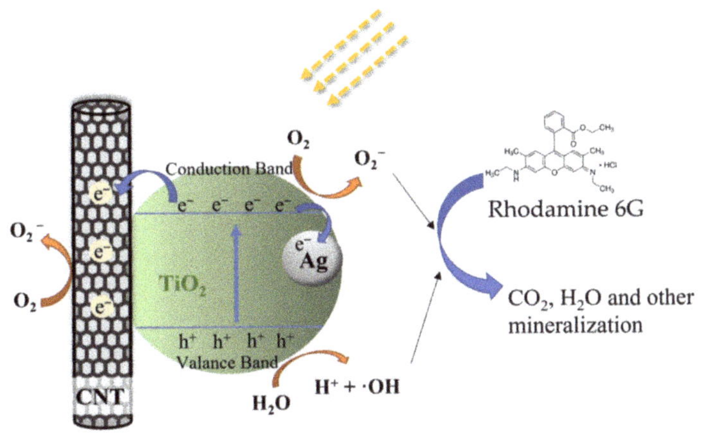

Figure 11. Possible schematic explaining the photocatalytic activity of the T-C-A ternary film.

3. Experimental

3.1. Experimental Process

The CNT suspensions (0.5 wt%) were prepared using 0.5 g of CNTs (Sigma-Aldrich, St. Louis, MO, USA, diameter: 110–170 nm and length: 5–9 μm) dispersed in 95 mL of deionized water with Triton X-100 as the dispersant. The Ag nanoparticle suspensions (0.1 wt%) were synthesized via co-precipitation [40]. Subsequently, 0.18 g of $AgNO_3$ (Nacalai Tesque, Inc., Tokyo, Japan) was dissolved into 100 mL of deionized water, and 2 mL of 0.01 g/mL of trisodium citric acid (Wako Pure Chemical Industries, Ltd., Osaka, Japan) aqueous solution was added. The mixture was then stirred at 100 °C for 1 h and cooled to ~25 °C. A solid suspension for spraying was prepared by mixing water, the CNT suspension, and the Ag nanoparticle suspension with a certain concentration.

Figure 12 shows the experimental setup consisting of different feeding systems [15,31]. A solid suspension (10 mL) was supplied by a syringe pump (YSP-301, YMC. Co., Ltd., Kyoto, Japan) with a capillary tube. A two-fluid nozzle was used to spray the solid suspension, with He gas flowing at a rate of 1000 sccm. The transport tube was preheated using a ribbon heater to dry the solid suspension. The obtained aerosol particles of CNT/Ag were supplied to the plasma cavity. On the other hand, TTIP (Tokyo Chemical Industry Co., Ltd., Tokyo, Japan) was vaporized in a bubbler (45 °C) with He gas at a flow rate of 50 sccm. Meanwhile, O_2 gas with a flow rate of 50 sccm and He gas with a flow rate of 400 sccm were employed to transfer the vaporized TTIP. The quartz tube (inner diameter: 25.4 mm; length: 300 mm) was wrapped and supported by a cavity. Furthermore, we designed a recessed part in the middle of the quartz tube to place a quartz mesh and hold a glass substrate, on which the films could be deposited. The plasma was produced by 190 W microwaves (2.45 GHz) in a quartz tube. A vacuum pump (Pascal 2025C1, Pfeiffer Vacuum, Inc., Asslar, Germany) was used to maintain all the equipment under vacuum pressure before feeding and at ~6 kPa after the materials were fed. The films were deposited on a quartz glass plate (1 × 1 cm) through the simultaneous feeding of the TTIP and CNT/Ag suspensions for 15 min. The obtained films were annealed at 600 °C and 700 °C in an N_2 atmosphere.

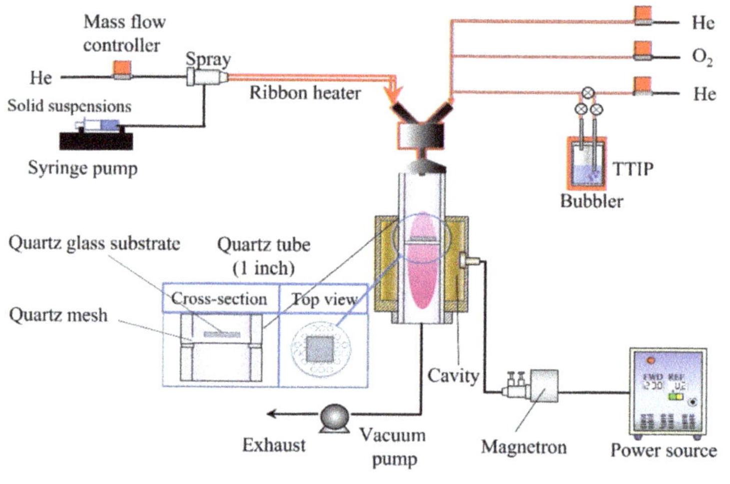

Figure 12. Experimental setup for PECVD.

3.2. Film Characterization

The morphology of the films was analyzed by SEM (S-5200, Hitachi High Technologies, Tokyo, Japan) and TEM (JEM-2010, JEOL, Tokyo, Japan), in conjunction with EDS (JED-2300T, JEOL, Tokyo, Japan). The elemental analysis was performed by XPS (ESCA-3400, SHIMADAZU, Kyoto, Japan). The crystal structures of the films were investigated by XRD (RINT-2100, Rigaku, Tokyo, Japan), using Cu-K_{α} radiation (λ = 1.5406 Å). The UV–Vis spectra of the films were recorded using a V-650 UV–Vis spectrophotometer (Jasco, Tokyo, Japan).

Further, the elemental analysis of the films was conducted by employing an ICP-OES instrument (SPS3000, Hitachi High Technologies, Tokyo, Japan). Each film was placed in a polypropylene sample tube and immersed in 1 mL of ethanol. Ethanol was used as a lubricant and collector for the scraped film powder. A micro grinder equipped with an electroplated diamond burr was used to scrape the film off the substrate. The obtained powder was dried and dissolved in HF. Excess HF was neutralized by H_3BO_4, and the solvents were then filtered and used for ICP-OES analysis.

3.3. Photocatalytic Activity

To investigate the photocatalytic activities of the films, the degradation of rhodamine 6G by the films under simulated solar-light irradiation was evaluated. The spectrum of the light used to activate the films is recorded by spectro multi-channel photo detector (MCPD-3000, Otsuka Electronics, Osaka, Japan), and the result is shown in Figure 13. Firstly, the films were immersed in 3 mL of a rhodamine 6G aqueous solution (5 mg/L), which was held in the cuvette cell and left in the dark for 30 min to reach adsorption equilibrium. Subsequently, the catalytic process was activated through irradiation with simulated solar light (300–800 nm). Using the equation for the absorbance and concentration of rhodamine 6G, the maximum absorbances of rhodamine 6G at different times were observed and converted into concentrations of rhodamine 6G. The concentrations before and after irradiation were determined through UV–Vis spectrophotometry (V-650, Jasco, Tokyo, Japan) every 10 min. Furthermore, the degradation efficiency was determined by C_t/C_0, and the rate constant k was calculated as follows:

$$\ln \frac{C_0}{C_t} = kt \tag{1}$$

where C_0 and C_t represent the concentrations of rhodamine 6G before and after irradiation, respectively, and t is the irradiation time.

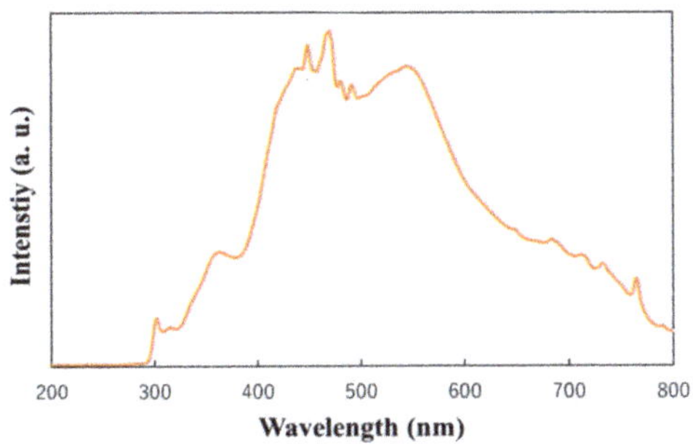

Figure 13. Light spectrum used to activate the films.

4. Conclusions

In this study, a T-C-A ternary composite film was successfully prepared using a novel and simple process with relatively good coverage. The XRD spectrum revealed that the structure of T-C-A had a pure anatase phase after annealing at 600 °C and an anatase–rutile mixed phase after annealing at 700 °C. The SEM images exhibited good coverage with the CNT embedded structure of the T-C-A film, and the thickness was approximately 1 μm. The TEM-EDS images indicated the presence of Ag. The XPS analysis proved that Ag existed in the metallic state. The UV–Vis spectrum confirmed that the absorbance of T-C-A extended to longer wavelength. The ICP-OES results indicated the presence of Ag. Moreover, the photocatalytic activity of the prepared T-C-A film was approximately 1.8 times higher than that of the prepared T film. The expected ternary structure and photocatalytic ability of T-C-A were determined using PECVD. We believe that the investigated process has immense potential for use in the preparation of films with different materials.

Author Contributions: Conceptualization, M.S.; methodology, J.L. and K.T.; formal analysis, J.L.; writing—J.L.; writing—Review and editing, J.L., M.S. and M.K.; supervision: M.S.; funding acquisition, M.S. All authors have read and agreed to the published version of the manuscript.

Funding: This work is partly supported by Japan Society for the Promotion of Science KAKENHI grant numbers 21K04750.

Data Availability Statement: Data are contained within the article.

Acknowledgments: We would like to thank M. Maeda for assistance with the TEM analysis.

Conflicts of Interest: There are no conflict of interest.

References

1. Zhu, X.; Lin, Y.; San Martin, J.; Sun, Y.; Zhu, D.; Yan, Y. Lead halide perovskites for photocatalytic organic synthesis. *Nat. Commun.* **2019**, *10*, 2843. [CrossRef] [PubMed]
2. Tobaldi, D.M.; Dvoranova, D.; Lajaunie, L.; Rozman, N.; Figueiredo, B.; Seabra, M.P.; Skapin, A.S.; Calvino, J.J.; Brezova, V.; Labrincha, J.A. Graphene-TiO_2 hybrids for photocatalytic aided removal of VOCs and nitrogen oxides from outdoor environment. *Chem. Eng. J.* **2021**, *405*, 126651. [CrossRef] [PubMed]
3. Yadav, A.A.; Hunge, Y.M.; Kulkarni, S.B. Synthesis of multifunctional $FeCo_2O_4$ electrode using ultrasonic treatment for photocatalysis and energy storage applications. *Ultrason. Sonochem.* **2019**, *58*, 104663. [CrossRef] [PubMed]
4. Wang, Z.; Lin, Z.; Shen, S.; Zhong, W.; Cao, S. Advances in designing heterojunction photocatalytic materials. *Chin. J. Catal.* **2021**, *42*, 710–730. [CrossRef]
5. He, S.; Rong, Q.; Niu, H.; Cai, Y. Platform for molecular-material dual regulation: A direct Z-scheme MOF/COF heterojunction with enhanced visible-light photocatalytic activity. *Appl. Catal. B Environ.* **2019**, *247*, 49–56. [CrossRef]
6. Zhu, Y.; Liu, Y.; Ai, Q.; Gao, G.; Yuan, L.; Fang, Q.; Tian, X.; Zhang, X.; Egap, E.; Ajayan, P.M.; et al. In Situ Synthesis of Lead-Free Halide Perovskite–COF Nanocomposites as Photocatalysts for Photoinduced Polymerization in Both Organic and Aqueous Phases. *ACS Mater. Lett.* **2022**, *4*, 464–471. [CrossRef]
7. Hieu, V.Q.; Phung, T.K.; Nguyen, T.-Q.; Khan, A.; Doan, V.D.; Tran, V.A.; Le, V.T. Photocatalytic degradation of methyl orange dye by Ti_3C_2–TiO_2 heterojunction under solar light. *Chemosphere* **2021**, *276*, 130154. [CrossRef]

8. Hao, R.; Wang, G.; Tang, H.; Sun, L.; Xu, C.; Han, D. Template-free preparation of macro/mesoporous g-C_3N_4/TiO_2 heterojunction photocatalysts with enhanced visible light photocatalytic activity. *Appl. Catal. B Environ.* **2016**, *187*, 47–58. [CrossRef]
9. Sobczyk-Guzenda, A.; Owczarek, S.; Szymanowski, H.; Volesky, L.; Walkowiak, B.; Miszczak, S.; Gazicki-Lipman, M. Iron doped thin TiO_2 films synthesized with the RF PECVD method. *Ceram. Int.* **2015**, *41*, 7496–7500. [CrossRef]
10. Li, D.; Bulou, S.; Gautier, N.; Elisabeth, S.; Goullet, A.; Richard-Plouet, M.; Choquet, P.; Granier, A. Nanostructure and photocatalytic properties of TiO_2 films deposited at low temperature by pulsed PECVD. *Appl. Surf. Sci.* **2019**, *466*, 63–69. [CrossRef]
11. Varshney, G.; Kanel, S.R.; Kempisty, D.M.; Varshney, V.; Agrawal, A.; Sahle-Demessie, E.; Varma, R.S.; Nadagouda, M.N. Nanoscale TiO_2 films and their application in remediation of organic pollutants. *Coord. Chem. Rev.* **2016**, *306*, 43–64. [CrossRef]
12. Rapsomanikis, A.; Apostolopoulou, A.; Stathatos, E.; Lianos, P. Cerium-modified TiO_2 nanocrystalline films for visible light photocatalytic activity. *J. Photochem. Photobiol. A Chem.* **2014**, *280*, 46–53. [CrossRef]
13. Lee, M.S.; Hong, S.S.; Mohseni, M. Synthesis of photocatalytic nanosized TiO_2-Ag particles with sol-gel method using reduction agent. *J. Mol. Catal. A Chem.* **2005**, *242*, 135–140. [CrossRef]
14. Yogi, C.; Kojima, K.; Wada, N.; Tokumoto, H.; Takai, T.; Mizoguchi, T.; Tamiaki, H. Photocatalytic degradation of methylene blue by TiO_2 film and Au particles-TiO_2 composite film. *Thin Solid Films* **2008**, *516*, 5881–5884. [CrossRef]
15. Lang, J.; Takahashi, K.; Kubo, M.; Shimada, M. Ag-Doped TiO_2 Composite Films Prepared Using Aerosol-Assisted, Plasma-Enhanced Chemical Vapor Deposition. *Catalysts* **2022**, *12*, 365. [CrossRef]
16. Lu, G.; Wang, X.; Wang, Y.; Shi, G.; Xie, X.; Sun, J. Anti-oxidative microstructure design of ultra-stable N-TiO_2 composite for the gaseous photodegradation reactions. *Chem. Eng. J.* **2021**, *408*, 127257. [CrossRef]
17. Chen, M.; Zhang, F.; Oh, W. Synthesis, characterization, and photocatalytic analysis of CNT/TiO_2 composites derived from MWCNTs and titanium sources. *New Carbon Mater.* **2009**, *24*, 159–166. [CrossRef]
18. Chaudhary, D.; Singh, S.; Vankar, V.D.; Khare, N. A ternary Ag/TiO_2/CNT photoanode for efficient photoelectrochemical water splitting under visible light irradiation. *Int. J. Hydrog. Energy* **2017**, *42*, 7826–7835. [CrossRef]
19. Zhou, W.; Pan, K.; Qu, Y.; Sun, F.F.; Tian, C.G.; Ren, Z.Y.; Tian, G.H.; Fu, H.G. Photodegradation of organic contamination in wastewaters by bonding TiO_2/single-walled carbon nanotube composites with enhanced photocatalytic activity. *Chemosphere* **2010**, *81*, 555–561. [CrossRef]
20. Banerjee, S.; Benjwal, P.; Singh, M.; Kar, K.K. Graphene oxide (rGO)-metal oxide (TiO_2/Fe_3O_4) based nanocomposites for the removal of methylene blue. *Appl. Surf. Sci.* **2018**, *439*, 560–568. [CrossRef]
21. Wang, S.; Gong, Q.M.; Zhu, Y.F.; Liang, J. Preparation and photocatalytic properties of silver nanoparticles loaded on CNTs/TiO_2 composite. *Appl. Surf. Sci.* **2009**, *255*, 8063–8066. [CrossRef]
22. Koo, Y.; Littlejohn, G.; Collins, B.; Yun, Y.; Shanov, V.N.; Schulz, M.; Pai, D.; Sankar, J. Synthesis and characterization of Ag–TiO_2–CNT nanoparticle composites with high photocatalytic activity under artificial light. *Compos. Part. B Eng.* **2014**, *57*, 105–111. [CrossRef]
23. Arabatzis, I.M.; Stergiopoulos, T.; Bernard, M.C.; Labou, D.; Neophytides, S.G.; Falaras, P. Silver-modified titanium dioxide thin films for efficient photodegradation of methyl orange. *Appl. Catal. B Environ.* **2003**, *42*, 187–201. [CrossRef]
24. Akhavan, O. Lasting antibacterial activities of Ag-TiO_2/Ag/a-TiO_2 nanocomposite thin film photocatalysts under solar light irradiation. *J. Colloid Interface Sci.* **2009**, *336*, 117–124. [CrossRef] [PubMed]
25. Liu, Y.; Wang, X.L.; Yang, F.; Yang, X.R. Excellent antimicrobial properties of mesoporous anatase TiO_2 and Ag/TiO_2 composite films. *Microporous Mesoporous Mater.* **2008**, *114*, 431–439. [CrossRef]
26. Suda, Y.; Kawasaki, H.; Ueda, T.; Ohshima, T. Preparation of high quality nitrogen doped TiO_2 thin film as a photocatalyst using a pulsed laser deposition method. *Thin Solid Films* **2004**, *453*, 162–166. [CrossRef]
27. Liu, Y.; Xu, C.; Feng, Z.D. Characteristics and anticorrosion performance of Fe-doped TiO_2 films by liquid phase deposition method. *Appl. Surf. Sci.* **2014**, *314*, 392–399. [CrossRef]
28. Lim, S.; Huang, N.M.; Lim, H.N.; Mazhar, M. Surface Modification of Aerosol-Assisted CVD Produced TiO_2 Thin Film for Dye Sensitised Solar Cell. *Int. J. Photoenergy* **2014**, *2014*, 1–12. [CrossRef]
29. Ebrahimi, I.; Gashti, M.P. Polypyrrole-MWCNT-Ag composites for electromagnetic shielding: Comparison between chemical deposition and UV-reduction approaches. *J. Phys. Chem. Solids* **2018**, *118*, 80–87. [CrossRef]
30. Alimohammadi, F.; Gashti, M.P.; Shamei, A.; Kiumarsi, A. Deposition of silver nanoparticles on carbon nanotube by chemical reduction method: Evaluation of surface, thermal and optical properties. *Superlattices Microstruct.* **2012**, *52*, 50–62. [CrossRef]
31. Kubo, M.; Taguchi, T.; Shimada, M. Preparation of nanoparticle-embedded thin films by simultaneous feeding of gaseous and solid raw materials in plasma-enhanced chemical vapor deposition process. *Thin Solid Films* **2017**, *632*, 55–65. [CrossRef]
32. Askari, M.B.; Banizi, Z.T.; Seifi, M.; Dehaghi, S.B.; Veisi, P. Synthesis of TiO_2 nanoparticles and decorated multi-wall carbon nanotube (MWCNT) with anatase TiO_2 nanoparticles and study of optical properties and structural characterization of TiO_2/MWCNT nanocomposite. *Optik* **2017**, *149*, 447–454. [CrossRef]
33. Choudhury, B.; Choudhury, A. Local structure modification and phase transformation of TiO_2 nanoparticles initiated by oxygen defects, grain size, and annealing temperature. *Int. Nano Lett.* **2013**, *3*, 55. [CrossRef]
34. Yu, J.G.; Yu, H.G.; Cheng, B.; Zhou, M.H.; Zhao, X.J. Enhanced photocatalytic activity of TiO_2 powder (P25) by hydrothermal treatment. *J. Mol. Catal. A Chem.* **2006**, *253*, 112–118. [CrossRef]
35. Zhao, C.; Guo, J.; Yu, C.; Zhang, Z.; Sun, Z.; Piao, X. Fabrication of CNTs-Ag-TiO_2 ternary structure for enhancing visible light photocatalytic degradation of organic dye pollutant. *Mater. Chem. Phys.* **2020**, *248*, 122873. [CrossRef]

36. Aazam, E.S. Visible light photocatalytic degradation of thiophene using Ag-TiO_2/multi-walled carbon nanotubes nanocomposite. *Ceram. Int.* **2014**, *40*, 6705–6711. [CrossRef]
37. Soltanieh, A.M.; Khavar, A.H.C.; Ganjidoust, H.; Mahjoub, A.R.; Khazaee, Z. Plasmon-induced charge separation by Ag nanoparticles between titanium dioxide and MWCNTs for natural sunlight-driven photocatalysis. *J. Iran. Chem. Soc.* **2021**. [CrossRef]
38. De la Flor, M.P.; Camarillo, R.; Martínez, F.; Jiménez, C.; Quiles, R.; Rincón, J. Synthesis and characterization of TiO_2/CNT/Pd: An effective sunlight photocatalyst for neonicotinoids degradation. *J. Environ. Chem. Eng.* **2021**, *9*, 106278. [CrossRef]
39. Wang, T.; Tang, T.; Gao, Y.; Chen, Q.; Zhang, Z.; Bian, H. Hydrothermal preparation of Ag-TiO_2-reduced graphene oxide ternary microspheres structure composite for enhancing photocatalytic activity. *Phys. E Low-Dimens. Syst. Nanostruct.* **2019**, *112*, 128–136. [CrossRef]
40. Gaafar, M.R.; Mady, R.F.; Diab, R.G.; Shalaby, T.I. Chitosan and silver nanoparticles: Promising anti-toxoplasma agents. *Exp. Parasitol.* **2014**, *143*, 30–38. [CrossRef]

catalysts

MDPI

Communication

Ag-Doped TiO_2 Composite Films Prepared Using Aerosol-Assisted, Plasma-Enhanced Chemical Vapor Deposition

Jianghua Lang [1], Kazuma Takahashi [1], Masaru Kubo [2] and Manabu Shimada [2,*]

1 Department of Chemical Engineering, Graduate School of Engineering, Hiroshima University, 1-4-1, Kagamiyama, Higashi-Hiroshima, Hiroshima 739-8527, Japan; d196006@hiroshima-u.ac.jp (J.L.); ka-864-@ezweb.ne.jp (K.T.)

2 Chemical Engineering Program, Graduate School of Advanced Science and Engineering, Hiroshima University, 1-4-1, Kagamiyama, Higashi-Hiroshima, Hiroshima 739-8527, Japan; mkubo@hiroshima-u.ac.jp

* Correspondence: smd@hiroshima-u.ac.jp

Abstract: TiO_2 is a promising photocatalyst, but its large bandgap restricts its light absorption to the ultraviolet region. The addition of noble metals can reduce the bandgap and electron-hole recombination; therefore, we prepared TiO_2-Ag nanoparticle composite films by plasma-enhanced chemical vapor deposition (PECVD) using a mixture of aerosolized $AgNO_3$, which was used as a Ag nanoparticle precursor, and titanium tetraisopropoxide, which acted as the TiO_2 precursor. Notably, the use of PECVD enabled a low process temperature and eliminated the need for pre-preparing the Ag nanoparticles, thereby increasing the process efficiency. The structures and morphologies of the deposited films were characterized by ultraviolet (UV)—visible spectroscopy, X-ray diffraction, X-ray photoelectron spectroscopy, scanning electron microscopy, transmission electron microscopy and energy-dispersive X-ray spectroscopy, and the effects of the $AgNO_3$ concentration on the photocatalytic activity of the deposited films were determined by assessing the degradation of methylene blue under UV light irradiation. The Ag ions were successfully reduced to metallic nanoparticles and were embedded in the TiO_2 film. The best photocatalytic activity was achieved for a 1 wt% Ag-loaded TiO_2 composite film, which was 1.75 times that of pristine TiO_2.

Keywords: metal nanoparticles; thin film; bandgap; photocatalytic activity

Citation: Lang, J.; Takahashi, K.; Kubo, M.; Shimada, M. Ag-Doped TiO_2 Composite Films Prepared Using Aerosol-Assisted, Plasma-Enhanced Chemical Vapor Deposition. *Catalysts* **2022**, *12*, 365. https://doi.org/10.3390/catal12040365

Academic Editors: Carolina Belver and Jorge Bedia

Received: 28 February 2022
Accepted: 20 March 2022
Published: 23 March 2022

1. Introduction

Since the first use of TiO_2 electrodes for the photocatalytic splitting of water in 1972 [1], TiO_2 has become one of the most promising photocatalysts. To date, TiO_2 photocatalysts have been used in the bulk, powdered, nanostructured, and thin-film forms. In particular, the synthesis of TiO_2 in the form of thin films allows it to be efficiently utilized, thereby expanding its applications [2]. Furthermore, the immobilization of catalyst thin films on a substrate is advantageous compared to the use of powdered catalysts, because it (1) eliminates the need for catalyst separation and filtration, (2) enables application in continuous flow systems, and (3) reduces agglomeration at various catalyst loadings [3]. However, despite the promise of thin-film TiO_2 photocatalysts, TiO_2 has an intrinsically low photocatalytic activity because of the large bandgap (3.2 eV) of the anatase phase, which allows absorption only in the ultraviolet (UV) region, thereby preventing the full utilization of solar energy (typically, only 3–5% of the total solar spectrum). In addition, the fast recombination of electron-hole pairs reduces the catalytic activity [4–6]. Therefore, metals, such as Ag, Au, Pt, and Cu, have been doped into TiO_2 films, which enhance the photocatalytic activity compared to that of undoped TiO_2 films. In this study, we focused on the doping of TiO_2 thin films with Ag.

Several methods have been developed for film deposition, among which chemical vapor deposition (CVD) is one of the most popular. It requires simple equipment, and yields good surface coverage and a high deposition rate [7], but it requires high temperatures. Therefore, to enable the use of lower temperatures, the plasma-enhanced CVD (PECVD) method has been proposed. In PECVD, films are deposited on a low-cost substrate by utilizing the reactive chemical species formed in a microwave-induced plasma [8]. Notably, thin TiO_2 films and metal-doped TiO_2 composite films have been deposited by PECVD [9–11], where titanium tetraisopropoxide (TTIP) was employed as a reactive TiO_2 precursor. During the reaction, TTIP was cracked by inelastic collisions with the ionized species in the plasma [12], enabling the use of a low overall process temperature. In addition, metal ions can be reduced to their metallic state by plasma treatment [13]. In fact, recently, we successfully fabricated TiO_2 composite films by adding a Ag nanoparticle suspension [14]; however, this required pre-preparing the nanoparticle mixture. Therefore, in this study, to increase the process efficiency, $AgNO_3$ was used as the Ag^+-ion source and TTIP was used as the TiO_2 precursor. Briefly, aqueous $AgNO_3$ was aerosolized and supplied to the PECVD reactor. In the plasma, Ag nanoparticles were obtained by the reduction of $AgNO_3$, and these were deposited on the substrate. Simultaneously, the TiO_2 film was deposited from the species formed by the cracking of the TTIP vapor introduced in the He and O_2 carrier gases. In particular, we investigated the influence of the $AgNO_3$ concentration on the structure and photocatalytic activity of the TiO_2-Ag composite films.

2. Results and Discussion

2.1. Ultraviolet (UV)—Visible (Vis) Spectra

Figure 1a demonstrates the UV-Vis spectra of the pristine TiO_2 film and the TiO_2-Ag composite films prepared using different concentrations of $AgNO_3$ (0.1–4 wt%). The pristine TiO_2 film exhibits a threshold wavelength of 358 nm, which does not fall in the visible light range. Increasing the concentration of $AgNO_3$ shifts the absorption edges of the TiO_2-Ag composite films toward longer wavelengths; that is, they shift toward the visible region. In particular, the spectra of the TiO_2-Ag composite films show a strong shoulder peak at 300–500 nm, which can be attributed to the surface plasmon absorption of spatially confined electrons in the Ag nanoparticles [15]. This suggests that doping TiO_2 with Ag can significantly enhance its visible-light absorption. The bandgap energy was calculated using the relation $(\alpha h\upsilon)^{1/\gamma} = B(h\upsilon - E_g)$, where α is the absorption coefficient, $h\upsilon$ is the incident photon energy, E_g is the bandgap energy, the γ-factor is equal to 1/2 for a direct bandgap, and B is a constant [16]. The $(\alpha h\upsilon)^2$ versus hυ plots, constructed based on the absorbance spectra, are shown in Figure 1b. The bandgap was estimated by extrapolating the tangent of the curve to the x-axis (Table 1). The bandgap of pristine TiO_2 was 3.54 eV, while those of the TiO_2-Ag composite films decreased from 3.21 to 2.61 eV with increasing $AgNO_3$ concentration. The difference in the bandgaps of these samples may originate from the different surface microstructures and compositions [17].

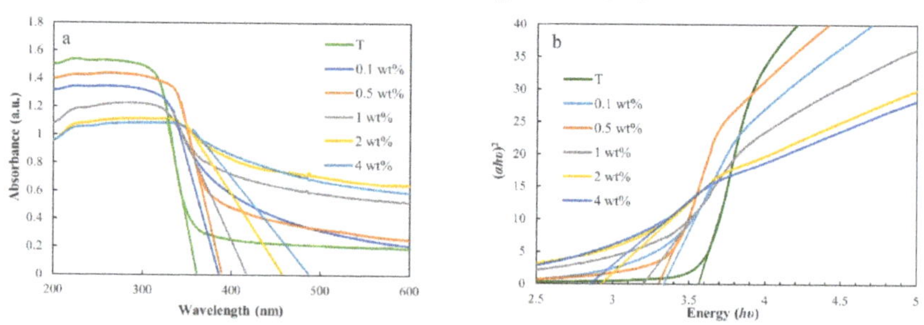

Figure 1. Optical properties of the pristine TiO_2 film (T) and TiO_2-Ag composite films deposited using $AgNO_3$ concentrations of 0.1–4 wt%: (**a**) absorbance spectra, (**b**) plot of $(\alpha h\upsilon)^2$ versus photon energy ($h\upsilon$).

Table 1. Bandgap energies (E_g) of the films.

Film	Threshold Wavelength (nm)	E_g (eV)
Pristine TiO_2	350	3.54
TiO_2-Ag (0.1 wt%)	386	3.21
TiO_2-Ag (0.5 wt%)	389	3.18
TiO_2-Ag (1 wt%)	410	3.02
TiO_2-Ag (2 wt%)	442	2.80
TiO_2-Ag (4 wt%)	457	2.61

2.2. X-ray Diffraction (XRD) Analysis

The effect of $AgNO_3$ concentration on the crystal phase of the prepared films was analyzed by XRD (Figure 2) using JCPDS cards of TiO_2, Ag, and SiO_2 as the reference. Regardless of the $AgNO_3$ concentration, the prepared films exhibited peaks only for the anatase phase at 25.82°, 37.53°, 38.34°, 39.16°, 48.56°, 54.44°, and 55.56° correspond to the (101), (112), (103), (004), (200), (105), and (211) planes, respectively. In particular, no peaks consistent with those of metallic Ag are observed, possibly because of the low loading of Ag or overlap with peaks corresponding to TiO_2 [18]. In addition, the peak near 23° may be attributed to the SiO_2 of the substrate. However, we were unable to identify the peaks near 35° and 43°, which may originate from unknown contamination.

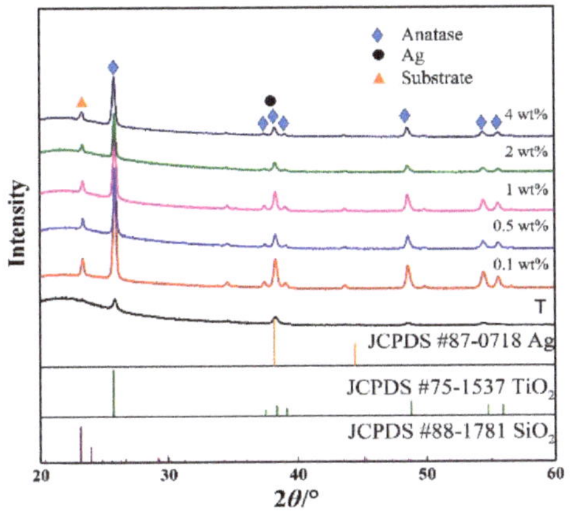

Figure 2. X-ray diffraction patterns of the pristine TiO_2 film and TiO_2-Ag composite films deposited using different concentrations of $AgNO_3$.

2.3. X-ray Photoelectron Spectroscopy (XPS) Analysis

Next, XPS measurements were performed to examine the chemical species and their oxidation states in the prepared 1 wt% $AgNO_3$-loaded TiO_2-Ag film, which was selected as a representative sample to confirm the conversion of $AgNO_3$ to metallic Ag. Figure 3a illustrates the XPS survey profile of this film, which contains three peaks corresponding to O 1s, Ti 2p, and Ag 3d. Figure 3b demonstrates the typical Ag 3d spectrum, with Ag $3d_{5/2}$ and Ag $3d_{3/2}$ peaks at binding energies of 368.2 and 374.1 eV, respectively; these peaks are characteristic of metallic Ag. The two peaks corresponding to the Ti $2p_{3/2}$ and Ti $2p_{1/2}$ peaks (459.0 eV and 464.7 eV, respectively) are shown in Figure 3c, and they confirm the presence of Ti^{4+}; this is consistent with the use of TiO_2 in the prepared films. The O 1s spectrum could be deconvoluted into three peaks at 530.2, 532.2, and 534.3 eV (Figure 3d), corresponding to the lattice oxygen in TiO_2 and the dissociated oxygen and hydroxyl-like groups [19,20], respectively.

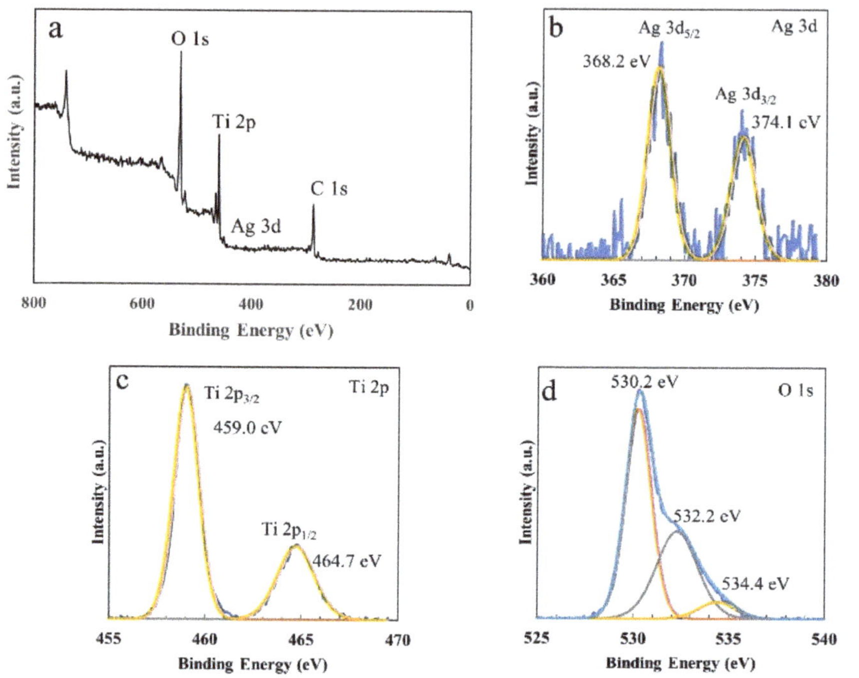

Figure 3. X-ray photoelectron profiles of the TiO_2-Ag (1 wt%) film: (**a**) survey, (**b**) Ag 3d, (**c**) Ti 2p, and (**d**) O 1s spectra.

2.4. Morphological Structure

Figure 4 demonstrates the scanning electron microscopy (SEM) images on the surface microstructures and cross-section of the TiO_2-Ag composite films deposited using different $AgNO_3$ concentrations. The pristine TiO_2 film has a dense surface structure. However, the prepared composite films contain secondary spherical particles formed by the aggregation of tiny particles; these secondary particles show a wide range of sizes. Generally, additives such as Ag nanoparticles provide nucleation centers, thereby inducing the formation of agglomerates or dense heterogeneities during deposition [21]. In addition, higher $AgNO_3$ concentrations resulted in a more severe agglomeration on the surface of the films, as previously reported [22,23]. However, the aggregated particles are tightly packed together, suggesting that the PECVD method produces films with good substrate coverage. The layer thickness was estimated using cross-section images, and all the films showed a thickness of approximately 0.7 μm. Furthermore, repeated experiments yielded a relatively smooth surface for the TiO_2 films and a rough surface for the TiO_2-Ag composite films, indicating that our film deposition process is reproducible.

A microgrid was employed to scratch the surface of the synthesized TiO_2-Ag composite films, which were subsequently investigated using transmission electron microscopy energy dispersive X-ray (TEM-EDS). The results reflect only a very small part of the surface of the film. The images of the 1 wt% $AgNO_3$-containing film (Figure 5) shows that spherical Ag nanoparticles of 40–60 nm diameter are attached to the very small TiO_2 particles. The large size of the Ag nanoparticles may cause the aggregation on the surface of the film as shown in the SEM images (b–f). Figure 5b shows the high-resolution TEM (HRTEM) images. The interplanar spacings of the (111) plane of Ag and (101) plane of anatase were determined to be 0.24 and 0.35 nm, respectively. Thus, the Ag nanoparticles were tightly coated by the TiO_2 particles. Furthermore, the EDS analysis of the region enclosed in a yellow square in Figure 5a confirmed the presence of the Ag species in the film.

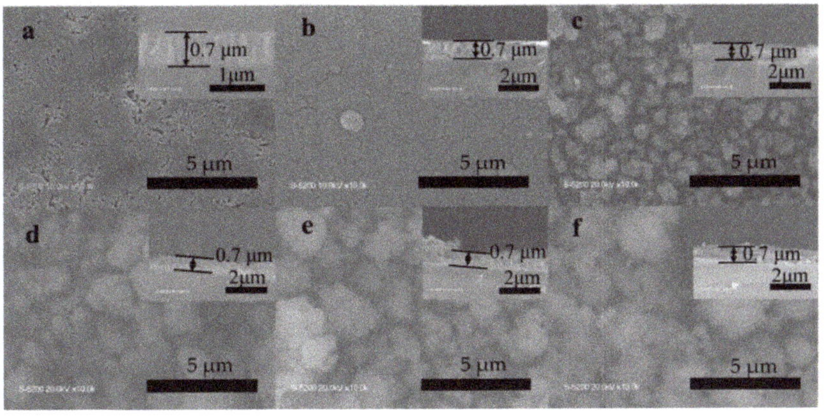

Figure 4. Scanning electron microscopy images of the surface view and cross-section of (**a**) pristine TiO_2 film and TiO_2-Ag composite films deposited using different $AgNO_3$ concentrations: (**b–f**) 0.1, 0.5, 1, 2, and 4 wt%, respectively.

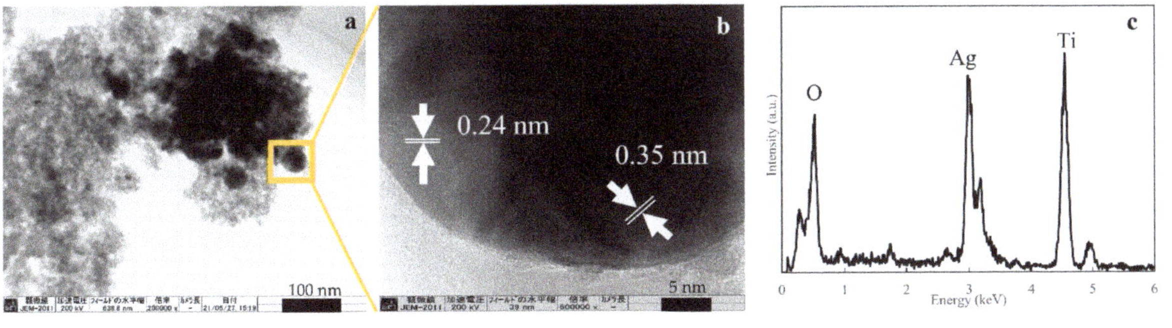

Figure 5. Structure and component of the TiO_2-Ag (1 wt%) film: (**a**) transmission electron microscopy (TEM) image, (**b**) high-resolution TEM images, and (**c**) the corresponding energy dispersive X-ray profile of the region in the yellow square.

2.5. Photocatalytic Activity

The photocatalytic activities of the samples were evaluated based on the decomposition of methylene blue (MB) under UV irradiation. Figure 6a shows the degradation of MB (C_t/C_0) with respect to time (t) for the different films under UV light irradiation. All the films exhibited photocatalytic activities under the same conditions. However, the photocatalytic activities of the TiO_2-Ag composite films were higher compared to that of the pristine TiO_2 film. A possible mechanism has been previously reported [19,24]. The percent MB degradation by the different films after 3 h of irradiation was calculated using the formula $[(C_0 - C_t)/C_0] \times 100\%$, where C_0 and C_t are the concentrations of MB at time $t = 0$ and after irradiation for time t, respectively. The TiO_2-Ag (1 wt%) composite film exhibited the best photocatalytic activity, yielding an MB degradation of approximately 41%, whereas the pristine TiO_2 film showed an activity of approximately 11%. As mentioned, the rate constant, k, for MB degradation by the different films was calculated using Equation (1), and the results are shown in Figure 6b. The resulting k values also confirm that the maximum photocatalytic activity was exhibited by the TiO_2-Ag (1 wt%) composite film, which was almost 1.7 times that of pristine TiO_2. Thus, 1 wt% was determined to be the optimal concentration because higher loadings (e.g., 2 wt% and 4 wt%) resulted in lower activities. Previously, adding excess Ag has been shown to reduce the photocatalytic activity because of the increased reflection of incident light [25,26]. In addition, electron-hole recombination can occur on the Ag nanoparticles, thereby hindering or preventing electron transfer to the surface and reducing the photocatalytic activity [27].

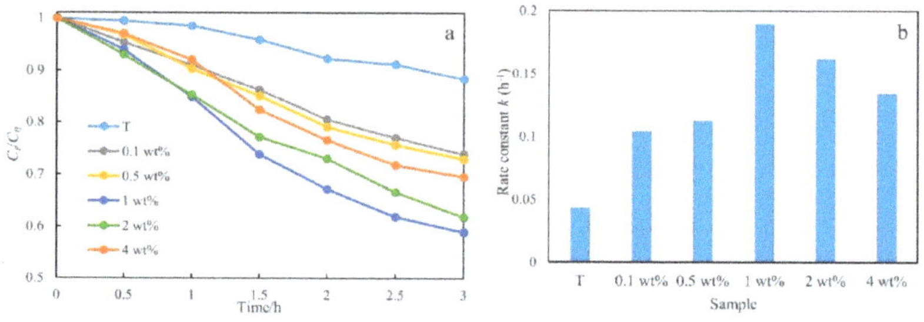

Figure 6. Photocatalytic activity for the degradation of methylene blue over different films: (**a**) degradation with respect to time and (**b**) rate constants.

3. Experimental Setup and Characterization

3.1. Experimental Setup

Figure 7 shows a schematic of the PECVD reactor consisting of the material feeding systems, a ribbon heater, a quartz tube with a substrate holder, the plasma generation system, and a vacuum pump [14,28]. The plasma was produced when the gaseous species was ionized by 2.45 GHz microwaves generated using a magnetron (MH2000S-212BB, Muegge, Co., Ltd., Reichelsheim, Germany). The temperature and pressure of the system were controlled using the ribbon heater and vacuum pump (Pascal 2015C1, Pfeiffer Vacuum, Inc., Asslar, Germany), respectively. On the left-hand side of the system, aqueous $AgNO_3$ (99.8%, Nacalai Tesque, Inc., Tokyo, Japan) solutions with concentrations ranging from 0 to 4 wt% were supplied by a syringe pump (YSP-301, YMC. Co., Ltd., Kyoto, Japan). To aerosolize the aqueous $AgNO_3$ solution, a two-fluid nozzle was used. The carrier gas was He gas and was introduced at a rate of 1500 standard cubic centimeters per minute (sccm). TTIP (Tokyo Chemical Industry Co., Ltd., Tokyo, Japan) was vaporized and fed into the system with He and O_2 gases at a flow rate of 50 sccm. The plasma gas was generated inside a glass tube with an inner diameter and length of 25.4 mm and 300 mm, respectively, and the materials were fed into this tube. As $AgNO_3$ and TTIP were simultaneously fed into the system over 15 min, the films were deposited onto the glass substrate. Subsequently, the obtained films were annealed at 600 °C for 12 h under N_2 gas flow at 200 sccm.

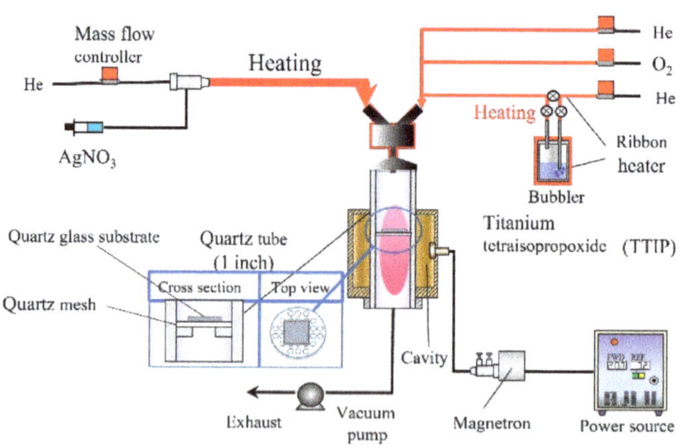

Figure 7. Schematic describing the fabrication process of the plasma-enhanced chemical vapor deposition method.

3.2. Characterization

The optical properties of the films were investigated using a V-650 spectrophotometer (JASCO, Tokyo, Japan). The crystal structures of the films were determined by XRD (RINT2100, Rigaku, Tokyo, Japan), using Cu-K_{α} radiation (λ = 1.5406 Å). The elemental composition and their oxidation states were determined using XPS (ESCA-3400, SHIMADAZU, Kyoto, Japan). The recorded XPS spectra were fitted using the C 1s peak at 284.6 eV as the reference; this peak corresponded to adventitious carbon on the sample surface. The morphologies of the films were visualized using SEM (S-5200, Hitachi High-Technologies, Tokyo, Japan) and TEM (JEM-2010, JEOL, Tokyo, Japan), in conjunction with energy-dispersive EDS (JED-2300T, JEOL, Tokyo, Japan).

The photocatalytic activities of the films were measured in a cuvette using 3 mL of MB (Kanto Chemical Co. Inc., Tokyo, Japan). To attain the adsorption equilibrium, the cuvette was left in the dark for 30 min. Then, the photocatalytic reaction was initiated by irradiating the sample with UV light (365 nm, 1407 μW/cm). The absorption spectrum of the sample was recorded on a V-650 spectrophotometer at an interval of 30 min. MB mainly absorbed in the range of 550–700 nm, with an absorption maximum at 665 nm. The measurement was conducted in a thermostatically controlled system maintained at 25 °C. The concentration of MB was determined using the Beer-Lambert law, and the percent degradation of MB was calculated as $[(C_0 - C_t)/C_0] \times 100\%$, where C_0 and C_t are the concentrations of MB at time t = 0 and after irradiation for time t, respectively. Further, the reaction rate constant, k, was calculated to evaluate the photocatalytic ability of the different films, as expressed by Equation (1).

$$\ln \frac{C_0}{C_t} = kt \tag{1}$$

4. Conclusions

In this work, pristine TiO_2 films and TiO_2-Ag composite films were successfully prepared with the PECVD method. The films were annealed at 600 °C after PECVD, and they exhibited a typical anatase structure. The addition of Ag nanoparticles significantly altered the morphology of the films due to particle aggregation. Notably, the use of the Ag dopant significantly improved the light absorption ability of the films; that is, compared to pristine TiO_2, the TiO_2-Ag composite films show significantly enhanced photocatalytic activities. Thus, the optimal film (in terms of the photocatalytic activity) was prepared using 1 wt% $AgNO_3$. Overall, the results demonstrate the success of PECVD for doping TiO_2 thin films with metal nanoparticles, thereby providing a simple synthetic process for preparing metal-doped films with high photocatalytic activity.

Author Contributions: Conceptualization, M.S.; methodology, J.L. and K.T.; formal analysis, J.L.; writing, J.L.; writing—review and editing, J.L., M.S. and M.K.; supervision: M.S.; funding acquisition, M.S. All authors have read and agreed to the published version of the manuscript.

Funding: This work was partly supported by the Japan Society for the Promotion of Science KAKENHI grant number 21K04750.

Data Availability Statement: Data are contained within the article.

Acknowledgments: The authors are grateful to M. Maeda for assistance with the TEM analysis.

Conflicts of Interest: The authors declare no conflict of interest.

References

1. Fujishima, A.; Honda, K. Electrochemical photolysis of water at a semiconductor electrode. *Nature* **1972**, *238*, 37–38. [CrossRef] [PubMed]
2. Xi, B.J.; Verma, L.K.; Li, J.; Bhatia, C.S.; Danner, A.J.; Yang, H.; Zeng, H.C. TiO_2 Thin Films Prepared via Adsorptive Self-Assembly for Self-Cleaning Applications. *ACS Appl. Mater. Interfaces* **2012**, *4*, 1093–1102. [CrossRef] [PubMed]
3. Arabatzis, I.M.; Stergiopoulos, T.; Bernard, M.C.; Labou, D.; Neophytides, S.G.; Falaras, P. Silver-modified titanium dioxide thin films for efficient photodegradation of methyl orange. *Appl. Catal. B* **2003**, *42*, 187–201. [CrossRef]

4. Diaz-Uribe, C.; Viloria, J.; Cervantes, L.; Vallejo, W.; Navarro, K.; Romero, E.; Quinones, C. Photocatalytic Activity of Ag-TiO_2 Composites Deposited by Photoreduction under UV Irradiation. *Int. J. Photoenergy* **2018**, *2018*, 6080432. [CrossRef]
5. Sen, S.; Mahanty, S.; Roy, S.; Heintz, O.; Bourgeois, S.; Chaumont, D. Investigation on sol-gel synthesized Ag-doped TiO_2 cermet thin films. *Thin Solid Film.* **2005**, *474*, 245–249. [CrossRef]
6. Tang, B.; Chen, H.Q.; Peng, H.P.; Wang, Z.W.; Huang, W.Q. Graphene Modified TiO_2 Composite Photocatalysts: Mechanism, Progress and Perspective. *Nanomaterials* **2018**, *8*, 205. [CrossRef]
7. Rasoulnezhad, H.; Kavei, G.; Ahmadi, K.; Rahimipour, M.R. Combined sonochemical/CVD method for preparation of nanostructured carbon-doped TiO_2 thin film. *Appl. Surf. Sci.* **2017**, *408*, 1–10. [CrossRef]
8. Li, D.; Goullet, A.; Carette, M.; Granier, A.; Zhang, Y.; Landesman, J.P. Structural and optical properties of RF-biased PECVD TiO_2 thin films deposited in an O_2/TTIP helicon reactor. *Vacuum* **2016**, *131*, 231–239. [CrossRef]
9. Li, D.; Bulou, S.; Gautier, N.; Elisabeth, S.; Goullet, A.; Richard-Plouet, M.; Choquet, P.; Granier, A. Nanostructure and photocatalytic properties of TiO_2 films deposited at low temperature by pulsed PECVD. *Appl. Surf. Sci.* **2019**, *466*, 63–69. [CrossRef]
10. Li, D.; Gautier, N.; Dey, B.; Bulou, S.; Richard-Plouet, M.; Ravisy, W.; Goullet, A.; Choquet, P.; Granier, A. TEM analysis of photocatalytic TiO_2 thin films deposited on polymer substrates by low-temperature ICP-PECVD. *Appl. Surf. Sci.* **2019**, *491*, 116–122. [CrossRef]
11. Sobczyk-Guzenda, A.; Owczarek, S.; Szymanowski, H.; Volesky, L.; Walkowiak, B.; Miszczak, S.; Gazicki-Lipman, M. Iron doped thin TiO_2 films synthesized with the RF PECVD method. *Ceram. Int.* **2015**, *41*, 7496–7500. [CrossRef]
12. Ehlbeck, J.; Schnabel, U.; Polak, M.; Winter, J.; von Woedtke, T.; Brandenburg, R.; von dem Hagen, T.; Weltmann, K.D. Low temperature atmospheric pressure plasma sources for microbial decontamination. *J. Phys. D Appl. Phys.* **2011**, *44*, 013002. [CrossRef]
13. Zou, J.-J.; Zhang, Y.-P.; Liu, C.-J. Reduction of Supported Noble-Metal Ions Using Glow Discharge Plasma. *Langmuir* **2006**, *22*, 11388–11394. [CrossRef] [PubMed]
14. Lang, J.; Takahashi, K.; Kubo, M.; Shimada, M. *Preparation of TiO_2-CNT-Ag Ternary Composite Film with Enhanced Photocatalytic Activity via Plasma-Enhanced Chemical Vapor Deposition*; Hiroshima University: Hiroshima, Japan, 2022; *to be submitted*.
15. Lei, X.F.; Xue, X.X.; Yang, H. Preparation and characterization of Ag-doped TiO_2 nanomaterials and their photocatalytic reduction of Cr(VI) under visible light. *Appl. Surf. Sci.* **2014**, *321*, 396–403. [CrossRef]
16. Makula, P.; Pacia, M.; Macyk, W. How To Correctly Determine the Band Gap Energy of Modified Semiconductor Photocatalysts Based on UV-Vis Spectra. *J. Phys. Chem. Lett.* **2018**, *9*, 6814–6817. [CrossRef]
17. Yu, J.G.; Xiong, J.F.; Cheng, B.; Liu, S.W. Fabrication and characterization of Ag-TiO_2 multiphase nanocomposite thin films with enhanced photocatalytic activity. *Appl. Catal. B* **2005**, *60*, 211–221. [CrossRef]
18. Shi, H.; Yu, Y.; Zhang, Y.; Feng, X.; Zhao, X.; Tan, H.; Khan, S.U.; Li, Y.; Wang, E. Polyoxometalate/TiO_2/Ag composite nanofibers with enhanced photocatalytic performance under visible light. *Appl. Catal. B Environ.* **2018**, *221*, 280–289. [CrossRef]
19. Demirci, S.; Dikici, T.; Yurddaskal, M.; Gultekin, S.; Toparli, M.; Celik, E. Synthesis and characterization of Ag doped TiO_2 heterojunction films and their photocatalytic performances. *Appl. Surf. Sci.* **2016**, *390*, 591–601. [CrossRef]
20. Yu, J.-G.; Yu, H.-G.; Cheng, B.; Zhao, X.-J.; Yu, J.C.; Ho, W.-K. The Effect of Calcination Temperature on the Surface Microstructure and Photocatalytic Activity of TiO_2 Thin Films Prepared by Liquid Phase Deposition. *J. Phys. Chem. B* **2003**, *107*, 13871–13879. [CrossRef]
21. Chen, C.Y.; Tuan, W.H. Effect of silver on the sintering and grain-growth behavior of barium titanate. *J. Am. Ceram. Soc.* **2000**, *83*, 2988–2992. [CrossRef]
22. Tomas, S.A.; Luna-Resendis, A.; Cortes-Cuautli, L.C.; Jacinto, D. Optical and morphological characterization of photocatalytic TiO_2 thin films doped with silver. *Thin Solid Film.* **2009**, *518*, 1337–1340. [CrossRef]
23. Singh, J.; Sahu, K.; Pandey, A.; Kumar, M.; Ghosh, T.; Satpati, B.; Som, T.; Varma, S.; Avasthi, D.K.; Mohapatra, S. Atom beam sputtered Ag-TiO_2 plasmonic nanocomposite thin films for photocatalytic applications. *Appl. Surf. Sci.* **2017**, *411*, 347–354. [CrossRef]
24. Zhang, H.; Liang, C.; Liu, J.; Tian, Z.; Wang, G.; Cai, W. Defect-Mediated Formation of Ag Cluster-Doped TiO_2 Nanoparticles for Efficient Photodegradation of Pentachlorophenol. *Langmuir* **2012**, *28*, 3938–3944. [CrossRef] [PubMed]
25. Yu, B.Y.; Leung, K.M.; Guo, Q.Q.; Lau, W.M.; Yang, J. Synthesis of Ag-TiO_2 composite nano thin film for antimicrobial application. *Nanotechnology* **2011**, *22*, 115603. [CrossRef] [PubMed]
26. Ko, S.; Banerjee, C.K.; Sankar, J. Photochemical synthesis and photocatalytic activity in simulated solar light of nanosized Ag doped TiO_2 nanoparticle composite. *Compos. Part B* **2011**, *42*, 579–583. [CrossRef]
27. Yang, D.; Sun, Y.Y.; Tong, Z.W.; Tian, Y.; Li, Y.B.; Jiang, Z.Y. Synthesis of Ag/TiO_2 Nanotube Heterojunction with Improved Visible-Light Photocatalytic Performance Inspired by Bioadhesion. *J. Phys. Chem. C* **2015**, *119*, 5827–5835. [CrossRef]
28. Kubo, M.; Taguchi, T.; Shimada, M. Preparation of nanoparticle-embedded thin films by simultaneous feeding of gaseous and solid raw materials in plasma-enhanced chemical vapor deposition process. *Thin Solid Film.* **2017**, *632*, 55–65. [CrossRef]

catalysts

MDPI

Article

Influence of Synthesis Approach on Controlled Microstructures and Photocatalytic Properties of Ag/AgBr-Activated Carbon Composites on Visible Light Degradation of Tetracycline

Saheed O. Sanni [1,2,*], Hendrik G. Brink [1] and Elvera L. Viljoen [2]

1 Department of Chemical Engineering, Faculty of Engineering, Built Environment and Information Technology, University of Pretoria, Pretoria 0028, South Africa; deon.brink@up.ac.za
2 Biosorption and Wastewater Treatment Research Laboratory, Department of Chemistry, Faculty of Applied and Computer Sciences, Vaal University of Technology, Vanderbijlpark 1900, South Africa; elverav@vut.ac.za
* Correspondence: saheed.sanni@up.ac.za

Abstract: The influence of the synthesis approach (thermal polyol and deposition–precipitation) regarding the dispersion of Ag/AgBr nanoparticles dispersed on activated carbon prepared from chemical impregnated pinecone (TP-AABR-ACK, and DP-AABR-ACK) was studied, to increase their photocatalytic efficiency on the degradation of tetracycline (TC). The physicochemical characterization evidenced the significance of the ACK catalyst promoter in enhancing controlled microstructures (morphologies and particle size distributions), synergistic interface interaction between AABR NPs and the carbonaceous support, and efficient photogenerated charge carriers separation within TP-AABR-ACK, and DP-AABR-ACK composites. The results revealed 92% removal of TC within 180 min under the LED visible light irradiation, which was achieved using TP-AABR-ACK when compared to DP-AABR-ACK composite and other catalysts in this study. Such superior results achieved with TP-AABR-ACK composite were attributed to controlled morphologies, reduced particle size and agglomeration, improved absorptivity, and superior cooperative effect between the AABR and ACK catalyst promoter as evidenced from SEM, EDX, TEM, UV-DRS, and electrochemical characterizations, respectively. Furthermore, enhanced TOC removal and abundance of reactive superoxide anion generation were achieved with the TP-AABR-ACK composite in this study.

Keywords: silver/silver bromide; activated carbon; controlled microstructures; tetracycline degradation; reactive oxygen species

Citation: Sanni, S.O.; Brink, H.G.; Viljoen, E.L. Influence of Synthesis Approach on Controlled Microstructures and Photocatalytic Properties of Ag/AgBr-Activated Carbon Composites on Visible Light Degradation of Tetracycline. *Catalysts* **2021**, *11*, 1396. https://doi.org/10.3390/catal11111396

Academic Editors: Jorge Bedia and Carolina Belver

Received: 15 October 2021
Accepted: 17 November 2021
Published: 18 November 2021

1. Introduction

Wastewater effluents from pharmaceutical wastes has been identified as a crucial environmental issue, requiring urgent attention for removal [1]. The persistent overuse and abuse of pharmaceutical antibiotics, in the society nowadays have adverse effects on both aquatic and terrestrial organisms in the environment [2,3]. Tetracycline (TC) antibiotics are commonly employed for health purposes, due to their antibacterial and antimicrobial influence, which are very challenging to biodegrade in living organisms [3]. The removal of such organic compounds from water using conventional remediation methods still has limitations, and adequate outcomes are not achieved; hence, the need for cost-effective treatment method is required [4]. Host of water treatment technologies such as adsorption, biodegradation, and photocatalysis have been applied to overcome this environmental issue [3–5]. Semiconductor photocatalysis under the advanced oxidation processes (AOPs) is considered an effective environmentally friendly and economic treatment approach in addressing threats of pharmaceutical TC antibiotics in the environment [3,5–7]. Therefore, it is urgent for the design and development of efficient and suitable photocatalysts that can harvest the visible-light potentials of the solar spectrum, in achieving enhance removal of TC antibiotic from wastewater.

With this fact, a host of visible light active photocatalyst semiconductors have been reported [8–12]. Among these visible light photocatalysts, the nanostructures from plasmonic silver/silver halide (Ag/AgX, X = Br, Cl, or I) materials have attracted attention owing to their excellent light harvesting through Ag NPs surface plasmon resonance (SPR), and electronic attributes that advance reactive oxygen species (ROS) production [13]. In the past years, silver/silver bromide (Ag/AgBr) photocatalyst have had vast applications for environmental remediation and energy generation purposes [14,15]. Specifically, a rapid formation of AgBr nanoparticles (NPs) occurs during the synthetic process, which significantly influence the faster sprouting of the metallic silver nanoparticles (Ag NPs) at the photoinduced reduction stage [3]: hence, irregular morphologies, large particle size, and an evidence of agglomeration associated with this process; thus, lower photocatalytic activity along with lower efficiency in the production of ROS by the Ag/AgBr photocatalyst [3]. However, several synthetic processes have been utilized in controlling the morphologies and reducing the agglomeration of Ag/AgBr NPs [12,16–18]. However, a rapid recombination of the photoexcited electron–hole pairs through interaction with visible light with preceding works [3,17,18], during the photocatalysis process still persist, thus limiting their potential applications. An alternative approach is the immobilization of Ag/AgBr (AABR) NPs on an inert, and appropriate catalyst support, that controls the microstructures of AABR NPs and expediates the separation of photogenerated charge carriers.

Specifically, three-dimensional (3D) architectures are a well known catalyst support, which have attracted considerable attention owing to their unique hierarchical porous network, which found vast applications in photocatalysis [19], supercapacitors [20], and sensing [21]. The 3D network materials aid in efficient dispersion of photocatalyst NPs, enhanced visible light absorptivity, and promote fast separation of photogenerated charge carriers in semiconductor photocatalysis applications [22–24]. Particularly, the high surface area activated carbon (AC) derived from low cost, reproducible, and abundance biomass materials has formed ordered nanostructures with a host of photocatalyst materials. The 3D AC also increased intimate contact between the pollutant and active sites of photocatalyst, thus facilitating the efficient separation of charge carriers with high catalytic activities [22,25,26]. Hence, the application of activated carbon synthesized from microwave pyrolysis of pinecone chemically impregnated with potassium hydroxide (ACK), will be employed as a potential catalyst promoter/support for this study, in the controlled morphologies of AABR NPs.

Studies have shown that preparation conditions (reaction temperature, structuring directing agents, and ammonium hydroxide volume) significantly influence the unique AABR nanostructures formation during the glycothermal process [17]. As such, the AABR-ACK composites were synthesized through two methods (thermal polyol—TP-AABR-ACK, and deposition–precipitation—DP-AABR-ACK) in this study [10,11], which has never been reported so far.

In this present work, the synthetic influence of the AABR-ACK composites (analyzed using SEM, XRD, FTIR, UV-DRS, and electrochemical studies) from two different methods were carried out. The visible light photocatalytic degradation of TC antibiotic was conducted. Furthermore, the generation rate of superoxide anion radical from the as-prepared catalyst, as well comparison of active photocatalysts with other previously existing studies on TC degradation, were evaluated.

2. Results and Discussions

The crystal structures of TP-AABR-ACK, DP-AABR-ACK, DP-AABR, and TP-AABR are comparatively presented as shown in Figure 1. Furthermore, the center cubic phase of AgBr depicts the peaks at about 26.6°, 31.1°, 44.3°, 54.8°, 64.8°, 73.4° that aligns with the diffraction of the (111), (200), (220), (222), (400), (420) planes of AgBr with JPCDS file: 079-0148. The Ag NPs (JCPDS file: 071-3762) [27] revealed four diffraction peaks at around 38.2°, 44.2°, 64.4°, and 77.8°, which were all assigned to (111), (200), (220), and (311) planes of Ag, [28] respectively. In addition, the peaks of metallic Ag and AgBr, for the AABR-ACK nanocomposites, shifted by 0.2°, thus being attributed to the interfacial interaction between the AABR nanoparticles and ACK catalyst support [29–31]. These attributes for TP-AABR-ACK and DP-AABR-ACK nanocomposites will further promote fast separation of photogenerated charge carriers, thus resulting in high catalytic activities on the removal of TC antibiotics.

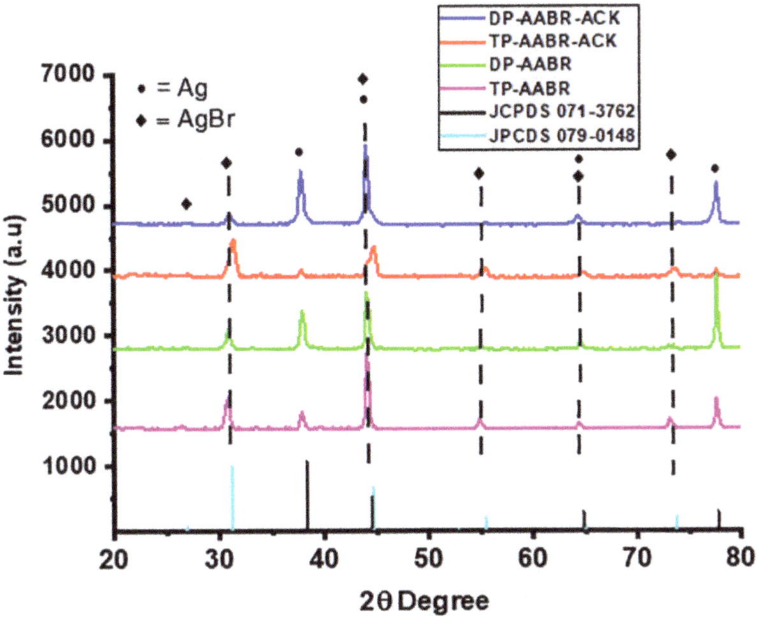

Figure 1. XRD patterns of DP-AABR-ACK, TP-AABR-ACK nanocomposites, DP-AABR, and TP-AABR.

The SEM micrographs of plasmon TP-AABR, TP-AABR-ACK, DP-AABR, and DP-AABR-ACK catalysts are presented in Figure 2a–f; herein, the morphologies of TP-AABR, and DP-AABR are irregular morphologies, agglomerated with large particle size distribution (1.2–1.7 μm) as presented in Figure 2a,d. From the SEM image for TP-AABR-ACK, and DP-AABR-ACK in Figure 2b,e, ordered nanospheres (TP-AABR-ACK) and near spheres (DP-AABR-ACK) of AABR particles are homogeneously dispersed on ACK with a significant reduction in particle size distribution to around 200–350 nm, respectively. The TP-AABR-ACK composite evidenced no agglomeration and a lower particle size diameter (Figure 2b,c) compared with DP-AABR-ACK with some slight agglomeration (Figure 2e,f). The TP-AABR-ACK composite with a minuscule particle size diameter and nanospheres morphology will thus possess a speedier charge carrier transfer [32].

The elements distribution of TP-AABR-ACK and DP-AABR-ACK composites were investigated through the energy-dispersive X-ray spectroscopy (EDX) measurement. The three elements Ag, Br, and C were distributed in the as-prepared samples, which confirmed the co-existence of AABR nanoparticles (NPs) and carbon in the samples as depicted in Figure 3a–h. In addition, a more than 1:1 molar ratio of Ag^+: Br^- in the DP-AABR-ACK sample was observed in EDX (Figure 3a,e), which confirmed the creation of more metallic Ag NPs in the composite, compared to TP-AABR-ACK. The excessive Ag content on the surface of AgBr for DP-AABR-ACK might results in reduced photocatalytic activity [33], while TP-AABBR-ACK with low content and order dispersion of AABR nanospheres will enhance TC removal efficiently in this study [33,34]. The three-dimensional network of the ACK structure favored for uniform distribution of AABR nanoparticles in the formed nanocomposites. Both TP-AABR-ACK and DP-AABR-ACK composites (Figure 3b–h) showed homogeneous distribution of the main elements that promotes improved photogenerated charge carriers transfer efficiency [35], and thus significantly enhances their photocatalytic activities on TC removal, as further discussed in Figure 7.

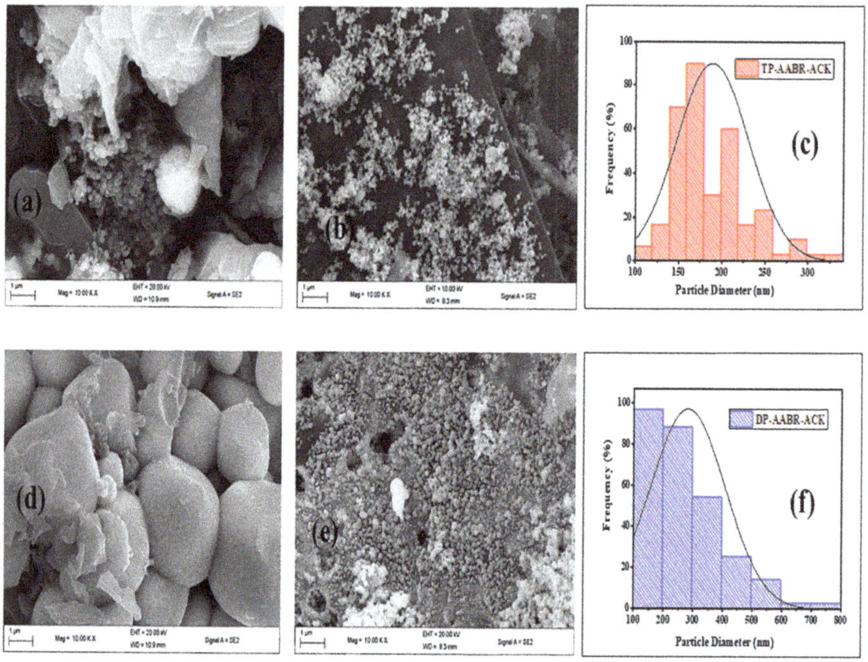

Figure 2. SEM images plasmon: (**a**) TP-AABR, (**b**) TP-AABR-ACK, (**d**) DP-AABR, (**e**) DP-AABR-ACK, and (**c**,**f**) particle size distribution of TP-AABR-ACK and DP-AABR-ACK photocatalysts.

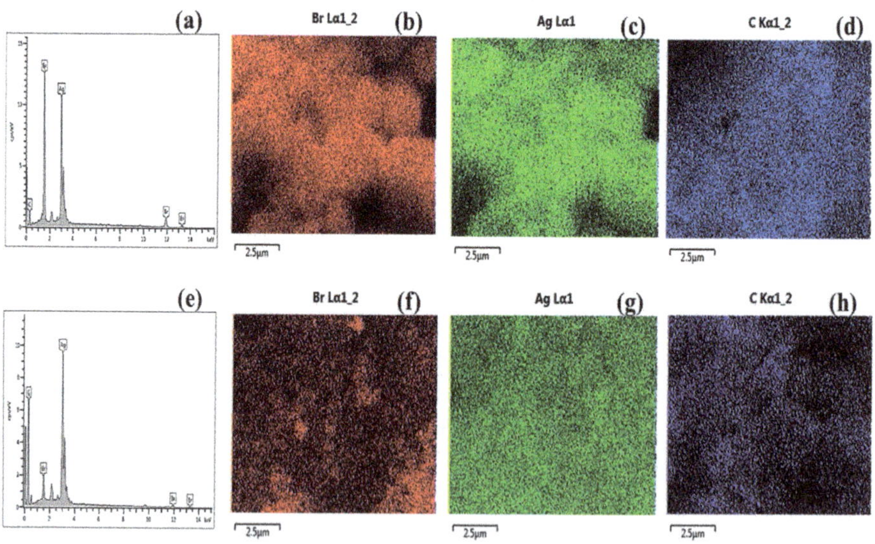

Figure 3. EDX spectrum and elemental mapping of TP-AABR-ACK (**a**–**d**) and DP-AABR-ACK (**e**–**h**).

The TEM image of TP-AAABR-ACK composite as presented in Figure 4a is nanospheres particles with particle size diameter around 3 nm (Figure 4c). The emergence of near spheres particles with the DP-AABR-ACK is evidenced by the TEM image (Figure 4d), with particle size distribution at about 6 nm (Figure 4f). Figure 4b,e from the HRTEM image reveals d-spacing of 0.237 nm, attributed to (111) planes of metallic Ag [36], and d-spacing of 0.286 nm is assigned to (200) planes of silver bromide [37]. The particle size diameter of TP-AABR-ACK is smaller compared to the DP-AABR-ACK composite (Figure 4c,f). The TEM images for both TP-AAABR-ACK, and DP-AAABR-ACK remarkably evidenced the metallic Ag and AgBr within the nanocomposites, as their direct contact with ACK is desirable in the development of a strong heterostructure for enhanced charge carrier separation.

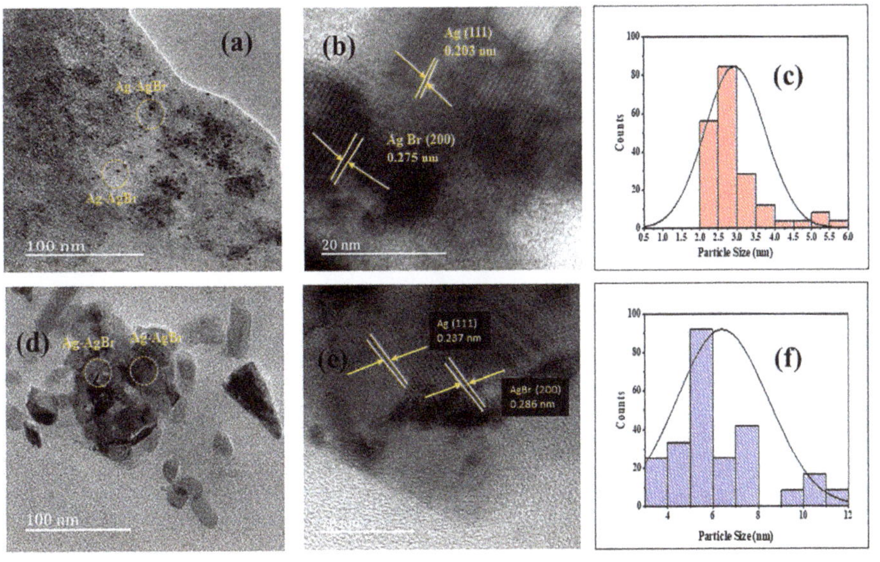

Figure 4. (**a**,**d**) Transmission electron micrographs, (**b**,**d**) high-resolution Transmission electron micrographs, and (**c**,**f**) particle diameter of the TP-AABR-ACK and DP-AABR-ACK photocatalysts.

The composites (TP-AAABR-ACK and DP-AABR-ACK) functionals groups and the presence of carbonaceous material in the composites were confirmed by FTIR measurement and presented in Figure 5. The FTIR weak band at 3405 cm^{-1} associated with weak O-H stretching vibrations of the ACK [38], while the carbonyl (-C=O) stretching band of the carboxylic groups was also observed, 1692 cm^{-1} [39], and the peak at 1575 cm^{-1} is attributed to the -C=C vibrational bond of the aromatic ring [25]. The characteristic bands of C-OH and C-O-C were also detected at 1288, 1127, and 1026 cm^{-1}, respectively, revealed the presence of ACK catalysts support within the AABR-ACK composites. However, the carbonyl stretching bands shifted to a high intensity for the synthesized photocatalysts as presented in Figure 5, which further affirms the formation of firm bonding as observed with previously reported studies [40,41], utilizing carbonaceous support in the hybridization of plasmon AABR composites. In addition, the aliphatic bands of the capped surfactants (PVP and HTAB) at 2912 and 2842 cm^{-1} appear with high and low intensity for TP-AAABR-ACK and DP-AABR-ACK [31,42]. The high intensity of aliphatic bands in TP-AAABR-ACK highlights the formation of ordered nanospheres AABR [43] with small particle diameter (as evidenced from SEM and TEM analysis in Figures 2b and 4a). Overall, reduction in the rapid growth of AABR particles, agglomeration prevention, and decrease in particle size distribution is evident for the active TP-AABR-ACK composite based on FTIR analysis (Figure 5).

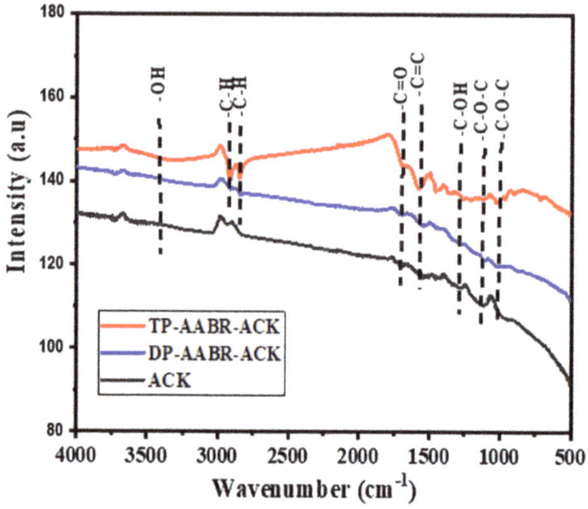

Figure 5. FTIR spectrum of ACK, DP-AABR-ACK, and TP-AABR-ACK composites.

The thermal behavior and the stability of ACK, TP-AABR-ACK, and DP-AABR-ACK/4) were investigated by thermal gravimetric analyses (TGA), and the results are presented in Figure S1. For ACK, three mass losses are evident between 30–150 °C, 150–450 °C, and 500–700 °C; which are ascribed to decomposition of adsorbed water, thermal degradation of cellulose backbone of carbon, and decomposition of extra residual carbonaceous skeleton [25,44]. The DP-AABR-ACK composite also undergoes three weight losses, as the first masses is attributed to the loss of water in the synthesized catalyst, whilst the next two occur around 230–700 °C, corresponding, respectively, to decomposition of inorganic matrixes present and extra residual carbonaceous skeleton. For TP-AABR-ACK, two main sharp decline of mass losses can be identified around 30–150 °C and 230–300 °C, as further temperature beyond 300 °C results in the sample stability, and no additional mass loss was observed. This evidenced that the TP-AABR-ACK from TP synthesis approach is critical in the controlled microstructures of AABR NPs on ACK.

The as-prepared catalysts' (TP-AABR-ACK, DP-AABR-ACK, TP-AABR, and DP-AABR) absorption strength in the visible region was investigated as presented in Figure 6a. The utilization of visible light was evident with the as-prepared catalyst, owing to the surface plasmon resonance (SPR) of metallic Ag NPs [45]. Notably, the absorption intensity for TP-AABR-ACK and DP-AABR-ACK composites increased with the addition of ACK. The 3D hierarchical structure of ACK also aids light transmittance within the TP-AABR-ACK and DP-AABR-ACK composites, thus enhancing the absorption edge of AABR for higher activity compared to TP-AABR and DP-AABR catalysts, similar to other reports [10,46,47]. However, the TP-AABR-ACK composite showed better absorption in the visible region of the solar spectrum compared to the DP-AABR-ACK catalyst, which can be ascribed to controlled microstructures (morphologies and particle size) of AABR NPs. The excessive Ag content in DP-AABR-ACK (from EDX analysis in Figure 3) will cover up the actives sites of the material, and thus reduces the visible light absorption for the generation of ROS [48,49]. The enhanced visible light response of TP-AABR-ACK with controlled microstructures will promote solar spectrum utilization in the rapid formation of photogenerated charge pairs [50], thus resulting in higher photocatalytic activity than other catalysts in this study.

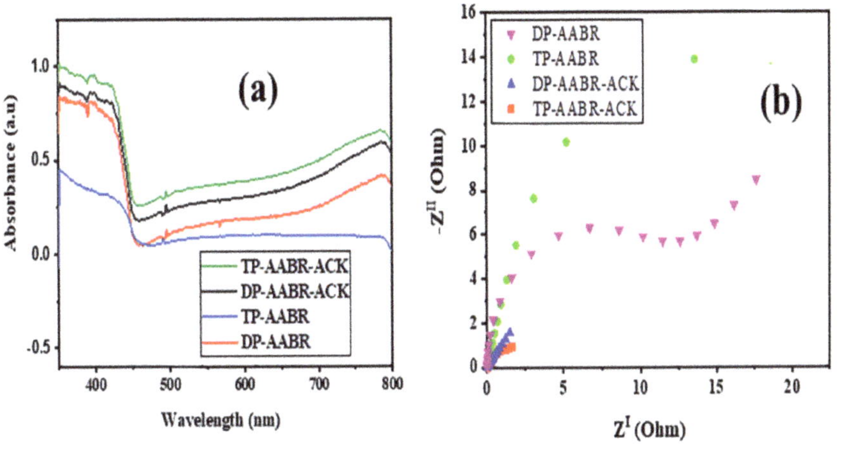

Figure 6. (**a**) UV–vis absorption spectra; and (**b**) EIS Nyquist plot of the synthesized catalysts.

The electrochemical impedance spectra (EIS) were evaluated to demonstrate the conducting attributes of DP-AABR, TP-AABR, DP-AABR-ACK, and TP-AABR-ACK catalysts as presented in Figure 6b. The corresponding EIS Nyquist plot for TP-AABR-ACK displays the smallest arc radius in comparison to other catalysts in this study (Figure 6b), which signifies a lower charge transfer resistance, thus promoting increased interfacial separation of charge carriers [35]. The ACK support significantly acts as a transport channel for the efficient separation of photo-generated charge carriers away from the AABR NPs, thus further boosting the catalytic attributes of TP-AABR-ACK in this study. TP-AABR-ACK with lower charge transfer also results in faster generation of ROS, thus enhancing TC degradation, which is ascribed to controlled morphologies, particle size distribution, and a superior cooperative effect between the AABR and ACK catalyst promoter.

The photocatalytic activities of the synthesized catalysts were assessed by the photocatalytic degradation of TC in aqueous under LED visible light irradiation as presented in Figure 7a. The TC solution without the photocatalyst degradation rate reached 14.66%. The TP-AABR and DP-AABR catalysts showed around 59.72 and 57.10% photoreduction of TC after 180 min of visible light irradiation. The TP-AABR-ACK composite degradation rate (92.08%) is better compared to DP-AABR-ACK (81.12%), which is attributed to the regular morphologies of AABR NPs with reduced particle diameter. The dispersion of regular shaped AABR NPs on ACK improves the degradation rate significantly. The excessive metallic Ag content in DP-AABR-ACK compared to TP-AABR-ACK (as evidenced from EDX analysis in Figure 3) can be a recombination center and hasten the recombination of charge carrier pair, resulting in reduced activity [49,51]. Overall, the TP-AABR-ACK photocatalytic performance was higher than that of DP-AABR-ACK, TP-AABR, DP-AABR NPs, and photolysis in the degradation of TC. The superior photocatalytic activity of the TP-AABR-ACK composite can be assigned to the formation of an effective heterojunction among the components [52]. High activity for the composite is also ascribed to the uniform dispersion of AABR nanospheres on ACK, thus increasing the contact area to visible light, and the promotion of the interfacial charge transfer process [53]. In addition, the photocatalytic degradation of CIP antibiotic and RhB dye was also carried out, as presented in Figure S2a,b. The results of CIP (10 mg/L; 150 mL) and RhB dye (5 mg/L; 150 mL) degradation under the same conditions clearly evidenced that TP-AABR-ACK composite had enhanced photocatalytic activities than other as-prepared catalysts in this study.

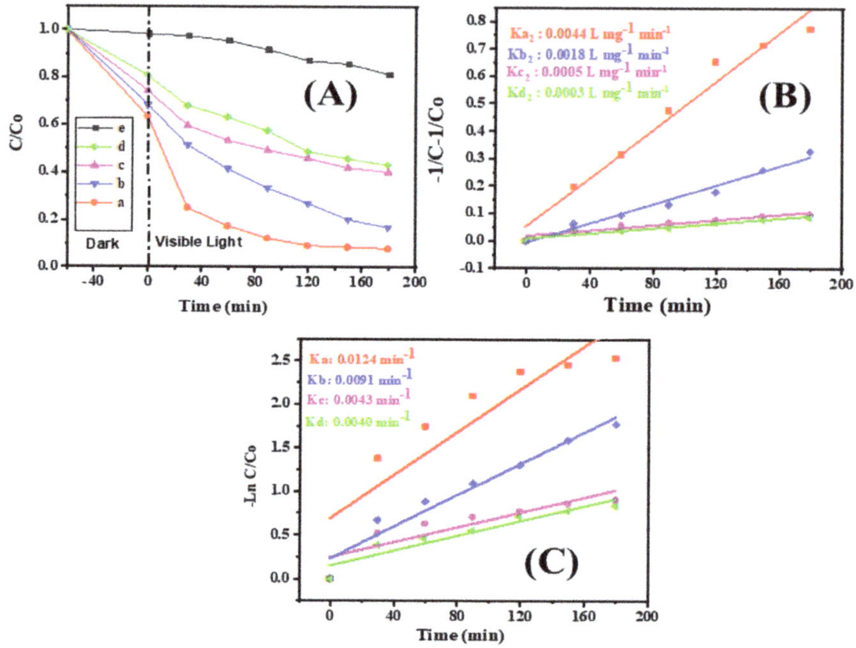

Figure 7. (**A**) Photocatalytic degradation of TC under LED visible light; (**a**) TP-AABR-ACK, (**b**) DP-AABR-ACK, (**c**) TP-AABR, (**d**) DP-AABR, and (**e**) photolysis; (**B**) second and (**C**) first order kinetics.

The photocatalytic degradation kinetics was also investigated to describe the photocatalytic rate and controlling factor for TC degradation. The kinetic data were estimated using the linearized pseudo first-order (Equation (1)) [54] and pseudo second-order models (Equation (2)) as presented below:

$$In\frac{C}{C_O} = -kt \quad (1)$$

$$\frac{1}{C} - \frac{1}{C_O} = -k_2 t \quad (2)$$

Here, t is reaction time, Co, and C are the TC concentrations at a reaction time of 0, and t min, k, and $k2$ represent the apparent rate constants for the pseudo-first order and pseudo second-order kinetics.

The pseudo second order fits the degradation of TC better compared to the pseudo first-order kinetic (Figure 7b,c). The k_{a2} value for TP-AABR-ACK (Figure 7b) is largest compared to DP-AABR-ACK, TP-AABR, and DP-AABR catalysts, respectively, in this work.

Figure 8a,b depict the UV-Vis absorption spectra of photodegraded TC over the active TP-AABR-ACK and DP-AABR-ACK composites, as a function of the time under LED visible light irradiation. The characteristic absorption band of TC antibiotic at 376 nm declines with increasing reaction time, highlighting the destruction of the aromatic ring in TC, as the intermediates were formed [42] and subsequently degraded. This analysis shows that the photocatalytic degradation efficiency of the TP-AABR-ACK is better compared to DP-AABR-ACK.

The TP-AABR-ACK and DP-AABR-ACK composites exhibit excellent photocatalytic degradation of TC in comparison with previously reported silver based photocatalysts in recent literature, as depicted in Table 1, thus justifying its potential applications in an industrial photocatalytic process.

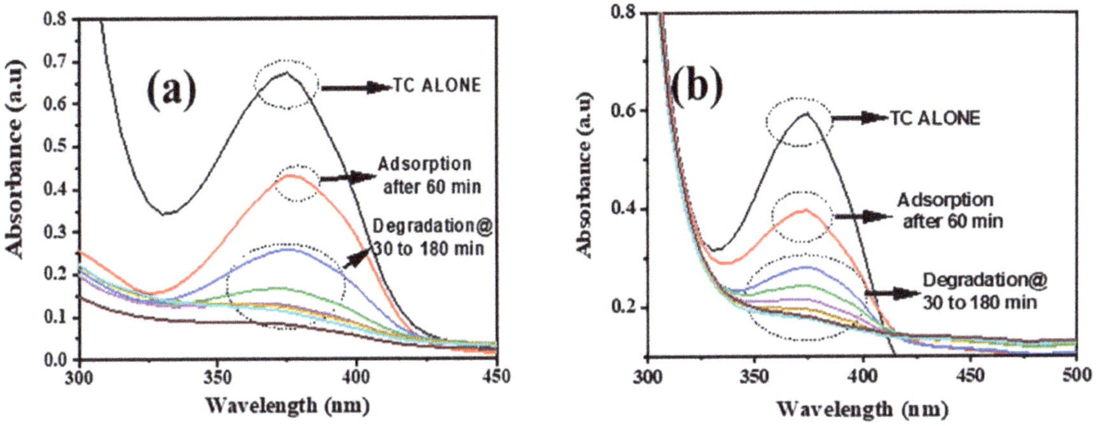

Figure 8. UV-vis absorption spectra of photodegraded TC solution using (**a**) TP-AABR-ACK and (**b**) DP-AABR-ACK.

Table 1. Comparison of degradation efficiency of different heterostructures with AABR-ACK composites on tetracycline removal.

Catalyst	Catalyst Mass (g), TC Concentration (ppm)	Light Source	% Degradation	References
$Ag/AgBr/AgIn(MoO_4)_2$	0.1, 10	500 W Xe lamp	42 after 40 min	[55]
$AgI\text{-}WO_3$	0.04, 35	300 W Xe lamp	75 after 60 min	[56]
$Ag\text{-}K_2Ta_2O_6$	0.1, 20	300 W Xe lamp	50 after 270 min	[57]
Ag/Bi_3TaO_7	0.05, 10	250 W Xe lamp	85 after 60 min	[58]
$Ag_2CO_3/Ag/WO_3$	0.1, 10	300 W Xe lamp	81 after 90 min	[59]
graphene-like BN/BiOBr	0.05, 20	300 W Xe lamp	75 after 80 min	[60]
GO/CN/BiOI	0.03, 20	35 W LED track light	74, after 100 min	[61]
Bi_4NbO_8Cl perovskite	0.01, 20	18-W LED bulb	79, after 60 min	[62]
TP-AABR-ACKDP-AABR-ACK	0.045, 15	36 W Visible LED Light	92 after 180 min 81 after 180 min	This Study

The recovery capacity of the active AABR-ACK nanocomposites was also researched under the LED powered visible light illumination up to five cycles, as presented in Figure S3. The degradation slightly diminishes from 91.9 to 83.27% (TP-AABR-ACK) and 81.12 to 74.14% (DP-AABR-ACK) after repeated five cycles., thus highlighting its reusability. Besides, the XRD pattern of active AABR-ACK nanocomposites after 5 cycles also affirms the stability of the synthesized composites (Figure S3b). The characteristic peaks of the photocatalysts remain unchanged, showing the good reusability of active AABR-ACK nanocomposites after the photocatalytic reaction process.

TOC Teledyne Tekmar analyzer evaluated the mineralization ability of the prepared catalysts, and their results are presented in Figure 9. The carbon content of degraded TC molecules decreased significantly, as the TP-AABR-ACK composite gave a total TOC removal efficiency of 87.5% after 180 min, which is higher compared to DP-AABR-ACK, TP-AABR, and DP-AABR (78.5%, 19%, and 15.4%) under the same conditions. This further indicates that the TP-AABR-ACK composite presents enhanced mineralization ability in TC antibiotic degradation, indicating the formation of inorganic ions, and CO_2. Overall, the AABR NPs uniformly decorated onto the 3D porous network ACK in the TP-AABR-ACK composite enhanced the mineralization process five-fold.

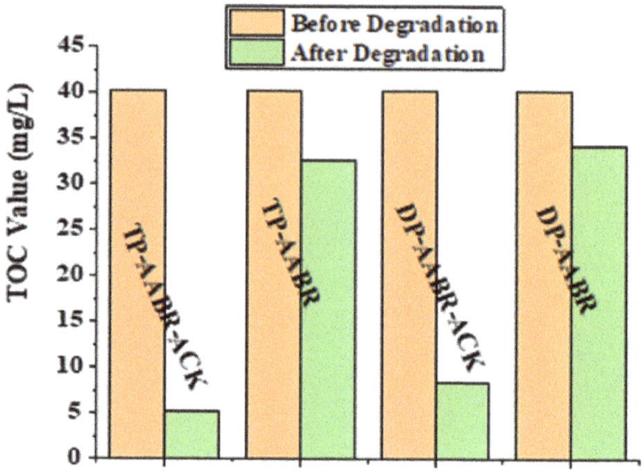

Figure 9. TOC changes of TC before and after photocatalytic reaction using as-prepared samples.

In the scavenging experiment, the degradation rate of TC significantly inhibited around 9–12%, and 29–35%, with the addition of BQ and EDTA-Na_2, respectively (Figure 10). The addition of IPA had a minimal contribution, while the mixture of all the scavengers highlights the ROS contribution in TC degradation in this study. The $^{\bullet}O_2^-$ significantly enhanced the TC degradation, which aligned well with previous studies [30,63]; thus, the significant role of superoxide anion radical is further explored in this study.

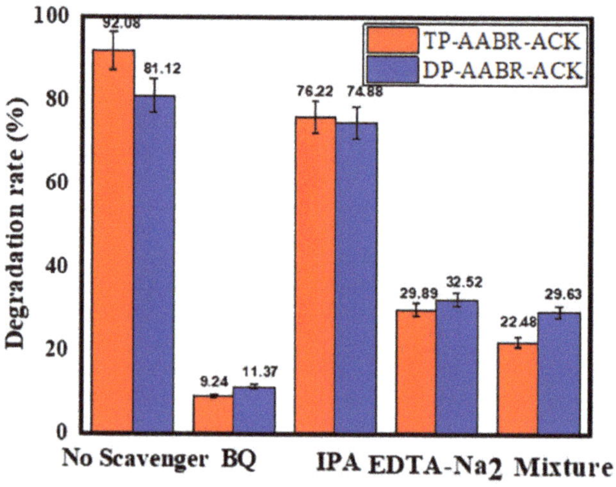

Figure 10. Photocatalytic degradation of TC solution over active AABR-ACK nanocomposites with scavenger agents.

The formation rate of superoxide anion radical ($^{\bullet}O_2^-$) through the nitroblue tetrazolium (NBT) photocatalytic degradation conversion into diformazan [64] under visible light irradiation was conducted. The production rate of $^{\bullet}O_2^-$ measured through the reduction in absorbance band of NBT molecules @259 nm is presented in Figure 11a–d. A higher amount of reactive $^{\bullet}O_2^-$ is generated more with TP-AABR-ACK composite (Figure 11d) with a rapid reduction in absorbance wavelength of NBT, which aligns with the enhanced degradation rate of TC antibiotic compared to DP-AABR-ACK, TP-AABR, and DP-AABR catalysts. Among the photocatalysts employed, the highest generation rate of a superoxide radical based on the highest NBT disappearance follows the order TP-AABR-ACK > DP-

AABR-ACK > TP-AABR > DP-AABR in this study. The enhanced production of ROS is significantly attributed with the efficient separation of the photogenerated charge carriers of the TP-AABR-ACK (as evidenced from EIS Nyquist spectra in Figure 6).

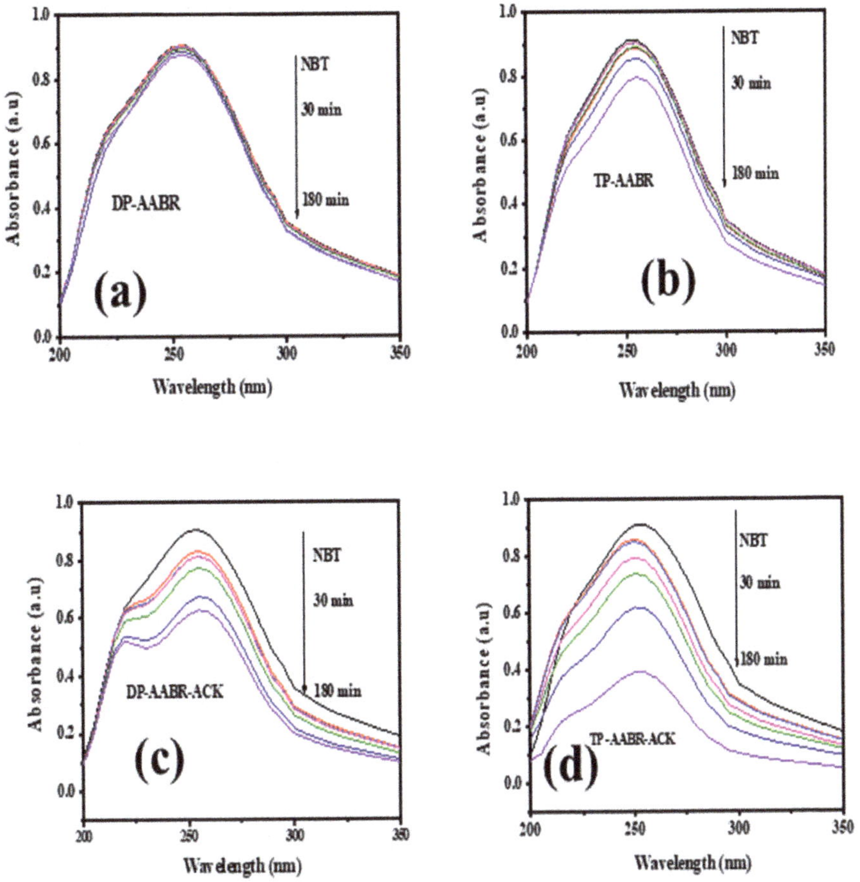

Figure 11. UV-Vis absorption spectra of NBT degraded by (**a**) DP-AABR, (**b**) TP-AABR, (**c**) DP-AABR-ACK, and (**d**) TP-AABR-ACK.

From the discussion described above, the ACK catalyst promoter/support significantly reduced aggregation and agglomeration AABR NPs, along with controlled morphologies of AABR-ACK composites, and thus promotes high interfacial separation of photogenerated charge carriers. These attributes also result in abundance generation of ROS with these composites, thus resulting in higher photocatalytic activity towards oxidative reduction of TC as depicted in Figure 7.

3. Experimental Section

3.1. Materials

Ethylene glycol (EG, 99%, Acros), polyvinylpyrrolidone (0.26 g, PVP, M.W. 58,000, K 29–32 Acros), hexadecyltrimethylammonium bromide (HTAB, Acros, 99%), silver nitrate ($AgNO_3$, Merck, Johannesburg, South Africa, 98%), nitroblue tetrazolium solution (Sigma Aldrich, St. Louis, MO, USA, 99%), *p*-benzoquinone (BQ), disodium ethylenediaminetetraacetic acid (EDTA-Na_2), and isopropanol (IPA) were used as received.

3.2. Activated Carbon (ACK)

Activated carbon produced from the microwave heating of impregnated pinecone biomass has been reported [25,30].

3.2.1. Preparation of TP-AABR-ACK through Thermal Polyol Route

TP-AABR-ACK was synthesized by a modified thermal polyol process [17,30]: 18 mL of EG was added into a250 mL round-bottom flask, and further subjected to heating at 60 °C for 30 min. Polyvinylpyrrolidone (0.26 g, PVP) and 0.48 g of HTAB were sequentially added to the stirred solution, and then, 0.03 g of ACK was further added to the reaction mixture. Two hundred mg $AgNO_3$ dissolved in 3 mL EG solution in the dark was then added drop-wise into the stirred solution. The suspension was maintained at 60 °C for another 30 min, and thereafter heated to 140 °C. After reaching 140 °C, the reaction was stirred further for 18 min. Thereafter, precipitate was collected by centrifugation, washed with ethanol three times, and dried at 60 °C in an oven. TP-AABR was prepared through the above-described route, without the addition of ACK.

3.2.2. Preparation of DP-AABR-ACK through Deposition-Precipitation Route

AABR NPs composite was synthesized onto ACK catalyst support via the deposition–precipitation method [11]. In brief, 12 mL EG solution was added into a 250 mL round-bottom flask, and heated at a temperature of 65 °C for 30 min. Then, 0.26 g HTAB and 0.09 g ACK were added to the magnetically stirred solution. A defined amount of $AgNO_3$, dissolved in 1 M ammonium water 2.5 mL, was added dropwise to the stirred solution. The suspension was magnetically stirred for another 6 h under ambient light, required for the production of Ag NPs. The precipitate was collected by centrifugation, further washed with ethanol and deionized water three times, and dried in an oven at 60 °C overnight. DP-AABR was synthesized using the same route without ACK in the synthesis process.

3.3. Characterization

X-ray diffraction (XRD) patterns were obtained using a powder X-ray diffractometer (Shimadzu X-ray 700) at 40 kV and 15 mA with a Cu Kα radiation source (0.1504 nm). The morphology and elemental mapping of the synthesized photocatalyst samples were observed by the scanning electron microscopy (SEM, ZEISS Ultra/Plus FEG-SEM) with an accelerating voltage around 10–20 kV. The microstructure attributes were conducted by transmission electron microscope (TEM, JEOL JEM-2010) with an accelerating voltage of 200 kV. Fourier transform infrared spectrometer (FTIR) spectra of the samples were recorded on a Perkin Elmer spectrum 400, in the range of 600–4000 cm^{-1}. PerkinElmer STA 6000 was employed for thermal studies at a heating rate of 10 °C min^{-1} in the temperature range from 30 to 900 °C under nitrogen purge stream. Ultraviolet-visible (UV–vis) diffuse reflectance spectra were performed on a Maya 2000, Ocean Optics spectrometer using $BaSO_4$ as the reference sample. A Biologic SP 240 potentiostat workstation was utilized for electrochemical impedance spectroscopy (EIS) tests under a typical three-electrode system. A platinum wire, glassy carbon electrode, an Ag/AgCl electrode served as the counter, working, and reference electrodes, respectively.

3.4. Measurement of Photocatalytic Activity

Photocatalytic activities of the synthesized photocatalyst were evaluated by the removal efficiency of TC antibiotic with visible light irradiation. One hundred and fifty mL aqueous suspensions, containing 0.3 g/L catalyst loading of the photocatalysts, and 15 mg/L of TC, were prepared in 500 mL beaker. Before irradiation, the above-mixed solution was subjected to a dark reaction for 60 min to establish an adsorption–desorption equilibrium. Afterwards, the prepare suspension was irradiated by LED Visible light (36 W) with continuous magnetic stirring. During the photoreaction, a 3.0 mL reaction solution was collected at 30 min regular intervals and centrifuged to separate the photocatalyst. The residual TC solution was measured via a UV-vis spectrophotometer at the characteristic absorption peak of 376 nm. Moreover, the influence of sacrificial agents comprising of *p*-benzoquinone (BQ) for superoxide ($^{\cdot}O_2^-$), disodium ethylenediaminetetraacetic acid (EDTA-Na_2) for the hole (h^+), and isopropanol (IPA) for hydroxyl ($^{\cdot}OH$) at 1 mmole, on the degradation rate of TC over most active TP-AABR-ACK and DP-AABR-ACK, was also evaluated in this study.

The recyclability experiment of our active AABR-ACK nanocomposites was examined for five cycles of TC photocatalytic degradation, utilizing 150 mL of the TC antibiotic (15 mg/L). After the initial first cycle, the photocatalyst nanocomposites were detached from the photodegraded TC solution by centrifugation and washed with distilled water to eliminate any adsorbed TC. The washed photocatalyst was then dried at room temperature, and afterward reused in the following TC degradation cycle. This process was repeated four times to acquire the five reusability cycles.

Similar to the photodegradation experiments, the measurement of superoxide radicals generated was carried out, using nitroblue tetrazolium molecules. The nitroblue tetrazolium (NBT) can be specifically reduced by the superoxide ion to form the insoluble purple formazan in the aqueous solution. A 20 mg/L Nitroblue Tetrazolium solution (Sigma Aldrich, 99%) with 0.3 g/L of photocatalysts was degraded under LED visible light, and the change in concentration was measured at a maximum absorbance wavelength of 259 nm. Total Organic Carbon (TOC) analysis was performed to determine the mineralization degree of the TC reached after the photocatalytic process.

4. Conclusions

The morphologies and particle size distributions of AABR-ACK composites (TP-AABR-ACK and DP-AABR-ACK) were controlled in this work, under the influence of two methods (thermal polyol and deposition-precipitation). TP-AABR-ACK had unique nanospheres morphologies, lower particle size, no agglomeration, thus resulting in restrained recombination of photogenerated charge carriers. The TP-AABR-ACK composite activities on TC degradation were superior to DP-AABR-ACK and corresponding AABR NPs under visible light irradiation, its exceptional morphological and conductivity properties, thus further evidenced in the TOC tests. The ACK catalyst support with interfacial interaction with AABR NPs improved the crystallinity of AABR NPs through shifting of the peaks. The ACK also boosted the light harvesting attributes of AABR NPs into the visible region for more production of ROS, and enhanced the separation of photogenerated charge carrier in the catalytic process. In addition, the TP-AABR-ACK nanocomposite is capable of producing abundant ROS (superoxide anions radicals), compared to DP-AABR-ACK, occurring through the intrinsic interactive reaction of nitroblue tetrazolium molecules with the active TP-AABR-ACK.

Supplementary Materials: The following are available online at https://www.mdpi.com/article/10.3390/catal11111396/s1, Figure S1: Thermogravimetric analysis of ACK, DP-AABR-ACK, and TP-AABR-ACK, Figure S2: Photocatalytic degradation efficiencies of (A) ciprofloxacin, and (B) rhodamine B as a function of irradiation time for (a) TP-AABR-ACK, (b) DP-AABR-ACK, (c) TP-AABR, (d) DP-AABR and (e) photolysis, Figure S3: Reusability test AABR-ACK nanocomposites for the degradation of TC under LED visible light illumination, and (b) XRD patterns of the AABR-ACK nanocomposites prior, and then afterward test.

Author Contributions: Conceptualization, S.O.S.; methodology, S.O.S., E.L.V.; writing—original draft preparation, S.O.S., E.L.V.; writing—review and editing, S.O.S., E.L.V., H.G.B.; resources, S.O.S., H.G.B.; supervision, S.O.S., E.L.V., H.G.B.; funding acquisition, S.O.S., H.G.B. All authors have read and agreed to the published version of the manuscript.

Funding: This research was supported by the Sasol University Collaboration Program, National Research Foundation of South Africa through the grant No. 99330, and 111330, and the Vaal University of Technology. We acknowledge the Built Environment and Information Technology Department, University of Pretoria, for their enormous financial support.

Data Availability Statement: The data presented in this study are openly available in the University of Pretoria Research Data Repository at DOI: 10.25403/UPresearchdata.17038907.

Conflicts of Interest: The authors declare no conflict of interest.

References

1. Zhang, H.; Liu, P.; Feng, Y.; Yang, F. Fate of antibiotics during wastewater treatment and antibiotic distribution in the effluent-receiving waters of the Yellow Sea, northern China. *Mar. Poll. Bull.* **2013**, *73*, 282–290. [CrossRef]
2. Fernandes, A.; Oliveira, C.; Pacheco, M.J.; Ciríaco, L.; Lopes, A. Anodic oxidation of oxytetracycline: Influence of the experimental conditions on the degradation rate and mechanism. *J. Electrochem. Sci. Eng.* **2014**, *4*, 203–213. [CrossRef]
3. Zhu, Y.; Han, Z.; Zhao, S.; Zhang, Q.; Shen, X.; Lv, H.; Liu, J.; Li, B. In-situ growth of Ag/AgBr nanoparticles on a metal organic framework with enhanced visible light photocatalytic performance. *Mater. Sci. Semicond. Process.* **2021**, *133*, 105973. [CrossRef]
4. Chao, Y.; Zhu, W.; Yan, B.; Lin, Y.; Xun, S.; Ji, H.; Wu, X.; Li, H.; Han, C. Macroporous polystyrene resins as adsorbents for the removal of tetracycline antibiotics from an aquatic environment. *J. Appl. Polym. Sci.* **2014**, *131*, 40561. [CrossRef]
5. Yao, Y.; Zhang, Y.; Shen, M.; Li, W.; Xia, W. The facile synthesis and enhanced photocatalytic properties of ZnO@ ZnS modified with Ag0 via in-situ ion exchange. *Colloids. Surf. A Physicochem. Eng. Aspects* **2020**, *591*, 124556. [CrossRef]
6. Fang, Y.; Li, Y.; Zhou, F.; Gu, P.; Liu, J.; Chen, D.; Li, N.; Xu, Q.; Lu, J. An Efficient Photocatalyst Based on Black TiO_2 Nanoparticles and Porous Carbon with High Surface Area: Degradation of Antibiotics and Organic Pollutants in Water. *Chem. Plus Chem.* **2019**, *84*, 474–480.
7. Zhao, X.; Lu, Z.; Wei, M.; Zhang, M.; Dong, H.; Yi, C.; Ji, R.; Yan, Y. Synergetic effect of carbon sphere derived from yeast with magnetism and cobalt oxide nanochains towards improving photodegradation activity for various pollutants. *Appl. Catal. B Environ.* **2018**, *220*, 137–147. [CrossRef]
8. Ye, Y.; Zang, Z.; Zhou, T.; Dong, F.; Lu, S.; Tang, X.; Wei, W.; Zhang, Y. Theoretical and experimental investigation of highly photocatalytic performance of CuInZnS nanoporous structure for removing the NO gas. *J. Catal.* **2018**, *357*, 100–107. [CrossRef]
9. Bhachu, D.S.; Moniz, S.J.; Sathasivam, S.; Scanlon, D.O.; Walsh, A.; Bawaked, S.M.; Mokhtar, M.; Obaid, A.Y.; Parkin, I.P.; Tang, J. Bismuth oxyhalides: Synthesis, structure and photoelectrochemical activity. *Chem. Sci.* **2016**, *7*, 4832–4841. [CrossRef]
10. Bano, Z.; Saeed, R.Y.; Zhu, S.; Xia, M.; Mao, S.; Lei, W.; Wang, F. Mesoporous CuS nanospheres decorated rGO aerogel for high photocatalytic activity towards Cr (VI) and organic pollutants. *Chemosphere* **2020**, *246*, 125846. [CrossRef] [PubMed]
11. Zhu, M.; Chen, P.; Liu, M. High-performance visible-light-driven plasmonic photocatalysts Ag/AgCl with controlled size and shape using graphene oxide as capping agent and catalyst promoter. *Langmuir* **2013**, *29*, 9259–9268. [CrossRef]
12. Yan, T.; Zhang, H.; Luo, Q.; Ma, Y.; Lin, H.; You, J. Controllable synthesis of plasmonic Ag/AgBr photocatalysts by a facile one-pot solvothermal route. *Chem. Eng. J.* **2013**, *232*, 564–572. [CrossRef]
13. Errokh, A.; Cheikhrouhou, W.; Ferraria, A.M.; do Rego, A.M.B.; Boufi, S. Cotton decorated with Cu_2O-Ag and Cu_2O-Ag-AgBr NPs via an in-situ sacrificial template approach and their antibacterial efficiency. *Colloids Surf. B Biointerface* **2021**, *200*, 111600. [CrossRef] [PubMed]
14. Cao, D.; Wang, Q.; Liu, Z.; Zhang, H.; Wang, Y.; Jin, R.; Gao, S. Enhanced the photoelectrocatalytic performance of TiO_2 nanotube arrays by the synergistic sensitization of Ag–AgBr nanospheres. *Spectrochim. Acta Part A Mol. Biomol. Spectrosc.* **2020**, *227*, 117674. [CrossRef] [PubMed]
15. Xiao, J.-Q.; Lin, K.-S.; Yu, Y. Novel Ag@ AgCl@ AgBr heterostructured nanotubes as high-performance visible-light photocatalysts for decomposition of dyes. *Catal. Today* **2018**, *314*, 10–19. [CrossRef]
16. Li, B.; Wang, H.; Zhang, B.; Hu, P.; Chen, C.; Guo, L. Facile synthesis of one dimensional AgBr@ Ag nanostructures and their visible light photocatalytic properties. *ACS Appl. Mater. Interface* **2013**, *5*, 12283–12287. [CrossRef] [PubMed]
17. Chen, G.; Li, F.; Huang, Z.; Guo, C.-Y.; Qiao, H.; Qiu, X.; Wang, Z.; Jiang, W.; Yuan, G. Facile synthesis of Ag/AgBr/RGO nanocomposite as a highly efficient sunlight plasmonic photocatalyst. *Catal. Commun.* **2015**, *59*, 140–144. [CrossRef]
18. Xu, X.; Shen, X.; Zhou, H.; Qiu, D.; Zhu, G.; Chen, K. Facile microwave-assisted synthesis of monodispersed ball-like Ag@ AgBr photocatalyst with high activity and durability. *Appl. Catal. A Gen.* **2013**, *455*, 183–192. [CrossRef]
19. Parale, V.G.; Kim, T.; Phadtare, V.D.; Yadav, H.M.; Park, H.-H. Enhanced photocatalytic activity of a mesoporous TiO_2 aerogel decorated onto three-dimensional carbon foam. *J. Mol. Liq.* **2019**, *277*, 424–433. [CrossRef]
20. Li, G.; Huang, B.; Pan, Z.; Su, X.; Shao, Z.; An, L. Advances in three-dimensional graphene-based materials: Configurations, preparation and application in secondary metal (Li, Na, K, Mg, Al)-ion batteries. *Energy Environ. Sci.* **2019**, *12*, 2030–2053. [CrossRef]
21. Ananthanarayanan, A.; Wang, X.; Routh, P.; Sana, B.; Lim, S.; Kim, D.H.; Lim, K.H.; Li, J.; Chen, P. Facile synthesis of graphene quantum dots from 3D graphene and their application for Fe^{3+} sensing. *Adv. Funct. Mater.* **2014**, *24*, 3021–3026. [CrossRef]
22. Che, H.; Che, G.; Zhou, P.; Liu, C.; Dong, H. Yeast-derived carbon sphere as a bridge of charge carriers towards to enhanced photocatalytic activity of 2D/2D Cu_2WS_4/g-C_3N_4 heterojunction. *J. Colloid Interface Sci.* **2019**, *546*, 262–275. [CrossRef] [PubMed]
23. Zhou, X.; Zhou, S.; Ma, F.; Xu, Y. Synergistic effects and kinetics of rGO-modified TiO_2 nanocomposite on adsorption and photocatalytic degradation of humic acid. *J. Environ. Manag.* **2019**, *235*, 293–302. [CrossRef] [PubMed]
24. Xu, P.; Cen, C.; Zheng, M.; Wang, Y.; Wu, Z.; Teng, Z. A facile electrostatic droplets assisted synthesis of copper nanoparticles embedded magnetic carbon microspheres for highly effective catalytic reduction of 4-nitrophenol and Rhodamine B. *Mater. Chem. Phys.* **2020**, *253*, 123444. [CrossRef]

25. Sanni, S.; Viljoen, E.; Ofomaja, A. Three-dimensional hierarchical porous carbon structure derived from pinecone as a potential catalyst support in catalytic remediation of antibiotics. *RSC Adv.* **2020**, *10*, 8717–8728. [CrossRef]
26. Fu, Y.; Xu, P.; Huang, D.; Zeng, G.; Lai, C.; Qin, L.; Li, B.; He, J.; Yi, H.; Cheng, M. Au nanoparticles decorated on activated coke via a facile preparation for efficient catalytic reduction of nitrophenols and azo dyes. *Appl. Surf. Sci.* **2019**, *473*, 578–588. [CrossRef]
27. Zhu, Q.; Wang, W.-S.; Lin, L.; Gao, G.-Q.; Guo, H.-L.; Du, H.; Xu, A.-W. Facile synthesis of the novel Ag_3VO_4/AgBr/Ag plasmonic photocatalyst with enhanced photocatalytic activity and stability. *J. Phys. Chem. C* **2013**, *117*, 5894–5900. [CrossRef]
28. Lou, Z.; Huang, B.; Qin, X.; Zhang, X.; Cheng, H.; Liu, Y.; Wang, S.; Wang, J.; Dai, Y. One-step synthesis of AgCl concave cubes by preferential overgrowth along <111> and <110> directions. *Chem. Commun.* **2012**, *48*, 3488–3490. [CrossRef]
29. Lin, Z.; Xiao, J.; Yan, J.; Liu, P.; Li, L.; Yang, G. Ag/AgCl plasmonic cubes with ultrahigh activity as advanced visible-light photocatalysts for photodegrading dyes. *J. Mater. Chem. A* **2015**, *3*, 7649–7658. [CrossRef]
30. Sanni, S.; Viljoen, E.; Ofomaja, A. Accelerated electron transport and improved photocatalytic activity of Ag/AgBr under visible light irradiation based on conductive carbon derived biomass. *Catal. Lett.* **2019**, *149*, 3027–3040. [CrossRef]
31. Sanni, S.; Viljoen, E.; Ofomaja, A. Tailored synthesis of Ag/AgBr nanostructures coupled activated carbon with intimate interface interaction for enhanced photodegradation of tetracycline. *Process Saf. Environ. Prot.* **2021**, *146*, 20–34. [CrossRef]
32. He, J.; Yang, J.; Jiang, F.; Liu, P.; Zhu, M. Photo-assisted peroxymonosulfate activation via 2D/2D heterostructure of Ti_3C_2/g-C_3N_4 for degradation of diclofenac. *Chemosphere* **2020**, *258*, 127339. [CrossRef]
33. Guo, H.; Niu, C.-G.; Wen, X.-J.; Zhang, L.; Liang, C.; Zhang, X.-G.; Guan, D.-L.; Tang, N.; Zeng, G.-M. Construction of highly efficient and stable ternary AgBr/Ag/$PbBiO_2Br$ Z-scheme photocatalyst under visible light irradiation: Performance and mechanism insight. *J. Colloid Interface Sci.* **2018**, *513*, 852–865. [CrossRef]
34. Zhang, Y.; Tang, Z.-R.; Fu, X.; Xu, Y.-J. Nanocomposite of Ag–AgBr–TiO_2 as a photoactive and durable catalyst for degradation of volatile organic compounds in the gas phase. *App. Catal. B Environ.* **2011**, *106*, 445–452. [CrossRef]
35. He, F.; Wang, Z.; Li, Y.; Peng, S.; Liu, B. The nonmetal modulation of composition and morphology of g-C_3N_4-based photocatalysts. *App. Catal. B Environ.* **2020**, *269*, 118828. [CrossRef]
36. Madigasekara, I.H.K.; Perera, H.C.S.; Kumari, J.M.K.W.; Senadeera, G.K.R.; Dissanayake, M.A.K.L. Photoanode modification of dye-sensitized solar cells with Ag/AgBr/TiO_2 nanocomposite for enhanced cell efficiency. *Sol. Energy* **2021**, *230*, 59–72. [CrossRef]
37. Sui, Y.; Su, C.; Yang, X.; Hu, J.; Lin, X. Ag-AgBr nanoparticles loaded on TiO_2 nanofibers as an efficient heterostructured photocatalyst driven by visible light. *J. Mol. Catal. A Chem.* **2015**, *410*, 226–234. [CrossRef]
38. López-Peñalver, J.J.; Sánchez-Polo, M.; Gómez-Pacheco, C.V.; Rivera-Utrilla, J. Photodegradation of tetracyclines in aqueous solution by using UV and UV/H2O2 oxidation processes. *J. Chem. Technol. Biotechnol.* **2010**, *85*, 1325–1333. [CrossRef]
39. Feng, Y.; Zhou, H.; Liu, G.; Qiao, J.; Wang, J.; Lu, H.; Yang, L.; Wu, Y. Methylene blue adsorption onto swede rape straw (Brassica napus L.) modified by tartaric acid: Equilibrium, kinetic and adsorption mechanisms. *Bioresour. Technol.* **2012**, *125*, 138–144. [CrossRef] [PubMed]
40. Petroski, J.; El-Sayed, M.A. FTIR study of the adsorption of the capping material to different platinum nanoparticle shapes. *J. Phys. Chem. A* **2003**, *107*, 8371–8375. [CrossRef]
41. Zhu, M.; Chen, P.; Liu, M. Graphene oxide enwrapped Ag/AgX (X = Br, Cl) nanocomposite as a highly efficient visible-light plasmonic photocatalyst. *ACS Nano* **2011**, *5*, 4529–4536. [CrossRef] [PubMed]
42. Sanni, S.; Viljoen, E.; Ofomaja, A. Design of ordered Ag/AgBr nanostructures coupled activated carbon with enhanced charge carriers separation efficiency for photodegradation of tetracycline under visible light. *J. Mol. Liq.* **2020**, *299*, 112032. [CrossRef]
43. Attia, Y.A.; Mohamed, Y.M.A. Silicon-grafted Ag/AgX/rGO nanomaterials (X = Cl or Br) as dip-photocatalysts for highly efficient p-nitrophenol reduction and paracetamol production. *Appl. Organomet. Chem.* **2019**, *33*, e4757. [CrossRef]
44. Singh, Y.D.; Mahanta, P.; Bora, U. Comprehensive characterization of lignocellulosic biomass through proximate, ultimate and compositional analysis for bioenergy production. *Renew. Energy* **2017**, *103*, 490–500. [CrossRef]
45. Jiang, J.; Li, H.; Zhang, L. New insight into daylight photocatalysis of AgBr@ Ag: Synergistic effect between semiconductor photocatalysis and plasmonic photocatalysis. *Chem. A Eur. J.* **2012**, *18*, 6360–6369. [CrossRef] [PubMed]
46. Sanni, S.; Modise, S.; Viljoen, E.; Ofomaja, A. Enhanced degradation of dye mixtures: Physicochemical and electrochemical properties of titania dispersed on clinoptilolite, synergistic influence. *SN Appl. Sci.* **2020**, *2*, 1–10. [CrossRef]
47. Wang, H.; Liu, H.; Wang, S.; Li, L.; Liu, X. Influence of tunable pore size on photocatalytic and photoelectrochemical performances of hierarchical porous TiO_2/C nanocomposites synthesized via dual-Templating. *Appl. Catal. B Environ.* **2018**, *224*, 341–349. [CrossRef]
48. Bhatt, D.K.; Patel, U.D. Mechanism underlying visible-light photocatalytic activity of Ag/AgBr: Experimental and theoretical approaches. *J. Phys. Chem. Solids* **2019**, *135*, 109118. [CrossRef]
49. Deng, F.; Zhao, L.; Luo, X.; Luo, S.; Dionysiou, D.D. Highly efficient visible-light photocatalytic performance of Ag/$AgIn_5S_8$ for degradation of tetracycline hydrochloride and treatment of real pharmaceutical industry wastewater. *Chem. Eng. J.* **2018**, *333*, 423–433. [CrossRef]
50. Soltani, T.; Tayyebi, A.; Lee, B.-K. Photolysis and photocatalysis of tetracycline by sonochemically heterojunctioned BiVO4/reduced graphene oxide under visible-light irradiation. *J. Environ. Manag.* **2019**, *232*, 713–721. [CrossRef]
51. Hu, P.; Cao, Y.J.D.T. A new chemical route to a hybrid nanostructure: Room-temperature solid-state reaction synthesis of Ag@ AgCl with efficient photocatalysis. *Dalton Trans.* **2012**, *41*, 8908–8912. [CrossRef]

52. Ren, Y.; Dong, T.; Ding, S.; Liu, X.; Zheng, H.; Gao, L.; Hu, J. AgBr Nanoparticles Anchored on CdS Nanorods as Photocatalysts for H2 Evolution. *ACS Appl. Nano Mater.* **2021**, *4*, 9274–9282. [CrossRef]
53. De Moraes, N.P.; Valim, R.B.; da Silva Rocha, R.; da Silva, M.L.C.P.; Campos, T.M.B.; Thim, G.P.; Rodrigues, L.A. Effect of synthesis medium on structural and photocatalytic properties of ZnO/carbon xerogel composites for solar and visible light degradation of 4-chlorophenol and bisphenol A. *Colloids Surf. A Physicochem. Eng. Asp.* **2020**, *584*, 124034. [CrossRef]
54. Xu, J.; Wang, W.; Shang, M.; Sun, S.; Ren, J.; Zhang, L. Efficient visible light induced degradation of organic contaminants by Bi_2WO_6 film on SiO_2 modified reticular substrate. *Appl. Catal. B Environ.* **2010**, *93*, 227–232. [CrossRef]
55. Yan, X.; Wang, X.; Gu, W.; Wu, M.; Yan, Y.; Hu, B.; Che, G.; Han, D.; Yang, J.; Fan, W. Single-crystalline AgIn $(MoO_4)_2$ nanosheets grafted Ag/AgBr composites with enhanced plasmonic photocatalytic activity for degradation of tetracycline under visible light. *Appl. Catal. B Environ.* **2015**, *164*, 297–304. [CrossRef]
56. Xu, D.; Liu, K.; Shi, W.; Chen, M.; Luo, B.; Xiao, L.; Gu, W. Ag-decorated $K_2Ta_2O_6$ nanocomposite photocatalysts with enhanced visible-light-driven degradation activities of tetracycline (TC). *Ceram. Int.* **2015**, *41*, 4444–4451. [CrossRef]
57. Yang, S.-F.; Niu, C.-G.; Huang, D.-W.; Zhang, H.; Zeng, G.-M. Ag/AgCl nanoparticles-modified $CdSnO_3{\cdot}3H_2O$ nanocubes photocatalyst for the degradation of methyl orange and antibiotics under visible light irradiation. *J. Colloid Interface Sci.* **2017**, *505*, 96–104. [CrossRef] [PubMed]
58. Yuan, X.; Jiang, L.; Chen, X.; Leng, L.; Wang, H.; Wu, Z.; Xiong, T.; Liang, J.; Zeng, G. Highly efficient visible-light-induced photoactivity of Z-scheme $Ag_2CO_3/Ag/WO_3$ photocatalysts for organic pollutant degradation. *Environ. Sci. Nano* **2017**, *4*, 2175–2185. [CrossRef]
59. Wang, T.; Quan, W.; Jiang, D.; Chen, L.; Li, D.; Meng, S.; Chen, M. Synthesis of redox-mediator-free direct Z-scheme AgI/WO 3 nanocomposite photocatalysts for the degradation of tetracycline with enhanced photocatalytic activity. *Chem. Eng. J.* **2016**, *300*, 280–290. [CrossRef]
60. Di, J.; Xia, J.; Ji, M.; Wang, B.; Yin, S.; Zhang, Q.; Chen, Z.; Li, H. Advanced photocatalytic performance of graphene-like BN modified BiOBr flower-like materials for the removal of pollutants and mechanism insight. *Appl. Catal. B Environ.* **2016**, *183*, 254–262. [CrossRef]
61. Wang, Q.; Li, Y.; Huang, L.; Zhang, F.; Wang, H.; Wang, C.; Zhang, Y.; Xie, M.; Li, H. Enhanced photocatalytic degradation and antibacterial performance by GO/CN/BiOI composites under LED light. *Appl. Surf. Sci.* **2019**, *497*, 143753. [CrossRef]
62. Majumdar, A.; Pal, A. Optimized synthesis of Bi_4NbO_8Cl perovskite nanosheets for enhanced visible light assisted photocatalytic degradation of tetracycline antibiotics. *J. Environ. Chem. Eng.* **2020**, *8*, 103645. [CrossRef]
63. Zhang, T.; Yin, Q.; Zhang, M.; Zhang, S.; Shao, Y.; Fang, L.; Yan, Q.; Sun, X. Enhanced photocatalytic activity of AgBr photocatalyst via constructing heterogeneous junctions with reduced graphene. *J. Mater. Sci. Mater. Electron.* **2021**, *32*, 15331–15342. [CrossRef]
64. Obregón, S.; Zhang, Y.; Colón, G. Cascade charge separation mechanism by ternary heterostructured $BiPO_4/TiO_2/gC_3N_4$ photocatalyst. *Appl. Catal. B Environ.* **2016**, *184*, 96–103. [CrossRef]

catalysts MDPI

Review

Tailoring Structure: Current Design Strategies and Emerging Trends to Hierarchical Catalysts

Virginia Venezia [1], Giulio Pota [1], Brigida Silvestri [2], Aniello Costantini [1], Giuseppe Vitiello [1,3] and Giuseppina Luciani [1,*]

1 Department of Chemical, Materials and Production Engineering, University of Naples Federico II, Piazzale Tecchio 80, Fuorigrotta, 80125 Naples, Italy
2 Department of Civil, Architectural and Environmental Engineering, University of Naples Federico II, Via Claudio 21, Fuorigrotta, 80125 Naples, Italy
3 CSGI—Center for Colloid and Surface Science, Via Della Lastruccia 3, 50019 Florence, Italy
* Correspondence: luciani@unina.it; Tel.: +39-0817682433

Abstract: Nature mimicking implies the design of nanostructured materials, which can be assembled into a hierarchical structure, thus outperforming the features of the neat components because of their multiple length scale organization. This approach can be effectively exploited for the design of advanced photocatalysts with superior catalytic activity for energy and environment applications with considerable development in the recent six years. In this context, we propose a review on the state of the art for hierarchical photocatalyst production. Particularly, different synthesis strategies are presented, including template-free structuring, and organic, inorganic, and hybrid templating. Furthermore, emerging approaches based on hybrid and bio-waste templating are also highlighted. Finally, a critical comparison among available methods is carried out based on the envisaged application.

Keywords: hierarchical photocatalysts; environment applications; inorganic structures; organic structures; hybrid structures; template free structures; environmental and energetic applications

Citation: Venezia, V.; Pota, G.; Silvestri, B.; Costantini, A.; Vitiello, G.; Luciani, G. Tailoring Structure: Current Design Strategies and Emerging Trends to Hierarchical Catalysts. *Catalysts* **2022**, *12*, 1152. https://doi.org/10.3390/catal12101152

Academic Editor: Jorge Bedia

Received: 3 September 2022
Accepted: 27 September 2022
Published: 1 October 2022

1. Introduction

"Learning from Nature" is always a winning strategy for materials scientists. Indeed, natural, and biological systems exhibit a multitude of properties and functions, often tailored for a specific application. Thus, they may be source of imitation for the design of high performance materials, able to meet rising ambitious industrial needs, following a biomimetic approach [1–4]. What makes bioavailable materials so interesting is that they often share a hierarchical structure made by a self-organization of molecular building blocks that are assembled with other phases, which in turn are self- organized at increasing size levels ranging from the nanoscale up to the macroscale [2,5–7]. The complex interplay between structure, morphology, and surface chemistry leads to a wide range of outstanding properties including superhydrophobicity (ex-lotus leaves [8,9]), improved mechanical performances (ex-bird feathers [10,11]), and unique optical behavior (ex-butterfly wings [12,13]). The appealing perspective to translate nature examples into industrial and technological devices steers research towards the synthesis of hierarchical nanomaterials. Indeed, intriguing features of nanostructured systems can be further improved if they are built up into hierarchically structured materials (HSPM), which exhibit a porous architecture, consisting of interconnected pores of different length scales from micro- (<2 nm), meso- (2–50 nm), to macropores (>50 nm), following bimodal or even trimodal pore size distributions [14]. Morphological features such as interconnected multilevel hierarchical porosity, high surface area, and large accessible space are responsible for enhanced light harvesting, electron and ion transport, mass loading, and diffusion [14,15].

Such outstanding properties confer to HSPM a pivotal role in a great number of technological fields including energy storage and conversion [16], catalysis [17] and photocatalysis [18], water remediation [19], separation [20], sensing [21], and biomedicine [22]. Indeed, the variety of physico-chemical characteristics exhibited by these nature-inspired materials can be modified by tuning surface chemistry and morphological properties such as surface area and pore distribution. Notably, a three-dimensional multimodal extended pore structure would be ideal for application in heterogeneous catalysis. The combination of different levels of intraparticle porosity, as well as the presence of mesopores, improve the diffusion of reactants and products towards the active sites and the external surface of the catalyst, respectively, and reduce the length of pore channels, resulting in decreased diffusion barriers in the solid volume [23–26]. Moreover, the large surface area (from 100 to thousands m^2/g [26,27]) allows for better accessibility and uniformity of active sites in the catalysts, causing an improvement in the surface reaction kinetics [15]. Building up a deeply extended multimodal porous architecture in semiconductors paves the way to the design of more efficient photocatalysts [28] by overcoming the typical limits that obstacle their performances such as short lifetime of photogenerated electron-hole pairs [29], partial utilization of solar light [30], low safety, high cost, and poor chemical stability [31]. Widespread application of photocatalysis accounts for huge technological potential of hierarchical photocatalysts, prompting scientific interest towards the design and use of these systems. Accordingly, a great number of outstanding studies has been published on this subject in the last years, which have been overviewed in different reviews [15,32–38].

Indeed, the number of articles on hierarchical photocatalysts have been constantly growing so far and so has the sensitivity towards environmental energetic issues and sustainability [39–43].

This has prompted the flourishing of several systems produced according to sustainable logics, some of them with exotic structures and unusual compositions, in the attempt to improve photocatalytic performance compared to state of the art [43–46].

Thus, we believe that a review summarizing the most recent advances and strategies in the design of hierarchical photocatalysts is timely and could enable more conscious development of these systems. First, the advantages of hierarchical photocatalysts are shortly illustrated. Then, the most promising strategies for their fabrication are systematically reviewed focusing on the most outstanding studies published in the period between 2015 and 2021. Particularly, as a strong element of novelty with respect to previous reviews published on this subject, emerging manufacturing approaches are highlighted, including the use of hybrid templates, such as metallorganic frameworks (MOFs). Furthermore, special emphasis is placed on the use of bioavailable structure directing agents, even derived from bio-waste valorization, according to a circular approach. Notably, the mechanism accounting for improved photocatalytic activity has been described to highlight the strength of each synthesis approach and provide scientists involved in this field with clear indication on the key parameters that can be easily manipulated to optimize catalytic performance. To this purpose, modification strategies, including doping and co-doping, the use of co-catalysts, and heterostructures formation have been illustrated. In addition, the use of emerging non-conventional energy sources in the synthesis of hierarchical nanostructures has been shortly overviewed. Furthermore, main applications in environmental and energetic fields, devoted to pollutants removal and hydrogen production, are reviewed and tables reporting collected results on photocatalytic performance were created to enable straight comparison among systems produced using similar or different synthesis strategies. These have been analyzed highlighting pros and cons to get interested scientists more conscious with the opportunities offered by each approach. Moreover, main unaddressed issues and future challenges for both fundamental investigation and technological application have been envisaged. We hope that the review can be useful to stimulate more focused and fruitful research in this field and can open new routes in the development of more effective photocatalytic systems.

2. Hierarchical Photocatalysts

The expression hierarchical photocatalysts usually refers to nanostructured semiconductors having multidimensional domains at different size levels and multimodal pore structure [16].

Since photocatalysis involves a series of complicated interconnected phenomena including light absorption, charge excitation/separation, charge migration, transport and recombination, and charge utilization, the overall efficiency of a photocatalyst is calculated as the product of the individual efficiency of each phenomenon [15]:

$$\eta_c = \eta_{abs} \cdot \eta_{cs} \cdot \eta_{cmt} \cdot \eta_{cu}, \quad (1)$$

It appears evident that a decrease in any singular efficiency would contribute to the decay of the overall efficiency. Hierarchical photocatalysts exhibit a series of advantages with respect to traditional photocatalysts, which are summarized in Figure 1 and include higher surface area, enhanced molecular diffusion/transfer, and improved light absorption, among others (Figure 1).

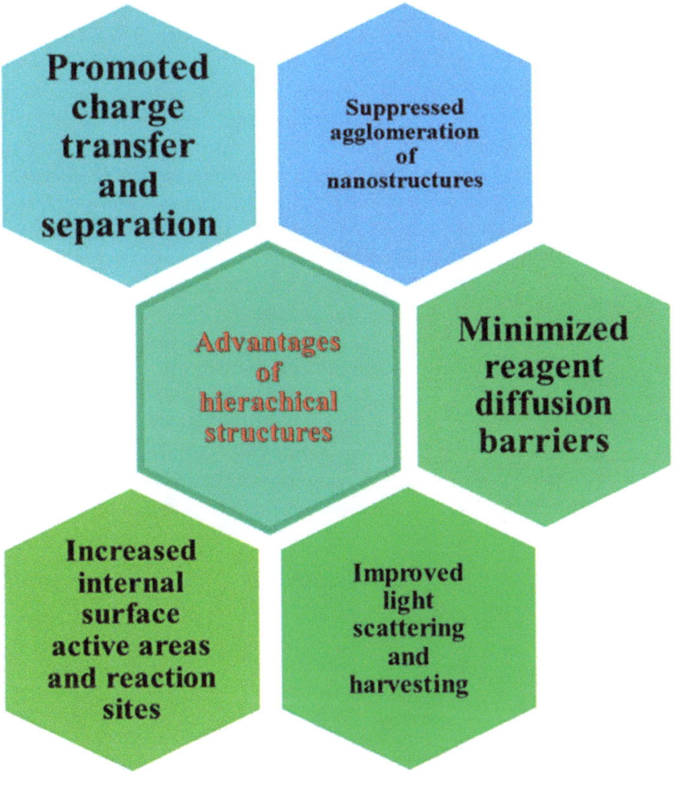

Figure 1. Main advantages of hierarchical porous structures in photocatalysis.

Indeed, properly tuning the pore hierarchy can increase the number of light traveling paths within the catalyst, resulting in multiple reflections as in hollow structures or enhanced light scattering phenomena, thus improving the absorption efficiency (Figure 2) [15,47].

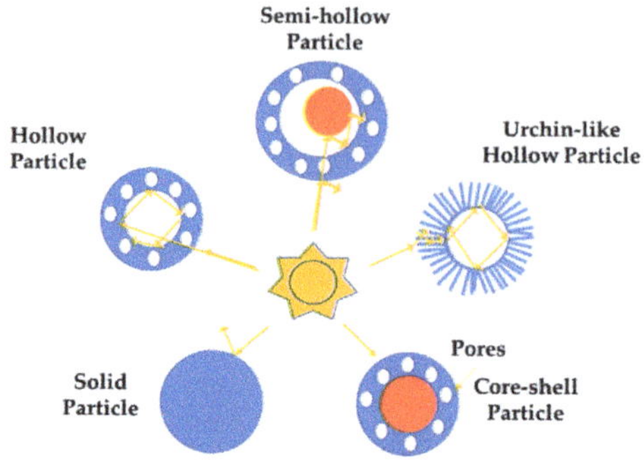

Figure 2. Schematic illustration of the light-harvesting behavior of heterogeneous photocatalysts with different nanostructures [47].

In addition, high surface area enables heterojunction formation and promotes surface reactions as well as charge separation. Furthermore, nanosized building blocks reduce migration distance of charge carriers (Figure 1). To date, pristine and doped TiO_2 [28,48–55], ZnO [56–61], CeO_2 [62,63], and graphene-blended hierarchical photocatalysts [64–66] have been synthesized in multiple shapes and morphologies such as nanospheres [67,68], nanoflowers [69–76], nanosheets [77–81], urchin-like nanostructures [68,82–87], and nanoflakes [88,89]. The effort to produce efficient hierarchical nanostructured photocatalysts materialized into a heterogeneous framework of synthesis strategies so far, all of them groupable into two macro-categories: templating strategies and template-free methods. Herein, we present an overview of the latest produced hierarchically nanostructured photocatalysts, focusing on the different synthesis strategies which have been implemented to develop them. Our sincere opinion is that this review can shed light on the most promising approaches that could be exploited to design cutting-edge photocatalysts with impressive performance in a huge range of energy and environmental applications, including H_2 production, water remediation, and organic pollutant degradation.

Synthesis Approaches to Hierarchical Structures

The main synthesis strategies to hierarchical semiconductors are overviewed in Figure 3 and can be grouped into three wide categories: templating approaches, templated free methods, and post-synthesis approaches.

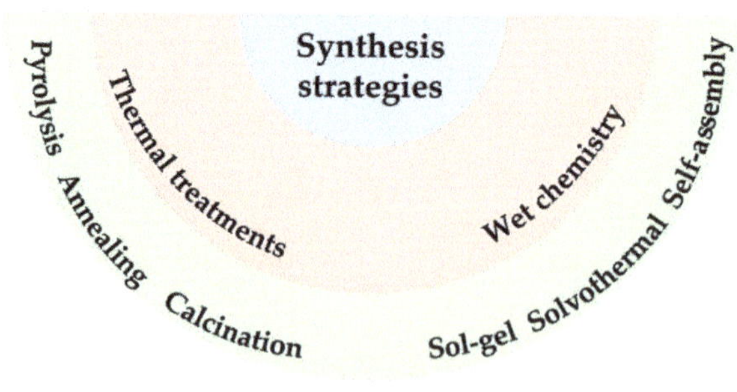

Figure 3. Main Synthesis Methods to realize hierarchical nanostructures.

Templated approaches are the most employed methods to obtain hierarchical materials, offering good reproducibility, large scale application, as well as tailored structures, due to the huge range of available templates. These approaches can be further distinguished into hard templating, soft templating, and bio-templating strategies.

In hard templating methods, also known as nanocasting processes, either inorganic or organic structures with rigid shape are employed as sacrificial templates to build up hierarchical framework [90]. These are subsequently removed through either selective leaching or thermal decomposition to achieve the final porous structure.

As major drawbacks, the use of corrosive and harmful leaching agents or high energy demand hamper large scale applications of these processes.

Contrarily to hard templates, soft templates are highly deformable and more easily removable. They include block copolymers (e.g., Pluronics P123, F127, PEG), microemulsions (micelles and vesicles), ionic liquids [91] hydrogels, surfactants (e.g., sodium dedecyl sulfate (SDS) and cetyl trimethylammonium bromide (CTAB)), and even gas bubbles [92]. On the other hand, bioavailable moieties are emerging as a cheaper and more sustainable alternative.

3. Inorganic Templates

Inorganic template-assisted synthesis is widely used for generation of multi-modal and hierarchical pores in the desired products. Generally, during this synthesis strategy, the catalyst precursors are filled into preexisting templates with suitable pore size, followed by subsequent removal of sacrificial framework thus generating a stable porous structure with uniform pore distribution. The final porous architecture can be controlled tuning the pore size and the wall thickness of hard templates. Most used templates include silica and ice as hard pore-directing agents (Table 1). Worthy of note is a bi-templated method, proposed to produce both $SrTiO_3/TiO_2/C$ (STC) (Figure 4) [93] and $PbTiO_3/TiO_2$/carbon (PTC) [94] heterostructures with a tri-modal micro-/meso-/macro-porosity. This was obtained by combining freeze casting process with the use of silica colloid as the hard template, then followed by an appropriate pyrolysis treatment. Particularly, ice and silica templates were used to obtain macro- and meso-pores, while pyrolysis introduced micropores. The proposed strategy was successful in tuning microstructure of both synthesized STC and PTC architectures leading to excellent photocatalytic performances in the degradation of methylene blue under UV light irradiation. In addition to the organic dye degradation, the photochemical reactivity of the STC samples was further investigated in the photocatalytic hydrogen production from water splitting resulting in a H_2 production rate as high as 2.52 mmol h^{-1} g^{-1} under UV irradiation.

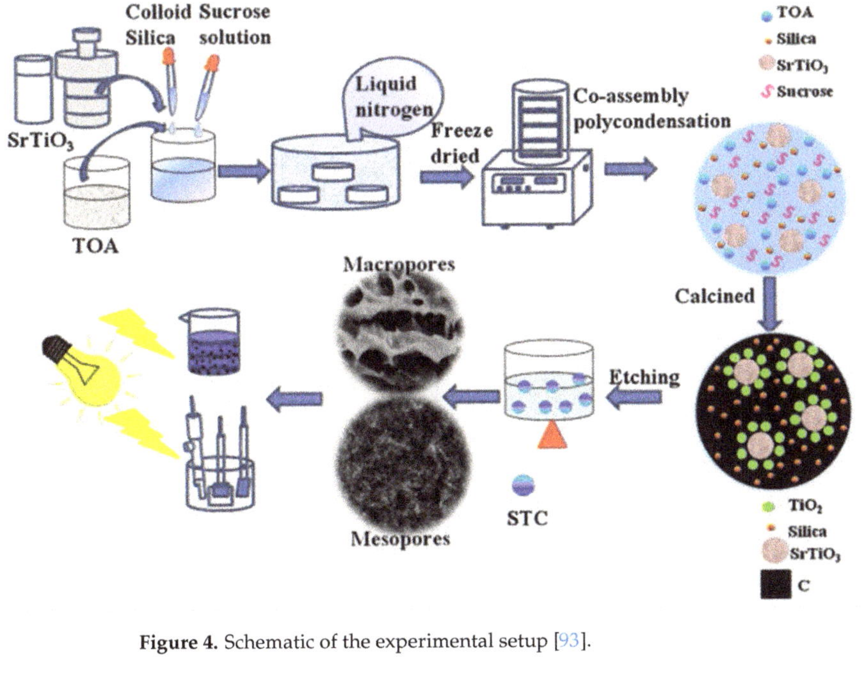

Figure 4. Schematic of the experimental setup [93].

Table 1. Hierarchical catalysts produced by using inorganic template(s).

Composition	Synthesis Strategy	Morphology	SSA (m^2/g)	Application	Activity	Irradiation	Template(s)	Reference
$SrTiO_3/TiO_2$@carbon	hydrothermal	heterostructure with tri-modal (micro-, meso-, macro-) pores	93–417	photocatalytic hydrogen production from water splitting, methylene blue degradation	2.52 mmol h^{-1} g^{-1} 100% Degradation	UV	KOH	[93]
$PbTiO_3/TiO_2$/carbon	dual template	quasi-1D nanoneedle	277–374	photocatalytic and photoelectrochemical performances	MB degradation r 100% Under UV, 75% Under Visible+ ultrasound radiation the rates of hydrogen generation are 2360 and 9.6 μmol^{-1} g^{-1}	UV-Vis	ice/silica hard templates	[94]
N–TiO_2/C	dropping	flower-like in hierarchical porous structure	217–407	MB degradation and hydrogen production	2.0 and 48 times of benchmark P25 for MB degradation: 95% under UV and 92% under Vis irradiation and the hydrogen production rates are as high as 2.832 and 0.038 mmol g^{-1} h^{-1}	UV-Vis	ice/silica hard templates	[95]
Metal oxide/C	Dual-template method followed by heat treatment	hollow "dragon-bone" structure	14–375	methylene blue degradation	100% (UV) 90% (Vis)	UV-Vis	ice/silica hard templates	[96]
TiO_2	hydrothermal	3D hierarchical porous core-shell	13.5–42.3	Adsorption and mineralization of tetracycline hydrochloride	up to 70%	UV-Vis	silica	[97]

A similar procedure was also employed to fabricate porous N doped TiO_2/C nanocomposites [95]. In this case, the ice/silica hard templates and urea pyrolysis were used to obtain flower-like hierarchical pores at micro-, meso-, and macro-scale and the resulting materials showed highly improved photocatalytic activity for both methylene blue degradation and hydrogen production under both UV and visible light irradiation. Furthermore, this approach was exploited to synthetize other metal oxide systems (TiO_2 and ZrO_2) [96], and in all cases the corresponding hierarchical carbon-based composites showed superior photocatalytic activity towards methylene blue degradation compared with the control samples. Silica, as mesopore structuring agent, was also exploited to produce hierarchical structure of porous core-shell homojunction constructed by crystalline and amorphous TiO_2 [97]. Combining the amorphous TiO_2 shell with the mesoporous rutile crystals resulted in a simultaneously enhancement of adsorption ability, removal rate, and mineralization efficiency, under UV irradiation, of Tetracycline hydrochloride (TCH) chosen as antibiotic model. The systems developed are shown in Table 1.

4. Organic Templates

A great number of different templating agents have been proposed for the synthesis of hierarchically porous materials and a widespread attention has been recently focused on the use of organic (Table 2) and/or bioavailable soft templates (Table 3). Biogenic-templated catalyst which owns intrinsic hierarchical, multi-dimensional, and multi-level structure holds promising potency to enhance light-harvesting and photocatalytic performance.

Hierarchically ordered macro–mesoporous anatase TiO_2 was successfully produced by Zhao et al. [98] using triblock copolymer P123 and natural pearl oyster shell as dual templates. Inexpensive and environment-friendly pearl oyster shell is exploited as bio-template for the fabrication of macroporous structures and at the same time P123 is employed as a mesopore-directing agent to produce highly ordered mesopores. The authors investigated the obtained materials after calcination at different temperatures and demonstrated that the resultant crystalline anatase structure is made of both macropores and mesopores which are well-preserved after calcination at 350 °C or 450 °C. The photocatalytic activities of the produced samples were monitored through the photodegradation of an aqueous Rhodamine B (RhB) solution under UV irradiation. The highest photocatalytic activity was obtained in the case of TiO_2 sample calcined at 450 °C, with a degradation of RhB of about 90%, because of the presence of hierarchical macro–mesoporous structures, high specific surface areas, and the anatase phase. Among bioavailable sources, bio-wastes are emerging as abundant and cheap templates to build up photoactive hierarchical structures, following a waste to wealth approach. Indeed, because of intrinsic lightness, some bio-wastes can drive formation of water floating photocatalysts, which exhibit remarkable advantages over powder suspensions, in terms of efficient recycling as well as high light harvesting efficiency due to their proximity to air-solution interface [99].

As a proof of concept, bio-waste poplar catkin was combined with triazine-based porous organic polymer to obtain a water-floatable photocatalyst for water remediation, exhibiting relevant sunlight activity towards Cr(VI) and methylene blue decontamination (Figure 5) [100].

Among biowastes, rice husk has attracted great research interest because of its unique and hierarchically porous structure with excellent mechanical property and extremely high light-harvesting and exploitation efficiency.

Rice husk was used by Chen et al. [101] as main raw material to form hybrid silica-carbon bio-template to fabricate porous TiO_2 anatase.

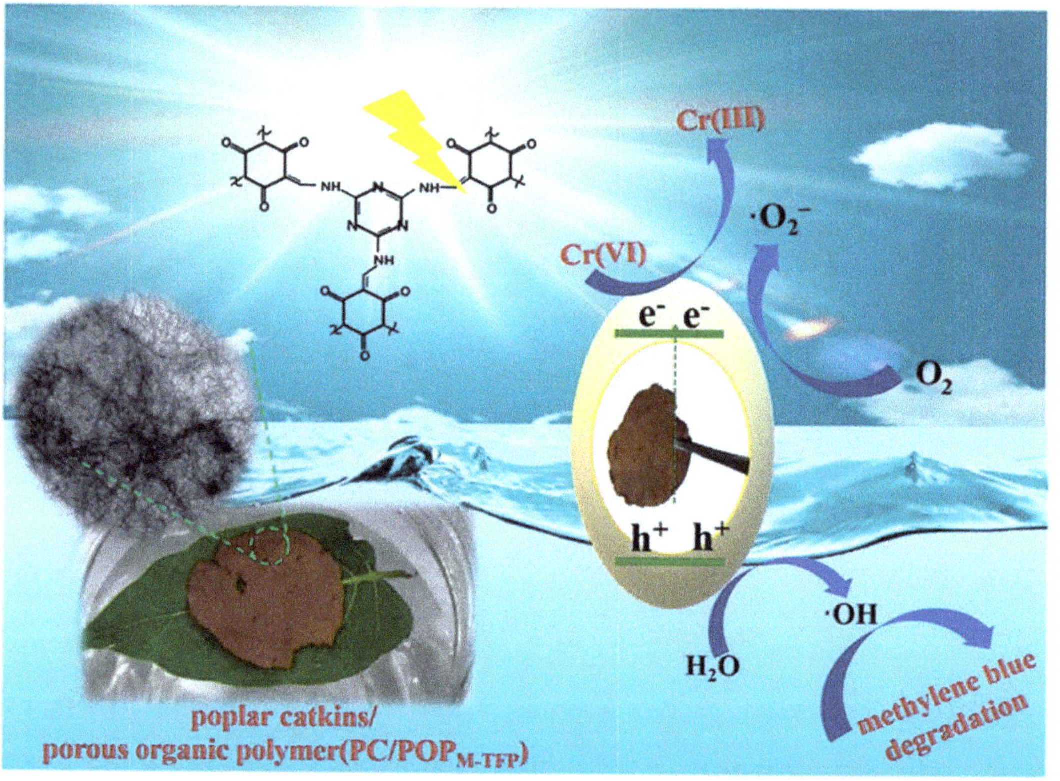

Figure 5. Scheme of floating triazine–based porous organic polymer (POP) photocatalyst grown on poplar catkins and used for water multipurpose water decontamination [100].

A detailed study was reported on the influence of calcination temperature and impregnation times on the hierarchical porous structures and photocatalytic activity of the synthesized materials. Porous structures with 1D nanostructures network connection showed efficient light-harvesting and photocatalytic degradation of Rh B (120 mg L^{-1}) which was completely removed in just 180 min. Furthermore, the observed high thermal stability opened new perspectives in the use of these materials in a wide range of applications including photocatalysis, catalysis, solar-cell, separation, and purification processes.

Biotemplated synthesis of hierarchical α-Fe_2O_3 fibers was proposed by Chen et al. [102], using cotton as biological template. The original structure of cotton was preserved in the final iron oxide fibers which also exhibit certain unexpected magnetic behavior; in fact, common α-Fe2O3 is not magnetic. In addition, superior photocatalytic decoloring performances were observed under visible light irradiation, up to 96.8% methylene blue degraded in only 200 min. Moreover, no loss of catalytic activity was appreciated after three uses, proving high stability. Indeed, this study evidenced the key role of hierarchical porosity in catalytic activity, with fibrous structures acting as effective catalytic centers for organic dye elimination. Recently, electrospinning fiber formation technology has emerged as a powerful and straightforward tool for tailoring material structures and assembling hierarchical architectures [103,104]. There is plenty of scope for improving photocatalytic properties if this methodology is combined with more conventional processes including solvothermal and microemulsion synthesis.

In this context, hierarchical titanium dioxide nanofibers with distinctive microstructures exhibiting mixed rutile/anatase crystalline phase were successfully fabricated via microemulsion electrospinning approach by Zhang et al. [105]. More in detail, in order to make the microemulsion, paraffin was used as oil phase, while tetrabutyl titanate was dissolved in an alcoholic acid solution (containing a mixture of CTAB and PVP) as continuous phase.

The structure of porous nanofibers was regulated by changing the ratio of continuous phase and oil phase. Different morphologies such as multi-channel, hollow, irregular hollow TiO_2 nanofibers were obtained by varying the ratio between the TiO_2 precursor, tetrabutyl titanate (TBT), and paraffin oil [105]. In addition, the authors demonstrated that brittle nanostructure with abundant mesopores were formed when the ratio of TBT/paraffin oil decreased to 1. Compared with solid TiO_2 nanofibers, all samples prepared by microemulsion electrospinning had improved photocatalytic performances towards the MB degradation UV irradiation. These results can be attributed to a combination of multiple structure dependent effects: first, porosity plays a vital role in absorbing photoelectron transition from rutile to anatase. In addition, photocatalyst with higher porosity can absorb more oxygen moieties on the surface which react with photoelectron, thus avoiding recombination and resulting in more electron and holes available for photocatalytic process (Figure 6). Finally, mixed anatase and rutile crystalline structure are expected to enhance charge separation ultimately improving photocatalytic performance

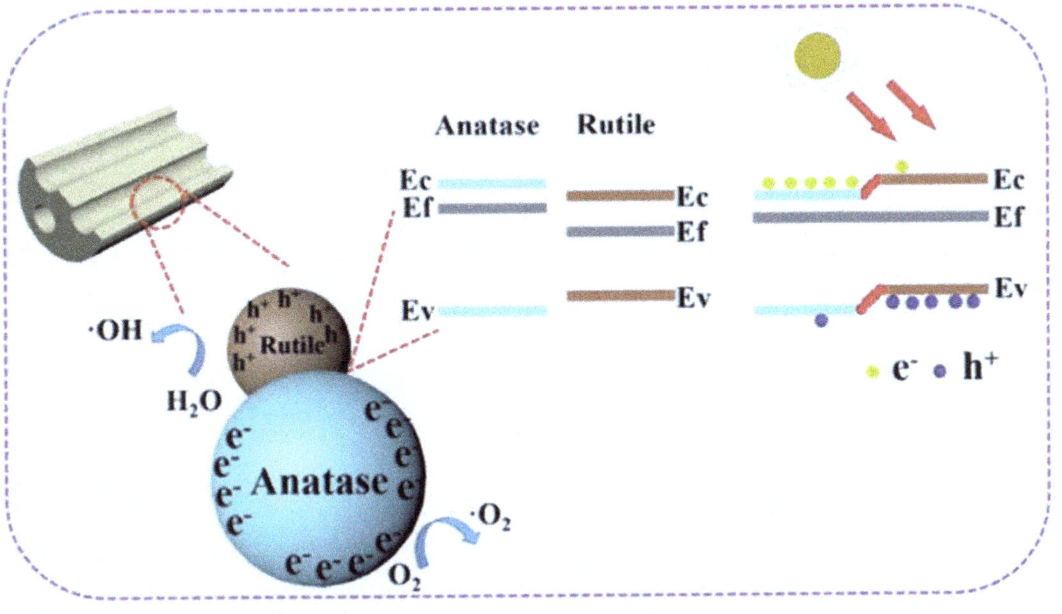

Figure 6. The mechanism of improved photocatalytic performances for hierarchical TiO_2 nanofibers [105].

A soft template approach was exploited by Zhou et al. [106] to produce three-dimensional (3D) flower-like β-Bi_2O_3/$Bi_2O_2CO_3$ heterojunction photocatalyst. A composite soft template composed of DL-aspartic acid (DLAA) and Pluronic F123 was employed to fabricate a 3D flower-like Bi containing micro/nanomaterial, in which DLAA acted as both coordination- and structure-directing agent to control the hierarchical structure, while F127 acted as a capping agent.

For the first time, the β-Bi_2O_3/$Bi_2O_2CO_3$ p–n heterojunction photocatalyst was applied in the simulated-sunlight-driven photodegradation of antibiotic agent tetracycline (TC). The photocatalyst obtained after calcination at 290 °C showed relevant photocatalytic performance with 98.79% TC degradation being achieved within 60 min of irradiation. The obtained results confirm that the combination between the narrow band gap, heterojunction structure, and 3D hierarchical structure results in excellent photocatalytic performances. In particular, •OH and h+ proved to be the main active species in TC photodegradation process. Finally, the β-Bi_2O_3/$Bi_2O_2CO_3$ heterojunction catalyst was not photo-corroded after six consecutive cycles, suggesting very high photostability (Figure 7).

Among bioavailable polymers, cellulose appears to be a promising green bio-template for porous hierarchical structure synthesis, because of its intrinsic hierarchical structure [107,108]. It was successfully exploited to produce $H_3PW_{12}O_{40}$/TiO_2 nanocomposite with remarkable activity towards organic pollutant removal under sunlight [107].

A non-solvent induced phase separation (NIPS) technique was explored by Sun et al. [108]. One-pot route towards active TiO_2 doped hierarchically porous cellulose provided for highly efficient photocatalysts for methylene blue degradation [108]. Ethyl acetate was chosen as the non-solvent with the aim to induce the phase separation of cellulose into a cellulose/LiCl/N,N-dimethylacetamide solution containing TiO_2 nanoparticles. The subsequent solvent exchange/freeze-drying treatment allowed to obtain cellulose/TiO_2 composite monoliths featuring large surface area and hierarchically porous structures with two kinds of interconnected macropores. The cellulose/TiO_2 monoliths showed high efficiency of photocatalytic activity in the decomposition of methylene blue dye up to 99% within 60 min under UV light. This behavior is probably due to high adsorption properties of the material itself because of the hierarchical porous structure which improves the contact frequency between photocatalysts and MB. In addition, after 10 cycles, the monoliths retained 90% of the photodegradation efficiency.

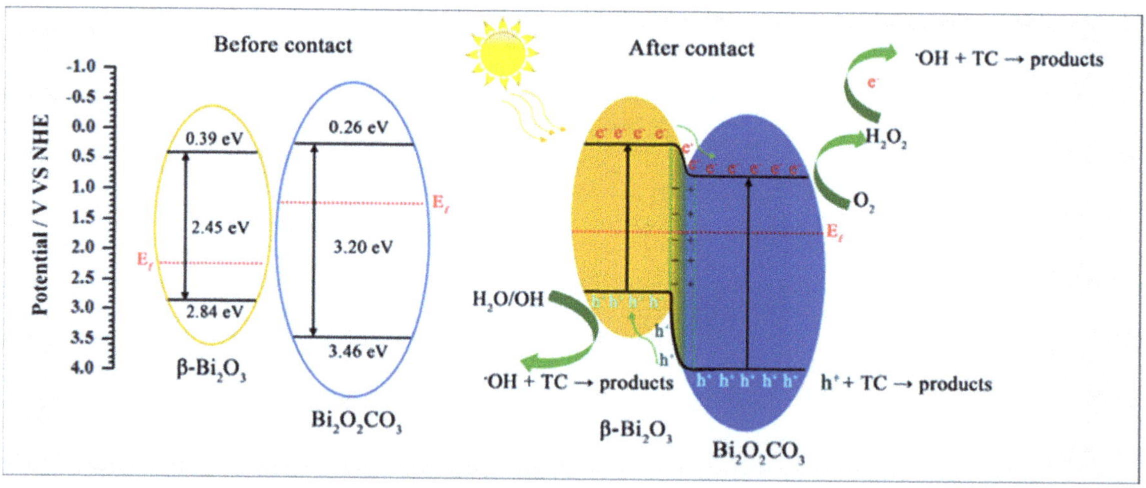

Figure 7. Schematic representation of the band structure of the β-Bi_2O_3/$Bi_2O_2CO_3$ heterojunction and the migration of photogenerated charges under simulated sunlight irradiation [106].

Table 2. Hierarchical catalysts produced by using synthesis templates.

Composition	Synthesis Strategy	Morphology	SSA (m^2/g)	Application	Activity	Irradiation	Template(s)	Template Nature	Reference
Anatase TiO_2	Sol-gel	Hierarchical ordered macro-meso	125.91	photodegradation of aqueous Rhodamine B (RhB)	90%	365 nm	copolymer P12/oyster shell	Soft/hard	[98]
Dimethylglyoxime (DMG)/TiO_2/ polyacrylonitrile (PAN) nanofiber	Electrospinning	Nanofibers	50–60	Photocatalytic MB degradation efficiency	97%	Visible Light	PAN/PVP	Soft	[103]
Bi_2WO_6/ WO_3/PAN	Electrospinning and solvothermal process	Nanofibrous membrane	-	Degradation of cationic pollutants	85% of RhB, 87.8% for BQ and 95.7% for IPA, 96% of MB, 77.4% of chlortetracycline hydrochloride and 90.37% of tetracycline hydrochloride	UV-Vis	Triton-X	Soft	[104]
Anatase/rutile TiO_2	microemulsion electrospinning (ME-ES)/pyrolysis	multi-channel irregular mesoporous TiO_2 nanofibers (hundreds of nm in diameter)	/	Photodegradation of aqueous methylene Blue	100%	365 nm	CTAB	soft	[105]
β-Bi_2O_3/$Bi_2O_2CO_3$	Sol-gel with reflux	Mesoporous Micrometric flower-like structures	27.78	Degradation of tetracycline	98.79%	Sunlight	DL-aspartic acid (DLAA)/Pluronic F123	soft	[106]

Table 3. Hierarchical catalysts produced by using bioavailable templates (biowaste).

Composition	Synthesis Strategy	Morphology	SSA (m^2/g)	Application	Activity	Irradiation	Template(s)	Template Nature	Reference
PC/POP	hydrothermal	Self-supporting film-like structures	122.5	Cr(IV) photoreduction/methylene blue photodegradation	100% Cr (IV) reduction/100% methylen blue photodegradation	Visible light	Poplar catckins	Hard	[100]
α-Fe_2O_3	Sol-gel	hierarchical porous and fibrous structures (15 µm diameter, hundreds µm length)	51.3	Photodegradation of aqueous methylene Blue	96.8%	Visible light	Cotton	hard	[102]
$H_3PW_{12}O_{40}/TiO_2$	Sol-gel & calcination	nanotubes	80.7	Methylene blue photodegradation	95% methylene blue degradation	UV light	Cellulose	soft	[107]
Cellulose-TiO_2	non-solvent induced phase separation (NIPS) technique	Macro-Mesoporous composite monoliths	16.96	photodegradation of aqueous methylene Blue	99%	UV lamp	Cellulose	soft	[108]
	-	-							

5. Hybrid Templates

At the boundary between organic and inorganic structure directing agents, metal-organic frameworks (MOF) are hugely porous crystalline hybrid structures made of metal clusters and organic binders. They are emerging as powerful and versatile hard templates to obtain hierarchical photocatalysts with tailored structure as well as porosity and function [109] (Table 4).

Indeed, upon decomposition of organic moieties, obtained materials inherit MOF large mesoporous structure allowing high active site exposure and enabling reactant and product exchange as well as fast charge migration to exposed reactive positions, with limited recombination phenomena [109,110]. As a result, MOF-templated photocatalysts are more active than their bulk counterparts and can be exploited in a huge range of energy and environmental applications, including H_2 production [111,112], CO_2 reduction [110,113,114], as well as organic pollutant degradation [115–119]. The main explored technological uses of MOF derived hierarchical photocatalysts are reported in Table 4.

MOF templated approach is extremely versatile and can be exploited to produce different semiconductor compositions. Porous metal oxides are easily obtained by MOF thermal decomposition [109]. Particularly, a Ti-MOF upon calcination leads to TiO_2, composed by both anatase and rutile nanostructured phases, whose intimate interaction suppresses electron-hole recombination and promote photocatalytic activity, which results even better than P25 [120].

Following a similar strategy, Co MOF with 1,4-naphtalenedicarboxylic acid as ligand has been decomposed to obtain Co_3O_4 nanosheets with uniform size distribution and significant CO_2 reduction conversion to CO, with high selectivity (77%) [110]. Moreover, metal oxides produced by MOF decomposition can be easily converted into metal sulfides by sulphuration process [121,122]. This strategy was exploited to obtain hierarchical CdS structures, with higher photocatalytic activity than both nanostructured and bulk counterparts, towards H_2 production by water splitting [109]. Following a similar procedure, ZnCoS solid solution was fabricated, which exhibited high stability and activity because of a wide light absorption range, a great number of catalytic sites, and fast electron migration [122] (Table 4).

However, MOF templated method is time consuming since it is built on two steps, synthesis and calcination, which are difficult to be carried out on a large scale. Furthermore, the criteria to adequately choose MOF precursors must be further investigated to create the desired composition exhibiting hierarchical structure [109]. Indeed, selection of MOFs based on the same metallic cation, but with different structures, can deeply affect the morphology, structure, and overall performance of obtained catalysts. As a general rule, thermal stability is strictly demanded to avoid structure collapse during thermal treatment. Among available composition, MIL(53) meets this requirement, thus it holds huge promise to build controlled multilevel porous structures [109].

A promising strategy to obtain unique hierarchical structures relies on the use of defect-rich MOF [111]. This approach was effectively exploited to obtain hollow C doped CuO structures through pyrolytic degradation of Cu-MOF. In this study, not only did MOF act as a template to tune the structure of the newborn catalytic phase, but if pyrolysis occurred in reducing atmosphere, they could also supply for carbon atoms to modify the catalyst lattice. Both C-doped and C-containing hybrid composites can be produced following this approach and exhibiting relevant photocatalytic activity towards H_2 production as well as pollutants degradation, due to peculiar porous and electronic structure [123,124] (Table 4).

Indeed, due to the controlled high surface area, MOF derived hierarchical structures are suitable platforms to be loaded with dopants that act as co-catalysts and can further improve photocatalytic properties. Apart from doping with C atoms, modification options include metal nanoparticles, encompassing noble metals (Pd, Pt) [114], and also cheaper transition metals (Ni, W, Cu) [125–127] as well as metal sulfides [128].

Indeed, doping can be carried out through "extra-situ" approach; in that case, MOF thermal decomposition can provide for either the catalyst lattice (Figure 8) or doping moieties for the preformed catalytic network. This can be further decorated with a proper dopant through hydrothermal treatment (Figure 8) [122,129,130].

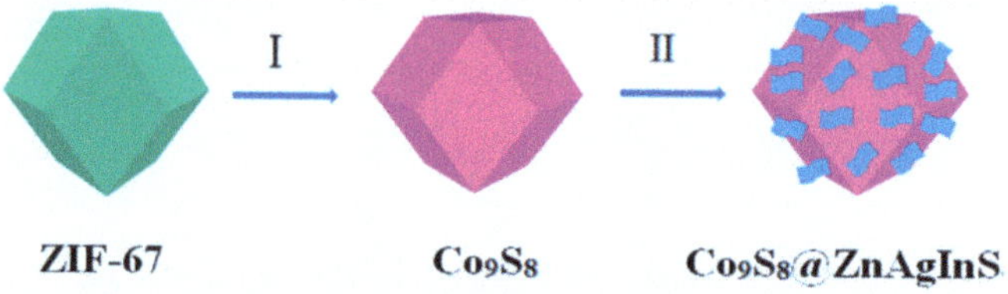

Figure 8. Scheme for hierarchical heterostructured Co_9S_8@98ZnAgInS photocatalyst: fabrication through extra situ approach through (I) sulfidation reaction and thermal treatment in a nitrogen atmosphere and (II) deposition of ZnAgInS nanosheets [115].

Alternatively, in-situ methodology can be also explored, wherein the formation of catalyst backbone from MOF decomposition and dopants growth in the lattice concurrently occur during solvothermal process [126]. Because of MOF huge surface area and controlled porosity, high dopant dispersion and tunable morphology are usually afforded, producing a significant improvement of both catalytic performance and stability [14] thanks to the presence of multifunctional sites and the improved charge carrier separation [127]. Furthermore, concurrent modification with several co-catalysts can be easily carried out to achieve high photocatalytic performance. In this regard, TiO_2 was combined with conductive bimetallic NiCoS-porous carbon shell, which afforded fast charge transport and prompt reaction for hydrogen production [129].

Moreover, MOF templates offer a simple, straightforward, and largely effective strategy in creating heterojunctions within hierarchical structures.

This approach is also successful for sulfide compounds, which usually require multistep preparation (Figure 8). The large number of heterojunctions promote charge separation and migration and consequently much higher photocatalytic activity than conventional catalysts [131] for a huge range of applications spanning from organic pollutant removal to hydrogen production through water splitting [112,132]. MOF tunable composition and porosity offer the chance to prepare a plethora of hybrid catalysts combining hierarchical structures with p-n heterojunctions for photocatalytic processes (Table 4) [133]. Following this approach, creative and unique morphologies and structures can be achieved, including hierarchical hollow heterostructural cages combined with 2d nanosheets (Figure 9), hollow caps [134], boxes, polyhedra [135], and capsules which can be further modified through hydrothermal approach, reaching outstanding catalytic performance even under visible light.

Figure 9. SEM micrographs of some hierarchical photocatalysts obtained with hybrid templates [131,133,134].

To exploit advantages of combining different semiconductors, interface must be accurately designed to match band gaps and enable fast charge migrations as well as their easy transfer to reactants. To this purpose an appropriate distribution of reactive sites must be achieved. All these requirements can be easily met through MOF templated approach [136]. As a proof of concept for the efficacy of this strategy, hepitaxially grown MOF-on-MOF heterostructure was exploited as a precursor to produce N-doped C encapsulated pagoda-like CuO–In_2O_3 (In_2O_3/CuO@N-C) micro-rods. They exhibited improved light-absorption efficacy due to peculiar lamellar structure, enhanced separation efficacy of charge carriers thanks to CuO–In_2O_3 p–n heterojunction, and separated reduction and oxidation sites to promote charge carrier transfer to reactants. These features confer outstanding activity for cross-dehydrogenative coupling (CDC) reaction [136].

Due to high surface and reactivity, combination of MOF obtained photocatalysts into heterojunctions can be easily carried out through solvothermal synthesis [137,138]. This strategy was exploited to produce p-n heterostructure $NiFe_2O_4$/$CuInSe_2$ catalysts with high activity towards bisphenol-A and resorcinol degradation [139].

Indeed, solvothermal approach appears as a straightforward production route, since moderate temperature during the process get organic ligands degraded, thus straightly leading to the final desired p-n semiconductor [137]. Following this method, even Z-scheme heterojunctions can be easily obtained [138], such as CoSx@CdS polyhedrons, provoking a marked improvement of charge transfer and enabling a high sensitivity toward Hg^{2+} ions detection (Figure 10) [138]. As an alternative, p-n heterojunction can be obtained by thermal treatment of bi-metallic MOF [112].

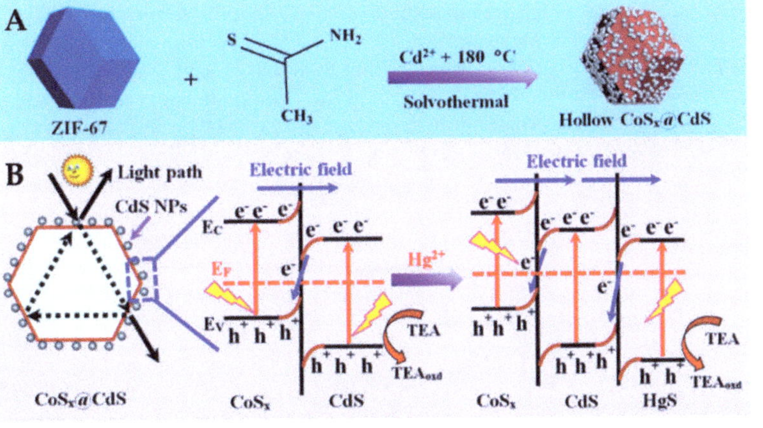

Figure 10. Scheme of (**A**) the synthetic route of hollow CoSx polyhedrons and CoSx@CdS composites; (**B**) Z scheme charge migration mechanism CoSx@CdS composites and double CoSx@CdS/HgS heterojunction under the visible-light illumination [138].

Moreover, one of the highest strengths of MOF templated approach lies in the opportunity to design complex hierarchical architectures even with multi-level hierarchy. To this regard, MOF can be effectively used to obtain 2D hierarchical nanoarrays, by assembling 2D structures and conductive sheets, which exhibit outstanding catalytic activity because of high catalytic site exposure and fast charge transport [113,117]. As a proof of concept, in-situ MOF derived approach was successfully employed to produce hierarchical Co-Co layered double hydroxide/Ti_3C_2TX nanosheets (Co-Co LDH/TNS) nanoarray through solvothermal process and evidencing relevant activity towards CO_2 photoreduction under visible light (Figure 11) [113].

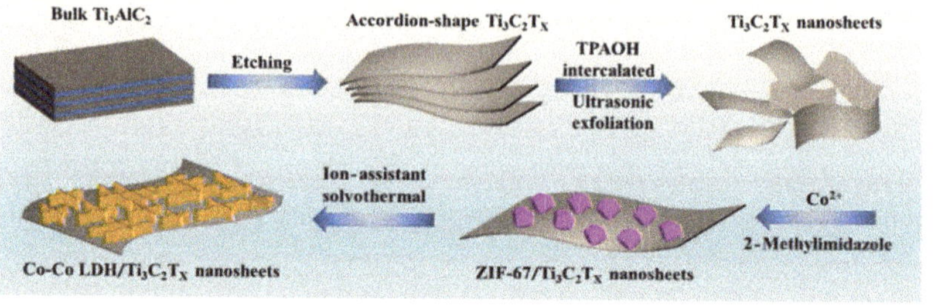

Figure 11. Synthesis scheme of heterostructure Co-Co LDH/TNS nanosheets [113].

Highly effective heterojunctions can be also obtained by combining MOF and C_3N_4 sheets via thermal treatment. Following this route, N doped carbon C_3N_4 composite was produced exhibiting relevant activity towards bisphenol A Degradation (Table 4) [117]. Finally, MOF derived multilevel composites exhibit multiple heterojunctions and consequently enhanced catalytic performance even in selective oxidation [116]. Similarly, one-pot calcination process of Zn-Fe mixed MOF produced $ZnO/ZnFe_2O_4$ hierarchical heterostructures with high photocatalytic activity towards dye degradation and which can be easily recovered from solution because of its ferromagnetic features [119].

Indeed, due to the high versatility of MOF templates, they provide the opportunity to engineer the photocatalyst structure, by combining heterostructure formation with co-catalyst surface modification [121,140]. This occurred in TiO_2 based photocatalyst, which was modified by Co_3O_4 and Ni to promote oxidation and reduction, respectively. Obtained ternary $TiO_2/Co_3O_4/Ni$ photocatalyst disclosed an impressive hydrogen production 8.7 times higher than that of neat TiO_2 under UV-visible irradiation [140]. Because of the high versatility of MOF templated approach, tuning of electronic properties, through heterojunction formation and co-catalyst introduction, is accompanied by fine morphology control. In this regard, hollow CdS/TiO_2 nanohybrids modified with NiS cocatalyst exhibited very good photocatalytic performance towards H_2 production under visible light conditions (Figure 12) [121].

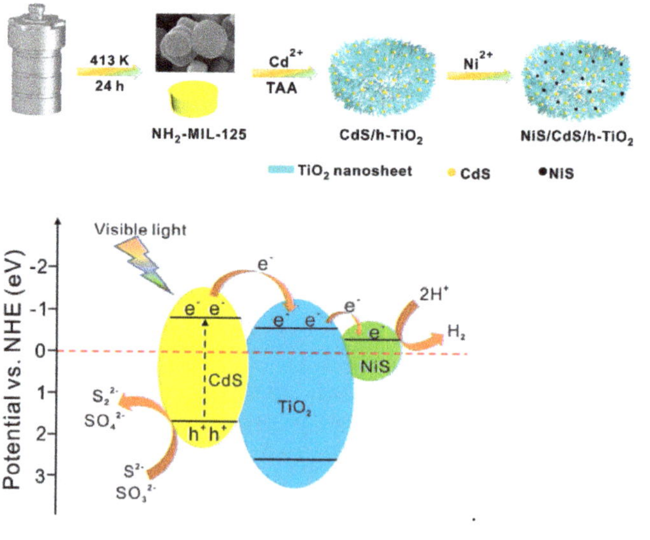

Figure 12. Schematic illustration of the synthetic procedure of NiS/CdS/h-nanocomposites with schematic of the photocatalytic mechanism of H_2 production on NiS/CdS/h-TiO_2 photocatalysts [121].

Table 4. Hierarchical catalysts produced by using hybrid templates.

Composition	Synthesis Strategy	Morphology	SSA (m^2/g)	Application	Activity	Irradiation	Template(s)	Reference
Rutile/anatase TiO_2	Sol-gel	Submicrometric parallelepiped with rounded corners	19	Photocatalytic reduction of water	1394 μmol h^{-1} g^{-1} H_2 production rate	UV-vis	MIL-125	[120]
Co_3O_4	Oil bath method	2D hierarchical nanosheets	24.96	Efficient photocatalytic conversion of CO_2	39.70 μmol h^{-1} CO generation	Visible light	Co/1,4-H_2NDC	[110]
NiS/CdS/TiO_2	Hydrolysis + sulfidation	Porous/hollow structure		Photocatalytic reduction of water	H_2 production rate of 2149.15 μmol g^{-1} h^{-1}	Visible light	NH_2-MIL-125	[121]
CdS	Impregnation + pyrolysis	Microporous nanoparticles	119	Photocatalytic reduction of water	634.0 μmol g^{-1} h^{-1}	Visible light	MIL-53 (Al)	[109]
C-CuO	pyrolysis	Hollow spheres	72.8	Photocatalytic reduction of water	67.3 mmol/g/h H_2 production	Visible light	Cu/benzoic acid/1,4 dicarboxybenzene	[111]
C-ZnO	Pyrolysis under N2 atmosphere	ZnO crystals embedded within a porous carbonaceous matrix	500	Adsorption and photodegradation of RhB	100% adsorption efficiency	//	MOF-5	[123]
Pd-Anatase TiO_2	pyrolysis	Submicron TiO2 tablets		Photocatalytic reduction of water	979.7/112.7 mol h^{-1} H_2 production	UV/solar light	NH_2-MIL-125	[114]
Ni/g-C_3N_4	Pyrolysis under Ar atmosphere	layered platelets with curled edges	64.9	Photocatalytic reduction of water	H_2 production rate of 2989.5 μmol g^{-1} h^{-1}	Visible light	2D Co-MOF	[125]
W/Co_3S_4	Hydrothermal treatment	10 mm square sheet arrangement	10.93	Photocatalytic reduction of water	85.7 μmol/h H_2 production rate	Visible light	Co-ZIF-9	[126]
Cu/TiO_2	Photolysis under N_2 atmosphere	Mesoporous core-shell Cu/TiO_2 hybrid nanoplatforms	-	Photocatalytic reduction of water	334 μmol g^{-1} h^{-1} H_2 production	Simulated light	Cu-MOF	[127]
TiO_2–NiCoS-PC	Annealing under Ar atmpsphere	Spherical porous carbon shell Embedding TiO_2	93.5	Photocatalytic reduction of water	1.29 mmol h^{-1} g^{-1}	Visible light	NiCo-ZIF	[129]
$MoS_2-Zn_{0.5}Co_{0.5}S$	Hydrothermal	Hollow rhombic dodecahedra	57	Photocatalytic reduction of water	15.47 mmol h^{-1} g^{-1}	UV-Vis	Zn_xCo_{1-x}-ZIF	[122]
MnS/In_2S_3	Solvothermal	Sub-micro rods	/	photocatalytic CO_2 reduction	58 μmol g^{-1} h^{-1} CO production rate	Xenon lamp 300 W (credo UV-vis)	MIL-68 (In)	[131]

Table 4. *Cont.*

Composition	Synthesis Strategy	Morphology	SSA (m^2/g)	Application	Activity	Irradiation	Template(s)	Reference
NiO/CeO_2	Hydrothermal and calcination	Porous microsphere	31.1	Dye photodegradation/water splitting	100% MO and MB degradation/71.5 μmol g^{-1} H_2 production rate	UV	Ni/Ce mixed-metal MOF	[112]
$Cu/CuO\text{-}TiO_2$	Sol-gel + calcination	nanoparticles	45.3	Photocatalytic reduction of water	286 mmol g^{-1} h^{-1} H_2 production rate	Solar (Sun)	Cu-MOF	[132]
CdS/ $ZnxCo_3\text{-}xO_4$	Hydrothermal method	Hollow polyhedra	29.6	Photocatalytic reduction of water	3978.6 μmol g^{-1} h^{-1} H_2 production rate	Visible	ZnCo-ZIF	[135]
$In_2O_3/CuO@N\text{-}C$	calcination	Pagoda-like heteroepitaxial micro-rods	147.1	cross-dehydrogenative coupling (CDC)	88–99% yield of reaction	Blue light	MIL-68-In	[136]
$CuInSe_2/NiFe_2O_4$	hydrothermal	Irregular-shaped nanoparticles	27.54	Bisphenol-A and resorcinol degradation	95–95% removal rate	Visible light	Ni–Fe MOF	[137]
CoSx/CdS	solvothermal	CdS NPs on the surface of CoSx polyhedrons	/	Hg^{2+} detection	0.010 to 1000 nM Hg^{2+} detection	Visible light	ZIF-67	[138]
Co-Co LDH/Ti_3C_2TX	solvothermal	3D nanosheets nanoarray	/	CO_2 photoreduction	1.25 × 104 μmol h^{-1} g^{-1} CO production rate	Visible light	ZIF-67	[113]
ZIF-NC/g-C_3N_4	Thermal treatment	ZIF-NC dodecahedra deposited onto g-C3N4 layers	/	bisphenol A degradation	95% removal	Visible light	ZIF-8	[117]
$MoS_2@Cu/Cu_2O@C$	Pyrolysis	Hierarchical rough polyhedra	94	Selective oxidation of cyclohexane to KA oil	1.31% conversion rate; 98% selectivity	Visible light	Cu-MOF	[116]
$ZnO/ZnFe_2O_4$	calcination	Spherical assemble of flake-like nanosheets	87.74	RhB and MB degradation	100% degradation	UV-vis	Zn(Fe)-MOF	[119]

6. Template Free Approach

Using a templating agent proved to be effective in building up hierarchical structures, yet it is a labor consuming approach, because of the operations required to remove the backbone.

On the other hand, self-assembly of primary particles into larger structures by oriented aggregation occurring during solvothermal processes can be a more straightforward and greener strategy, since it can potentially avoid toxic chemicals [141]. In these processes, it appears that solvents play a key role in hierarchical morphology formation [142]; indeed, their molecules or, in the case of alcohols, polymers obtained by condensation reactions can adsorb on the particle surface and tune their growth [143]. Solvothermal route has been successfully exploited to obtain hierarchically structured conventional semiconductor oxides including ZnO [60,142] TiO_2 [144–146], CeO_2 [147], Cu_2O [148], and more unusual compositions, such as WO_3 [149] and mesoporous Bi_2WO_6 [150]. Furthermore, doping of the main lattice with various ions can be carried out concurrently with its formation [151]. Moreover, a relevant number of heterostructured composites has been easily produced so far through one pot solvothermal treatment; these include combinations of semiconductor oxides (TiO_2, ZnO, CeO_2) with other semiconductors [62] including metal sulfides (SnS_2) [152], carbon nitride [133], or with carbonaceous species (CS). This class encompasses ZnO-graphene [141,153] and TIO_2-CS particles [154] as well as TiO_2 nanotubes wrapped by carbon coatings [155], among others. Furthermore, CoS/CdS heterojunctions [156] and even less conventional compositions, including $BiVO_4/Bi_2WO_6$ [157] systems, can be easily fabricated. Indeed, the presence of different morphology nanostructures, such as g-C_3N_4 nanosheets and ZnO nanoparticles, in the same batch during the solvothermal treatment appears to be a good strategy promoting self-assembly into hierarchical structures to decrease surface energy [133]. Moreover, synthesis parameters and particularly reactant molar ratio, solvent and precursor nature, as well as temperature and treatment time [145] can tune morphological and microstructural features. Accordingly, a significant range of morphologies can be easily fabricated, including hollow microspheres [141,142,149,155,158,159], nanoplates [156], as well as more exotic arrangements, such as broom-like [151], urchin-like [160], and nest-like [149] structures (Figure 13).

Figure 13. SEM micrographs of some hierarchical photocatalysts obtained without templates [149,151,158,160].

As a matter of fact, photocatalytic activity of hierarchical ZnO obtained by solvothermal process decrease according to the following order of employed precursor: zinc sulfate > zinc nitrate > zinc acetate > zinc chloride [60]. Furthermore, the use of pure ethanol as solvent during solvothermal treatment of $TiOSO_4$ precursor produces nanoparticle morphology, whereas the addition of ethyl ether and ethanol mixture has morphology changed to microrods with huge surface area (244.4 m^2 g^{-1}), promoting light scattering and reactivity [155].

Solvothermal hierarchical structures usually feature an interconnected porosity with proper pore size to enable mass transfer as well as improved light harvesting and even fast charge carrier mobility; these properties concur to achieve higher photocatalytic performance than conventional systems [150]. Obtained systems exhibit relevant activity as well as photoelectrochemical performances under UV [141,146] and even visible [145,148,156,157] and full solar irradiation [155] for energy and environmental photocatalysis. Applications range from environmental to energetic field, spanning from pollutant degradation in water and gas phase [144], including dye [62,133,145,146,157] and nitric oxide removal (Table 5) [150,161] to H_2 production [156].

The use of small organic molecules as structure directing agents can aid self-assembly and significant enhance photocatalytic performance, due to the formation of peculiar morphologies, with high degree of crystallinity, high light-harvesting ability, as well as efficient separation of photogenerated carriers [159,161,162]. At the same time, their decomposition can provide a source of doping species for the catalyst including carbon atoms which can also give rise to organized porous frameworks hosting the photoactive phase and contributing to a significant improvement of its performance under UV-irradiation [163]. To this regard, citric acid has been exploited to synthetize urchin-like $LaPO_4$ hollow spheres, exhibiting marked activity towards CO_2 reduction under near UV [159].

Similarly, the use of N-Acetyl-d-Proline amino acid during hydrothermal treatment produced multi-shelled ZnO microspheres more active for NO oxidation under UV irradiation than other hierarchical ZnO samples [161]. Indeed, low molecular weight compounds can play multiple roles during the hydrothermal process, behaving not only as morphological agents but even as reducing compounds [164,165]. As a matter of fact, L-cysteine and glucose were employed during the hydrothermal treatment to obtain a TiO_2-graphene-MoS_2 composite as a high performance and stable catalyst for CO_2 reduction (Figure 14). Notably, L-cysteine acted as a reducing agent for graphene oxide, whereas glucose played a key role to control morphology, since it formed amorphous carbon which inhibited the TiO_2 and MoS_2 growth (Table 5 and Figure 14) [165].

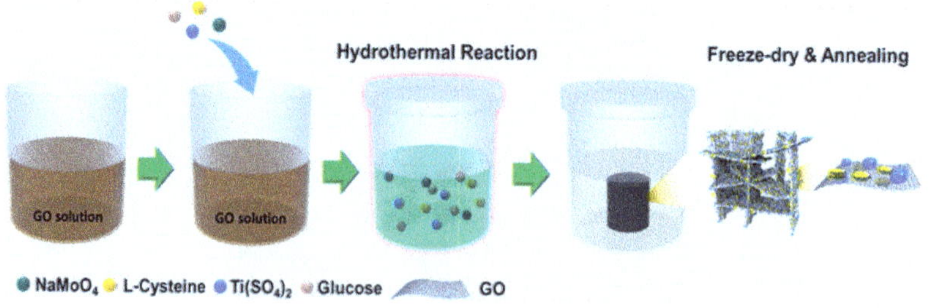

Figure 14. Schematization of one-pot solvothermal synthesis of TiO_2-graphene-MoS_2 composite [165].

Template free solvothermal processes can be carried out sequentially [162] or straightforwardly combined with wet impregnation or deposition procedures to have surface of produced catalyst decorated with other active moieties [161,163]. These often include plasmonic metal nanoparticles [163,166] as well as other semiconductors including both inorganic and organic systems [162,167]. The former improve catalytic activity through surface plasmon resonance [163]. The latter enable formation of p-n [161] as well as Z scheme heterostructures (Figure 15) [156,162,167,168], which exhibit superior visible light photocatalytic activity coupled with relevant stability and reusability [168,169].

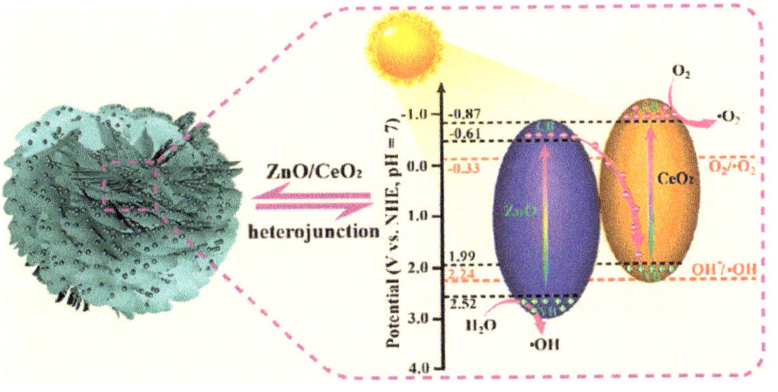

Figure 15. Schematic representation of ZnO/CeO_2 composites and charge transfer mechanism according to Z-scheme heterojunction [162].

Among wet chemistry routes, precipitation represents a facile low temperature approach to make hierarchical structures [159,170]. This method was effectively employed to produce $CdMoO_4$ hollow microspheres which exhibited higher photocatalytic activity towards Rhodamine-B aqueous solution under the UV irradiation than commercial TiO_2 [170].

In some cases, self-assembly into a hierarchical architecture can be promoted by electrostatic aggregation [171]. To this regard, $g\text{-}C_3N_4/ZnO$ microspheres were easily produced starting from a physical dispersion of the two components and exploiting their opposite surface charge. Z-scheme obtained by band matching between the two semiconductors as well as enhanced light absorption because of multiple scattering were responsible for a marked improvement of photocatalytic performance towards CO_2 reduction under visible light (Figure 16) [171].

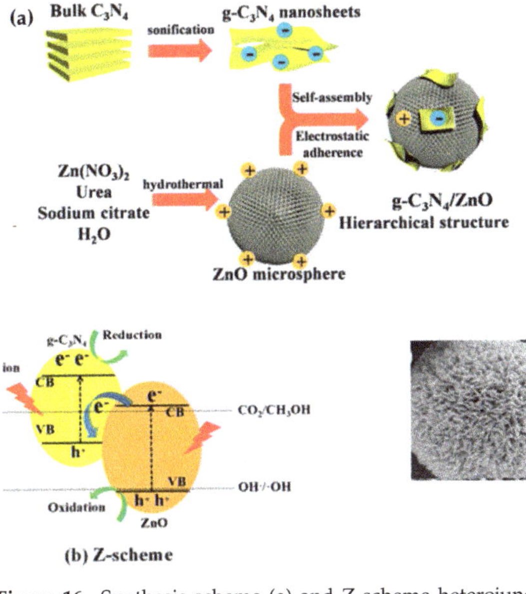

Figure 16. Synthesis scheme (**a**) and Z-scheme heterojunction (**b**) for $g\text{-}C_3N_4/ZnO$ hierarchical microspheres [171].

Other template free chemistry routes than solvothermal synthesis can be exploited to obtain hierarchical architectures. These include solid phase processes based on either solid or solid-gas reactions. Notably, solid-state reactions followed by hydrolysis at room temperature proved to be a simple route to make bismuth oxoiodide (BiO) with surface heterojunction between (0 0 1) and (1 1 0) surfaces promoting photocatalytic performance under visible light [172]. Solid-gas phase reactions usually involve pyrolysis [173], decomposition [174], or oxidation [175] and occur at moderate temperatures (higher than 150 °C). In these processes, temperature plays a key role in defining the final structure and overall photocatalytic performance, accordingly [174]. Moreover, either metal or non-metal heteroatom doping [175,176] as well as semiconductor heterostructures [171] can be easily achieved during the thermal process leading to superior photocatalytic performances [175,176]. Following this approach, nanosheet flower-like SnS_2 nanostructures have been obtained and used for Cr(IV) photocatalytic reduction under visible light with high performance as stability [173], even though these methods have been more extensively applied to make C_3N_4 catalysts. As a matter of fact, a facile pyrolysis process with urea was exploited to produce Yttrium-doped graphitic carbon nitride (Y/g-C_3N_4) catalysts, which feature high surface area, poor recombination phenomena, as well as wide spectral absorption accounting for enhanced photocatalytic activity towards rhodamine B degradation under visible light [176]. Alternatively, thermal oxidation and exfoliation, followed by curling condensation were carried out on bulk C_3N_4. Obtained O-doped carbon nitride (g-C_3N_4) nanotubes exhibit outstanding photocatalytic activity towards CO_2 reduction to synthetic fuels, sensitively higher than massive g-C_3N_4 [175].

Indeed, obtaining porous g-C_3N_4 is a challenging task, most of available processes to fabricate being time-consuming and energy intensive. A green and facile alternative consists in a fast quenching from high temperature, which produces an 3D interconnected porous assembly of g-C_3N_4 nano-scrolls, featuring high light absorption, poor recombination phenomena, and enhanced carrier mobility, accounting for improved catalytic performance toward CO_2 reduction [177].

Hierarchical heterostructures between semiconductors can be straightforwardly obtained by simultaneous decomposition of their precursors [171]. Following this route, calcination of tetrabutyl titanate and melamine precursors produced Z-scheme TiO_2/g-C_3N_4 heterojunction with unique features even due to concurrent C and N doping [171].

Template free processes also include combustion synthesis, a simple one-pot method exploiting fuel-oxidizer solutions, such as those containing sorbitol-nitrate mixtures, to fabricate mixed phase hierarchical structures, including bismuth rich-bismuth oxychlorides nanocatalysts, which exhibited high photocatalytic activity towards antibiotic ofloxacin degradation under solar light irradiation [178].

Moreover, non-conventional energy sources can be exploited to build self-templated hierarchical structures. To this regard, pulsed laser deposition (PLD) followed by thermal treatment had hierarchical nanostructured hematite coated onto glass surface and exhibiting enhanced and stable catalytic performance because of high availability of active sites, improved interaction with reactants as well as limited recombination phenomena [179].

Alternatively, ultrasound microwave and ultrasound irradiation can assist wet chemistry processes to promote hierarchical structure formation. Notably, microwave irradiation of copper acetate and thiourea aqueous solutions led to hierarchical CuS structures with tunable porosity and morphology, which evidenced higher activity than commercial CuS catalyst in dye degradation processes under visible light [180,181]. As an alternative, ultrasound irradiation is understood to be a facile and fast yet extremely efficient method to assist particles self-assembly processes into hierarchical architectures [160]. It proved effective in producing Urchin-shaped iron oxide nanoparticles, with huge surface area, up to 282.7 m^2/g and which evidenced 25 times faster removal of As(V) and Cr(VI) than commercial Fe_2O_3 (Figure 17) [160].

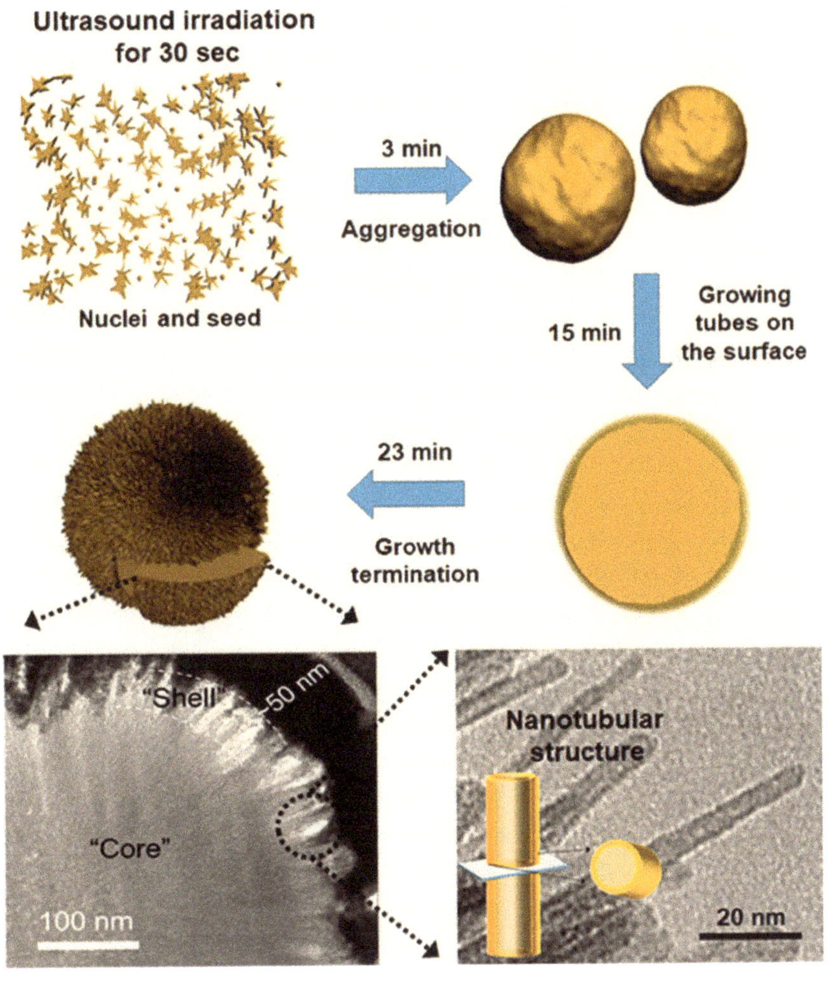

Figure 17. Formation mechanism and TEM images of urchin-shaped iron oxide nanostructures (Fe_2O_3) produced through ultrasound irradiation [160].

Similarly, sonochemical treatment of Melamine in water and its further calcination produced hierarchical graphitic-C_3N_4, which was tested for rhodamine B degradation and H_2 production and exhibited 16.7 and 8.7 higher visible light photo-catalytic activity than bulk homologous composition [182].

On the other hand, reactive magnetron sputtering deposition appears to be a facile methodology for large scale fabrication of hierarchical films [183]. This approach was exploited to fabricate two level porosity AgO nanorod arrays, which exhibited excellent photocatalytic properties under UV and visible-light conditions [183].

Table 5. Hierarchical catalysts produced without templates.

Composition	Synthesis Strategy	Morphology	SSA (m^2/g)	Application	Activity	Irradiation	Ref
ZnCuCo layered double hydroxide	Hydrothermal	3D flower-like hierarchical morphologies with in-terlaced petal-like nanosheet	86–179	photocatalytic H_2 production and degradation of SMZ	H_2 production rate (3700 μmol g^{-1} h^{-1}, 95% sulfamethazine (SMZ) degradation	Visible light	[43]
Ti_3C_2/Bi_2WO_6	hydrothermal	2D/2D hetero junction i	33.5–58.3	Photocatalysis degradation of tetracycline hydrochloride (TC-HCl)	0.430 min^{-1}	Visible light	[44]
$ZnFe_2O_4$ modified Cu_2S	in-situ self-assembly method	Dendritic fractal structure similar to a snowflake	-	photocatalytic reduction of nitrobenzene with and degradation of methyl orange and methylene blue dyes proficiently	98% yield of degradation of nitrobenzene, 94.3 and 86% fordegradation of methyl orange and methylene blue, respectively	Visible light	[45]
ZnO-graphene nanocomposite	Solution route	hollow microspheres	29.7–37.6	Adsorption/photocatalytic activity towards degradation of water-soluble organic pollutants (such as Rhodamine B, methyl orange, phenol)	90% adsorption capacity	UV (254 nm)	[141]
ZnO	Solvothermal route	Monodisperse microspheres	18	Photocatalytic activity for degradation of methylene blue	100%	UV-Vis	[142]
ZnO	Hydrothermal and calcination	Hierarchically porous microspheres composed of nanosheets	46–91	Photocatalytic RhB degradation activity	Up to 100%	solar	[60]
TiO_2	Hydrothermal route	(i) long and well oriented macrochannels (ii) surface macropores of 0.8–9.3 μm in size, and (iii) porous walls with pores mostly smaller than 0.5 μm.	82.9–216.1	Decomposition of methyl ethyl ketone (MEK) in a continuous flow photoreactor	Up to 37%	UV (254 nm)	[144]
TiO_2	Hydrothermal route	Partial spherical like structure	30–43.7	RhB photodegradation	Up to 98%	Visible light	[145]
TiO_2	Hydrothermal route	Nanoflowers with a spherical hierarchical structure	20–80	Aqueous methylene blue photo-oxidation	Up to 50%	solar	[146]

Table 5. *Cont.*

Composition	Synthesis Strategy	Morphology	SSA (m^2/g)	Application	Activity	Irradiation	Ref
CeO_2	Hydrothermal route	Mesoporous nanosphere	42.1–68.2	photocatalytic activity of rhodamine B (RhB) dye degradation	97.8–92% RhB dye degradation	UV-Vis and acidic condition	[147]
Cu_2O	Low temperature route	Spherical, cuboctahedral or cubic nanoparticles	7–13	Photocatalytic degradation of the antibiotic trimethoprim	Up to 48%	Visible light	[148]
WO_3	Hydrothermal route	Regular-shaped nanosheets with an average thickness of approximately 30–40 nm	2–18	Photocatalytic activity towards an aqueous solution of tetracycline (TC) and possess good stability and reusability	Up to 94%	Visible light	[149]
Bi_2WO_6	Hydrothermal method and calcining process	Mesoporous nanoplate multi-directional	53.5	Photocatalytic oxidation of NO	Up to 90%	Solar	[150]
Sm, Y, La and Nd-doped CeO_2	hydrothermal	broom-like hierarchical structure	-	BPA degradation and on CO_2 evolution from CH_3CHO decomposition	Up to 99%	UV	[151]
Yttria (Y_2O_3) nanosphere decorated ceria (CeO_2)	Hydrothermal route	Yttrium nanoparticles are anchored on the surface of CeO_2 nanorod with a particle size of 10 nm	-	Photocatalytic decomposition of aqueous Rhodamine B	Up to 96%	Solar light	[62]
SnS_2/TiO_2	Ultrasonic treatment	Ordered channels with a size about 1–3 μm were formed in the particles, and lots of holes appeared on the wall of the channels.	28	Photocatalytic degradation of Methyl Orange (MO)	Up to 90.9%	Solar light	[152]
$ZnO/g\text{-}C_3N_4$	Solvothermal	Porous microsphere with a size of about 700nm.	9.9–32-7	Photocatalytic degradation on rhodamine B and phenol	Up to 100%	Solar	[133]
ZnO/graphene	Solvothermal	Core-shell structure	65–201	Photocatalytic degradation of rhodamine B	Up to 98.5%	Visible light	[153]
TiO_2 Microspheres with Carbonaceous Species	Solvothermal	Porous structure	337	Photodegradation of rhodamine B	Up to 100%	Visible light	[154]
Carbon-coated TiO_2	solvothermal	Hierarchical nanotubes	Up to 244.4	Photocatalysts activity for water oxidation	Up to 705 $\mu mol\ h^{-1}\ g^{-1}$	Solar light	[155]
S-deficient CoS/CdS	Solvothermal	Hexagonal nanoplates	Up to 84.32	Photocatalytic water-splitting	Up to 14.5%	Visible light	[156]

Table 5. *Cont.*

Composition	Synthesis Strategy	Morphology	SSA (m^2/g)	Application	Activity	Irradiation	Ref
$BiVO_4/Bi_2WO_6$	Solvothermal	Self-assembled hierarchical $BiVO_4/Bi_2WO_6$ heterostructured composites	5.16	Photocatalytic activities for methylene blue (MB) degradation and photoelectrochemical performance	Up to 50%	Visible light	[157]
3D $LaPO_4$	Solution route using citric acid (CA)	Urchin-like hollow sphere	51–124	Photocatalytic CO_2-reduction performance	6.8-fold enhancement of the AQY	UV	[159]
ZnO	Hydrothermal-calcinatio n	Mesoporous multi-shelled ZnO microspheres	3–20	Photocatalysis for NO oxidation	Up to 77%	UV	[161]
ZnO/CeO_2	Hydrothermal/calcination	Spherically hierarchical structure	41–56	photocatalytic rhodamine B (RhB)	Up to 96%	Visible light	[162]
TiO_2/graphene/MoS_2	Hydrothermal	2D rGO sheets assembled into macroporous 3D structures	124	CO_2 Reduction Photocatalyst	Up to 97%	UV-Vis	[165]
Ag,Au/TiO_2	Solvothermal	Mesoporous spherical shape	153–173	Degradation of Textile Dyes	Up to 85%	Solar and visible light	[163]
Au-H-ZnO	Low temperature aqueous reaction and heat treatment	Nanosheets	-	Gas sensing and photocatalytic properties	Up to 94.8%	Visible light	[166]
PDA-modified Zn	Self-assembly	Rough microstructures	32–38	Photocatalytic CO_2 reduction	Up to 0.95 μmol h^{-1} g^{-1} of CH_3OH	UV	[167]
$ZnIn_2S_4$ marigold flower/Bi_2WO_6 (ZIS/BW)	Hydrothermal method followed by wet-impregnation	Hierarchical marigold flower and flower-like morphologies	14–73	Decomposition of metronidazole	Up to 56%	Visible light	[168]
BiOCl/$BiVO_4$	Coprecipitation-hydrothermal method	Micro-nanosheet	1.53–2.83	Photo-degradation rate of rhodamine B (RhB)	Up to 96%	Visible light	[169]
CuS	Microwave-assisted wet chemical process	Spherical monodispersed submicron particles	17.8–26.4	Decomposition of organic dyes including rhodamine B, methylene blue and malachite green	Up to 100%	Visible light	[180]

Table 5. *Cont.*

Composition	Synthesis Strategy	Morphology	SSA (m^2/g)	Application	Activity	Irradiation	Ref
CuS	Microwave irradiation	Aggregates of roughly spherical nanoparticles	14.74	Degradation of methylene blue, methyl orange, and 4-chlorophenol	Up to 100%	Solar light	[181]
3D- Fe_2O_3	Ultrasound irradiation method	Large quantity of 3D sea urchin-like structures combining a 1D rod-like structure on spherical support	129.4–282.7	Adsorption for heavy metals and photocatalytic activities toward the dyes (methylene blue and phenol)	Up to 100%	Solar light	[160]
g-C_3N_4	Ultrasound-assisted molecular rearrangement strategy	Hierarchical Rodlike	88.6	RhB degradation and H_2 evolution	Up to 23%	Visible light	[182]
$CdMoO_4$	Low temperature oil bath method	Uniform and porous spheres	-	Photocatalytic removal of mixed dye aqueous solutions	Up to 100%	UV-Vis	[170]
g-C_3N_4/ZnO	Electrostatic self-assembly method	nanosheets	156	Photocatalytic CO_2 reduction activity	0.64 µmol h^{-1} g^{-1}	Solar light	[171]
BiOI	Solid-state reaction with subsequent hydrolysis at room temperature.	Hierarchical microspheres assembled by nanoplates	3.1–13.8	Photocatalytic activity for phenol degradation.	Up to 100%	Visible light	[172]
SnS_2	Heating the mixture of SnCl2·2H2O and thiourea in air at 170 °C for 2	Porous flower-like hierarchical nanostructure	36.15–82.4	Adsorption and photocatalytic reduction of aqueous Cr(VI)	Up to 79.4%	Visible light	[173]
ZnO	Annealing of zinc oxalate.	Mesoporous nanostructured microlumps	1.7–29–9	Discoloration of the Methyl violet 2B	Up to 100%	UV	[174]
O-Doped g-C_3N_4	Successive thermal oxidation exfoliation and curling-condensation of bulk g-C_3N_4	Uniform porous network	36	Photocatalytic CO_2 Reduction Activity	Up to 0.88 µmol g^{-1} h^{-1}	Visible light	[175]
Yttrium-doped g-C_3N_4	Pyrolysis method	Sheet-like morphology with worm-like pores	39–106	Photocatalytic performance in rhodamine B degradation	Up to 100%	Visible light	[176]
3D g-C_3N_4	Cold quenching	Self-assembled nanoscrolls	76	Photocatalytic CO_2 reduction	Up to 11.2 µmol g^{-1} h^{-1}	UV-Vis	[177]

Table 5. *Cont.*

Composition	Synthesis Strategy	Morphology	SSA (m^2/g)	Application	Activity	Irradiation	Ref
Bismuth oxychlorides (BOC)	One-pot sorbitol-nitrate solution auto-combustion method	Mesoporous-mixed-phase of grain-like	93	Photocatalytic application in treatment of antibiotic effluents	Up to 80%	Visible light	[178]
Fe_2O_3	Pulsed Laser Deposition and thermal oxidation	Spherical particulates to an urchin-like structure with evolution of nanowires	-	Photocatalytic water purification	Up to 100%	H_2O_2 and visible light	[179]
AgO	Oxidizing solid Ag films in an environment of reactive magnetron sputtering deposition of NiO	Porous AgO nanorod films.	-	Photocatalyst and all solid-state thin film battery	Up to 100%	UV-Vis	[183]

7. Conclusions

Hierarchical structures have raised great interest for different photocatalytic applications, because of improved light harvesting, charge separation, as well as mass transport and adsorption efficacy, resulting in superior catalytic properties. Even though different fabrication strategies have been explored, as discussed in this review, some major challenges must be still addressed to allow easy synthesis of hierarchical photocatalysts and help to scale up process from laboratory to industrial production.

Among available methods, templated free wet synthesis approaches based on self-aggregation processes offer a facile route to obtain a wide choice of compositions and structures. Although most of them feature lower average surface area than systems obtained using inorganic templates, they usually exhibit comparable photocatalytic performance especially towards pollutant degradation, which is complete under UV irradiation and, for some composition reaches even 95% under visible light. Yet, some issues must be addressed mainly related to the difficulty to carry out large scale and high yield synthesis. Furthermore, post-process modification is usually required to further improve photocatalytic properties. Among templating approaches, inorganic backbones lead to the largest surface area, as high as 400 m^2/g, even though this feature does not necessarily result in the best photocatalytic performance, while organic templated systems exhibit comparable activity, allowing for organic pollutant complete degradation even under visible light. On the other hand, hybrid templating strategies, based on MOFs, are extremely versatile since they provide for a huge number of semiconductor compositions, with tunable morphology and structure. In addition, post-modification is usually not required, since catalyst functional modification including doping, hetero-junction formation, and loading with co-catalyst moieties can be easily carried out in-situ upon thermal treatments in dry or solvothermal conditions, concurrently with template decomposition. Carbonaceous residues resulting from this process can act themselves as doping species and sensitively improve photocatalytic properties. However, MOF templated method is expensive, time consuming, and difficult to be carried out on a large scale. Moreover, hierarchical structures obtained through this approach often show comparable performance to those obtained through self-templated methods, particularly for decontamination processes, thus the latter are usually preferred for these applications. On the other hand, more advanced technological routes involving hydrogen production are worth the effort of using MOF templated structures, which outperform with respect to conventional templates because of their extremely versatile features.

Nevertheless, the criteria to adequately choose MOF precursors must be further investigated to create the desired composition exhibiting hierarchical structure.

Bio-templates offer a sustainable and simple opportunity to design hierarchical photocatalytic nanomaterials. Among bioavailable compounds, bio-wastes are attracting great interest as abundant and cheap templates. However, their heterogeneity and poor knowledge of growth mechanisms make the process difficult and poor reproducible. Further investigation on building blocks growth and assembly is highly demanded to overcome these limitations and achieve rational control of the final hierarchical structure. More efforts should be taken towards the investigation of structure/function relationships and the mechanism underlying photocatalytic behavior.

Finally, a combination of the described strategies appears to be a promising way to combine benefits of each single approach. In this view, very recent studies evidence that integrating bio-templated design with MOF derived nanoparticles provide for a unique hierarchical structure, with efficient light absorption, superior charge separation efficacy, accounting for potent photocatalytic activity [184]. To this aim, this study could provide useful information to move towards more performant hierarchical nanostructures for advanced environmental and energy applications.

Author Contributions: G.L. drafted this review; V.V. and G.P. contributed equally to this work; B.S., A.C. and G.V. commented on the manuscript. All authors have read and agreed to the published version of the manuscript.

Funding: This research received no external funding.

Conflicts of Interest: The authors declare no conflict of interest.

References

1. Mukherjee, A. *Biomimetics Learning from Nature*; IntechOpen: London, UK, 2012; ISBN 9789533070254.
2. Bhushan, B. Biomimetics: Lessons from Nature—An overview. *Philos. Trans. R. Soc. A* **2009**, *367*, 1445–1486. [CrossRef] [PubMed]
3. Wegst, U.G.K.; Bai, H.; Saiz, E.; Tomsia, A.P.; Ritchie, R.O. Bioinspired structural materials. *Nat. Mater.* **2015**, *14*, 23–36. [CrossRef] [PubMed]
4. Guan, Q.F.; Yang, H.B.; Han, Z.M.; Ling, Z.C.; Yu, S.H. An all-natural bioinspired structural material for plastic replacement. *Nat. Commun.* **2020**, *11*, 5401. [CrossRef] [PubMed]
5. Martin-Martinez, F.J.; Jin, K.; López Barreiro, D.; Buehler, M.J. The rise of hierarchical nanostructured materials from renewable sources: Learning from nature. *ACS Nano* **2018**, *12*, 7425–7433. [CrossRef]
6. Fratzl, P.; Weinkamer, R. Nature's hierarchical materials. *Prog. Mater. Sci.* **2007**, *52*, 1263–1334. [CrossRef]
7. Dan, N. Synthesis of hierarchical materials. *Trends Biotechnol.* **2000**, *18*, 370–374. [CrossRef]
8. Koch, K.; Bhushan, B.; Jung, Y.C.; Barthlott, W. Fabrication of artificial Lotus leaves and significance of hierarchical structure for superhydrophobicity and low adhesion. *Soft Matter* **2009**, *5*, 1386–1393. [CrossRef]
9. Latthe, S.S.; Terashima, C.; Nakata, K.; Fujishima, A. Superhydrophobic surfaces developed by mimicking hierarchical surface morphology of lotus leaf. *Molecules* **2014**, *19*, 4256–4283. [CrossRef]
10. Lingham-Soliar, T. Microstructural tissue-engineering in the rachis and barbs of bird feathers. *Sci. Rep.* **2017**, *7*, 45162. [CrossRef]
11. Sullivan, T.N.; Wang, B.; Espinosa, H.D.; Meyers, M.A. Extreme lightweight structures: Avian feathers and bones. *Mater. Today* **2017**, *20*, 377–391. [CrossRef]
12. Zhu, Y.; Zhang, W.; Zhang, D. Fabrication of Sensor Materials Inspired by Butterfly Wings. *Adv. Mater. Technol.* **2017**, *2*, 1600209. [CrossRef]
13. Niu, S.; Li, B.; Mu, Z.; Yang, M.; Zhang, J.; Han, Z.; Ren, L. Excellent structure-based multifunction of morpho butterfly wings: A review. *J. Bionic Eng.* **2015**, *12*, 170–189. [CrossRef]
14. Su, B.-L.; Sanchez, C.; Yang, X.-Y. *Hierarchically Structured Porous Materials: From Nanoscience to Catalysis, Separation, Optics, Energy, and Life Science*; John Wiley & Sons: Hoboken, NJ, USA, 2012; ISBN 3527639594.
15. Li, X.; Yu, J.; Jaroniec, M. Hierarchical photocatalysts. *Chem. Soc. Rev.* **2016**, *45*, 2603–2636. [CrossRef] [PubMed]
16. Trogadas, P.; Ramani, V.; Strasser, P.; Fuller, T.F.; Coppens, M. Hierarchically structured nanomaterials for electrochemical energy conversion. *Angew. Chem. Int. Ed.* **2016**, *55*, 122–148. [CrossRef]
17. Shi, Y.; Ye, G.; Yang, C.; Tang, Y.; Peng, C.; Qian, G.; Yuan, W.; Duan, X.; Zhou, X. Pore engineering of hierarchically structured hydrodemetallization catalyst pellets in a fixed bed reactor. *Chem. Eng. Sci.* **2019**, *202*, 336–346. [CrossRef]
18. Hsu, M.-H.; Chang, C.-J. S-doped ZnO nanorods on stainless-steel wire mesh as immobilized hierarchical photocatalysts for photocatalytic H_2 production. *Int. J. Hydrog. Energy* **2014**, *39*, 16524–16533. [CrossRef]
19. Miao, Y.-E.; Wang, R.; Chen, D.; Liu, Z.; Liu, T. Electrospun self-standing membrane of hierarchical SiO_2@ γ-AlOOH (Boehmite) core/sheath fibers for water remediation. *ACS Appl. Mater. Interfaces* **2012**, *4*, 5353–5359. [CrossRef]
20. Brady, R.; Woonton, B.; Gee, M.L.; O'Connor, A.J. Hierarchical mesoporous silica materials for separation of functional food ingredients—A review. *Innov. Food Sci. Emerg. Technol.* **2008**, *9*, 243–248. [CrossRef]
21. Nayak, A.K.; Ghosh, R.; Santra, S.; Guha, P.K.; Pradhan, D. Hierarchical nanostructured WO_3-SnO_2 for selective sensing of volatile organic compounds. *Nanoscale* **2015**, *7*, 12460–12473. [CrossRef]
22. Gao, X.; Wang, Y.; Ji, G.; Cui, R.; Liu, Z. One-pot synthesis of hierarchical-pore metal–organic frameworks for drug delivery and fluorescent imaging. *CrystEngComm* **2018**, *20*, 1087–1093. [CrossRef]
23. Hartmann, M. Hierarchical zeolites: A proven strategy to combine shape selectivity with efficient mass transport. *Angew. Chemie Int. Ed.* **2004**, *43*, 5880–5882. [CrossRef] [PubMed]
24. Su, B.-L.; Sanchez, C.; Yang, X.-Y. *Zeolite Molecular Sieves Chemistry of Zeolites and Related Porous Materials Ordered Mesoporous Materials*; John Wiley & Sons: Chichester, UK, 2012; ISBN 9780470577578.
25. Holm, M.S.; Taarning, E.; Egeblad, K.; Christensen, C.H. Catalysis with hierarchical zeolites. *Catal. Today* **2011**, *168*, 3–16. [CrossRef]
26. Parlett, C.M.A.; Wilson, K.; Lee, A.F. Hierarchical porous materials: Catalytic applications. *Chem. Soc. Rev.* **2013**, *42*, 3876–3893. [CrossRef] [PubMed]
27. Costantini, A.; Venezia, V.; Pota, G.; Bifulco, A.; Califano, V.; Sannino, F. Adsorption of cellulase on wrinkled silica nanoparticles with enhanced inter-wrinkle distance. *Nanomaterials* **2020**, *10*, 1799. [CrossRef] [PubMed]
28. Tang, Y.; Wee, P.; Lai, Y.; Wang, X.; Gong, D.; Kanhere, P.D.; Lim, T.T.; Dong, Z.; Chen, Z. Hierarchical TiO_2 nanoflakes and nanoparticles hybrid structure for improved photocatalytic activity. *J. Phys. Chem. C* **2012**, *116*, 2772–2780. [CrossRef]

29. Gao, C.; Wei, T.; Zhang, Y.; Song, X.; Huan, Y.; Liu, H.; Zhao, M.; Yu, J.; Chen, X. A Photoresponsive Rutile TiO_2 Heterojunction with Enhanced Electron–Hole Separation for High-Performance Hydrogen Evolution. *Adv. Mater.* **2019**, *31*, 1806596. [CrossRef] [PubMed]
30. Yahya, N.; Aziz, F.; Jamaludin, N.A.; Mutalib, M.A.; Ismail, A.F.; Salleh, W.N.W.; Jaafar, J.; Yusof, N.; Ludin, N.A. A review of integrated photocatalyst adsorbents for wastewater treatment. *J. Environ. Chem. Eng.* **2018**, *6*, 7411–7425. [CrossRef]
31. Li, X.; Yu, J.; Low, J.; Fang, Y.; Xiao, J.; Chen, X. Engineering heterogeneous semiconductors for solar water splitting. *J. Mater. Chem. A* **2015**, *3*, 2485–2534. [CrossRef]
32. Fang, M.; Dong, G.; Wei, R.; Ho, J.C. Hierarchical Nanostructures: Hierarchical Nanostructures: Design for Sustainable Water Splitting (Adv. Energy Mater. 23/2017). *Adv. Energy Mater.* **2017**, *7*, 1770135. [CrossRef]
33. Stein, A.; Rudisill, S.G.; Petkovich, N.D. Perspective on the influence of interactions between hard and soft templates and precursors on morphology of hierarchically structured porous materials. *Chem. Mater.* **2014**, *26*, 259–276. [CrossRef]
34. Wang, P.; Xuan, J.; Zhang, R.; Zhang, H.; Wang, Q.; Wang, H.; Liu, H.; Zhang, L. Hierarchically Structured Components: Design, Additive Manufacture, and Their Energy Applications. *Adv. Mater. Technol.* **2022**, *7*, 2100672. [CrossRef]
35. Wang, S.; Wang, Y.; Zang, S.; Lou, X.W. Hierarchical hollow heterostructures for photocatalytic CO_2 reduction and water splitting. *Small Methods* **2020**, *4*, 1900586. [CrossRef]
36. Sun, C.; Yang, J.; Xu, M.; Cui, Y.; Ren, W.; Zhang, J.; Zhao, H.; Liang, B. Recent intensification strategies of SnO_2-based photocatalysts: A review. *Chem. Eng. J.* **2022**, *427*, 131564. [CrossRef]
37. Jiang, L.; Zhou, H.; Yang, H.; Sun, N.; Huang, Z.; Pang, H. Applications of hierarchical metal–organic frameworks and their derivatives in electrochemical energy storage and conversion. *J. Energy Storage* **2022**, *55*, 105354. [CrossRef]
38. Janani, R.; Preethi, V.R.; Singh, S.; Rani, A.; Chang, C.-T. Hierarchical ternary sulfides as effective photocatalyst for hydrogen generation through water splitting: A review on the performance of $ZnIn_2S_4$. *Catalysts* **2021**, *11*, 277. [CrossRef]
39. Meng, S.; Wu, H.; Cui, Y.; Zheng, X.; Wang, H.; Chen, S.; Wang, Y.; Fu, X. One-step synthesis of 2D/2D-3D NiS/$Zn_3In_2S_6$ hierarchical structure toward solar-to-chemical energy transformation of biomass-relevant alcohols. *Appl. Catal. B Environ.* **2020**, *266*, 118617. [CrossRef]
40. Tahir, M.B.; Asiri, A.M.; Nabi, G.; Rafique, M.; Sagir, M. Fabrication of heterogeneous photocatalysts for insight role of carbon nanofibre in hierarchical WO_3/$MoSe_2$ composite for enhanced photocatalytic hydrogen generation. *Ceram. Int.* **2019**, *45*, 5547–5552. [CrossRef]
41. Tan, H.; Li, J.; He, M.; Li, J.; Zhi, D.; Qin, F.; Zhang, C. Global evolution of research on green energy and environmental technologies: A bibliometric study. *J. Environ. Manage.* **2021**, *297*, 113382. [CrossRef]
42. Jiang, Y.; Liao, J.-F.; Xu, Y.-F.; Chen, H.-Y.; Wang, X.-D.; Kuang, D.-B. Hierarchical $CsPbBr_3$ nanocrystal-decorated ZnO nanowire/macroporous graphene hybrids for enhancing charge separation and photocatalytic CO_2 reduction. *J. Mater. Chem. A* **2019**, *7*, 13762–13769. [CrossRef]
43. Gholami, P.; Khataee, A.; Ritala, M. Template-free hierarchical trimetallic oxide photocatalyst derived from organically modified ZnCuCo layered double hydroxide. *J. Clean. Prod.* **2022**, *366*, 132761. [CrossRef]
44. Han, Z.; Zhang, X.; Zuo, Y.; Dong, H.; Ren, H. Decorating 2D Ti_3C_2 on flower-like hierarchical Bi_2WO_6 for the 2D/2D heterojunction construction towards photodegradation of tetracycline antibiotics. *Sep. Purif. Technol.* **2022**, *299*, 121715. [CrossRef]
45. Kaushik, B.; Rana, P.; Solanki, K.; Rawat, D.; Yadav, S.; Naikwadi, D.R.; Biradar, A.V.; Sharma, R.K. In-situ synthesis of 3-D hierarchical $ZnFe_2O_4$ modified Cu_2S snowflakes: Exploring their bifunctionality in selective photocatalytic reduction of nitroarenes and methyl orange degradation. *J. Photochem. Photobiol. A Chem.* **2022**, *433*, 114165. [CrossRef]
46. Li, Q.; Zhou, J.; Fu, L.; Chen, C.; Mao, S.; Pu, Z.; Yang, J.; Shi, J.-W.; Wu, K. Fabrication of heterostructural Ru-$SrTiO_3$ fibers through in-situ exsolution for visible-light-induced photocatalysis. *J. Alloys Compd.* **2022**, *925*, 166747. [CrossRef]
47. Li, X.; Yu, J.; Jaroniec, M. Hierarchical porous photocatalysts. *Interface Sci. Technol.* **2020**, *31*, 63–102.
48. Shi, J.W.; Zong, X.; Wu, X.; Cui, H.J.; Xu, B.; Wang, L.; Fu, M.L. Carbon-doped Titania Hollow Spheres with Tunable Hierarchical Macroporous Channels and Enhanced Visible Light-induced Photocatalytic Activity. *ChemCatChem* **2012**, *4*, 488–491. [CrossRef]
49. Yu, J.; Zhang, L.; Cheng, B.; Su, Y. Hydrothermal preparation and photocatalytic activity of hierarchically sponge-like macro-/mesoporous Titania. *J. Phys. Chem. C* **2007**, *111*, 10582–10589. [CrossRef]
50. Wang, X.; Yu, J.C.; Ho, C.; Hou, Y.; Fu, X. Photocatalytic activity of a hierarchically macro/mesoporous titania. *Langmuir* **2005**, *21*, 2552–2559. [CrossRef] [PubMed]
51. Zheng, Z.; Huang, B.; Qin, X.; Zhang, X.; Dai, Y. Strategic synthesis of hierarchical TiO_2 microspheres with enhanced photocatalytic activity. *Chem. A Eur. J.* **2010**, *16*, 11266–11270. [CrossRef]
52. Yu, J.; Su, Y.; Cheng, B. Template-free fabrication and enhanced photocatalytic activity of hierarchical macro-/mesoporous titania. *Adv. Funct. Mater.* **2007**, *17*, 1984–1990. [CrossRef]
53. Zhang, L.; Yu, J.C. A sonochemical approach to hierarchical porous titania spheres with enhanced photocatalytic activity. *Chem. Commun.* **2003**, *3*, 2078–2079. [CrossRef]
54. Shao, G.S.; Wang, F.Y.; Ren, T.Z.; Liu, Y.; Yuan, Z.Y. Hierarchical mesoporous phosphorus and nitrogen doped titania materials: Synthesis, characterization and visible-light photocatalytic activity. *Appl. Catal. B Environ.* **2009**, *92*, 61–67. [CrossRef]
55. Dong, G.; Wang, Y.; Lei, H.; Tian, G.; Qi, S.; Wu, D. Hierarchical mesoporous titania nanoshell encapsulated on polyimide nanofiber as flexible, highly reactive, energy saving and recyclable photocatalyst for water purification. *J. Clean. Prod.* **2020**, *253*, 120021. [CrossRef]

56. Ta, Q.T.H.; Cho, E.; Sreedhar, A.; Noh, J.S. Mixed-dimensional, three-level hierarchical nanostructures of silver and zinc oxide for fast photocatalytic degradation of multiple dyes. *J. Catal.* **2019**, *371*, 1–9. [CrossRef]
57. Yukhnovets, O.; Semenova, A.A.; Levkevich, E.A.; Maximov, A.I.; Moshnikov, V.A. Zinc oxide hierarchical nanostructures for photocatalysis. *J. Phys. Conf. Ser.* **2018**, *993*, 012009. [CrossRef]
58. Kim, J.H.; Joshi, M.K.; Lee, J.; Park, C.H.; Kim, C.S. Polydopamine-assisted immobilization of hierarchical zinc oxide nanostructures on electrospun nanofibrous membrane for photocatalysis and antimicrobial activity. *J. Colloid Interface Sci.* **2018**, *513*, 566–574. [CrossRef]
59. Adhyapak, P.V.; Meshram, S.P.; Tomar, V.; Amalnerkar, D.P.; Mulla, I.S. Effect of preparation parameters on the morphologically induced photocatalytic activities of hierarchical zinc oxide nanostructures. *Ceram. Int.* **2013**, *39*, 7367–7378. [CrossRef]
60. Wang, S.; Kuang, P.; Cheng, B.; Yu, J.; Jiang, C. ZnO hierarchical microsphere for enhanced photocatalytic activity. *J. Alloys Compd.* **2018**, *741*, 622–632. [CrossRef]
61. Yin, Q.; Qiao, R.; Li, Z.; Zhang, X.L.; Zhu, L. Hierarchical nanostructures of nickel-doped zinc oxide: Morphology controlled synthesis and enhanced visible-light photocatalytic activity. *J. Alloys Compd.* **2015**, *618*, 318–325. [CrossRef]
62. Magdalane, C.M.; Kaviyarasu, K.; Priyadharsini, G.M.A.; Bashir, A.K.H.; Mayedwa, N.; Matinise, N.; Isaev, A.B.; Al-Dhabi, N.A.; Arasu, M.V.; Arokiyaraj, S. Improved photocatalytic decomposition of aqueous Rhodamine-B by solar light illuminated hierarchical yttria nanosphere decorated ceria nanorods. *J. Mater. Res. Technol.* **2019**, *8*, 2898–2909. [CrossRef]
63. Qian, J.; Chen, Z.; Sun, H.; Chen, F.; Xu, X.; Wu, Z.; Li, P.; Ge, W. Enhanced photocatalytic H_2 production on three-dimensional porous CeO_2/carbon nanostructure. *ACS Sustain. Chem. Eng.* **2018**, *6*, 9691–9698. [CrossRef]
64. Xiang, Q.; Cheng, F.; Lang, D. Hierarchical layered WS2/graphene-modified CdS nanorods for efficient photocatalytic hydrogen evolution. *ChemSusChem* **2016**, *9*, 996–1002. [CrossRef] [PubMed]
65. Wang, W.; Yu, J.; Xiang, Q.; Cheng, B. Enhanced photocatalytic activity of hierarchical macro/mesoporous TiO_2–graphene composites for photodegradation of acetone in air. *Appl. Catal. B Environ.* **2012**, *119*, 109–116. [CrossRef]
66. Luo, Q.-P.; Yu, X.-Y.; Lei, B.-X.; Chen, H.-Y.; Kuang, D.-B.; Su, C.-Y. Reduced graphene oxide-hierarchical ZnO hollow sphere composites with enhanced photocurrent and photocatalytic activity. *J. Phys. Chem. C* **2012**, *116*, 8111–8117. [CrossRef]
67. Sin, J.-C.; Lam, S.-M.; Satoshi, I.; Lee, K.-T.; Mohamed, A.R. Sunlight photocatalytic activity enhancement and mechanism of novel europium-doped ZnO hierarchical micro/nanospheres for degradation of phenol. *Appl. Catal. B Environ.* **2014**, *148*, 258–268. [CrossRef]
68. Chen, J.; Wang, H.; Huang, G.; Zhang, Z.; Han, L.; Song, W.; Li, M.; Zhang, Y. Facile synthesis of urchin-like hierarchical Nb_2O_5 nanospheres with enhanced visible light photocatalytic activity. *J. Alloys Compd.* **2017**, *728*, 19–28. [CrossRef]
69. Sang, Y.; Cao, X.; Dai, G.; Wang, L.; Peng, Y.; Geng, B. Facile one-pot synthesis of novel hierarchical Bi_2O_3/Bi_2S_3 nanoflower photocatalyst with intrinsic p-n junction for efficient photocatalytic removals of RhB and Cr(VI). *J. Hazard. Mater.* **2020**, *381*, 120942. [CrossRef]
70. Sharma, S.; Khare, N. Hierarchical Bi_2S_3 nanoflowers: A novel photocatalyst for enhanced photocatalytic degradation of binary mixture of Rhodamine B and Methylene blue dyes and degradation of mixture of p-nitrophenol and p-chlorophenol. *Adv. Powder Technol.* **2018**, *29*, 3336–3347. [CrossRef]
71. Deas, R.; Pearce, S.; Goss, K.; Wang, Q.; Chen, W.T.; Waterhouse, G.I.N. Hierarchical Au/TiO_2 nanoflower photocatalysts with outstanding performance for alcohol photoreforming under UV irradiation. *Appl. Catal. A Gen.* **2020**, *602*, 39–41. [CrossRef]
72. Cao, F.; Shi, W.; Zhao, L.; Song, S.; Yang, J.; Lei, Y.; Zhang, H. Hydrothermal synthesis and high photocatalytic activity of 3D wurtzite ZnSe hierarchical nanostruetures. *J. Phys. Chem. C* **2008**, *112*, 17095–17101. [CrossRef]
73. Zhang, H.; Hu, C. Effective solar absorption and radial microchannels of SnO_2 hierarchical structure for high photocatalytic activity. *Catal. Commun.* **2011**, *14*, 32–36. [CrossRef]
74. Li, D.; Fang, M.; Jiang, C.; Lin, H.; Luo, C.; Qi, R.; Huang, R.; Peng, H. Size-controlled synthesis of hierarchical bismuth selenide nanoflowers and their photocatalytic performance in the presence of H_2O_2. *J. Nanoparticle Res.* **2018**, *20*, 228. [CrossRef]
75. Song, J.M.; Mao, C.J.; Niu, H.L.; Shen, Y.H.; Zhang, S.Y. Hierarchical structured bismuth oxychlorides: Self-assembly from nanoplates to nanoflowers via a solvothermal route and their photocatalytic properties. *CrystEngComm* **2010**, *12*, 3875–3881. [CrossRef]
76. Liang, Y.; Ding, M.; Yang, Y.; Xu, K.; Luo, X.; Yu, T.; Zhang, W.; Liu, W.; Yuan, C. Highly dispersed Pt nanoparticles on hierarchical titania nanoflowers with {010} facets for gas sensing and photocatalysis. *J. Mater. Sci.* **2019**, *54*, 6826–6840. [CrossRef]
77. Malik, R.; Tomer, V.K.; Rana, P.S.; Nehra, S.P.; Duhan, S. Surfactant assisted hydrothermal synthesis of porous 3-D hierarchical SnO_2 nanoflowers for photocatalytic degradation of Rose Bengal. *Mater. Lett.* **2015**, *154*, 124–127. [CrossRef]
78. Wu, Y.; Wang, H.; Tu, W.; Liu, Y.; Wu, S.; Tan, Y.Z.; Chew, J.W. Construction of hierarchical 2D-2D $Zn_3In_2S_6$/fluorinated polymeric carbon nitride nanosheets photocatalyst for boosting photocatalytic degradation and hydrogen production performance. *Appl. Catal. B Environ.* **2018**, *233*, 58–69. [CrossRef]
79. Jin, Z.; Dong, W.; Yang, M.; Wang, J.; Gao, H.; Wang, G. One-Pot Preparation of Hierarchical Nanosheet-Constructed Fe_3O_4/MIL-88B(Fe) Magnetic Microspheres with High Efficiency Photocatalytic Degradation of Dye. *ChemCatChem* **2016**, *8*, 3510–3517. [CrossRef]
80. Jia, Y.; Ma, Y.; Tang, J.; Shi, W. Hierarchical nanosheet-based Bi_2MoO_6 microboxes for efficient photocatalytic performance. *Dalt. Trans.* **2018**, *47*, 5542–5547. [CrossRef]

81. Tanveer, M.; Cao, C.; Ali, Z.; Aslam, I.; Idrees, F.; Khan, W.S.; But, F.K.; Tahir, M.; Mahmood, N. Template free synthesis of CuS nanosheet-based hierarchical microspheres: An efficient natural light driven photocatalyst. *CrystEngComm* **2014**, *16*, 5290–5300. [CrossRef]
82. Taheri, M.; Abdizadeh, H.; Golobostanfard, M.R. Hierarchical ZnO nanoflowers and urchin-like shapes synthesized via sol-gel electrophoretic deposition with enhanced photocatalytic performance. *Mater. Chem. Phys.* **2018**, *220*, 118–127. [CrossRef]
83. Li, H.; Fei, G.T.; Fang, M.; Cui, P.; Guo, X.; Yan, P.; Zhang, L. De Synthesis of urchin-like Co_3O_4 hierarchical micro/nanostructures and their photocatalytic activity. *Appl. Surf. Sci.* **2011**, *257*, 6527–6530. [CrossRef]
84. Yang, Y.; Wu, J.; Xiao, T.; Tang, Z.; Shen, J.; Li, H.; Zhou, Y.; Zou, Z. Urchin-like hierarchical CoZnAl-LDH/RGO/g-C_3N_4 hybrid as a Z-scheme photocatalyst for efficient and selective CO_2 reduction. *Appl. Catal. B Environ.* **2019**, *255*, 117771. [CrossRef]
85. Liu, J.; Guo, Z.; Wang, W.; Huang, Q.; Zhu, K.; Chen, X. Heterogeneous ZnS hollow urchin-like hierarchical nanostructures and their structure-enhanced photocatalytic properties. *Nanoscale* **2011**, *3*, 1470–1473. [CrossRef]
86. Xiang, L.; Zhao, X.; Yin, J.; Fan, B. Well-organized 3D urchin-like hierarchical TiO_2 microspheres with high photocatalytic activity. *J. Mater. Sci.* **2012**, *47*, 1436–1445. [CrossRef]
87. Cheng, P.; Wang, Y.; Xu, L.; Sun, P.; Su, Z.; Jin, F.; Liu, F.; Sun, Y.; Lu, G. High specific surface area urchin-like hierarchical ZnO-TiO_2 architectures: Hydrothermal synthesis and photocatalytic properties. *Mater. Lett.* **2016**, *175*, 52–55. [CrossRef]
88. Xiao, X.; Xing, C.; He, G.; Zuo, X.; Nan, J.; Wang, L. Solvothermal synthesis of novel hierarchical $Bi_4O_5I_2$ nanoflakes with highly visible light photocatalytic performance for the degradation of 4-tert-butylphenol. *Appl. Catal. B Environ.* **2014**, *148*, 154–163. [CrossRef]
89. Ong, W.L.; Natarajan, S.; Kloostra, B.; Ho, G.W. Metal nanoparticle-loaded hierarchically assembled ZnO nanoflakes for enhanced photocatalytic performance. *Nanoscale* **2013**, *5*, 5568–5575. [CrossRef]
90. Sun, M.; Chen, C.; Chen, L.; Su, B. Hierarchically porous materials: Synthesis strategies and emerging applications. *Front. Chem. Sci. Eng.* **2016**, *10*, 301–347. [CrossRef]
91. Yang, Q.; Hao, J. Synthesis of metal sulfides via ionic liquid-mediated assembly strategy and their photocatalytic degradation of dyes in water. *Colloids Surf. A Physicochem. Eng. Asp.* **2022**, *633*, 127848. [CrossRef]
92. Sang, Y.; Ding, G.; Guo, Z.; Xue, Y.; Li, G.; Zhang, R. Facile synthesis of amorphous bimetallic hydroxide on Fe-doped Ni_3S_2 as an active electrocatalyst for oxygen evolution reaction. *J. Alloys Compd.* **2022**, *919*, 165855. [CrossRef]
93. Xu, T.; Wang, S.; Li, L.; Liu, X. Dual templated synthesis of tri-modal porous $SrTiO_3$/TiO_2@ carbon composites with enhanced photocatalytic activity. *Appl. Catal. A Gen.* **2019**, *575*, 132–141. [CrossRef]
94. Xu, T.; Liu, X.; Wang, S.; Li, L. Ferroelectric oxide nanocomposites with trimodal pore structure for high photocatalytic performance. *Nano-micro Lett.* **2019**, *11*, 37. [CrossRef] [PubMed]
95. Yao, L.; Wang, H.; Zhang, Y.; Wang, S.; Liu, X. Fabrication of N doped TiO_2/C nanocomposites with hierarchical porous structure and high photocatalytic activity. *Microporous Mesoporous Mater.* **2019**, *288*, 109604. [CrossRef]
96. Liu, H.; Liu, X.; Mu, S.; Wang, S.; Wang, S.; Li, L.; Giannelis, E.P. A novel fabrication approach for three-dimensional hierarchical porous metal oxide/carbon nanocomposites for enhanced solar photocatalytic performance. *Catal. Sci. Technol.* **2017**, *7*, 1965–1970. [CrossRef]
97. Lyu, J.; Shao, J.; Wang, Y.; Qiu, Y.; Li, J.; Li, T.; Peng, Y.; Liu, F. Construction of a porous core-shell homojunction for the photocatalytic degradation of antibiotics. *Chem. Eng. J.* **2019**, *358*, 614–620. [CrossRef]
98. Zhao, J.; Liao, C.; Chen, X.; Song, W. Hierarchically ordered macro–mesoporous anatase TiO_2 prepared by pearl oyster shell and triblock copolymer dual templates for high photocatalytic activity. *RSC Adv.* **2018**, *8*, 38461–38469. [CrossRef]
99. Rana, A.; Sudhaik, A.; Raizada, P.; Nguyen, V.-H.; Xia, C.; Khan, A.A.P.; Thakur, S.; Nguyen-Tri, P.; Nguyen, C.C.; Kim, S.Y. Graphitic carbon nitride based immobilized and non-immobilized floating photocatalysts for environmental remediation. *Chemosphere* **2022**, *297*, 134229. [CrossRef]
100. Xu, M.; Wu, J.; Wang, J.; Mao, Y.; Liu, M.; Yang, Y.; Yang, C.; Sun, L.; Du, Y.; Li, Y. The integration of Triazine-based porous organic polymer with bio-waste poplar catkin as water-floatable photocatalyst. *Appl. Surf. Sci.* **2022**, *581*, 152409. [CrossRef]
101. Chen, H.; Zhao, L.; He, X.; Huang, Z.; Wang, G.; Fang, W.; Li, W. Highly enhanced photocatalytic degradation of dye rhodamine B on the biogenic hierarchical porous TiO_2/SiO_2 with 1D/3D-chain. *Clean Technol. Environ. Policy* **2018**, *20*, 887–897. [CrossRef]
102. CHEN, F.; CHEN, X.; CHEN, Z. *Artificial Synthesis and Improved Photo Catalysis Performance of Bionic Hierarchical Fibers Alpha-Fe_2O_3*; 2nd International Seminar on Applied Physics, Optoelectronics and Photonics (APOP): Suzhou, China, 2017; ISBN 978-1-60595-522-3.
103. Lv, H.; Zhang, M.; Wang, P.; Xu, X.; Liu, Y.; Yu, D.-G. Ingenious construction of $Ni(DMG)_2$/TiO_2-decorated porous nanofibers for the highly efficient photodegradation of pollutants in water. *Colloids Surf. A Physicochem. Eng. Asp.* **2022**, *650*, 129561. [CrossRef]
104. He, D.; Ma, Y.; Zhao, R.; Qiu, J.; Sun, B.; Wang, H.; Yang, M.; Jia, X.; Li, X.; Li, Y. Complete-lifecycle-available, lightweight and flexible hierarchical structured Bi_2WO_6/WO_3/PAN nanofibrous membrane for X-Ray shielding and photocatalytic degradation. *Adv. Mater. Interfaces* **2021**, *8*, 2002131. [CrossRef]
105. Zhang, J.; Hou, X.; Pang, Z.; Cai, Y.; Zhou, H.; Lv, P.; Wei, Q. Fabrication of hierarchical TiO_2 nanofibers by microemulsion electrospinning for photocatalysis applications. *Ceram. Int.* **2017**, *43*, 15911–15917. [CrossRef]
106. Zhou, H.; Zhong, S.; Shen, M.; Yao, Y. Composite soft template-assisted construction of a flower-like β-Bi_2O_3/$Bi_2O_2CO_3$ heterojunction photocatalyst for the enhanced simulated sunlight photocatalytic degradation of tetracycline. *Ceram. Int.* **2019**, *45*, 15036–15047. [CrossRef]

107. Lin, Z.; Huang, J. A hierarchical $H_3PW_{12}O_{40}/TiO_2$ nanocomposite with cellulose as scaffold for photocatalytic degradation of organic pollutants. *Sep. Purif. Technol.* **2021**, *264*, 118427. [CrossRef]
108. Sun, X.; Wang, K.; Shu, Y.; Zou, F.; Zhang, B.; Sun, G.; Uyama, H.; Wang, X. One-pot route towards active TiO_2 doped hierarchically porous cellulose: Highly efficient photocatalysts for methylene blue degradation. *Materials* **2017**, *10*, 373. [CrossRef]
109. Xiao, J.; Jiang, H. Thermally Stable Metal-Organic Framework-Templated Synthesis of Hierarchically Porous Metal Sulfides: Enhanced Photocatalytic Hydrogen Production. *Small* **2017**, *13*, 1700632. [CrossRef]
110. Ren, J.-T.; Zheng, Y.-L.; Yuan, K.; Zhou, L.; Wu, K.; Zhang, Y.-W. Self-templated synthesis of Co_3O_4 hierarchical nanosheets from a metal–organic framework for efficient visible-light photocatalytic CO_2 reduction. *Nanoscale* **2020**, *12*, 755–762. [CrossRef]
111. Wang, F.; Xiao, L.; Chen, J.; Chen, L.; Fang, R.; Li, Y. Regulating the Electronic Structure and Water Adsorption Capability by Constructing Carbon-Doped CuO Hollow Spheres for Efficient Photocatalytic Hydrogen Evolution. *ChemSusChem* **2020**, *13*, 5711–5721. [CrossRef]
112. Li, P.; Zhang, M.; Li, X.; Wang, C.; Wang, R.; Wang, B.; Yan, H. MOF-derived NiO/CeO_2 heterojunction: A photocatalyst for degrading pollutants and hydrogen evolution. *J. Mater. Sci.* **2020**, *55*, 15930–15944. [CrossRef]
113. Chen, W.; Han, B.; Xie, Y.; Liang, S.; Deng, H.; Lin, Z. Ultrathin Co-Co LDHs nanosheets assembled vertically on MXene: 3D nanoarrays for boosted visible-light-driven CO_2 reduction. *Chem. Eng. J.* **2020**, *391*, 123519. [CrossRef]
114. Yan, B.; Zhang, L.; Tang, Z.; Al-Mamun, M.; Zhao, H.; Su, X. Palladium-decorated hierarchical titania constructed from the metal-organic frameworks NH2-MIL-125 (Ti) as a robust photocatalyst for hydrogen evolution. *Appl. Catal. B Environ.* **2017**, *218*, 743–750. [CrossRef]
115. Zhao, F.; Yin, D.; Khaing, K.K.; Liu, B.; Chen, T.; Deng, L.; Li, L.; Guo, X.; Wang, J.; Xiao, S. Fabrication of hierarchical Co_9S_8@ ZnAgInS heterostructured cages for highly efficient photocatalytic hydrogen generation and pollutants degradation. *Inorg. Chem.* **2020**, *59*, 7027–7038. [CrossRef] [PubMed]
116. Peng, D.; Zhang, Y.; Xu, G.; Tian, Y.; Ma, D.; Zhang, Y.; Qiu, P. Synthesis of multilevel structured MoS_2@ Cu/Cu_2O@ C visible-light-driven photocatalyst derived from MOF–guest polyhedra for cyclohexane oxidation. *ACS Sustain. Chem. Eng.* **2020**, *8*, 6622–6633. [CrossRef]
117. Gong, Y.; Zhao, X.; Zhang, H.; Yang, B.; Xiao, K.; Guo, T.; Zhang, J.; Shao, H.; Wang, Y.; Yu, G. MOF-derived nitrogen doped carbon modified g-C_3N_4 heterostructure composite with enhanced photocatalytic activity for bisphenol A degradation with peroxymonosulfate under visible light irradiation. *Appl. Catal. B Environ.* **2018**, *233*, 35–45. [CrossRef]
118. He, X.; Nguyen, V.; Jiang, Z.; Wang, D.; Zhu, Z.; Wang, W.-N. Highly-oriented one-dimensional MOF-semiconductor nanoarrays for efficient photodegradation of antibiotics. *Catal. Sci. Technol.* **2018**, *8*, 2117–2123. [CrossRef]
119. Xu, Y.; Wu, S.; Li, X.; Huang, Y.; Wang, Z.; Han, Y.; Wu, J.; Meng, H.; Zhang, X. Synthesis, characterization, and photocatalytic degradation properties of $ZnO/ZnFe_2O_4$ magnetic heterostructures. *New J. Chem.* **2017**, *41*, 15433–15438. [CrossRef]
120. Kampouri, S.; Ireland, C.P.; Valizadeh, B.; Oveisi, E.; Schouwink, P.A.; Mensi, M.; Stylianou, K.C. Mixed-phase MOF-derived titanium dioxide for photocatalytic hydrogen evolution: The impact of the templated morphology. *ACS Appl. Energy Mater.* **2018**, *1*, 6541–6548. [CrossRef]
121. Li, N.; Huang, H.; Bibi, R.; Shen, Q.; Ngulube, R.; Zhou, J.; Liu, M. Noble-metal-free MOF derived hollow CdS/TiO_2 decorated with NiS cocatalyst for efficient photocatalytic hydrogen evolution. *Appl. Surf. Sci.* **2019**, *476*, 378–386. [CrossRef]
122. Mao, Q.; Chen, J.; Chen, H.; Chen, Z.; Chen, J.; Li, Y. Few-layered 1T-MoS_2-modified ZnCoS solid-solution hollow dodecahedra for enhanced photocatalytic hydrogen evolution. *J. Mater. Chem. A* **2019**, *7*, 8472–8484. [CrossRef]
123. Yang, S.J.; Im, J.H.; Kim, T.; Lee, K.; Park, C.R. MOF-derived ZnO and ZnO@C composites with high photocatalytic activity and adsorption capacity. *J. Hazard. Mater.* **2011**, *186*, 376–382. [CrossRef]
124. Dolgopolova, E.A.; Shustova, N.B. Metal–organic framework photophysics: Optoelectronic devices, photoswitches, sensors, and photocatalysts. *MRS Bull.* **2016**, *41*, 890–896. [CrossRef]
125. Li, M.; Song, S.; Su, C.; Li, L.; Yan, Z.; Cao, X. MOF-templated in situ fabrication of surface-modified Ni/graphitic carbon nitride with enhanced photocatalytic hydrogen evolution. *Catal. Sci. Technol.* **2019**, *9*, 3828–3835. [CrossRef]
126. Wang, H. Rational design W-doped Co-ZIF-9 based Co_3S_4 composite photocatalyst for efficient visible-light-driven photocatalytic H 2 evolution. *Sustain. Energy Fuels* **2019**, *3*, 173–183. [CrossRef]
127. Mondal, I.; Gonuguntla, S.; Pal, U. Photoinduced fabrication of Cu/TiO_2 core–shell heterostructures derived from Cu-MOF for solar hydrogen generation: The size of the Cu nanoparticle matters. *J. Phys. Chem. C* **2019**, *123*, 26073–26081. [CrossRef]
128. Liu, H.; Li, J.; Chen, Y.; Sun, X.; Xu, X.; Qiu, L.; Duo, S.; Li, P. Ternary photocatalysts based on MOF-derived TO_2 co-decorated with $ZnIn_2S_4$ nanosheets and CdS nanoparticles for effective visible light degradation of organic pollutants. *New J. Chem.* **2022**, *46*, 7195–7201. [CrossRef]
129. Gao, L.-M.; Zhao, J.-H.; Li, T.; Li, R.; Xie, H.-Q.; Zhu, P.-L.; Niu, X.-Y.; Li, K. High-performance TO_2 photocatalyst produced by the versatile functions of the tiny bimetallic MOF-derived NiCoS-porous carbon cocatalyst. *CrystEngComm* **2019**, *21*, 3686–3693. [CrossRef]
130. Zhang, C.; Liu, B.; Cheng, X.; Guo, Z.; Zhuang, T.; Lv, Z. Noble-Metal-Free CdS Decorated Porous $Ni_xCo_{1-x}O$ Skeleton Derived from Metal–Organic Framework for Efficient Visible-Light H_2 Production. *ACS Appl. Energy Mater.* **2019**, *3*, 852–860. [CrossRef]
131. Tan, J.; Yu, M.; Cai, Z.; Lou, X.; Wang, J.; Li, Z. MOF-derived synthesis of MnS/In_2S_3 pn heterojunctions with hierarchical structures for efficient photocatalytic CO_2 reduction. *J. Colloid Interface Sci.* **2021**, *588*, 547–556. [CrossRef]

132. Mondal, I.; Pal, U. Synthesis of MOF templated Cu/CuO@ TO_2 nanocomposites for synergistic hydrogen production. *Phys. Chem. Chem. Phys.* **2016**, *18*, 4780–4788. [CrossRef] [PubMed]
133. Wu, S.; Zhao, H.-J.; Li, C.-F.; Liu, J.; Dong, W.; Zhao, H.; Wang, C.; Liu, Y.; Hu, Z.-Y.; Chen, L. Type II heterojunction in hierarchically porous zinc oxide/graphitic carbon nitride microspheres promoting photocatalytic activity. *J. Colloid Interface Sci.* **2019**, *538*, 99–107. [CrossRef]
134. Chen, X.; Li, J.-J.; Chen, X.; Cai, S.-C.; Yu, E.-Q.; Chen, J.; Jia, H. MOF-Templated Approach for Hollow NiO_x/Co_3O_4 Catalysts: Enhanced Light-Driven Thermocatalytic Degradation of Toluene. *ACS Appl. Nano Mater.* **2018**, *1*, 2971–2981. [CrossRef]
135. Chen, W.; Fang, J.; Zhang, Y.; Chen, G.; Zhao, S.; Zhang, C.; Xu, R.; Bao, J.; Zhou, Y.; Xiang, X. CdS nanosphere-decorated hollow polyhedral ZCO derived from a metal–organic framework (MOF) for effective photocatalytic water evolution. *Nanoscale* **2018**, *10*, 4463–4474. [CrossRef]
136. Yang, Y.; Sun, L.; Zhan, W.; Wang, X.; Han, X. Separated redox site strategies for engineering highly efficient photocatalysts: A pagoda-like In_2O_3/CuO heteroepitaxial structure coated with a N-doped C layer. *J. Mater. Chem. A* **2021**, *9*, 4310–4316. [CrossRef]
137. Jiang, Z.; Feng, L.; Zhu, J.; Liu, B.; Li, X.; Chen, Y.; Khan, S. Construction of a hierarchical $NiFe_2O_4/CuInSe_2$ (pn) heterojunction: Highly efficient visible-light-driven photocatalyst in the degradation of endocrine disruptors in an aqueous medium. *Ceram. Int.* **2021**, *47*, 8996–9007. [CrossRef]
138. Zhang, L.; Feng, L.; Li, P.; Chen, X.; Jiang, J.; Zhang, S.; Zhang, C.; Zhang, A.; Chen, G.; Wang, H. Direct Z-scheme photocatalyst of hollow CoSx@ CdS polyhedron constructed by ZIF-67-templated one-pot solvothermal route: A signal-on photoelectrochemical sensor for mercury (II). *Chem. Eng. J.* **2020**, *395*, 125072. [CrossRef]
139. Al-Hajji, L.A.; Ismail, A.A.; Bumajdad, A.; Alsaidi, M.; Ahmed, S.A.; Almutawa, F.; Al-Hazza, A. Construction of Au/TiO_2 Heterojunction with high photocatalytic performances under UVA illumination. *Ceram. Int.* **2020**, *46*, 20155–20162. [CrossRef]
140. Li, T.; Cui, J.-D.; Xu, M.-L.; Li, R.; Gao, L.-M.; Zhu, P.-L.; Xie, H.-Q.; Li, K. Engineering a hetero-MOF-derived TiO_2–Co_3O_4 heterojunction decorated with nickel nanoparticles for enhanced photocatalytic activity even in pure water. *CrystEngComm* **2020**, *22*, 5620–5627. [CrossRef]
141. Bera, S.; Pal, M.; Naskar, A.; Jana, S. Hierarchically structured ZnO-graphene hollow microspheres towards effective reusable adsorbent for organic pollutant via photodegradation process. *J. Alloys Compd.* **2016**, *669*, 177–186. [CrossRef]
142. Zhou, L.; Han, Z.; Li, G.-D.; Zhao, Z. Template-free synthesis and photocatalytic activity of hierarchical hollow ZnO microspheres composed of radially aligned nanorods. *J. Phys. Chem. Solids* **2021**, *148*, 109719. [CrossRef]
143. Bian, Z.; Zhu, J.; Li, H. Solvothermal alcoholysis synthesis of hierarchical TiO_2 with enhanced activity in environmental and energy photocatalysis. *J. Photochem. Photobiol. C Photochem. Rev.* **2016**, *28*, 72–86. [CrossRef]
144. Mamaghani, A.H.; Haghighat, F.; Lee, C.-S. Photocatalytic oxidation of MEK over hierarchical TiO_2 catalysts: Effect of photocatalyst features and operating conditions. *Appl. Catal. B Environ.* **2019**, *251*, 1–16. [CrossRef]
145. Hiremath, V.; Deonikar, V.G.; Kim, H.; Seo, J.G. Hierarchically assembled porous TiO_2 nanoparticles with enhanced photocatalytic activity towards Rhodamine-B degradation. *Colloids Surfaces A Physicochem. Eng. Asp.* **2020**, *586*, 124199. [CrossRef]
146. Harris, J.; Silk, R.; Smith, M.; Dong, Y.; Chen, W.-T.; Waterhouse, G.I.N. Hierarchical TiO_2 nanoflower photocatalysts with remarkable activity for aqueous methylene blue photo-oxidation. *ACS Omega* **2020**, *5*, 18919–18934. [CrossRef] [PubMed]
147. Singh, S.; Lo, S.-L. Single-phase cerium oxide nanospheres: An efficient photocatalyst for the abatement of rhodamine B dye. *Environ. Sci. Pollut. Res.* **2018**, *25*, 6532–6544. [CrossRef]
148. Sekar, K.; Chuaicham, C.; Balijapalli, U.; Li, W.; Wilson, K.; Lee, A.F.; Sasaki, K. Surfactant-and template-free hydrothermal assembly of Cu_2O visible light photocatalysts for trimethoprim degradation. *Appl. Catal. B Environ.* **2021**, *284*, 119741. [CrossRef]
149. Rong, R.; Wang, L. Synthesis of hierarchical hollow nest-like WO_3 micro/nanostructures with enhanced visible light-driven photocatalytic activity. *J. Alloys Compd.* **2021**, *850*, 156742. [CrossRef]
150. Wan, J.; Du, X.; Wang, R.; Liu, E.; Jia, J.; Bai, X.; Hu, X.; Fan, J. Mesoporous nanoplate multi-directional assembled $Bi2WO_6$ for high efficient photocatalytic oxidation of NO. *Chemosphere* **2018**, *193*, 737–744. [CrossRef]
151. Xu, B.; Yang, H.; Zhang, Q.; Yuan, S.; Xie, A.; Zhang, M.; Ohno, T. Design and Synthesis of Sm, Y, La and Nd-doped CeO_2 with a broom-like hierarchical structure: A photocatalyst with enhanced oxidation performance. *ChemCatChem* **2020**, *12*, 2638–2646. [CrossRef]
152. Dai, G.; Qin, H.; Zhou, H.; Wang, W.; Luo, T. Template-free fabrication of hierarchical macro/mesoporpous $SnS2/TiO_2$ composite with enhanced photocatalytic degradation of Methyl Orange (MO). *Appl. Surf. Sci.* **2018**, *430*, 488–495. [CrossRef]
153. Zhu, L.; Liu, Z.; Xia, P.; Li, H.; Xie, Y. Synthesis of hierarchical ZnO&Graphene composites with enhanced photocatalytic activity. *Ceram. Int.* **2018**, *44*, 849–856.
154. Liu, W.; Xu, Y.; Zhou, W.; Zhang, X.; Cheng, X.; Zhao, H.; Gao, S.; Huo, L. A facile synthesis of hierarchically porous TiO_2 microspheres with carbonaceous species for visible-light photocatalysis. *J. Mater. Sci. Technol.* **2017**, *33*, 39–46. [CrossRef]
155. Liang, Z.; Bai, X.; Hao, P.; Guo, Y.; Xue, Y.; Tian, J.; Cui, H. Full solar spectrum photocatalytic oxygen evolution by carbon-coated TiO_2 hierarchical nanotubes. *Appl. Catal. B Environ.* **2019**, *243*, 711–720. [CrossRef]
156. Li, Z.; Chen, H.; Li, Y.; Wang, H.; Liu, Y.; Li, X.; Lin, H.; Li, S.; Wang, L. Porous direct Z-scheme heterostructures of S-deficient CoS/CdS hexagonal nanoplates for robust photocatalytic H_2 generation. *CrystEngComm* **2022**, *24*, 404–416. [CrossRef]
157. Chen, L.; Meng, D.; Wu, X.; Wang, A.; Wang, J.; Yu, M.; Liang, Y. Enhanced visible light photocatalytic performances of self-assembled hierarchically structured $BiVO_4/Bi_2WO_6$ heterojunction composites with different morphologies. *RSC Adv.* **2016**, *6*, 52300–52309. [CrossRef]

158. Zhen, M.; Li, K.; Guo, S.-Q.; Li, H.; Shen, B. Template-free construction of hollow TiO_2 microspheres for long-life and high-capacity lithium storage. *J. Alloys Compd.* **2021**, *859*, 157761. [CrossRef]
159. Pan, B.; Zhou, Y.; Su, W.; Wang, X. Self-assembly synthesis of $LaPO_4$ hierarchical hollow spheres with enhanced photocatalytic CO_2-reduction performance. *Nano Res.* **2017**, *10*, 534–545. [CrossRef]
160. Lee, S.C.; Jeong, Y.; Kim, Y.J.; Kim, H.; Lee, H.U.; Lee, Y.-C.; Lee, S.M.; Kim, H.J.; An, H.-R.; Ha, M.G. Hierarchically three-dimensional (3D) nanotubular sea urchin-shaped iron oxide and its application in heavy metal removal and solar-induced photocatalytic degradation. *J. Hazard. Mater.* **2018**, *354*, 283–292. [CrossRef]
161. Chen, X.; Zhang, H.; Zhang, D.; Miao, Y.; Li, G. Controllable synthesis of mesoporous multi-shelled ZnO microspheres as efficient photocatalysts for NO oxidation. *Appl. Surf. Sci.* **2018**, *435*, 468–475. [CrossRef]
162. Zhu, L.; Li, H.; Xia, P.; Liu, Z.; Xiong, D. Hierarchical ZnO decorated with CeO_2 nanoparticles as the direct Z-scheme heterojunction for enhanced photocatalytic activity. *ACS Appl. Mater. Interfaces* **2018**, *10*, 39679–39687. [CrossRef] [PubMed]
163. Anjugam Vandarkuzhali, S.A.; Pugazhenthiran, N.; Mangalaraja, R.V.; Sathishkumar, P.; Viswanathan, B.; Anandan, S. Ultrasmall plasmonic nanoparticles decorated hierarchical mesoporous TiO_2 as an efficient photocatalyst for photocatalytic degradation of textile dyes. *ACS Omega* **2018**, *3*, 9834–9845. [CrossRef]
164. Cai, Y.; Song, J.; Liu, X.; Yin, X.; Li, X.; Yu, J.; Ding, B. Soft BiOBr@ TiO_2 nanofibrous membranes with hierarchical heterostructures as efficient and recyclable visible-light photocatalysts. *Environ. Sci. Nano* **2018**, *5*, 2631–2640. [CrossRef]
165. Jung, H.; Cho, K.M.; Kim, K.H.; Yoo, H.-W.; Al-Saggaf, A.; Gereige, I.; Jung, H.-T. Highly efficient and stable CO_2 reduction photocatalyst with a hierarchical structure of mesoporous TiO_2 on 3D graphene with few-layered MoS_2. *ACS Sustain. Chem. Eng.* **2018**, *6*, 5718–5724. [CrossRef]
166. Qu, X.; Yang, R.; Tong, F.; Zhao, Y.; Wang, M.-H. Hierarchical ZnO microstructures decorated with Au nanoparticles for enhanced gas sensing and photocatalytic properties. *Powder Technol.* **2018**, *330*, 259–265. [CrossRef]
167. Nie, N.; He, F.; Zhang, L.; Cheng, B. Direct Z-scheme PDA-modified ZnO hierarchical microspheres with enhanced photocatalytic CO_2 reduction performance. *Appl. Surf. Sci.* **2018**, *457*, 1096–1102. [CrossRef]
168. Jo, W.-K.; Lee, J.Y.; Natarajan, T.S. Fabrication of hierarchically structured novel redox-mediator-free $ZnIn_2S_4$ marigold flower/Bi_2WO_6 flower-like direct Z-scheme nanocomposite photocatalysts with superior visible light photocatalytic efficiency. *Phys. Chem. Chem. Phys.* **2016**, *18*, 1000–1016. [CrossRef] [PubMed]
169. Song, L.; Pang, Y.; Zheng, Y.; Chen, C.; Ge, L. Design, preparation and enhanced photocatalytic activity of porous BiOCl/$BiVO_4$ microspheres via a coprecipitation-hydrothermal method. *J. Alloys Compd.* **2017**, *710*, 375–382. [CrossRef]
170. Madhusudan, P.; Zhang, J.; Yu, J.; Cheng, B.; Xu, D.; Zhang, J. One-pot template-free synthesis of porous $CdMoO_4$ microspheres and their enhanced photocatalytic activity. *Appl. Surf. Sci.* **2016**, *387*, 202–213. [CrossRef]
171. Nie, N.; Zhang, L.; Fu, J.; Cheng, B.; Yu, J. Self-assembled hierarchical direct Z-scheme g-C_3N_4/ZnO microspheres with enhanced photocatalytic CO_2 reduction performance. *Appl. Surf. Sci.* **2018**, *441*, 12–22. [CrossRef]
172. He, R.; Zhang, J.; Yu, J.; Cao, S. Room-temperature synthesis of BiOI with tailorable (0 0 1) facets and enhanced photocatalytic activity. *J. Colloid Interface Sci.* **2016**, *478*, 201–208. [CrossRef]
173. Wei, H.; Hou, C.; Zhang, Y.; Nan, Z. Scalable low temperature in air solid phase synthesis of porous flower-like hierarchical nanostructure SnS_2 with superior performance in the adsorption and photocatalytic reduction of aqueous Cr (VI). *Sep. Purif. Technol.* **2017**, *189*, 153–161. [CrossRef]
174. Sedlak, J.; Kuritka, I.; Masar, M.; Machovsky, M.; Urbanek, P.; Bazant, P.; Janota, P.; Dvorackova, M. Contributions of morphological and structural parameters at different hierarchical morphology levels to photocatalytic activity of mesoporous nanostructured ZnO. *Appl. Surf. Sci.* **2020**, *513*, 145773. [CrossRef]
175. Fu, J.; Zhu, B.; Jiang, C.; Cheng, B.; You, W.; Yu, J. Hierarchical porous O-doped g-C_3N_4 with enhanced photocatalytic CO_2 reduction activity. *Small* **2017**, *13*, 1603938. [CrossRef]
176. Wang, Y.; Li, Y.; Bai, X.; Cai, Q.; Liu, C.; Zuo, Y.; Kang, S.; Cui, L. Facile synthesis of Y-doped graphitic carbon nitride with enhanced photocatalytic performance. *Catal. Commun.* **2016**, *84*, 179–182. [CrossRef]
177. Liu, M.; Wageh, S.; Al-Ghamdi, A.A.; Xia, P.; Cheng, B.; Zhang, L.; Yu, J. Quenching induced hierarchical 3D porous gC_3N_4 with enhanced photocatalytic CO_2 reduction activity. *Chem. Commun.* **2019**, *55*, 14023–14026. [CrossRef]
178. Shabani, M.; Haghighi, M.; Kahforoushan, D.; Haghighi, A. Mesoporous-mixed-phase of hierarchical bismuth oxychlorides nanophotocatalyst with enhanced photocatalytic application in treatment of antibiotic effluents. *J. Clean. Prod.* **2019**, *207*, 444–457. [CrossRef]
179. Edla, R.; Tonezzer, A.; Orlandi, M.; Patel, N.; Fernandes, R.; Bazzanella, N.; Date, K.; Kothari, D.C.; Miotello, A. 3D hierarchical nanostructures of iron oxides coatings prepared by pulsed laser deposition for photocatalytic water purification. *Appl. Catal. B Environ.* **2017**, *219*, 401–411. [CrossRef]
180. Hu, H.; Wang, J.; Deng, C.; Niu, C.; Le, H. Microwave-assisted controllable synthesis of hierarchical CuS nanospheres displaying fast and efficient photocatalytic activities. *J. Mater. Sci.* **2018**, *53*, 14250–14261. [CrossRef]
181. Nethravathi, C.; Rajamathi, J.T.; Rajamathi, M. Microwave-assisted synthesis of porous aggregates of CuS nanoparticles for sunlight photocatalysis. *ACS Omega* **2019**, *4*, 4825–4831. [CrossRef]
182. Huang, Z.; Yan, F.-W.; Yuan, G. Ultrasound-assisted fabrication of hierarchical rodlike graphitic carbon nitride with fewer defects and enhanced visible-light photocatalytic activity. *ACS Sustain. Chem. Eng.* **2018**, *6*, 3187–3195. [CrossRef]

183. Xu, W.; Wang, S.Q.; Zhang, Q.Y.; Ma, C.Y.; Wang, Q.; Wen, D.H.; Li, X.N. Hierarchically structured AgO films with nano-porosity for photocatalyst and all solid-state thin film battery. *J. Alloys Compd.* **2019**, *802*, 210–216. [CrossRef]
184. Zhong, G.; Liu, D. Biomorphic CdS/Cd Photocatalyst Derived from Butterfly Wing for Efficient Hydrogen Evolution Reaction. *Adv. Mater. Interfaces* **2022**, *9*, 2102423. [CrossRef]

catalysts

MDPI

Review

Advancements in Solar Desalination of Seawater by Various Ti_3C_2 MXene Based Morphologies for Freshwater Generation: A Review

Adem Sreedhar and Jin-Seo Noh *

Department of Physics, Gachon University, 1342 Seongnamdaero, Sujeong-gu, Seongnam-si 461-701, Gyeonggi-do, Korea; ademgu@gachon.ac.kr
* Correspondence: jinseonoh@gachon.ac.kr

Abstract: For a few years, we have been witnessing ubiquitous fresh and drinking water scarcity in various countries. To mitigate these problematic situations, many countries relied on non-conventional freshwater generation technologies through solar desalination of seawater. In this manner, we excel the ability of new class 2D Ti_3C_2 MXenes as a photothermal material (solar absorber) for freshwater generation via the solar desalination technique. In this review, the air–water interfacial interaction is highlighted for improving the evaporation efficiency. To provide the dependence of the desalination efficiency on the microstructure of the solar absorbers, we summarized various forms of 2D Ti_3C_2 MXenes (aerosol, films, foam, hydrogel, membrane, monolith and porous structure) and their characteristics. These microstructures prevailed ultrahigh photoconversion efficiency. In this aspect, we further explained key features such as light absorption, reflection, multiple internal reflection, hydrophilicity, lower thermal conduction, light-to-heat generation, and salt rejection for achieving efficient desalination output throughout the visible and broadband region. Specifically, we targeted to explore the self-floating and salt rejection nature of various state-of-the-art 2D Ti_3C_2 MXene structures. Further, we highlighted the long-term stability. Among the above morphologies, Ti_3C_2 MXene in the form of a membrane is believed to be a promising morphology which effectively desalinates seawater into freshwater. Finally, we highlighted the challenges and future perspectives, which can pave a potential path for advancing the sustainable solar desalination of seawater into freshwater.

Keywords: 2D Ti_3C_2 MXene; membrane; broadband; solar desalination; freshwater

Citation: Sreedhar, A.; Noh, J.-S. Advancements in Solar Desalination of Seawater by Various Ti_3C_2 MXene Based Morphologies for Freshwater Generation: A Review. *Catalysts* **2021**, *11*, 1435. https://doi.org/10.3390/catal11121435

Academic Editors: Carolina Belver and Jorge Bedia

Received: 29 October 2021
Accepted: 23 November 2021
Published: 25 November 2021

1. Introduction

Water is the most essential resource for life on the earth. Due to the enormous growth of population and industrialization, our ecological system is significantly facing freshwater scarcity and contamination. At the same time, the availability of freshwater on the earth is 3% only [1]. Additionally, yearly, one-third of the population is not in a position to access fresh drinking water [2]. Thus, the production of freshwater from abundantly available seawater is necessary under eco-friendly techniques because conventional wastewater purification methods consume high energy [3]. Reverse osmosis (RO) and pervaporation (PV) proved their importance for the desalination process [4–6]. However, RO requires high pressure to conquer the osmotic pressure of seawater [7]. Additionally, the processing of RO is cost-effective, and maintenance is associated with scaling, fouling, and degradation. This energy-intensive technique has a huge impact on the economy (high energy demand) and environment (CO_2 emissions). For a decade, the desalination process has been skyrocketing in various countries. There is another interesting technology, such as capacitive deionization (CDI), which yields desalination efficiency under low energy consumption under an ecofriendly process [8]. Specifically, water scarcity becomes ever more threatening. For this, many countries permanently rely on these techniques. Thus, it is the right time to develop an advanced, eco-friendly, 100% carbon neutral, harmless, and low-cost desalination technique. In such a way, the solar desalination process raised its importance through the harvesting of the abundantly available solar energy into steam (freshwater) under the strategic conversion of light into heat [9]. For example, solar desalination is a process for obtaining pure water from abundantly available seawater through the pressure-free water purification process. It is to be noted that renewable solar energy is highly beneficial, which is an alternative to the utilization of carbon emissive fossil fuels (oil, coal, and gas). Thus, we can nullify the emission of carbon byproducts by utilizing renewable solar energy. Specifically, challenges associated with the developed solar absorber during the water evaporation process are (i) extended light absorption throughout the solar spectrum, (ii) the hydrophilic nature, (iii) low thermal conductivity, and (iv) high electrical conductivity. On the other hand, we should focus on crystallization of salt on the absorber. This process suppresses the water transport channel, localization of heat, and freshwater production under reduced energy conversion. We should further focus on solar desalination not only on a sunny day, but also under complex conditions, such as cloudy, rainy, or night conditions.

Transition metal carbides are effectively used as electrocatalysts for oxygen evolution reactions [10]. Due to the conceivable photocatalytic activity of Ti_3C_2 MXene [11], porous structured novel 2D MXenes raised its importance in the field of desalination of seawater under strategic utilization of abundant solar electromagnetic spectrum. It should be noted that 2D Ti_3C_2 MXene can absorb near-infrared light for the photofixation of N_2 [12]. Additionally, Ti_3C_2 MXene proved superior photothermal conversion efficiency of 100% (theoretical), which paved a potential path for the absorption and conversion of electromagnetic radiation into heat [13], which is one of the required features for advancing the water evaporation process. These features guarantee the solar desalination of 2D Ti_3C_2 MXene under controlled interfacial interaction with various semiconducting materials. Therefore, the developing of highly surface-active Ti_3C_2 MXene-based binary or ternary composites effectively converts incident light into heat under solar and IR/NIR regions, which is urgently required. In addition, strengthened surface termination groups (-OH, -O, and -F) during the formation of Ti_3C_2 MXene from its precursor (Ti_3AlC_2 MAX phase) provide impressive optical properties [14], which further boost the overall solar desalination process.

Indeed, the morphology of the developed Ti_3C_2 MXene is considered a promising feature during the desalination process for the efficient absorption of seawater. Significant research is going on for controlling the surface morphological features of Ti_3C_2 MXene to provide the right water transportation channels and efficient saltwater desalination. Thus, it is necessary to estimate the desalination efficiency of Ti_3C_2 MXene under various surface morphological features. Many morphologies, such as aerosol [15], films [16], foam [17], hydrogel [18], membrane [19], monolith [20] and porous structure [21] have been extracted from layer structured 2D Ti_3C_2 MXenes. Among the large community of the MXene family, Ti_3C_2 MXene is skyrocketing in various energy conversion applications, such as hydrogen production [22], dye degradation [23], drug degradation [24], CO_2 reduction [25,26], N_2 reduction [12,27], triboelectric nanogenerator [28], supercapacitor [29], solar cells [30,31], etc.

Based on the above constructive features, we have comprehensively summarized the innovative desalination capacity of various Ti_3C_2 MXene morphologies. This review mainly involved the reliability of the Ti_3C_2 MXenes morphology, which greatly enhances the desalination ability. There is scarce research on the development of optimized photothermal Ti_3C_2 MXene materials and the design of prototypes. In such a way, we provided the effectiveness of the Ti_3C_2 MXene morphology for sustainable solar desalination by converting light into heat and then water vapor. Until now, there were very few review articles and research articles published based on Ti_3C_2 MXenes for solar desalination [1,32]. Thus, 2D Ti_3C_2 MXene as an advanced material provides the innovative paths and technological solutions for freshwater generation. Understanding the oxidation ability of Ti_3C_2 MXene during the desalination process also provides a benchmark for realizing long-term stability.

2. Desalination

The production of freshwater (low salinity) from the abundantly available seawater (high salinity) is known as desalination. Majorly, the water available on earth is about 97% and it is saline in nature because 70% of earth is covered by the oceans, which means we only get 3% of freshwater for daily and industrial purposes [33]. Desalination is a non-conventional freshwater generation process, which is currently fulfilling 1% of the world's fresh drinking water. During the thermal (distillation) or mechanical (RO) desalination process, high energy and pressure conditions are necessary. Currently, multi-stage flash distillation (thermal energy) and RO using membranes (mechanical energy) are majorly used for desalination. Here, RO requires less energy compared to multi-stage flash distillation. During the RO desalination process, high pressure pumps the seawater through the membrane, which results in low-pressure freshwater and high-pressure brine. Due to the high energy consumption during the above desalination processes, the development of a facile desalination process by utilizing a renewable energy resource (solar energy) is desirable. Despite the reliability of the RO process, significant efforts have been focusing on pressure-free and low energy consumption desalination processes. In this category, the generation of freshwater vapor from seawater using a photothermal material under abundant solar light is skyrocketing. Figure 1 presents the cost-effective schematic representation of solar desalination of seawater into freshwater under solar light, using novel 2D layer structured Ti_3C_2 MXenes.

Figure 1. Schematic representation of seawater to freshwater generation by the 2D layer structured Ti_3C_2 MXenes.

3. Key Factors for Achieving Superior Solar Desalination

It is necessary to develop a stable photothermal material which address the desalination ability throughout the solar spectrum. For this, we should conquer few key points during the desalination process. These features make a potential path to realize the solar-driven water purification process. Therefore, we have provided the following fundamental features which profoundly demonstrate the effectiveness of the solar desalination of seawater into freshwater.

3.1. Control on Heat Loss

To convert saline seawater into freshwater, the developed photothermal material should overcome three kinds of heat losses: (i) radiation, (ii) convection, and (iii) conduction. Specifically, heat localization is the underlying parameter for advancing the exceptional solar-to-vapor efficiency during the desalination process [34]. All these factors make standalone solar evaporators for productive freshwater generation. For instance, it is necessary to develop a low thermal conductive material to localize the heat at the interface of water–air during the desalination process. At suppressed heat loss toward the bulk water, the localized heat effect significantly improves the steam generation. Figure 2 presents the schematic representation and key factors, which greatly influence the solar desalination by the layer structured 2D Ti_3C_2 MXenes. Specifically, lowering the thermal conduction, radiation, and convection at improved internal reflection greatly improves the overall water evaporation rate.

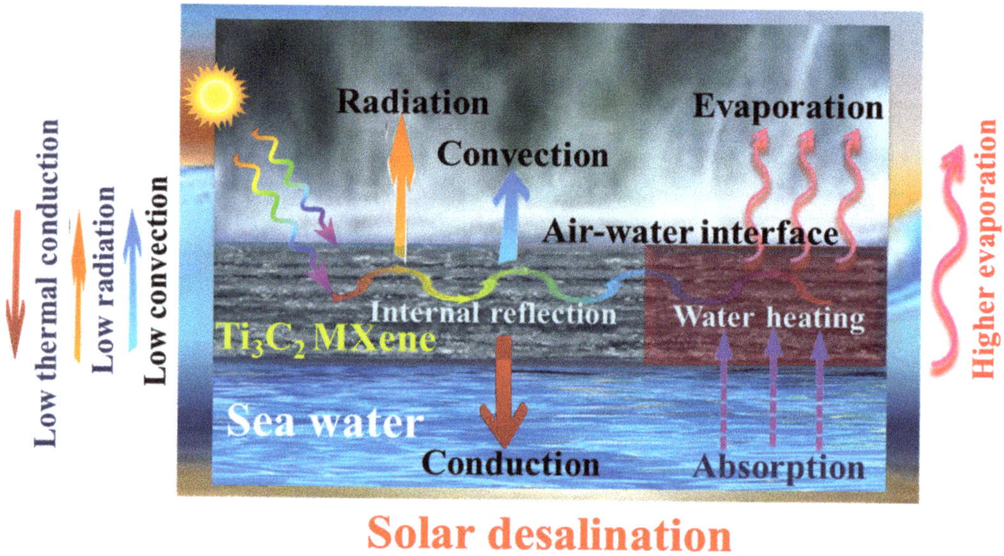

Figure 2. Prominent figures of merits of Ti_3C_2 MXenes during solar desalination process.

3.2. Reduced Light Reflection, Transmission, and Improved Broadband Light Absorption

In addition to the above features, light absorption, transmission, and reflection also determine the solar conversion efficiency. Reduced light reflection and improved absorption make more light-to-heat generation capacity by the photothermal materials (solar absorber). For this, we require various morphological and microstructure designs, which eventually improved the light absorption capacity under reduced reflection [35]. On the other hand, we should suppress the transmittance toward the bulk water. Thus, the development of porous structured solar absorbers is highly beneficial for reduced light reflection.

3.3. Localized Surface Plasmon Resonance (LSPR)

Conventional plasmonic metals, such as gold (Au), silver (Ag), and copper (Cu), behave like plasmonic absorbers. Similarly, metallic 2D $Ti_3C_2T_x$ MXenes prevails in its plasmonic resonance feature by absorbing the incident light from visible to the near-IR region [36]. Consequently, the LSPR significantly influences the light-to-heat conversion efficiency, which is ideal during solar desalination. Typically, plasmonic materials greatly absorb the solar light at about 95% and heat the water above the boiling point of water.

3.4. Multiple Internal Light Reflection

To be an effective solar evaporator, the photothermal material should prevail over the pronounced multiple light reflections within the developed specific microstructure. Thus, we should focus on the selective morphology to explain the importance of light-trapping and multiple light reflections by its inner microstructure. It is to be noted that the unabsorbed light is effectively trapped by the dense surroundings of the microstructure, which eventually creates multiple internal reflections through the reabsorption [37]. Moreover, we can achieve much lower light reflection and higher absorption at a low effective refractive index. The internally reflected light is transmitted through the absorber.

3.5. Hydrophilic Nature

The developed solar absorber should prevail in the high wettable property, which provides persistent water absorption and transportation channel from the bottom toward the surface (top of the absorber) within a short period. Additionally, the hydrophilic nature promotes the heat transfer at the evaporation surface, which strengthens the solar steam generation. In the case of Ti_3C_2 MXene, specific surface termination groups (–F, –O, or –OH) provide a hydrophilic nature [38]. This phenomenon is helpful for facilitating the rapid and required water supply through the water transportation channels during the desalination process. On the contrary, hydrophilic materials cannot provide a self-floating property when placed on the water. So, we need to assist additional layers, which reduces the heat conduction toward the bottom layer and provides a self-floating property. It should be noted that hydrophobic structures often yield lower solar evaporation efficiency, compared to hydrophilic structures [39]. Moreover, the hydrophilic wet layer provides more light absorption, compared to the dry layer counterparts.

3.6. Light-to-Heat Generation

The photothermal materials possess an intrinsic light absorption capacity at suppressed light reflection. In such a way, the incident light transforms into thermal energy (heat). Later, the heat evaporates the seawater into freshwater. Compared to the bulk water heating into vapor, interfacial solar heating through the solar absorber inhibits the heat loss toward the water [40]. Eventually, the light-to-heat generation process highlights the desalination of seawater into freshwater. Specific morphological features effectively convert the incident light into heat and strengthen the evaporation rate. Overall, this process should occur at lower heat loss to fulfill the light-to-vapor conversion.

3.7. Reduced Salt Blocking or Resistance

The solar absorber generates the vapor at the interface of the water. On the other hand, salt ions migrate from the surface of the absorber toward the bulk water. The blocking or clogging of salt on the surface of the solar absorber gradually increases the incident light reflection and reduces the water supply channels [41]. Thus, it is necessary to avoid salt blocking on the solar absorber to achieve stable desalination activity throughout the desalination process. On the other hand, we should enable long-term interfacial contact between the seawater and solar absorber (photothermal material) during the desalination process, which defines the salt resistance property. The developing of novel layer structured materials could dissolve the accumulated salt on the solar absorber and diffuse it toward the bulk water.

4. Equations Determines the Solar Desalination of Photothermal Material

During the steady desalination process, the following mathematical equations provide in-depth analysis to estimate the efficacy of the developed solar absorbers.

4.1. Solar-to-Vapor Conversion Efficiency (η)

Effective freshwater production efficiency can be defined as how much solar light is utilized and converted into vapor generation. The solar-to-vapor conversion efficiency (η) is the key parameter to determine the desalination process, which can be estimated by the following expression [42],

$$\eta = \frac{mh_{\mathrm{LV}}}{C_{\mathrm{opt}}P_0} \tag{1}$$

where m is the evaporation rate, h_{LV} is the change in the total enthalpy of the liquid-vapor phase, C_{opt} is the optical concentration and P_0 is the incident light intensity (1 sun = 100 mW/cm^2). In this situation, the evaporation efficiency of developed material should be higher than the natural evaporation rate of pure water.

4.2. Photothermal Conversion Ability (E)

The ability of evaporation rate (E) is quantitatively evaluated by the mass loss of water per unit area in a unit time, using the following equation [43],

$$E = \frac{\Delta_m}{AT} \tag{2}$$

where A is the area of the used container, Δ_m is the water mass loss and T is the light irradiation time.

4.3. Wettability of Photothermal Material

The wettability or hydrophilic nature (contact angle < 90°) of the developed photothermal material through the effective water flowing channels eventually improves the freshwater generation rate during solar desalination. The wettability can be estimated by the following equation [44],

$$\mathrm{H} = \frac{2\sigma COS\theta}{r\rho g} \tag{3}$$

where σ is the surface tension of water, ρ is the density of water, r is the diameter of the channel, θ is the developed contact angle between water and channel and g is the gravitational constant.

4.4. Salt Rejection Ratio

It is necessary to estimate the salt rejection capability of developed photothermal material, which greatly improves the long-term stability of the desalination process. The salt rejection (SR) ratio can be defined as the ratio of salt concentration (C_{out}) in the supernatant phase to the initial concentration of saltwater (C_0) [45],

$$SR = \frac{C_{out}}{C_0} \times 100\% \tag{4}$$

It is to be noted that C_{out} should be less than C_0 after the desalination process, which yields more freshwater generation.

4.5. Solar Absorption

The developed photothermal material should prevail over superior optical absorption throughout the solar spectrum. As a result, we can achieve continuous water evaporation. In such a way, the solar absorption can be evaluated by the following expression [46],

$$A(\lambda) = 1 - R(\lambda) - T(\lambda) \tag{5}$$

where R(λ) and T(λ) are the reflectance and transmittance of the developed sample at different wavelengths (λ). Here, we should obtain an absorption of about 95%, which results from sufficient interaction of light with photothermal material. Eventually, the water evaporation rate can be effectively achieved.

4.6. Thermal Conductivity

The developed solar absorber should prevail low thermal conductivity under minimized heat transfer toward the bulk water. As a result, we can confine the incident light within the surface of the material to achieve more heat generation. In other words, the material should prevail over the reduced thermal diffusion heat loss. The following expression gives the conduction loss during solar desalination [47],

$$Q = Cm\Delta T \quad (6)$$

where C represents the specific heat of water, m is the water weight, and ΔT is the improved water temperature within a certain time. Based on this equation, the conduction should be less than 2%, which endows the possible thermal conductive material with freshwater generation.

4.7. Evaporation Rate (v)

The ability of evaporation of the developed solar absorber can be estimated by using the following expression [48]:

$$v = \frac{\mathrm{d}m}{S\mathrm{d}t} \quad (7)$$

where S is the resultant surface area of the developed solar absorber and m is the weight of the evaporated fresh water under the specified time (t).

5. Importance of 2D Materials for Solar Desalination

State-of-the-art 2D materials prevail in their (i) distinct ultrathin morphology, (ii) response to the electromagnetic spectrum, (iii) active surface area, and (iv) hydrophilic nature. These factors guide the cost-effective desalination of seawater into freshwater. It should be noted that graphene, transition metal dichalcogenides (TMDs), and layered double hydroxide (LDH) are reliable for multifunctional applications. Firstly, the 2D material starting with graphene (semi-metallic) can absorb the broad range of the solar spectrum. Later, another 2D material MoS_2 raised its importance by providing optical absorption from visible to mid-infrared regions. Here, 2D MoS_2 shows a semiconducting-like behavior. Similarly, after the groundbreaking report on 2D material Ti_3C_2 MXene by Gogosti et al. [49], Ti_3C_2 MXenes were successfully utilized as a photothermal material in the solar desalination process. Nowadays, great interest is continuing for layer structured 2D Ti_3C_2 MXenes for freshwater generation. Additionally, the above-mentioned 2D ultra-thin layered materials provide a low toxic nature during the water purification process, which is primarily necessary for real-time environmental applications.

6. Why 2D Ti_3C_2 MXenes for Solar Desalination?

After inventing the layer structured 2D Ti_3C_2 MXene in the year 2011 by selective HF etching of aluminum layers in the Ti_3AlC_2 MAX phase [49], Ti_3C_2 MXenes have been widely used in various applications due to their superior chemical, electrical and mechanical properties. Figure 3 presents the Ti_3AlC_2 MAX phase transformation into multilayer and monolayer Ti_3C_2 MXene under selective HF etching of Al layers. After realizing the layered morphology, novel 2D Ti_3C_2 MXenes were explored for freshwater generation under the solar desalination process by satisfying the following conditions. For achieving stable solar desalination, the material should provide low thermal conductivity and superior photothermal conversion efficiency (PTCE). For this, metallic Ti_3C_2 MXene provides localized surface plasmon resonance (LSPR), which paved a potential pathway for the absorption of a wide solar spectrum (250–2500 nm). It should be pointed out that the PTCE and photothermal vaporization capability of Ti_3C_2 MXene are 100% and 84%, respectively [50]. Specifically, Ti_3C_2 dissipates the absorbed electromagnetic radiation into heat, which is a key factor during the solar desalination process. As a result, multiple internal light reflections occur by the absorption of electromagnetic radiation [13]. On the other hand, Ti_3C_2 MXene does not transfer the heat toward bulk water and dissipates the generated heat at the surface. Accordingly, generated heat converts the absorbed water into evaporation. Among all the MXenes, Ti- and C-rich $Ti_3C_2T_x$ MXenes received significant attention in water treatment technology. Remarkably, due to the intrinsic super-hydrophilic nature of Ti_3C_2 MXene [51], Ti_3C_2 MXene effectively absorbs and transports the bulk water through its water transportation channels. Additionally, Ti_3C_2 can recycle after the water purification process [16]. Through the above light-to-heat conversion efficiency, hydrophilicity, lower thermal conductivity, LSPR, and ideal absorption of broadband, Ti_3C_2 MXene raised its ability in freshwater generation and environmental remediation under an eco-friendly solar-driven desalination process.

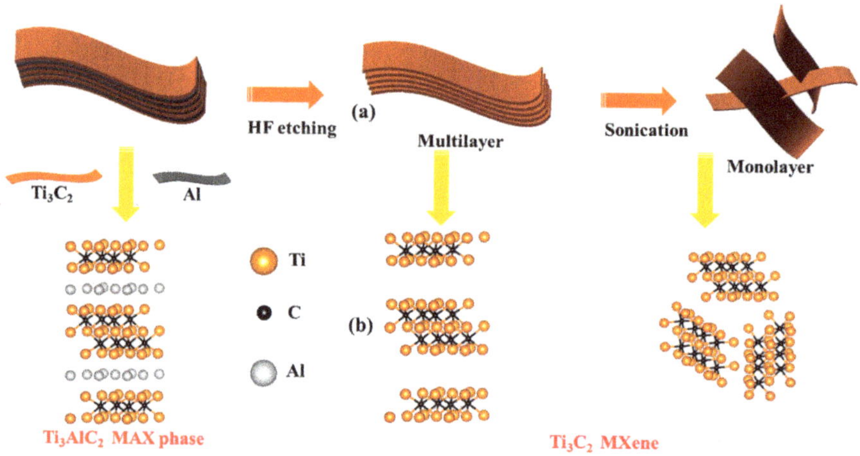

Figure 3. (**a**) Schematic representation of HF etching of Ti_3AlC_2 MAX phase and its transformation into Ti_3C_2 MXene multilayer and monolayer, and (**b**) corresponding chemical bondings in the Ti_3AlC_2 MAX phase and Ti_3C_2 MXene before and after etching.

7. Ti_3C_2 MXene-Based Composites

Considering the above constructive features, 2D Ti_3C_2 MXene-based composites are greatly dedicated to solar desalination of seawater in the broadband region. However, there is scarce research on solar desalination using 2D Ti_3C_2 MXenes. Thus, we addressed the status and ability of Ti_3C_2 MXene-based composites toward solar desalination by highlighting its intrinsic photothermal conversion under ideal light consumption, water absorption, and hydrophilic nature. Specifically, we explored the importance of various morphological features of Ti_3C_2 MXenes for seawater absorption and desalination. Thus, we provided the proof-of-concept of Ti_3C_2 MXenes in the form of aerogel, films, foam, hydrogel, membrane, monolith, and porous structures to reveal the importance of light absorption capacity and light-to-heat generation for efficient solar desalination and freshwater generation.

7.1. Aerogel

Aerogels prevail in their abundant porous network, low thermal conductivity, hydrophilic nature, and low mass density. Thus, the aerogels morphology is highly beneficial for solar thermal evaporation [52]. Specifically, aerogel features can be successfully achieved by 2D Ti_3C_2 MXenes, which can effectively apply in the solar desalination process.

Ming et al. [53] successfully developed the multilevel porous structured aerogel by integrating the 2D $Ti_3C_2T_x$ MXene and 3D graphene oxide (GO), which can weigh 500 times (500 g) higher than its own weight. Mainly, they explored heat localization, photothermal ability, solar-driven interfacial water evaporation in complex conditions (including summer and winter). For this, various MXene to GO (GMA) (1:1, 1:3, and 1:5) ratios were developed. Under constructive interfacial interaction between GO and $Ti_3C_2T_x$, self-enhanced photothermal conversion was achieved through heat localization on the top of the layer. The role of GO in the multilevel GMA aerogel was to prevent the oxidization and restacking of $Ti_3C_2T_x$ MXene, which results in the synergy between GO and $Ti_3C_2T_x$ MXene. Optimized GMA-3 (1:3) aerogel is satisfied with a hydrophilic nature, which is highly beneficial for rapid water absorption and transportation. It should be noted that the resultant GMA aerogel exhibited an ultra-lightweight feature due to the prominent multilevel porous structure. Figure 4a shows the digital image and morphology of GMA-3 during water evaporation and the microstructure with the internal gap, micron channels, and nano wrinkles for achieving interfacial evaporation. In such a way, GMA-3 achieved an evaporation efficiency of about ~90.7% at the evaporation rate of 1.27 kg/m^{2}h under 1 sun, compared to GO aerosol (~63.7% and 0.88 kg/m^{2}h) and $Ti_3C_2T_x$ MXene (~71.4% and 1.02 kg/m^{2}h). Due to the high thermal conduction of GO, the top portion of the GMA-3 aerogel generated heat, which converted into evaporation. Consequently, the sharp rise in the surface temperature (41.9 °C in 300 s) was observed. The resultant GMA composite achieved excellent photothermal activity by satisfying the self-reduction/oxidation between GO and $Ti_3C_2T_x$ MXene compared to pure GO and MXene. Interestingly, the internal light reflection, interfacial evaporation, and low thermal conductivity were predominately observed during the desalination process as shown in Figure 4b. Finally, the solar-driven interfacial evaporation process was highlighted through the (i) interfacial evaporation (~90.7%), (ii) lower thermal radiation (~3.42%), (iii) lower thermal convection (~2.50%), and (iv) lower heat convection (~1.50%). These factors motivated the development of a prototype for freshwater generation (Figure 4c). In comparison, solar evaporation during summer (1.27 kg/m^{2}h) is higher compared to the winter season (0.73 kg/m^{2}h) under peak sunlight (12:00 a.m. to 14:00 p.m.). Under these conditions, the GMA-3 was observed with 24 h stability.

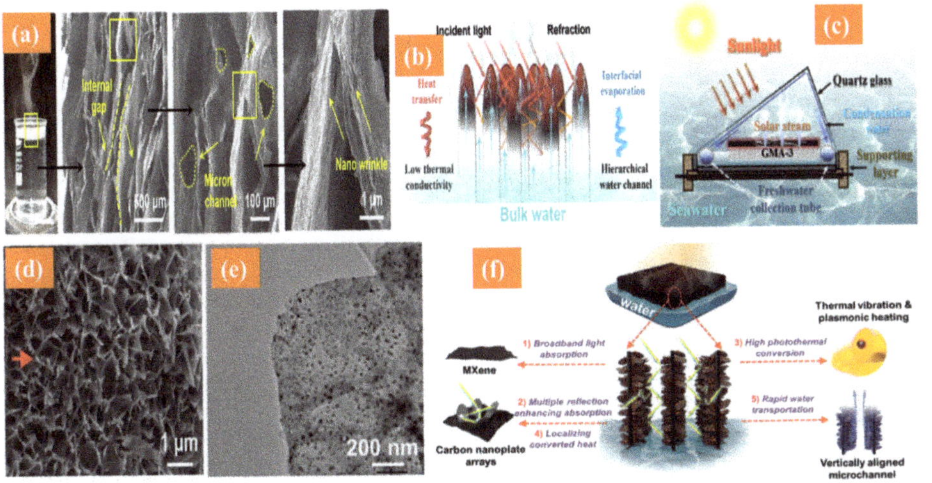

Figure 4. (**a**) Internal gap, microchannels, and nano wrinkles developed on the internal surface of GO supported Ti_3C_2 MXene aerogels for water supplying channels during solar desalination, (**b**) internal light refraction for superior light absorption and heat localization for interfacial evaporation at low thermal conductivity, and (**c**) schematic representation of solar desalination process by the GMA-3 (reprinted from Ref. [53], copyright with permission from Elsevier, 2020), (**d**) magnified SEM image of 3D Ti_3C_2 MXene nanosheets supported hierarchical cobalt nanoparticle–carbonaceous nanosheets/MXene foam, (**e**) corresponding TEM image with Co nanoparticles within the Co-CNS/M foam, and (**f**) key features which improved the solar desalination of Co-CNS/M foam. (Reprinted from Ref. [17], copyright with permission from Wiley Online Library, 2020).

Besides sunny days, it is necessary to achieve steam generation in all weather conditions and any complex environments. Thus, a strategic combination of photothermal and electro-thermal conversion effects is highly beneficial for all-weather steam generation. Specifically, during sunny days, batteries can store the electrical energy, which will be utilized for electro-thermal steam generation under low light and dark conditions. For this, Zaho et al. [15] strategically explored the groundbreaking report on continuous steam generation in complex environments (strong sunlight, weak light, and no light conditions) by developing the cross-linked MXene aerogel (CMA). The CMA was developed by assembling the GO with $Ti_3C_2T_x$ MXene (co-gelation strategy) at an optimized MXene content of 90% (CMA-90%). The cross-linked MXene aerogel loaded 1040 g, which is higher than previously reported GO-supported Ti_3C_2 MXene aerogel (500 g) [53]. Here, GO played a crucial role in assembling $Ti_3C_2T_x$ MXene nanosheets into crosslinked aerogels. It is to be noted that Ti_3C_2 MXene prevailed over LSPR for photo-thermal conversion of conventional plasmonic metals (Au and Ag). As a result, the light absorption capacity of CMA-90% is 99%. The surface temperature of aerogel changed from 80° (at an incident angle of 30°) to 235 °C with the increase in solar irradiation from 1 kWm^{-2} to 5 kWm^{-2} respectively, which suggests the superior photothermal conversion capability of CMA aerogel. The CMA sample prevailed as superior in its hydrophilic nature, which supported water absorption and transportation ability. In addition to photothermal conversion, CMA revealed the electro-thermal conversion ability. It should be noted that simultaneous electrical energy and solar illumination effects prevail in their increase in the surface temperature and reduced humidity of the overall system, respectively. Under both conditions (photothermal + electro-thermal) at a weak solar light intensity of 0.5 kWm^{-2} and electric supply of 2.5 V, CMA-90% achieved a superior evaporation rate of 1.624 kg/m^2h and energy conversion efficiency of 88.4%, which is higher than the sole 1 sun solar light (1.337 kg/m^2h and ~85%). CMA-90% aerogel achieved superior electrical conductivity, compared to the conventional graphene aerogel and carbon nanotube foam. Overall, a strategic combination of photothermal and electro-thermal effects was successfully studied by the $Ti_3C_2T_x$ MXene-based aerogel for the all-weather solar desalination process.

In another study, vertically aligned two-layer (hydrophobic/hydrophilic interface) Janus Ti_3C_2 MXene aerogel (VA-MXA) was developed for revealing the solar desalination process [54]. The specific features of the vertically aligned Janus Ti_3C_2 MXene succeeded the desalination process with a prominent (i) capillary water transport channel, (ii) light absorption capacity, (iii) multiple internal light reflection, and (iv) vapor escape ability compared to conventional MXene and Janus MXene. Specifically, the top layer provided a superior hydrophobic nature with a contact angle of 133.2° and the bottom layer acted as a hydrophilic surface (46.2°). The top layer benefited from the light absorption, heat confinement, reduced heat loss, and salt rejection by retaining salt ions from the bottom (hydrophilic surface), whereas the submerged bottom hydrophilic surface effectively pumped the water toward the top layer, where solar steam production occurred. Under these well-ordered features, they avoided the crystallization of salt on the absorber for improvement in the water evaporation. Interestingly, stable steam generation was observed at the air–water interface. The selective $VA\text{-}MXA_{15}$ aerogel with microchannel size of 15 μm (at Ti_3C_2-25 mg. mL^{-1}) achieved an energy conversion efficiency of ~87% and evaporation rate of 1.46 kg/m^2h (without any absorber-0.49 kg/m^2h) for 15 days under 1 sun following the LSPR [55]. The incident light (200–2500 nm) absorption capacity was maintained at about 96% after 12 h. Additionally, the scalable freshwater yield of ~6 L m^{-2} was achieved. The average surface temperature increased from 21.3 °C to 59.4 °C within 300 s. Here, the central part of the $VA\text{-}MXA_{15}$ aerogel achieved 98.5 °C compared to the bottom portion (29.4 °C), which suggested that the top layer of VA-MXA localized the generated heat and did not transfer toward the bottom following the heat localization effect and lower thermal conductivity. Thus, the hydrophobic/hydrophilic interface effectively confined the generated heat. Additionally, they mainly overcame the crystallization of saltwater on the absorber to prevail in stable freshwater production. Overall, vertically aligned Janus Ti_3C_2 MXene aerogel under opposite wettable conditions, heat localization, salt rejection, and improved light-to-heat conversion guided a new concept of freshwater generation.

Of these aerogels designed by the 2D Ti_3C_2 MXenes, the superior hydrophilic nature guided effective water transportation channels for water absorption and photothermal activity at low thermal conductivity. The improved solar absorption and multiple light reflection within the MXene aerogels contributed through the intrinsic LSPR property of Ti_3C_2 MXene. Specifically, 2D Ti_3C_2 MXenes aerogels contributed to their desalination performance in all-weather conditions, which proved the environmental viability of Ti_3C_2 MXene. Additionally, Ti_3C_2 MXene achieved good salt-resistance activity when it interacted with seawater during the desalination process. The salt resistance activity of the developed solar absorber provided long-term interfacial interaction with seawater.

7.2. Foam

As we mentioned before, incident light reflects and blocks water transportation channels due to salt blocking during the seawater desalination process. Thus, we should conquer the salt blocking ability and broadband absorption for continuous freshwater generation with high efficiency. For this, Fan et al. [17] combined the 3D Ti_3C_2 MXene nanosheets and cobalt-based metal-organic frameworks (Co-MOFs) as precursors for designing the hierarchical cobalt nanoparticle-carbonaceous nanosheets/MXene foam (Co-CNS/M foam). Here, the Co-MOF nanosheets are embedded with Co nanoparticles, which are vertically aligned on the MXene (Figure 4d,e). The vertically aligned microchannels in the foam exhibited a width of about 30–50 μm. The microchannel wall in MXene-based foam was observed with a lamellar structure. By developing these hierarchical structures, the following interesting features were observed: (i) multiple light reflection by the carbon nanoplate arrays and Co nanoparticles for improved light absorption, (ii) improved light-to-heat conversion by the thermal vibration of molecules and plasmonic localized heating, (iii) carbon nanoplate arrays on the MXene skeleton for reduced thermal conductivity and chemical stability for heat loss, (iv) broadband absorption of up to 2500 nm by the MXene nanosheets and carbon nanosheets, and (v) hydrophilic nature for enabling the water transportation channel. All these constructive features significantly improved the salt-rejection capacity throughout the solar desalination process as shown in Figure 4 (f). Specifically, Co-CNS/M foam revealed a thermal conductivity about 0.242 $Wm^{-1}K^{-1}$, which is less than pure MXene (0.345 $Wm^{-1}K^{-1}$) due to the localization of heat. These factors suggest that amorphous carbon nanoplates developed on MXene and the existence of a hierarchical porous architecture. The Co-CNS/M foam achieved superior solar absorbance of about 97.5% compared to pure MXene foam, (95%) up to 2000 nm. Most importantly, the surface temperature of Co-CNS/M was about 69.8°C within 1 min. Overall, Co-CNS/M foam achieved solar to thermal conversion efficiency at an average of 93.06% under 1 sun for 10 days (10 h per day) with a water evaporation rate of 1.393 kg/m^2h, compared to pure MXene-1.306 kg/m^2h.

7.3. Hydrogel

Hydrogels can hold water multiple times higher than its weight [56]. Considering this point, MXene-based photothermal hydrogels were developed for multifunctional solar water purification (photothermal) and the degradation of contaminants. Fan et al. [18] developed the mixed-dimensional $Ti_3C_2T_x$ MXene-based nanohetero structures ($Ti_3C_2T_x$/ $La_{0.5}Sr_{0.5}CoO_3$ (MLH)) for promoting solar water evaporation under excellent thermal insulation. Here, a facile hydrothermal reaction process was adopted to attach the 0D $La_{0.5}Sr_{0.5}CoO_3$ nanoparticles on $Ti_3C_2T_x$ MXene. Figure 5a,b shows the TEM image and interfacial interaction between $Ti_3C_2T_x$ MXene nanosheets (blue line) and $La_{0.5}Sr_{0.5}CoO_3$ nanoparticles (green lines). Selective weight ratio between $Ti_3C_2T_x$ MXene and LSC at 1:10 (MLH-2) yielded fruitful evaporation rate of 2.73 kg/m^2h, which was higher than pure water (0.55 kg/m^2h), MXene (2.26 kg/m^2h), and $La_{0.5}Sr_{0.5}CoO_3$ (2.24 kg/m^2h) with superior evaporation efficiency of 92.3% under 1 sun and average solar absorption (250–2500 nm) of 94.1% as shown in Figure 5c. Specifically, the MLH-2 hydrogel exhibited a microporous skeleton structure. At this condition, the saturated water content significantly increased within 60 s, which suggests the hydrophilic nature. The ability of cross-linked hydrogels was evaluated by measuring the mechanical stability, which suggested a higher storage modulus than loss modulus. It should be noted that the developed composite included negligible heat convection (~1.40%), radiation (~2.28%), and conduction (~0.8%), which supported improvement in the evaporation efficiency. The solar evaporation system designed based on the $Ti_3C_2T_x/La_{0.5}Sr_{0.5}CoO_3$ hydrogel is shown in Figure 5d. The low energy distribution of MLH-2 hydrogel with an enthalpy of 1175 J g^{-1} was observed compared to pure water (2250 J g^{-1}), which suggests efficient evaporation ability. Overall, the low thermal conduction ability of $Ti_3C_2T_x/La_{0.5}Sr_{0.5}CoO_3$ composite extensively absorbed the wide spectrum of solar light and saturated the water content in a shorter time, for achieving stable solar desalination process. The efficient mechanical and water storage capacity of the hydrogels suggests its importance in practical solar desalination of seawater into freshwater. Table 1 presents the solar desalination activity of Ti_3C_2 MXene-based aerogel, foam, and hydrogel morphologies.

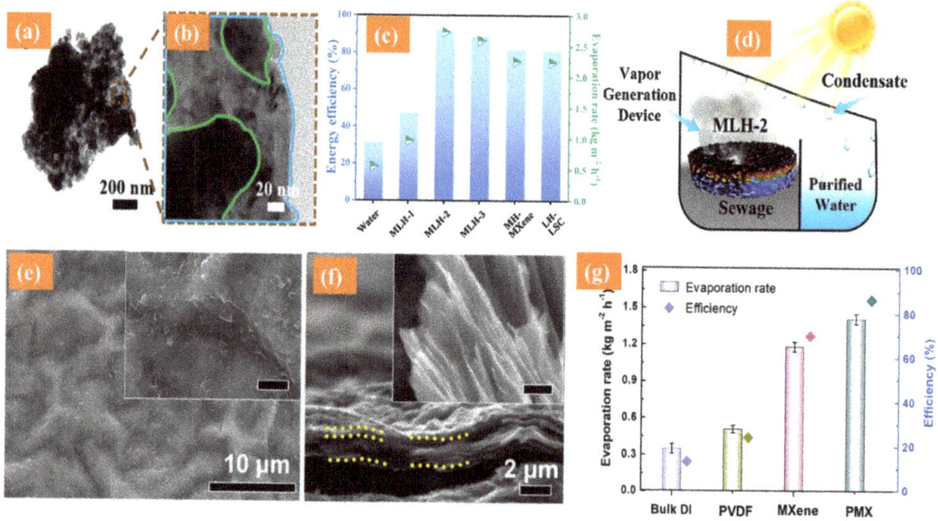

Figure 5. (**a**,**b**) TEM image and interfacial interaction between $Ti_3C_2T_x$ MXene nanosheets (blue line) and $La_{0.5}Sr_{0.5}CoO_3$ nanoparticles (green lines), (**c**) evaporation efficiency of $Ti_3C_2T_x$ MXene and $La_{0.5}Sr_{0.5}CoO_3$ at different weight ratios, and (**d**) designed prototype of solar-driven sewage purification system (Reprinted from Ref. [18], copyright with permission from Elsevier, 2021), (**e**,**f**) top and cross-section view of Janus PMX membrane, and (**g**) evaporation rate and efficiency of Janus PMX membrane compared to base components (reprinted from Ref. [57], copyright with permission from American Chemical Society, 2021).

Table 1. Summary of Ti_3C_2-based aerogel, foam and hydrogel morphologies for solar desalination.

Ti_3C_2-Based Aerogel, Foam and Hydrogel Morphologies for Solar Desalination							
Ti_3C_2-Based Composite (Optimized Condition)	**Ti_3C_2 Functional Groups (Etching Process)**	**Solar Evaporation Efficiency (%) (1 Sun)**	**Surface Temperature (°C)**	**Stability**	**Solar Desalination Rate (Kg/m^2h)**		**Ref.**
					Base Material	**Resultant Composite**	
Aerogel							
$GO/Ti_3C_2T_x$ (MXene to GO-1:3)	O, OH and/or F (HCl and LiF)	~90.7	~41.9	24 h (acid and alkaline environment)	GO-0.88, $Ti_3C_2T_x$-1.02	1.27	[53]
$GO/Ti_3C_2T_x$ ($Ti_3C_2T_x$-90%)	O, OH, and F (HCl and LiF)	~85.0	80	—	—	1.337	[15]
Ti_3C_2 (Micro-channel size of 15 μm (at Ti_3C_2-25 $mg.mL^{-1}$)	—	87	59.4 (300 s)	15 days (sea water)	—	1.46	[54]
Foam							
Cobalt nanoparticle-carbonaceous nanosheets/MXene	HCl and LiF	93.06	69.8 (60 s)	10 days	MXene foam-1.306	1.393	[17]
Hydrogel							
$Ti_3C_2T_x/La_{0.5}Sr_{0.5}CoO_3$ (Weight ratio of MXene to $La_{0.5}Sr_{0.5}CoO_3$-1:10)	NH_4F and HCl	92.3	40 (10 min)	10 cycles (Each cycle 60 min)	$Ti_3C_2T_x$-2.26, $La_{0.5}Sr_{0.5}CoO_3$-2.24	2.73	[18]

7.4. Membrane

Membrane-based water purification technology plays a significant role in producing fresh drinking water from seawater and wastewater. Nowadays, MXene-based composites in the form of the membrane have received much attention for solar desalination. Thus, we plan to reveal the ability of MXene-based composite membranes for freshwater generation, which are practically applicable in the development of advanced water treatment technologies. In this manner, various researchers particularly focused on Ti_3C_2 MXene membrane-related composites to achieve significant water purification.

During the desalination process, the salt blocking activity of the developed absorber plays a key role to achieve stable freshwater production. Generally, Ti_3C_2 MXene prevails in its superior hydrophilic nature and intrinsic light-to-heat conversion ability [39]. In contrast, the hydrophobic surface enables the salt-bocking feature on delaminated Ti_3C_2 MXene (10 mg) nanosheet membrane compared to the hydrophilic surface for efficient solar desalination [58]. Thus, they compared the desalination efficiency of the hydrophobic surface with hydrophilic membranes. Here, the hydrophobic Ti_3C_2 MXene membrane was achieved following the vacuum filtration method of Ding et al. [19]. It should be noted that hydrophobic delaminated Ti_3C_2 MXene successfully blocked the salt compared to the hydrophilic surface. The salt blocking strategy and hydrophobic nature of the -OH terminated Ti_3C_2 MXene membrane was effectively achieved by the addition of trimethoxy (1H,1H,2H,2H-per-fuorodecyl) silane (PFDTMS). These features are utilized for the development of commercial water filter membranes. Here, PFDTMS completely transformed the hydrophilic Ti_3C_2 MXene into a hydrophobic nature. In such a way, pure Ti_3C_2 MXene prevailed over the hydrophilic nature at a water contact angle of 38.8°. However, the PFDTMS-modified Ti_3C_2 MXene surface provided the hydrophobic nature with a water contact angle of 102.0°. The Ti_3C_2 MXene surface was covered with -CF_3 groups for achieving a hydrophilic nature. During the solar desalination process, the membrane gained a vapor temperature about 39 °C within 5–10 min. Due to the non-wettable feature of the Ti_3C_2 MXene surface, the developed heat effectively localized on the surface without transferring toward the bulk water. Under these strange surface features, the hydrophobic Ti_3C_2 MXene membrane achieved an evaporation rate of 1.31 kg/m^{2}h (pure seawater-0.42 kg/m^{2}h) with steam conversion efficiency of 71% and seawater desalination rate of 99.5% for 200 h under 1 sun. The conventional hydrophilic Ti_3C_2 MXene membrane achieved an evaporation rate of 1.41 kg/m^{2}h at a steam conversion efficiency of 74%. Additionally, negligible natural evaporation of about 0.19 kg/m^{2}h was also observed under a dark state due to the hydrophobic nature of the membrane. Overall, the hydrophilic membranes prevailed as the key feature of the salt blocking ability for the efficient and long-term solar desalination of seawater into freshwater.

In addition to the salt blocking, we should reduce the incident light reflection to improve the solar steam generation. In such a way, the first-time deformation of the hierarchical $Ti_3C_2T_x$ MXene (G_1) nanocoatings into a crumpled morphology succeeded for broadband light absorption [59]. In this study, they mainly highlighted the bioinspired $Ti_3C_2T_x$ MXene crumpled structures (such as West African Gaboon Viper), which strongly scattered and reflected the incident light multiple times within the MXene. Consequently, a superior light-to-heat phenomenon was observed. Specifically, broadband absorption (up to 2500 nm) was achieved under an improved equilibrium temperature. Additionally, improvement in the light absorption and decreased reflectance were successfully achieved. Such biomimetic $Ti_3C_2T_x$ MXene morphological features significantly achieved an evaporation rate of 1.33 kg/m^{2}h under 1 sun with broadband absorption up to 93.2% at a lower solar thermal loading of 0.32 mg cm^{-2}. The hydrophilic nature of MXene was also highlighted during the water supply process. Moreover, the $Ti_3C_2T_x$ MXene converted the incident light into heat at about 65.4°C. The planar MXene morphology was observed with reduced light absorption (46.8–64.0%) and light-to-heat generation (50.4–58.1 °C) compared to the crumpled morphology. Finally, mechanically deformed $Ti_3C_2T_x$ MXenes prevailed with the following prominent features: (i) thermal insulation, (ii) mechanical stretchability,

(iii) increased light absorption by scattering, (iv) multiple incident light reflection within the MXene, (v) negligible transmittance, and (vi) low reflection. On the other hand, bioinspired $Ti_3C_2T_x$ MXene features suggested the formation of novel stretchable solar steam generation devices.

As we discussed before, we need to conquer the light reflection losses and salt resistance at the interface of the solar absorber and liquid. In this manner, we can improve the incident light absorption capacity by the photothermal 2D materials. For this, Wang et al. [60] developed the Cu_3BiS_3 (CBS) on Ti_3C_2 MXene nanosheets (0.2 g) for the first time via the thermal-injection method in the ratio of 1:1 (CBS-Ti_3C_2). Later, this composition at different loadings (0.12, 0.24, 0.48, and 0.60 mg cm^{-2}) was placed on a hydrophilic porous filter membrane (PVDF) and vacuum dried overnight. As a result, the coatings were strongly attached to the PVDF membrane during the desalination process. Accordon-like Ti_3C_2 transformed into improved porosity and specific surface area in the CBS–Ti_3C_2 composite. These morphological features proved to have rapid water transportation channels and avoided the formation of high saltwater areas. Accordingly, the CBS–Ti_3C_2 (0.48 mg cm^{-2}) composite revealed steady light-to-heat generation of about 62.3°C (at 0° incident angle), which is higher than that of pure Ti_3C_2 (53.8°C). In such a way, CBS–Ti_3C_2 achieved superior photothermal evaporation efficiency of 91.9% under 1 sun, which is higher than other light intensities (3 sun—88.2%, and 5 sun—90.2%). The main -OH termination groups transformed into -O in the CBS–Ti_3C_2 compared to pure Ti_3C_2 under the unchanged chemical state of Ti_3C_2. These results suggest the efficient light-to-heat conversion ability of CBS–Ti_3C_2. In the same manner, CBS–Ti_3C_2 yielded an evaporation rate of 1.32 kg/m^{2}h (1 sun), 3.84 kg/m^{2}h (3 sun) and 6.26 kg/m^{2}h (5 sun) in 60 min. Due to the combination of the high absorption coefficient of Ti_3C_2 in the visible region and Cu_3BiS_3 in the near-infrared region, the solar light absorption capacity of CBS–Ti_3C_2 (0.48 mg cm^{-2}) was about 87.11% in the range of 300–2500 nm, which is higher than CBS–Ti_3C_2 at 0.12 mg cm^{-2} (85.56%).

In another study, the light-to-heat conversion efficiency of 100% was achieved on a self-floating 2D Ti_3C_2 MXene thin film membrane [39]. In this study, the light-to-heat conversion efficiency was revealed by the droplet laser heating process. During the experiment, the droplet temperature was accurately measured by a precalibrated IR camera. An aqueous droplet of Ti_3C_2 MXene (0.1 mg/mL) with 9.0 μL hung at the end of the PTFE pipet. Here, monochromatic wavelengths of 473 nm and 785 nm were illuminated at a power density of 82 mW and spot size of 0.85 nm in diameter. The laser point perfectly focused on the center of the droplet. During the light-to-heat conversion process, incident light energy was significantly absorbed by the MXene sheets and converted into heat. The scattered light is partially absorbed by the Ti_3C_2 MXene, which does not directly interact with the laser beam. During the water droplet evaporation, the droplet volume was slightly shrunken by about 0.2 mm. On the other hand, a certain amount of heat energy is converted into internal energy in the droplet system. Here, the heat energy gained by the MXene droplet was equal to the heat energy dissipation at equilibrium condition, which maintained the droplet temperatures constantly, compared to the carbon nanotubes. Thus, MXene sheets achieved a light-to-heat conversion efficiency of 100%. On the other hand, the self-floating nature, mechanical strength, and light-to-water evaporation efficiency were achieved by the interaction of MXene with the hydrophilic PVDF photothermal membrane (0.22 μm). The developed MXenes could be peeled off from the PVDF to create a self-floating activity. To decrease the surface energy and hydrophilic nature of MXene–PVDF, PDMS was grafted on MXene–PVDF to self-float on top of the water. The MXene–PVDF membrane achieved an equilibrium temperature of about 75°C (PVDF 30 °C). The PDMS maintained negligible light absorption and reflection of the membranes. Generally, superior water steam generation can be achieved under higher surface temperatures. In this situation, a light-to-water evaporation efficiency of 84% was achieved by the MXene sheets under one sun irradiation.

Very recently, Zhao et al. [61], for the first time, reported the self-assembly of 0D polydopamine microsphere (PDA) and 2D Ti_3C_2 MXene flakes (PDA@MXene microsphere) to realize the broadband light absorption capacity (250–1500 nm). The PDA@MXene microsphere was developed on a hydrophilic PVDF membrane. This composite prevailed with folding, refolding, and recovering features. The presence of -OH, -F, and -O termination groups on the Ti_3C_2 MXene created the negative charge. Specifically, key features, such as (i) reduced light reflection, (ii) minimized heat conduction towards bulk water, and (iii) convention and radiation losses were explored. Additionally, the PDA@MXene achieved rapid water transportation toward the surface and heat localization, compared to the PDA/MXene. Most importantly, PDA@MXene composite achieved a surface temperature of 80 °C, which enlightened the photothermal conversion capability under a stable light absorption capacity of 96%. Accordingly, the selective hydrophilic PDA@MXene (0.8 mg/cm^2) membrane composite succeeded the water evaporation ratio (1.276 kg/m^{2}h), solar-to-vapor conversion efficiency (85.2%), and salt rejection rate beyond 99%. These values are higher than those of pure PDA (1.157 kg/m^{2}h) and Ti_3C_2 MXene (1.095 kg/m^{2}h) under 1 sun. The oxidation stability and light absorption stability of the PDA@MXene membrane were also studied, which revealed no oxidation peaks (TiO_2) after 50 h of aging. Eventually, the PDA@MXene membrane was highlighted with significant light absorption capacity and negligible oxidation during solar desalination for efficient freshwater generation.

During the desalination process, we should mainly focus on the hydrophilic nature of the developed photothermal material. In contrast, Zhang et al. [57] developed a stable hydrophobic/hydrophilic fluorinated porphyrin–$Ti_3C_2T_x$ MXene (PMX) Janus membrane to reveal the superior photothermal desalination. Mainly, they addressed the good salt-resistance performance. Figure 5e,f presents the top and cross-sectional views of the PMX membrane. Previously, a vertically aligned Janus Ti_3C_2 MXene aerogel was also studied based on the hydrophobic/hydrophilic phenomena, which benefited multiple light reflections at an evaporation rate of 1.46 kg/m^{2}h [54]. The upper surface of the hydrophobic/hydrophilic Janus PMX membrane made a water contact angle of 143.9° (fluorinated porphyrin) and the bottom layer made one of a hydrophilic nature (72°). Among these, the hydrophobic surface greatly benefited the heat localization and salt resistance. Such a salt-blocking feature was also observed on the Ti_3C_2 MXene surface with a hydrophobic nature of 102.0° [58], whereas hydrophilic $Ti_3C_2T_x$ MXene effectively pumped the water toward the hot interface. As a result, the resultant PMX membrane contributed efficient vapor generation under 1 sun by generating a surface temperature of about 66.0 °C, which is higher than that of the pure MXene membrane (43.2 °C). Overall, the PMX Janus membrane achieved a stable photothermal evaporation rate of about 1.41 ± 0.04 kg/m^{2}h with an evaporation rate of 86.4% (Figure 5g). These results are almost similar to the previously studied another form of the Janus Ti_3C_2 MXene aerogel [54].

Similar to the PVDF membrane interfacial interaction with the Ti_3C_2 MXene [39,60,61], another study on the Ti_3C_2/PVDF membrane was explored by Peng et al. [62]. They revealed an efficient air-water interface during solar desalination. In this study, the Ti_3C_2/PVDF membrane achieved a hydrophobic nature with a water contact angle of 128.6°, which benefited the floating of the Ti_3C_2/PVDF membrane (light-absorbing material) on water due to surface tension. Here, they highlighted the air–water interface activity of Ti_3C_2 MXene under a superior solar absorption efficiency of 96.36%. In such a way, the surface temperature of the Ti_3C_2/PVDF membrane achieved 43.3 °C in 10 min after solar light illumination (48 °C, 1 h). This value was 11.8 °C higher than pure water. It should be noted that the temperature at the air–water interface without light condition was 21.2 °C. The difference in the temperature confirms the light-to-heat conversion effect by the self-floating Ti_3C_2/PVDF membrane. Specifically, Ti_3C_2 MXene is suitable for the heat absorption and air–water interface heating effect. The selective Ti_3C_2 content of 60 mg in Ti_3C_2/PVDF achieved an evaporation rate of 0.98 kg/m^{2}h (pure water, 0.35 kg/m^{2}h) under 2 sun and stable for 30 days. At this condition, the water evaporation rate was 0.01633 kg/m^{2}min. At a higher Ti_3C_2 content (70 mg), a decrement in the evaporation rate was observed (0.89 kg/m^{2}h) due to the suppressed vapor transport velocity of excessive Ti_3C_2.

The above Ti_3C_2 membrane-based composites effectively explained the self-floating nature and salt-blocking effect for boosting the sweater desalination activity. Specifically, the internal light reflection and hydrophobic nature of the top layer significantly increased the light-to-heat generation and localization of heat for efficient water evaporation. Thus, the Ti_3C_2 membrane-based composites paved a potential path toward freshwater generation.

7.5. Monoliths

In addition to the solar desalination (photothermal effect), voltage difference (Seebeck effect) was also explored through the temperature difference between the top and bottom layer of the absorber under solar light. This phenomenon can be explained as the light-to-heat-to-electricity effect. Following the above concept, Zhang et al. [20] realized the novel temperature difference–induced electricity by the combination of the bilayer Ti_3C_2 MXene-based monolith (B-MXM) (hydrophobic and hydrophilic) with n- and p-type silicon slices. In this study, the bottom portion of Ti_3C_2 MXene submerged in water prevailed with a stable hydrophilic nature and a water contact angle of 43.5°, which continuously led to the rapid dissociation of salt and avoided salt crystallization. The hydrophobic nature of the Ti_3C_2 MXene upper layer localized the heat as shown in Figure 6a. Zhang et al. [57] also reported such a hydrophobic/hydrophilic phenomenon on the $Ti_3C_2T_x$ MXene Janus membrane for solar desalination. The upper hydrophobic surface layer generated a temperature of about 98 °C than the lower B-MXM surface (30 °C), which suggests that heat effectively localized on the top of the surface. Consequently, thermal conduction was reduced toward the bulk water. Accordingly, B-MXM achieved an evaporation rate of about 1.39 kg/m^2h under outdoor natural sunlight (0.91 kWm^{-2}) with a solar thermal efficiency of ~82.9% and sustained voltage of ~0.3 V in the afternoon 12:00 p.m. (noon) as shown in Figure 6b. It should be noted that this study further elevated the light-to-heat concept into light-to-heat-to-electricity. Figure 6c presents the simultaneous solar desalination and voltage measurements, using hydrophobic and hydrophilic mediated B-MXM composites.

7.6. Porous

Ti_3C_2 MXenes in the form of tightly packed 2D membranes create high reflection losses and restrict the water from flowing, which suppresses the overall evaporation rate. To overcome this, macroporous 3D Ti_3C_2 MXene interfaced carbonized melamine foam (CMF@d-Ti_3C_2) was developed for improving the rapid flow of water and escaping the water vapor [63]. In this study, CMF@d–Ti_3C_2 was used as a solar absorber and evaporator. For floating the porous CMF@d–Ti_3C_2 composite, thermal insulated polystyrene foam was used. For water transportation under the capillary effect, nonwoven fabric was used. The 2D to 3D morphologic transformation of Ti_3C_2 in the resultant porous structured CMF@d–Ti_3C_2 elongated the optical path and improved the absorption of incident solar light. Additionally, CMF@d–Ti_3C_2 sustained an excellent hydrophilic nature. The absorber temperature gradually reached 40 °C after 10 min with a vapor temperature of 42 °C. The porous surface of the CMF@d–Ti_3C_2 composite greatly reduced the specular reflection losses. Under these constructive 3D architectures, CMF@d–Ti_3C_2 significantly achieved an evaporation rate of 1.60 kg/m^2h, which is higher than that of 2D Ti_3C_2 (1.41 kg/m^2h), with thermal evaporation efficiency of 84.6% (13 h) under 1 sun.

In a similar way, the morphological transformation of 2D $Ti_3C_2T_x$ into 3D was successfully explored for the solar desalination activity under strategic self-floating and direct contact with bulk water [21]. Here, the microporous hydrophilic 3D $Ti_3C_2T_x$ MXene architecture (3DMA) was developed on the melamine foam (MF) skeleton under the adhesion force of PVA. The strong capillary effect was observed by the hydrophilic 3DMA and MF. Specifically, continuous water flow was observed from MF to 3DMA. Moreover, the 2D to 3D morphological evolution of $Ti_3C_2T_x$ MXene significantly improved the light absorption capacity from the UV to NIR region. Additionally, 3DMA achieved ~98% absorbance (350–1500 nm). However, a lower surface temperature (39 °C) was observed on the surface while floating on the water. The unique resultant 3D architecture enhanced the extended light path, multiple light scattering, and decreased light scattering at the liquid-to-material interface, which eventually improved the light absorption capacity. As a result, 3DMA achieved the evaporation rate of 1.41 kg/m^{2}h under 1 sun at solar steam efficiency of 88.7%. Additionally, the developed $Ti_3C_2T_x$ MXene-based porous structure was highly stable for 10 cycles at 30 min of each cycle. In this study, direct contact of 3DMA on water lost its light-to-heat towards the bulk water. Thus, expandable polyethylene foam (insulating layer) was wrapped around the architecture to minimize heat loss and improve the photothermal conversion efficiency.

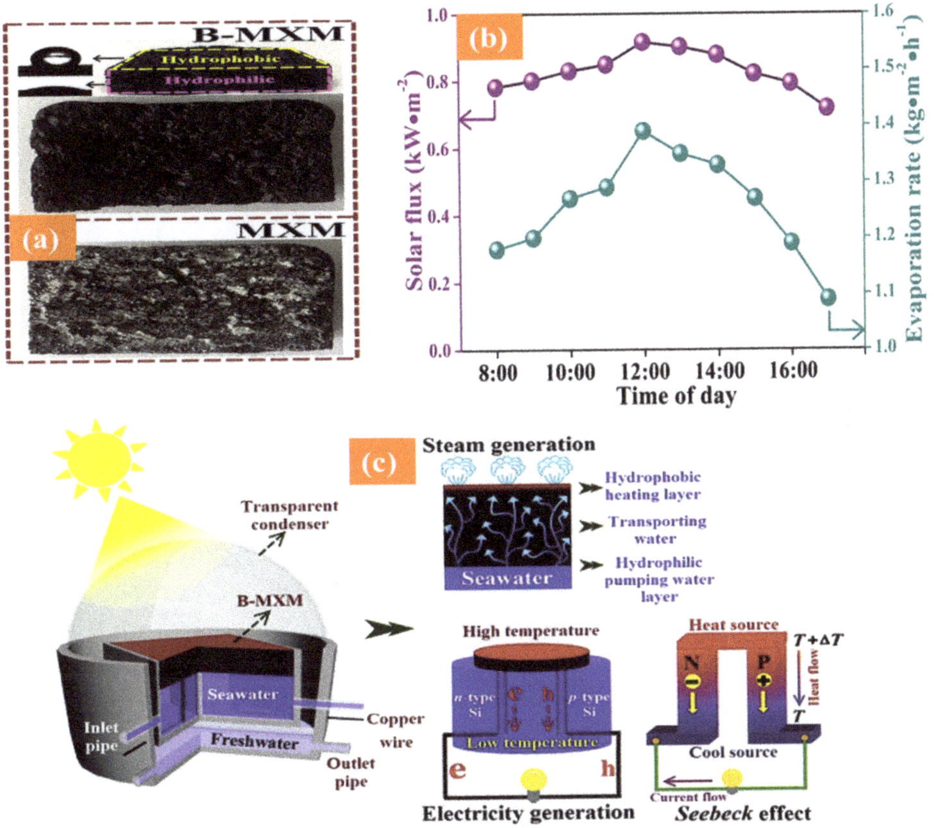

Figure 6. (**a**) Hydrophobic/hydrophilic bilayers of Ti_3C_2 MXene-based monolith with heat localization by hydrophobic nature, (**b**) corresponding evaporation rate at various time and solar intensities, and (**c**) implementation of simultaneous solar desalination and voltage measurements through temperature difference, using hydrophobic and hydrophilic Ti_3C_2 MXene layers (reprinted from Ref. [20], copyright with permission from Elsevier, 2020).

7.7. Nanocomposite

In addition to the above pure 2D Ti_3C_2 MXene, the developing of 2D/2D hierarchical nanocomposites also raised its importance in solar steam generation. Very recently, Xu et al. [64] interrelated the Ti_3C_2 and MoS_2 for the development of an accordion-like-layer-structured 2D/2D Ti_3C_2/MoS_2 composite via the hydrothermal method. This composite provided broadband light absorption between 200 nm and 2500 nm. Figure 7a presents the surface morphology of the Ti_3C_2/MoS_2 composite at evenly distributed MoS_2 layers on accordion-like Ti_3C_2. Interestingly, both Ti_3C_2 and MoS_2 act as photothermal materials. Under this situation, a selective mass fraction of Ti_3C_2 (65 wt%) in the Ti_3C_2/MoS_2 composite (TM-3) achieved superior evaporation rate, compared to the other conditions of MoS_2 (35, 50, and 80%). Additionally, an optical absorption capacity of 92.4% was achieved, which is higher than other counterparts as shown in Figure 7b. Moreover, TM-3 addressed lower transmission of light (3.77%) and reflection (3.78%). On the other hand, the achieved morphological features and large gaps between layers succeeded in multiple light reflections and refractions, which eventually boosted the improvement in the light absorption. Under the superior light absorption ability of TM-3, light-to-heat conversion was observed in the range of 28.5–55.3 °C within a short time of 4 min and reached 60.8 °C in 56 min. Under these constructive morphological features, TM-3 composite yielded the evaporation rate of 1.36 kg/m^{2}h, photothermal conversion efficiency of 87.2%, and strong light absorption capacity of 92.4% (Figure 7c).

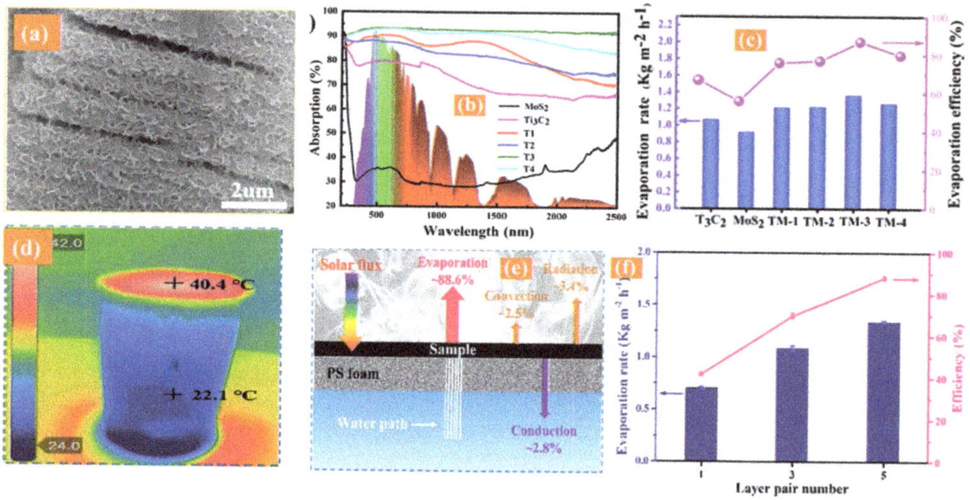

Figure 7. (**a**) Surface morphology of Ti_3C_2/MoS_2 composite at evenly distributed MoS_2 layers on accordion-like Ti_3C_2, (**b**) solar absorption behavior of pure Ti_3C_2, MoS_2, and various mass fractions of MoS_2 in Ti_3C_2/MoS_2, and (**c**) corresponding water evaporation rate and solar evaporation efficiency (reprinted from Ref. [64], copyright with permission from Elsevier, 2021), (**d**) variation in the temperature change on the surface and bottom of five times layer-by-layer grown $Ti_3C_2T_x$ MXene/CNT composite, (**e**) evaporation, convection, radiation, and thermal conduction rate during solar desalination, and (**f**) corresponding evaporation rate and light conversion efficiency of $Ti_3C_2T_x$ MXene/CNT at different layer growths (1, 3, and 5) (reprinted from Ref. [65], copyright with permission from Elsevier, 2021).

7.8. Other Morphologies

In addition to the above constructive synthesis of various morphologies, HF-free (LiF/HCl) etched Ti_3C_2 (-OH,-F and -O) MXene nanosheets (pH of 5) were developed on carbon nanotube (CNT)–coated cotton fabric (MC) by the layer-by-layer assembly to highlight the solar evaporation under an intimate interfacial interaction [65]. Here, they found key electrostatic interaction and hydrogen bonding between $Ti_3C_2T_x$ MXene (negatively charged) and CNT (positively charged). The -OH, -F and -O functional groups raised on the Ti_3C_2 surface and -NH_2 developed on CNT, which provided the hydrophilic nature with cotton fabric (polar substrate). Additionally, hydrophilic MXenes were well dispersed in water. These features rapidly wetted the composite fabric for water transportation during the evaporation process. In such a way, selective layer-by-layer grown (five times) $Ti_3C_2T_x$ MXene/CNT composite $(MC)_5$ cotton fabric achieved the light absorption capacity of 93.5% (dry state) and 97.7% (wet state) throughout the solar spectrum (250–2500 nm). Specifically, the porous structure benefited from the light absorption and air–water interfacial area. Thus, the $(MC)_5$ cotton fabric achieved a surface temperature of 40.4°C under 1 sun in 10 min (Figure 7d). As a result, $(MC)_5$ achieved the evaporation rate of 1.35 kg/m^{2}h, which is 3.29 times higher than pure water (0.41 kg/m^{2}h) with an energy conversion evaporation efficiency of 88.2 ± 0.9% (convection ~2.5%, radiation ~3.4% and conduction ~2.8%) as shown in Figure 7e,f. The lower thermal conducting behavior of $(MC)_5$ during the desalination process boosted the localization of heat on the surface, which eventually reduced the heat loss toward the surrounding environments. In other words, the thermal conductivity of pure water was 0.6 $Wm^{-1}K^{-1}$, which was higher than $(MC)_5$ at 0.12 $Wm^{-1}K^{-1}$. These results suggest the lower thermal conductivity of $Ti_3C_2T_x$ MXene-based composites, which boosted the water evaporation rate. Thus, $(MC)_5$ gained improvement in the evaporation rate compared to pure water. Overall, $Ti_3C_2T_x$ MXene/CNT cotton fabric composite achieved an efficient solar evaporation rate based on the following interesting factors: (i) low reflection, (ii) low thermal conductivity to reduce the convection heat to ambient (air), (iii) high optical absorption, and (iv) photothermal conversion. Table 2 presents the solar desalination activity of Ti_3C_2 MXene-based membrane, monolith, and porous morphologies.

Table 2. Summary of Ti_3C_2-based membrane, monolith, and porous morphologies for solar desalination.

Ti_3C_2-Based Membrane, Monolith, and Porous Morphologies for Solar Desalination							
Ti_3C_2-Based Composite (Optimized Condition)	Ti_3C_2 Functional Groups (Etching Process)	Solar Evaporation Efficiency (%) (1 Sun)	Surface Temperature (°C)	Stability	Solar Desalination Rate (Kg/m^2h)		Ref
					Base Material	Resultant Composite	
Membrane							
Ti_3C_2	HCl and LiF	99.5	39 (10 min)	200 h	1.41 (hydrophilic)	1.31	[58]
$Ti_3C_2T_x$	HCl and LiF	—	65.4 (5 min)	—	—	1.33	[59]
Cu_3BiS_3/Ti_3C_2 (1:1)	(O and OH) HCl and LiF	87.11	62.3	—	—	1.32	[60]
Ti_3C_2	HF	84	75	—	—	—	[39]
PDA@ MXene	HCl and LiF	85.2	80	—	PDA-1.157, $Ti_3C_2T_x$-1.095	1.276	[61]
Fluorinated porphyrin–$Ti_3C_2T_x$ MXene	HCl and LiF	86.4	66	—	—	1.41	[57]
Ti_3C_2/PVDF	HF	—	43.3 (10 min)	—	—	0.98	[62]
Monolith							
Ti_3C_2 MXene	—	82.9	98 (upper surface), 30 (lower surface)	—	—	1.39	[20]
Porous							
Carbonized melamine foam (CMF)@d-Ti_3C_2	HCl and LiF	84.6	38.5 (10 min)	—	Ti_3C_2-1.41	1.60	[63]
$Ti_3C_2T_x$	HCl and LiF	88.7	39	—	—	1.41	[21]
Nanocomposite							
Ti_3C_2/MoS_2	—	87.2	55.3 (4 min)	—	—	1.36	[64]
Other morphologies							
MXene nanosheets on carbon nanotube coated cotton fabric	-OH, -F and -O (HCl and LiF)	88.2±0.9	40.4	—	—	1.35	[65]

8. Conclusions and Future Perspectives

The current review comprehensively summarized the potentiality of various morphological Ti_3C_2 MXene-based composites for the effective solar desalination of seawater for freshwater generation. It is concluded that the desalination process of Ti_3C_2 MXene greatly depends on the morphological features, which include aerogel, foam, hydrogel, membrane, monolith, and porous structures. The evolution of Ti_3C_2 MXene morphologies extended the water absorption capacity and light absorption toward the broadband region, thanks to the exceptional photothermal conversion ability of the layer-structured 2D Ti_3C_2 MXenes. To date, the desalination of seawater using Ti_3C_2 MXene is still in its infancy. The Ti_3C_2 MXenes are significant for explaining the light-to-heat generation and heat localization on the surface for the successful evaporation of seawater into freshwater. Additionally, incident light was greatly absorbed inside the MXene through multiple light-reflection phenomena. Specifically, the intrinsic hydrophilic nature of the Ti_3C_2 MXene makes a potential path for water transportation channel toward the top of the surface. It was found that Ti_3C_2 MXenes prevailed with low radiation, low convection, and low thermal conduction, which eventually increased the light absorption and evaporation. The above-discussed concepts reveal that Ti_3C_2 MXenes are eco-friendly during the solar desalination of seawater. Due to scarce studies on the desalination ability of Ti_3C_2 MXenes, there are many constructive suggestions and perspectives:

1. The intrinsic hydrophilic nature and water absorption ability of layer-structured Ti_3C_2 MXenes determine its applicability in the field of eco-friendly solar desalination of seawater into freshwater.
2. In the contest of long-term stability, researchers should pay more attention toward membrane-based Ti_3C_2 MXenes for understanding efficient solar desalination. Thus, further research is necessary to understand the efficiency of Ti_3C_2 MXene membranes as a forerunner in solar desalination.
3. Due to the limited research on the Ti_3C_2 MXenes for solar desalination, it is further required to develop the wide band gap semiconductor material interaction with Ti_3C_2 MXenes for widening the solar light absorption. More research needs to be carried out in this direction.
4. The long-term stability of the layer-structured Ti_3C_2 MXenes is also a major challenge. However, surface modification through the selective termination groups shows great attention on the air–water interface for long-term stability.
5. Regarding the all-weather and complex conditions, there is infinite scope and opportunity for researchers to expand the potentiality of Ti_3C_2 MXenes for freshwater generation.
6. The developing of self-floating Ti_3C_2 MXenes with a hydrophilic nature has not been investigated. Thus, ceaseless efforts are required to accomplish the self-floating nature of Ti_3C_2 MXenes.
7. In addition to the self-floating nature, the salt-blocking nature during the desalination process significantly boosts the light absorption capacity and water transportation toward the top of the absorber. Such configurations are ideal for highly efficient solar desalination and long-term durability.
8. The oxidation stability of MXene during interaction with water under solar light illumination needs to be clearly explained.

In a word, this review provides new ideas for the design of novel 2D Ti_3C_2 MXenes-based composites for superior photothermal conversion and the solar desalination of seawater into freshwater.

Author Contributions: Conceptualization, A.S.; Writing-original draft preparation, A.S.; Writing-review and editing, J.-S.N.; supervision, J.-S.N.; funding acquisition, J.-S.N. All authors have read and agreed to the published version of the manuscript.

Funding: This work was funded by the National Research Foundation of Korea (NRF) by the Korea government (MSIT) (No. 2019R1A2C1008746).

Acknowledgments: This work was supported by the National Research Foundation of Korea (NRF) grant funded by the Korea government (MSIT) (No. 2019R1A2C1008746).

Conflicts of Interest: The authors declare no conflict of interest.

References

1. Wang, Y.; Niu, B.; Zhang, X.; Lei, Y.; Zhong, P.; Ma, X. Review—$Ti_3C_2T_x$ MXene: An emerging two-dimensional layered material in water treatment. *ECS J. Solid State Sci. Technol.* **2021**, *10*, 47002. [CrossRef]
2. Hodges, B.C.; Cates, E.L.; Kim, J.-H. Challenges and prospects of advanced oxidation water treatment processes using catalytic nanomaterials. *Nat. Nanotechnol.* **2018**, *13*, 642–650. [CrossRef] [PubMed]
3. Elimelech, M.; Phillip, W.A. The future of seawater desalination: Energy, technology, and the environment. *Science* **2011**, *333*, 712–717. [CrossRef]
4. Prathapaneni, D.R.; Detroja, K. Optimal design of energy sources and reverse osmosis desalination plant with demand side management for cost-effective freshwater production. *Desalination* **2020**, *496*, 114741. [CrossRef]
5. Eljaddi, T.; Mendez, D.L.M.; Favre, E.; Roizard, D. Development of new pervaporation composite membranes for desalination: Theoretical and experimental investigations. *Desalination* **2021**, *507*, 115006. [CrossRef]
6. Liu, G.; Shen, J.; Liu, Q.; Liu, G.; Xiong, J.; Yang, J.; Jin, W. Ultrathin two-dimensional MXene membrane for pervaporation desalination. *J. Memb. Sci.* **2018**, *548*, 548–558. [CrossRef]
7. Davenport, D.M.; Deshmukh, A.; Werber, J.R.; Elimelech, M. High-pressure reverse osmosis for energy-efficient hypersaline brine desalination: Current status, design considerations, and research needs. *Environ. Sci. Technol. Lett.* **2018**, *5*, 467–475. [CrossRef]
8. Yu, F.; Wang, L.; Wang, Y.; Shen, X.; Cheng, Y.; Ma, J. Faradaic reactions in capacitive deionization for desalination and ion separation. *J. Mater. Chem. A* **2019**, *7*, 15999–16027. [CrossRef]
9. Lv, B.; Gao, C.; Xu, Y.; Fan, X.; Xiao, J.; Liu, Y.; Song, C. A self-floating, salt-resistant 3D Janus radish-based evaporator for highly efficient solar desalination. *Desalination* **2021**, *510*, 115093. [CrossRef]
10. Wang, Y.; Wu, Q.; Zhang, B.; Tian, L.; Li, K.; Zhang, X. Recent advances in transition metal carbide electrocatalysts for oxygen evolution reaction. *Catalysts* **2020**, *10*, 1164. [CrossRef]
11. Ta, Q.T.H.; Tran, N.M.; Noh, J.-S. Rice crust-like $ZnO/Ti_3C_2T_x$ MXene hybrid structures for improved photocatalytic activity. *Catalysts* **2020**, *10*, 1140. [CrossRef]
12. Hou, T.; Li, Q.; Zhang, Y.; Zhu, W.; Yu, K.; Wang, S.; Xu, Q.; Liang, S.; Wang, L. Near-infrared light-driven photofixation of nitrogen over $Ti_3C_2T_x/TiO_2$ hybrid structures with superior activity and stability. *Appl. Catal. B Environ.* **2020**, *273*, 119072. [CrossRef]
13. Shahzad, F.; Alhabeb, M.; Hatter, C.B.; Anasori, B.; Man Hong, S.; Koo, C.M.; Gogotsi, Y. Electromagnetic interference shielding with 2D transition metal carbides (MXenes). *Science* **2016**, *353*, 1137–1140. [CrossRef]
14. Jiang, X.; Kuklin, A.V.; Baev, A.; Ge, Y.; Ågren, H.; Zhang, H.; Prasad, P.N. Two-dimensional MXenes: From morphological to optical, electric, and magnetic properties and applications. *Phys. Rep.* **2020**, *848*, 1–58. [CrossRef]
15. Zhao, X.; Peng, L.M.; Tang, C.Y.; Pu, J.H.; Zha, X.J.; Ke, K.; Bao, R.Y.; Yang, M.B.; Yang, W. All-weather-available, continuous steam generation based on the synergistic photo-thermal and electro-thermal conversion by MXene-based aerogels. *Mater. Horizons* **2020**, *7*, 855–865. [CrossRef]
16. Xie, X.; Chen, C.; Zhang, N.; Tang, Z.-R.; Jiang, J.; Xu, Y.-J. Microstructure and surface control of MXene films for water purification. *Nat. Sustain.* **2019**, *2*, 856–862. [CrossRef]
17. Fan, X.; Yang, Y.; Shi, X.; Liu, Y.; Li, H.; Liang, J.; Chen, Y. A MXene-based hierarchical design enabling highly efficient and stable solar-water desalination with good salt resistance. *Adv. Funct. Mater.* **2020**, *30*, 2007110. [CrossRef]
18. Fan, D.; Lu, Y.; Zhang, H.; Xu, H.; Lu, C.; Tang, Y.; Yang, X. Synergy of photocatalysis and photothermal effect in integrated 0D perovskite oxide/2D MXene heterostructures for simultaneous water purification and solar steam generation. *Appl. Catal. B Environ.* **2021**, *295*, 120285. [CrossRef]
19. Ding, M.; Xu, H.; Chen, W.; Yang, G.; Kong, Q.; Ng, D.; Lin, T.; Xie, Z. 2D laminar maleic acid-crosslinked MXene membrane with tunable nanochannels for efficient and stable pervaporation desalination. *J. Memb. Sci.* **2020**, *600*, 117871. [CrossRef]
20. Zhang, Q.; Chen, S.; Fu, Z.; Yu, H.; Quan, X. Temperature-difference-induced electricity during solar desalination with bilayer MXene-based monoliths. *Nano Energy* **2020**, *76*, 105060. [CrossRef]
21. Zhao, X.; Zha, X.-J.; Pu, J.-H.; Bai, L.; Bao, R.-Y.; Liu, Z.-Y.; Yang, M.-B.; Yang, W. Macroporous three-dimensional MXene architectures for highly efficient solar steam generation. *J. Mater. Chem. A* **2019**, *7*, 10446–10455. [CrossRef]
22. Su, T.; Hood, Z.D.; Naguib, M.; Bai, L.; Luo, S.; Rouleau, C.M.; Ivanov, I.N.; Ji, H.; Qin, Z.; Wu, Z. Monolayer $Ti_3C_2T_x$ as an effective co-catalyst for enhanced photocatalytic hydrogen production over TiO_2. *ACS Appl. Energy Mater.* **2019**, *2*, 4640–4651. [CrossRef]
23. Cheng, X.; Zu, L.; Jiang, Y.; Shi, D.; Cai, X.; Ni, Y.; Lin, S.; Qin, Y. A titanium-based photo-Fenton bifunctional catalyst of mp-MXene/TiO_{2-x} nanodots for dramatic enhancement of catalytic efficiency in advanced oxidation processes. *Chem. Commun.* **2018**, *54*, 11622–11625. [CrossRef]
24. Shahzad, A.; Rasool, K.; Nawaz, M.; Miran, W.; Jang, J.; Moztahida, M.; Mahmoud, K.A.; Lee, D.S. Heterostructural $TiO_2/Ti_3C_2T_x$ (MXene) for photocatalytic degradation of antiepileptic drug carbamazepine. *Chem. Eng. J.* **2018**, *349*, 748–755. [CrossRef]
25. Low, J.; Zhang, L.; Tong, T.; Shen, B.; Yu, J. TiO_2/MXene Ti_3C_2 composite with excellent photocatalytic CO_2 reduction activity. *J. Catal.* **2018**, *361*, 255–266. [CrossRef]
26. Chen, Y.-H.; Qi, M.-Y.; Li, Y.-H.; Tang, Z.-R.; Wang, T.; Gong, J.; Xu, Y.-J. Activating two-dimensional $Ti_3C_2T_x$-MXene with single-atom cobalt for efficient CO_2 photoreduction. *Cell Rep. Phys. Sci.* **2021**, *2*, 100371. [CrossRef]

27. Li, M.; Liu, X.; Wang, L.; Hou, F.; Dou, S.X.; Liang, J. Rational design on photo(electro)catalysts for artificial nitrogen looping. *EcoMat* **2021**, *3*, e12096. [CrossRef]
28. Zhu, J.; Zhu, M.; Shi, Q.; Wen, F.; Liu, L.; Dong, B.; Haroun, A.; Yang, Y.; Vachon, P.; Guo, X.; et al. Progress in TENG technology—A journey from energy harvesting to nanoenergy and nanosystem. *EcoMat* **2020**, *2*, e12058. [CrossRef]
29. Zhang, M.; Héraly, F.; Yi, M.; Yuan, J. Multitasking tartaric-acid-enabled, highly conductive, and stable MXene/conducting polymer composite for ultrafast supercapacitor. *Cell Rep. Phys. Sci.* **2021**, *2*, 100449. [CrossRef]
30. Saranin, D.; Pescetelli, S.; Pazniak, A.; Rossi, D.; Liedl, A.; Yakusheva, A.; Luchnikov, L.; Podgorny, D.; Gostischev, P.; Didenko, S.; et al. Transition metal carbides (MXenes) for efficient NiO-based inverted perovskite solar cells. *Nano Energy* **2021**, *82*, 105771. [CrossRef]
31. Bati, A.S.R.; Sutanto, A.A.; Hao, M.; Batmunkh, M.; Yamauchi, Y.; Wang, L.; Wang, Y.; Nazeeruddin, M.K.; Shapter, J.G. Cesium-doped $Ti_3C_2T_x$ MXene for efficient and thermally stable perovskite solar cells. *Cell Rep. Phys. Sci.* **2021**, *2*, 100598. [CrossRef]
32. Ihsanullah, I. Potential of MXenes in water desalination: Current status and perspectives. *Nano-Micro Lett.* **2020**, *12*, 72. [CrossRef] [PubMed]
33. Li, L.; Wang, C.; Feng, K.; Huang, D.; Wang, K.; Li, Y.; Jiang, F. Kesterite Cu_2ZnSnS_4 thin-film solar water-splitting photovoltaics for solar seawater desalination. *Cell Rep. Phys. Sci.* **2021**, *2*, 100468. [CrossRef]
34. Morciano, M.; Fasano, M.; Boriskina, S.V.; Chiavazzo, E.; Asinari, P. Solar passive distiller with high productivity and marangoni effect-driven salt rejection. *Energy Environ. Sci.* **2020**, *13*, 3646–3655. [CrossRef]
35. Zhu, Z.; Xu, Y.; Luo, Y.; Wang, W.; Chen, X. Porous evaporators with special wettability for low-grade heat-driven water desalination. *J. Mater. Chem. A* **2021**, *9*, 702–726. [CrossRef]
36. Chaudhuri, K.; Alhabeb, M.; Wang, Z.; Shalaev, V.M.; Gogotsi, Y.; Boltasseva, A. Highly broadband absorber using plasmonic titanium carbide (MXene). *ACS Photonics* **2018**, *5*, 1115–1122. [CrossRef]
37. Zhang, Y.; Xiong, T.; Nandakumar, D.K.; Tan, S.C. Structure architecting for salt-rejecting solar interfacial desalination to achieve high-performance evaporation with in situ energy generation. *Adv. Sci.* **2020**, *7*, 1903478. [CrossRef]
38. Ghidiu, M.; Lukatskaya, M.R.; Zhao, M.-Q.; Gogotsi, Y.; Barsoum, M.W. Conductive two-dimensional titanium carbide 'clay' with high volumetric capacitance. *Nature* **2014**, *516*, 78–81. [CrossRef]
39. Li, R.; Zhang, L.; Shi, L.; Wang, P. MXene Ti_3C_2: An effective 2D light-to-heat conversion material. *ACS Nano* **2017**, *11*, 3752–3759. [CrossRef]
40. Liu, H.; Huang, Z.; Liu, K.; Hu, X.; Zhou, J. Interfacial solar-to-heat conversion for desalination. *Adv. Energy Mater.* **2019**, *9*, 1900310. [CrossRef]
41. Liu, G.; Chen, T.; Xu, J.; Yao, G.; Xie, J.; Cheng, Y.; Miao, Z.; Wang, K. Salt-Rejecting solar interfacial evaporation. *Cell Rep. Phys. Sci.* **2021**, *2*, 100310. [CrossRef]
42. Han, X.; Zang, L.; Zhang, S.; Dou, T.; Li, L.; Yang, J.; Sun, L.; Zhang, Y.; Wang, C. Hydrophilic polymer-stabilized porous composite membrane for water evaporation and solar desalination. *RSC Adv.* **2020**, *10*, 2507–2512. [CrossRef]
43. Yao, J.; Yang, G. An efficient solar-enabled 2D layered alloy material evaporator for seawater desalination. *J. Mater. Chem. A* **2018**, *6*, 3869–3876. [CrossRef]
44. Thoai, D.N.; Hoai Ta, Q.T.; Truong, T.T.; Van Nam, H.; Van Vo, G. Review on the recent development and applications of three dimensional (3D) photothermal materials for solar evaporators. *J. Clean. Prod.* **2021**, *293*, 126122. [CrossRef]
45. Fengler, C.; Arens, L.; Horn, H.; Wilhelm, M. Desalination of seawater using cationic poly(acrylamide) hydrogels and mechanical forces for separation. *Macromol. Mater. Eng.* **2020**, *305*, 2000383. [CrossRef]
46. Ying, L.; Zhu, H.; Li, H.; Zhu, Z.; Sun, S.; Wang, X.; Lu, S.; Du, M. Heterostructure design of Cu_2O/Cu_2S core/shell nanowires for solar-driven photothermal water vaporization towards desalination. *Sustain. Energy Fuels* **2020**, *4*, 6023–6029. [CrossRef]
47. Yang, Y.; Zhao, R.; Zhang, T.; Zhao, K.; Xiao, P.; Ma, Y.; Ajayan, P.M.; Shi, G.; Chen, Y. Graphene-based standalone solar energy converter for water desalination and purification. *ACS Nano* **2018**, *12*, 829–835. [CrossRef]
48. Kospa, D.A.; Ahmed, A.I.; Samra, S.E.; Ibrahim, A.A. High efficiency solar desalination and dye retention of plasmonic/reduced graphene oxide based copper oxide nanocomposites. *RSC Adv.* **2021**, *11*, 15184–15194. [CrossRef]
49. Naguib, M.; Kurtoglu, M.; Presser, V.; Lu, J.; Niu, J.; Heon, M.; Hultman, L.; Gogotsi, Y.; Barsoum, M.W. Two-dimensional nanocrystals: Two-dimensional nanocrystals produced by exfoliation of Ti_3AlC_2 (Adv. Mater. 37/2011). *Adv. Mater.* **2011**, *23*, 4207. [CrossRef]
50. Xie, Z.; Peng, Y.-P.; Yu, L.; Xing, C.; Qiu, M.; Hu, J.; Zhang, H. Solar-inspired water purification based on emerging 2D materials: Status and challenges. *Sol. RRL* **2020**, *4*, 1900400. [CrossRef]
51. Sreedhar, A.; Noh, J.-S. Interfacial engineering insights of promising monolayer 2D Ti_3C_2 MXene anchored flake-like ZnO thin films for improved PEC water splitting. *J. Electroanal. Chem.* **2021**, *883*, 115044. [CrossRef]
52. Biener, J.; Stadermann, M.; Suss, M.; Worsley, M.A.; Biener, M.M.; Rose, K.A.; Baumann, T.F. Advanced carbon aerogels for energy applications. *Energy Environ. Sci.* **2011**, *4*, 656. [CrossRef]
53. Ming, X.; Guo, A.; Zhang, Q.; Guo, Z.; Yu, F.; Hou, B.; Wang, Y.; Homewood, K.P.; Wang, X. 3D macroscopic graphene oxide/MXene architectures for multifunctional water purification. *Carbon N.Y.* **2020**, *167*, 285–295. [CrossRef]
54. Zhang, Q.; Yi, G.; Fu, Z.; Yu, H.; Chen, S.; Quan, X. Vertically aligned janus MXene-based aerogels for solar desalination with high efficiency and salt resistance. *ACS Nano* **2019**, *13*, 13196–13207. [CrossRef]

55. Lin, H.; Wang, X.; Yu, L.; Chen, Y.; Shi, J. Two-dimensional ultrathin MXene ceramic nanosheets for photothermal conversion. *Nano Lett.* **2017**, *17*, 384–391. [CrossRef] [PubMed]
56. Guo, Y.; Yu, G. Engineering hydrogels for efficient solar desalination and water purification. *Accounts Mater. Res.* **2021**, *2*, 374–384. [CrossRef]
57. Zhang, B.; Gu, Q.; Wang, C.; Gao, Q.; Guo, J.; Wong, P.W.; Liu, C.T.; An, A.K. Self-assembled hydrophobic/hydrophilic porphyrin-$Ti_3C_2T_x$ MXene janus membrane for dual-functional enabled photothermal desalination. *ACS Appl. Mater. Interfaces* **2021**, *13*, 3762–3770. [CrossRef] [PubMed]
58. Zhao, J.; Yang, Y.; Yang, C.; Tian, Y.; Han, Y.; Liu, J.; Yin, X.; Que, W. A hydrophobic surface enabled salt-blocking 2D Ti_3C_2 MXene membrane for efficient and stable solar desalination. *J. Mater. Chem. A* **2018**, *6*, 16196–16204. [CrossRef]
59. Li, K.; Chang, T.; Li, Z.; Yang, H.; Fu, F.; Li, T.; Ho, J.S.; Chen, P. Biomimetic MXene textures with enhanced light-to-heat conversion for solar steam generation and wearable thermal management. *Adv. Energy Mater.* **2019**, *9*, 1901687. [CrossRef]
60. Wang, Z.; Yu, K.; Gong, S.; Mao, H.; Huang, R.; Zhu, Z. Cu_3BiS_3/MXenes with excellent solar–thermal conversion for continuous and efficient seawater desalination. *ACS Appl. Mater. Interfaces* **2021**, *13*, 16246–16258. [CrossRef]
61. Zhao, X.; Zha, X.-J.; Tang, L.-S.; Pu, J.-H.; Ke, K.; Bao, R.-Y.; Liu, Z.; Yang, M.-B.; Yang, W. Self-assembled core-shell polydopamine@MXene with synergistic solar absorption capability for highly efficient solar-to-vapor generation. *Nano Res.* **2020**, *13*, 255–264. [CrossRef]
62. Peng, H.; Zhu, K.; Li, C.; Xiao, Y.; Ye, M. Ti_3C_2/PVDF membrane for efficient seawater desalination based on interfacial solar heating. *Water Supply* **2021**, *21*, 918–926. [CrossRef]
63. Ju, M.; Yang, Y.; Zhao, J.; Yin, X.; Wu, Y.; Que, W. Macroporous 3D MXene architecture for solar-driven interfacial water evaporation. *J. Adv. Dielectr.* **2019**, *9*, 1950047. [CrossRef]
64. Xu, R.; Wei, N.; Li, Z.; Song, X.; Li, Q.; Sun, K.; Yang, E.; Gong, L.; Sui, Y.; Tian, J.; et al. Construction of hierarchical 2D/2D Ti_3C_2/MoS_2 nanocomposites for high-efficiency solar steam generation. *J. Colloid Interface Sci.* **2021**, *584*, 125–133. [CrossRef]
65. Wang, Y.; Qi, Q.; Fan, J.; Wang, W.; Yu, D. Simple and robust MXene/carbon nanotubes/cotton fabrics for textile wastewater purification via solar-driven interfacial water evaporation. *Sep. Purif. Technol.* **2021**, *254*, 117615. [CrossRef]

MDPI
St. Alban-Anlage 66
4052 Basel
Switzerland
Tel. +41 61 683 77 34
Fax +41 61 302 89 18
www.mdpi.com

Catalysts Editorial Office
E-mail: catalysts@mdpi.com
www.mdpi.com/journal/catalysts

www.ingramcontent.com/pod-product-compliance
Lightning Source LLC
Chambersburg PA
CBHW061817210926
51850CB00006B/731
* 9 7 8 3 0 3 6 5 8 3 4 0 2 *